Genome Stability
DNA Repair and Recombination

For Melissa, Deborah, and Susan

Genome Stability
DNA Repair and Recombination

James E. Haber

GS Garland Science
Taylor & Francis Group

NEW YORK AND LONDON

Garland Science
Vice President: Denise Schanck
Editor: Summers Scholl
Senior Editorial Assistants: William Sudry and Kelly O'Connor
Production Editor: Ioana Moldovan
Layout: Georgina Lucas
Illustrator, Cover, and Text Designer: Matthew McClements, Blink Studio, Ltd.
Development Editor: Elizabeth Zayatz
Copyeditor: Sally Huish
Proofreader: Chris Purdon
Indexer: Medical Indexing Ltd.

ISBN 978-0-8153-4485-8

Preface
Images of normal chromosomes and those from tumor cells courtesy of Molecular Cytogenetics of
Common Epithelial Cancers, Cancer Genomics Program, University of Cambridge. With permission
of Paul Edwards, University of Cambridge.

Library of Congress Cataloging-in-Publication Data
Haber, James E., author.
  Genome stability : DNA repair and recombination / James E. Haber.
    p. ; cm.
Summary: "Genome Stability: DNA Repair and Recombination describes the various mechanisms
of repairing DNA damage by recombination, most notably the repair of chromosomal breaks. The
text presents a definitive history of the evolution of molecular models of DNA repair, emphasizing
current research. The book introduces the central players in recombination. An overview of the four
major pathways of homologous recombinational repair is followed by a description of the several
mechanisms of nonhomologous end-joining. Designed as a textbook for advanced undergraduate
and graduate students with a molecular biology and genetics background, researchers and
practitioners, especially in cancer biology, will also appreciate the book as a reference"--Provided
by publisher.
  ISBN 978-0-8153-4485-8 (pbk : alk. paper)
  I. Title.
  [DNLM: 1. Genomic Instability. 2. DNA Repair. 3. Models, Molecular.
4. Recombinational DNA Repair. QS 677]
  QH467
  572.8'6459--dc23
                        2013031803

Published by Garland Science, Taylor & Francis Group, LLC,
an informa business,
711 Third Avenue, New York, NY, 10017, USA,
and 3 Park Square, Milton Park, Abingdon, OX14 4RN, UK.

Printed in the United States of America

15 14 13 12 11 10 9 8 7 6 5 4 3 2 1

Garland Science
Taylor & Francis Group

Visit our website at http://www.garlandscience.com

PREFACE

The primordial tumorigenic cell [...] is, according to my hypothesis, a cell that harbours a specific faulty assembly of chromosomes as a consequence of an abnormal event.

Theodore Boveri (1914)
Translated by Henry Harris

The factors responsible for fusions of broken ends or for the healing of a broken end are not understood but are probably related to the method by which the chromosome becomes broken and to the physiological conditions surrounding the broken end.

Barbara McClintock (1941)

One of the most striking molecular aspects of cancers cells is their shocking departure from the normal chromosome number and arrangement. DNA replication is over 99% accurate, but the task of replicating six billion base pairs of human DNA in every cell is still precarious, both in terms of simple mutations and—more dangerously—in the creation of double-strand breaks (DSBs) that must be repaired. This textbook explains how genome stability is maintained.

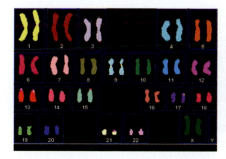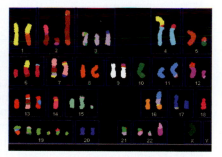

In contrast to normal chromosomes (above left), chromosomes from tumor cells (above right) exhibit dozens of alterations—truncations, translocations, duplications, and amplifications of chromosome segments, as well as gains and losses of whole chromosomes.

Cells have evolved two key processes to deal with broken chromosomes. First, they have elaborated a variety of different mechanisms to repair these breaks, most often using an intact sister chromatid or an homologous chromosome as the template to patch up the break. Much of this book will deal with understanding in detail how these largely error-free repair mechanisms function. These *homologous recombination* mechanisms are backed up by other, less precise *nonhomologous end-joining* pathways that can join broken ends together, with little regard for their origin. When the more accurate DNA repair processes fail, these alternative mechanisms take over, creating the rearrangements that we see in tumor cells. Repair of chromosome breaks is enhanced by a second process, termed the DNA damage checkpoint, which operates initially to prevent cells with chromosome breaks from entering mitosis, thus providing more time for repair to take place. If this restraint fails, then a second aspect of the DNA damage checkpoint is to destroy cells with unrepaired DNA damage by triggering apoptosis. Nearly all tumor cells have lost their ability to repair DSBs by homologous recombination and/or have lost the DNA damage checkpoint response.

Two exceptionally thoughtful books initially influenced my own thinking and prompted my wish to contribute a more molecular perspective. The first is H.L.K. Whitehouse's suggestive *Towards an Understanding of the Mechanism of Heredity* (1969); the second is Frank Stahl's inventive *Genetic Recombination: Thinking About It in Phage and Fungi* (1979). Both of these books preceded the explosion of molecular biological and genetic techniques that have made it possible to dissect the mechanisms of DNA repair in great detail, most especially in bacteria and yeasts, but increasingly in metazoans. I have jokingly said that this textbook is the sequel to Stahl's, but "thinking about it in fungi and mice." I have included a number of examples and concepts derived from studies of bacterial recombination and a smaller number from the emerging world of Archaea, but the focus is on eukaryotic chromosomes and their repair and recombination. Much of this textbook concentrates on chromosomal DSBs, the most dangerous type of DNA lesions. Some types of DNA repair—nucleotide excision repair or base-excision repair—are mentioned tangentially, but the focus is on repairing a completely broken chromosome.

This text is for advanced undergraduate and graduate students in molecular biology, genetics, and biochemistry. It is also intended as a reference for researchers and practitioners, especially in cancer biology. In writing this textbook I have assumed that the reader will have had some basic knowledge of genetics and molecular biology, knowing roughly how DNA replication proceeds. Consequently, the book begins with the problem of re-starting DNA replication at sites of damage or breakage (Chapter 1), as a way of introducing some of the basic mechanisms that are revisited in more detail in later chapters. The focus is on homologous recombination, driven by RecA and Rad51 recombination proteins, but Chapter 15 addresses nonhomologous end-joining in its several guises. After an overview of the various DSB repair mechanisms in Chapter 2, we begin with a review of the key recombination proteins RecA and Rad51 and how they work (Chapter 3). Then we turn to how DNA ends are processed to enable recombinase proteins to be loaded and to begin the search for homologous sequences with which repair can be effected (Chapter 4). Chapters 5 through 9 deal with different types of homologous recombination to repair a broken chromosome: single-strand annealing, mitotic gene conversion, and break-induced replication. A mixture of genetic and molecular biological evidence is presented to support our current understanding of the molecular mechanisms that underlie these processes. But homologous recombination is also a tool in modern genetics, so Chapters 10 and 11 examine gene targeting and site-specific recombination in detail. Only then do we confront recombination as it was initially studied a century ago—in meiosis (Chapters 12 and 13)—because meiotic recombination has elaborated and differentiated the basic mechanisms of DSB repair to ensure the accurate completion of generating recombined haploid germ cells from a diploid.

I have been forced to choose among many experiments to illustrate the important concepts in the book and have not mentioned numerous critical findings that led up to the selected experiments. Each of these experiments is cited in the relevant figure legends and each chapter includes suggested reading. Many other possible citations are absent, but they are available to the reader in two ways. First, I have added as an Appendix to this book (available online) a history of the evolution of molecular mechanisms of recombination, which has about 250 references that give full credit to the brave pioneers who launched the studies we continue today. Second, a combination of PubMed and Google searches will quickly bring an interested student to the relevant literature. Sprinkled throughout the book are brief boxes on nomenclature, perspectives, and measurement. The book also contains over 300 images and illustrations that, I hope, provide a way to visualize the processes that occur inside the human cell, a world too small to see.

ONLINE RESOURCES FOR STUDENTS AND INSTRUCTORS

Accessible from www.garlandscience.com/genomestability, the Student and Instructor Resource Websites provide learning and teaching tools created for *Genome Stability*. This book presents molecular models of recombination based on our present understanding, reflecting genetic, molecular biological, and biochemical approaches. But these models slowly evolved from ideas that first emerged soon after the discovery of the structure of DNA. Many clever ideas were postulated by creative scientists whose fundamentally important contributions are often overlooked as we focus on our present knowledge of these processes. An historical account of the way that our present models of DSB-mediated recombination evolved is presented in an Appendix in PDF format entitled "Evolution of Models of Homologous Recombination." This Appendix is available on both the Student and Instructor Resource Sites.

For Students: The Student Resource Site is open to everyone, and users have the option to register in order to use book-marking and note-taking tools.

For Instructors: All of the images from the book are available in two convenient formats: Microsoft PowerPoint® and JPEG. They have been optimized for display on a computer. Figures are searchable by figure number, figure name, or by keywords used in the figure legend from the book. The Instructor Resource Site requires registration; and access is available to instructors who have assigned the book to their course. To access the Instructor Resource Site, please contact your local sales representative or e-mail science@garland.com. You can also access the resources available for other Garland Science titles.

ACKNOWLEDGMENTS

I could not have even imagined writing a book on DNA repair and recombination without the invaluable help of many people. Jeffrey Hall taught me much of what I know about classical *Drosophila* genetics and I am especially grateful to Barbara McClintock who encouraged me early on in our discussions of her studies of genome instability in maize. A visit by Isamu Takano to Harlyn Halvorson's lab, coupled with my own interest in how mating type genes controlled meiosis and gene expression in yeast, prompted my investigation of homothallic *MAT* switching. Innumerable conversations with colleagues all over the world and collaborations with more than 75 different labs have also contributed to this endeavor. I owe particular thanks to my role models and colleagues, some of whom also reviewed individual chapters: Maury Fox, Frank Stahl, Jean-Luc Rossignol, Francis Fabre, Matt Meselson, Yasuji Oshima, Jack Szostak, Rod Rothstein, Richard Kolodner, Scott Hawley, Tom Petes, Shirleen Roeder, Nancy Kleckner, Maria Jasin, Scott Keeney, Neil Hunter, Doug Bishop, Michael Lichten, Dana Carroll, Bill Holloman, Lorraine Symington, Bill Holloman, Phil Hastings, Susan Rosenberg, Susan Lovett, John Wilson, Tony Carr, Benoit Arcangioli, Bernard Dujon, Steve West, Steve Kowalczykowski, Michael Cox, Hannah Klein, Patrick Sung, Akira Shinohara, Fred Alt, Shunichi Takeda, Ted Weinert, Virginia Zakian, Gerry Smith, Wolf Heyer, Simon Boulton, Roland Kanaar, Bernard Lopez, Vincenzo Costanzo, Ralph Scully, Marco Foiani, Doug Koshland, Angelika Amon, and Titia de Lange. I am certain I have forgotten to mention others and apologize in advance. I have deep, abiding memories of Nick Cozzarelli, Seymour Fogel, Fred Sherman, Ira Herskowitz, and my mentors Dan Koshland, Harlyn Halvorson, and Alan Wilson.

Most especially my understanding of DSB repair has come from the day-to-day discussions and the exceptional creativity of more than 80 postdocs and graduate students (plus several devoted technicians and a number of undergraduates) with whom I have been blessed to work. Here is a list of the people from my own lab with whom I have so far published: Peter Wejksnora (my first grad student), Paloma Liras (first postdoc), Sim Gek Kee (first undergrad), Ellen Kraig, Susan Remer, Nancy Pearson, Anne Comeau, Deborah Wygal Mascioli, Dave Rogers, Sue Stewart, Barbara Garvik, Jeanne George, Pat Thorburn, Elaine Sugarman, Norah Rudin,

Lance Davidow, Michael Lichten, John McCusker, Sandra Harris, Mark Hearn, Barbara Weiffenbach, Bryan Ray, Jacqueline Fishman-Lobell, Charles White, Wai-Ying Leung, Ed Louis, Rhona Borts, Kate Kramer, Neal Sugawara, Genya Ivanov, Shalini Anand, Songqing Na, Xiaochun Zhou, Xiaohua Wu, Zixhi Zhou, Cherry Wu, Kaiming Sun, Giovanni Bosco, Ayelet Arbel-Eden, Eliyahu Kraus, Kent Moore, Mónica Colaiácovo, Guy-Franck Richard, Allyson Holmes, Frédéric Pâques, Allison Landman, Taya Feldman, Masha German, Maria-Ai Naylor, Moreshwar Vaze, Sang Eun Lee, Xuan Wang, Ellen Lipstein, Maria Valencia Burton, Ulricke Sattler, Anna Malkova, Jung-Ae Kim, Nicolas Tanguy Le Gac, Gregory Ira, Mercedes Gallardo, Eric Coïc, André Walther, Debra Bressan, Jake Harrison, Farokh Dotiwala, Vikram Ranade, Sue Yen Tay, Minlee Kim, Taehyun Ryu, Miyuki Yamaguchi, Wade Hicks, Suvi Jain, Susannah Gordon-Messer, Zach Lipkin-Moore, John Lydeard, Jin Li, Quiqin Wu, Kihoon Lee, Cheng-Sheng Lee, Michael Tsabar, Vinay Eapen, Ranjith Anand, and Anuja Mehta. They provided many of the insights and creative experimental approaches that my lab has developed over a 40-year period. Anuja Mehta deserves special mention for her eagle-eyed proofreading that removed many errors that had escaped my sight.

I am also grateful for a number of exceptionally helpful reviewers of the individual chapters, including: Thorsten Allers (University of Nottingham, UK), Willy M. Baarends (Erasmus University Medical Center, the Netherlands), Jiri Bartek (Danish Cancer Society Research Center), Rodrigo Bermejo (Institute for Functional Biology and Genomics, Spain), Grant W. Brown (University of Toronto), Phillip Carpenter (University of Texas Health Science Center at Houston), Richard Fishel (Ohio State University), Thanos Halazonetis (University of Geneva, Switzerland), Jim Hu (Texas A&M University), Pablo Huertas (University of Seville, Spain), Hiroshi Iwasaki (Tokyo Institute of Technology, Japan), Scott Keeney (Memorial Sloan-Kettering Cancer Center), Anthony Schwacha (University of Pittsburgh), Agata Smogorzewska (The Rockefeller University), Marcus Bustamante Smolka (Cornell University), Jeremy M. Stark (Beckman Research Institute, City of Hope), Kiyoe Ura (Osaka University School of Medicine, Japan), Dik C. van Gent (Erasmus University Medical Center, the Netherlands), Marcel A.T.M. van Vugt (University of Groningen, the Netherlands), Zhao-Qi Wang (Leibniz Institute for Age Research, Germany), and Shan Zha (Columbia University). Their comments helped improve the accuracy and scope of the book, but the remaining faults are entirely mine. Were I to start over I would have included a few more topics. I am also immensely grateful to Matthew McClements, who has taken my rough-drawn figures as well as disparate images from many publications and turned them into a coherent set of illustrations. Development Editor, Elizabeth Zayatz, provided invaluable guidance in making chapters more accessible to a broad readership. Kelly O'Connor, William Sudry, and especially Ioana Moldovan facilitated the editorial process. And none of this would have happened without the encouragement and enthusiasm of my editor, Summers Scholl.

I first started working on a version of this book as a John Simon Guggenheim Fellow in 2000, but that effort soon faltered as I realized how little we really knew about the mechanism of strand exchange, the resolution of Holliday junctions, and much more. However the following decade has been remarkable in unveiling many of these mysteries. I didn't resume serious work on this book until I was a Fellow of the Radcliffe Institute for Advanced Study in 2008. Sabbatical visits with Steve West (London Research Institute, UK), Nevan Krogan (University of California, San Francisco), and Geneviéve Almouzni (Institut Curie) moved things along. Brandeis University has been a wonderful home for over 40 years and I have had the benefit of teaching bright students and working with stimulating faculty colleagues. I am particularly indebted to Susan Lovett, my DNA repair colleague, and to Jeffrey Hall, who—when I first started to teach—introduced me to the world of *Drosophila* and maize genetics, to chromosome segregations, and especially to the work of Barbara McClintock and other pioneers in the study of chromosome instability. Research from my own lab has been generously supported by the National Institutes of Health, the National Science Foundation, the US Department of Energy, and the American Cancer Society.

CONTENTS

DETAILED CONTENTS

CHAPTER 5   SINGLE-STRAND ANNEALING

CHAPTER 6   GENE CONVERSION

CHAPTER 7 "IN VIVO BIOCHEMISTRY": RECOMBINATION IN YEAST

CHAPTER 8 BREAK-INDUCED REPLICATION

CHAPTER 9 SISTER CHROMATID REPAIR

CHAPTER 10 GENE TARGETING

CHAPTER 11   SITE-SPECIFIC RECOMBINATION

CHAPTER 12   CYTOLOGY AND GENETICS OF MEIOSIS

CHAPTER 13   MOLECULAR EVENTS DURING MEIOTIC RECOMBINATION

CHAPTER 14   HOLLIDAY JUNCTION RESOLVASES AND CROSSING OVER

CHAPTER 1

RESTARTING DNA REPLICATION BY RECOMBINATION

For almost 100 years, scientists have been fascinated with the mechanisms that promote crossing over in meiosis. It is much more recently that we have begun to understand that the mechanisms that underlie meiotic recombination also account for the ways organisms from bacteria to people are able to cope with damaged and broken chromosomes in all types of cells. Genetic recombination is in fact essential for the viability of human cells, and defects in recombination and repair mechanisms are frequently associated with the loss of genetic integrity that is found in cancer cells. The fundamental mechanisms of genetic recombination and repair are conserved in bacteria, archaea, and eukaryotes.

This book examines a variety of recombination mechanisms that are used in different settings to repair broken and damaged chromosomes. Although many of the concepts on which the subject of recombinational repair is based have indeed come from the study of meiosis, this book begins by introducing some important ideas and the molecular cast of characters by focusing instead on the way cells cope with intrinsic problems that arise every time DNA is replicated. Errors of replication occur even in the absence of damage caused by UV- or ionizing radiation or by alkylation or oxidation of DNA bases. The DNA replication machinery is surprisingly quite fragile: in mammalian cells, failure of replication leads to perhaps a dozen broken chromatids, which must be repaired if each daughter cell is to inherit a faithfully copied and intact set of chromosomes. Many DNA replication problems arise because the DNA replication machinery stalls in regions that are difficult to traverse, whereupon cells must find ways to restart replication. So, although we will eventually arrive at the most highly evolved recombination mechanisms—those that are used in meiosis—we will begin at a place that most of us find familiar, the problem of replicating DNA.

1.1 DNA BREAKS OCCUR FREQUENTLY DURING REPLICATION

The idea that recombination mechanisms are essential for cellular life is illustrated by the consequences of depleting vertebrate cells of a protein essential for the repair of DNA damage. When expression of Rad51—the key recombinase protein in eukaryotes—is repressed, cells arrest within a single replication cycle and have 10 or more chromatid breaks (**Figure 1.1**), where one sister is intact and the other is broken, the consequence of not repairing the lesions. Without repair, vertebrate cells die. When the Rad51 protein is deleted in budding yeast cells, most cells do not die, but that is simply because yeasts have much smaller genomes

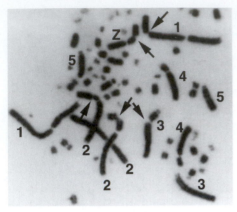

Replicated sister chromatids,
one of which has a DSB

Figure 1.1 Accumulation of double-strand breaks (DSBs) in one o f two sister chromatids after depletion of the Rad51 recombination enzyme in chicken DT40 cells. Above left, Rad51-deficient cells undergoing mitosis are spread on a slide and examined microscopically. Numbers in the figure indicate the major avian chromosomes including the Z sex chromosome, some of which are present in three copies. Arrows indicate sites of DSBs on one of the two replicated chromatids. The chromatid break is illustrated below left, showing one completely replicated chromatid and one chromatid with a DSB. (Adapted from Sonoda E, Sasaki MS, Buerstedde JM et al. [1998] *EMBO J* 17:598–608. With permission from Macmillan Publishers, Ltd.)

than vertebrates; consequently the probability that there will be one break on one of yeast's chromosomes is much less than 1. In fact, the rate of spontaneous chromosome breakage per megabase (Mb) of DNA is roughly the same for yeasts or fruit flies or mice.

A related indication of the number of lesions arising in chromosomes is the frequency of spontaneous sister chromatid crossing over. When a lesion arises during replication on one DNA double helix, it can be repaired by copying over the damaged information, using the undamaged sister chromatid as a template. How this occurs in detail is the subject of this book. Suffice it to say for now that some of these repair events lead to a crossover between the sister chromatids (**Figure 1.2A**). *Sister chromatid exchanges* (SCEs) can be visualized after labeling cells for one generation with BrdU (bromodeoxyuridine), a base analog of deoxythymidine, and then allowing the cells to replicate again in the absence of BrdU. At the end of this process, a pair of just-replicated sister chromatids will have only a single strand of labeled DNA. However, if during the completion of the second round of DNA replication—that which occurs in the absence of BrdU—there is an exchange (a crossover) between the sister chromatids, then the line of BrdU labeling along the chromatid will be visibly interrupted and the labeled DNA will continue along the sister chromatid. These characteristic interruptions occur roughly 10 times in every replicating cell; that is, there must have been some sort of DNA repair event leading to a sister chromatid exchange at least 10 times every time the cell copied its DNA. The number of SCEs likely gives us an underestimate of the number of repair events, because only a fraction of the repair processes actually lead to a crossover. But whether the actual number of repair events is 10 or 100, these observations suggest that the enzymatic machinery to assure that such lesions are repaired must be active in every cell during every replication cycle.

An important way in which the proteins involved in DNA repair have been identified is by the genetic and cellular phenotypes of cells or organisms that lack a particular function. For example, the number of sister chromatid exchanges is much greater than normal in cells of people with a defect in the Bloom's syndrome DNA helicase, BLM (**Figure 1.2B**); not surprisingly, people with Bloom's syndrome are cancer-prone. The cells lacking functional BLM could have a defect that leads to an increased number of DNA insults or—as now seems to be the case—the same number of lesions are more likely to be accompanied by sister chromatid crossing over. We will examine the roles of BLM and its homologs in other organisms, and many other key proteins, as we go along.

To begin to understand the molecular basis for repair of DNA by recombination, we will first look at the ways that replication damage can be repaired by enzymes that recognize broken chromosomes—*double-strand breaks* (DSBs)—created during DNA replication. This will provide an introduction both to the way DNA is processed and to the key proteins that enable such repair to occur.

(A)

(B)

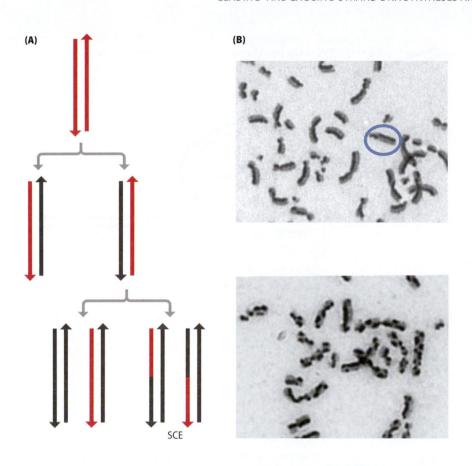

SCE

Figure 1.2 Sister chromatid crossing over revealed by BrdU labeling. (A) Cells are grown in a medium containing bromodeoxyuridine (BrdU) until their DNA has incorporated BrdU (*red*) into both strands (*top*). Cells are then transferred to normal medium. After one generation (*middle*), each duplex has one labeled and one unlabeled (*dark gray*) strand. After a second round of DNA replication (*bottom*), each pair of sister chromatids has one labeled strand. If there is no sister chromatid exchange (SCE), then, as shown at left, staining for BrdU-containing DNA reveals one continuous labeled strand. If an SCE has occurred, the label is discontinuous, with BrdU residing on a different part of each sister chromatid. (B) One of at least five SCE events in a normal cell is circled in *blue* (*top*). In a cell from a patient suffering from Bloom's syndrome (*bottom*), there are many SCE events on each pair of sister chromatids. (B, courtesy of Shriparna Sarbajna and Stephen West, Cancer Research UK.)

1.2 LEADING- AND LAGGING-STRAND DNA SYNTHESES ARE COORDINATED AT THE REPLICATION FORK

Replication of double-stranded DNA (dsDNA) requires the coordinated synthesis of leading and lagging strands by a pair of DNA polymerases that are coupled together and moving in the same direction along DNA (or, more likely, DNA is pulled through a stationary replication complex). The components of the eukaryotic replication fork are shown in **Figure 1.3**. Synthesis of the lagging strand is initiated by an RNA primer, created by the primase complex. The polymerases are preceded by a DNA helicase complex that opens up the double helix. In prokaryotes, the DnaB helicase complex travels along the lagging strand, whereas in eukaryotes, the Cdc45–GINS–MCM (CGM) helicase complex migrates along the leading strand.

In prokaryotes, a single DNA polymerase (DNA PolIII) carries out all the functions that are divided up among three eukaryotic DNA polymerases: Polα, Polδ, and Polε. Polα and the primase complex initiate both leading-strand synthesis and the synthesis of each Okazaki fragment of lagging-strand synthesis by copying a short RNA segment from the template. DNA synthesis of the leading strand depends on DNA Polε, while the extension of each Okazaki fragment involves transfer from Polα to Polδ. Each Okazaki fragment must be trimmed at its 5′ end, to remove at a minimum the short RNA primer, but it is likely that the entire short stretch of DNA copied by Polα is removed because Polα does not have the proofreading 3′ to 5′ exonuclease activities that make Polδ and Polε more accurate. Removal of these sequences by a flap endonuclease (FEN-1) or by a combination of a helicase and endonuclease leaves a gap in the DNA that can be filled in by a DNA polymerase. The filled-in

Figure 1.3 Components of the eukaryotic DNA replication fork. The eukaryotic replication fork is shown copying DNA from left to right. The newly synthesized strands (*red*) and the short RNA primers that initiate Okazaki fragment synthesis are shown (*wavy lines*). Leading-strand synthesis is carried out by DNA polymerase ε, which is stabilized as it progresses by its association with the PCNA clamp proteins that are loaded and unloaded from the DNA by the RFC (replication factor C) complex. The replication proteins are preceded by a DNA helicase complex that encircles the leading-strand template and opens up the duplex DNA in advance of the replication machinery, exposing single-stranded DNA (ssDNA) that is bound by a trimer of RPA proteins (replication protein A). Lagging-strand synthesis is shown progressing in the same direction as leading-strand synthesis; this requires that the lagging strand be looped around in what is often called a trombone configuration. Okazaki fragment synthesis begins with DNA polymerase α and its associated primase that lay down the initial RNA primer and the beginning of DNA copying before synthesis continues using DNA polymerase δ, which also associates with the replication clamp protein, PCNA. Regions of the lagging strand that have not yet been copied are coated with RPA. Once an Okazaki fragment is extended to reach the previous Okazaki fragment the RNA primer is removed by the flap endonuclease, FEN-1, and other nucleases; then the fragments are joined together by DNA ligase 1.

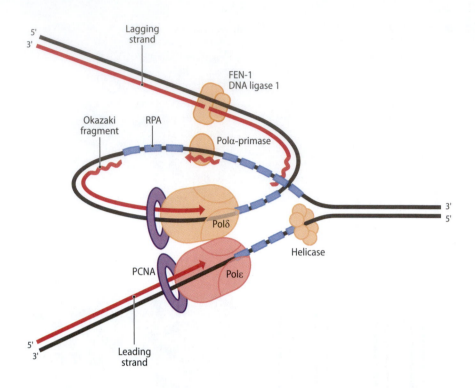

regions can then be ligated by DNA ligase 1. A failure to fill in and ligate a pair of adjacent Okazaki fragments would leave a nick in one strand.

Replication of a DNA molecule with a single-strand nick (ss-nick) left on the lagging strand (**Figure 1.4A**) will—in the next round of synthesis—be encountered by the leading-strand polymerase. Precisely what occurs *in vivo* is not well established, but it seems likely that replication of the leading strand is blocked at the position of the nick on its template (Figure 1.4A), leaving one intact DNA duplex and one that is truncated at the location of the nick. As we shall see, the broken end can be used to re-initiate DNA replication. Alternatively, a second replication fork, converging from the other direction, could copy the remaining part of the template (**Figure 1.4B**), leading to one completely replicated molecule and one with a double-strand break. In this case the DSB can be repaired by recombination with its sister chromatid.

1.3 REPLICATION FORK STALLING MAY OCCUR IN SEVERAL DIFFERENT WAYS

The progression of the replication fork can be blocked or stalled in different ways. DNA sequences such as those encoding transfer RNA (tRNA) or certain trinucleotide repeats such as CTG can form strong secondary or tertiary structures of self-complementary sequences when DNA is unwound ahead of the replication fork. Regions of heterochromatic DNA also impair fork progression. Stalling DNA replication can be visualized

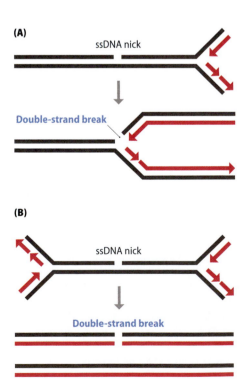

Figure 1.4 Creation of a DSB by replication through a single-strand nick.
(A) A replication fork moves from right to left toward a nick in the DNA. When the fork reaches the nick, the leading-strand synthesis is blocked because its template is broken; this results in a single broken end and an intact chromatid that can be used as a template to restart replication. (B) If two converging replication forks approach a nick in the template, one of the two chromatids will have a true two-ended double-strand break.

by two-dimensional Brewer–Fangman gel electrophoresis, which separates branched molecules on the basis of both size and shape (**Figure 1.5**). Normally the arc of replication intermediates is a continuous curve, but if replication pauses for a significant time at particular sequences, the arc is punctuated by a concentration of hybridization at the site where more molecules of a particular size and shape have accumulated. In budding yeast, where this phenomenon has been best studied, there are significant delays in fork progression at heterochromatic (transcriptionally silent) regions, at centromeres, and at tRNA genes (presumably because each strand of the DNA, when unwound, can form elaborate secondary structures similar to those formed by the tRNA itself) (**Figure 1.5D**). Absence of a 5′ to 3′ DNA helicase, Rrm3, exacerbates fork stalling and results in broken chromosomes (designated BR in the figure).

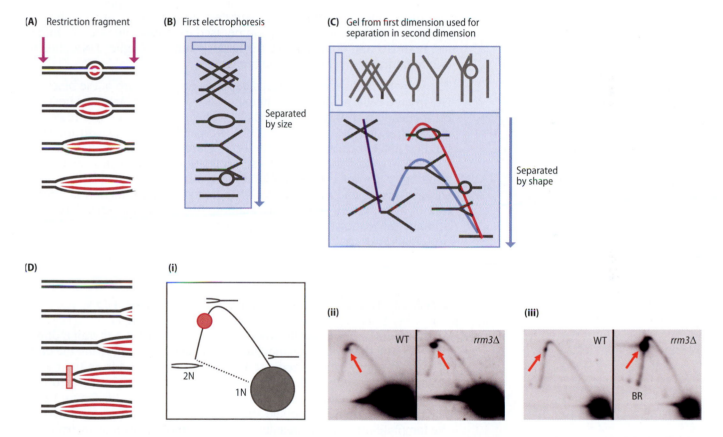

Figure 1.5 Visualization of DNA replication intermediates by two-dimensional gel electrophoresis. (A) DNA replication creates a set of different sized restriction fragments. If replication is initiated within this region, some of the fragments will contain different sized "bubbles," but as replication proceeds the fragments will become Y-shaped with two replicated arms and one as-yet unreplicated region. DNA replication that is initiated outside the restriction fragment will produce a series of different sized Y-shaped intermediates. (B) The different sized molecules can be separated as a function of their total mass by gel electrophoresis. (C) The gel containing the size-separated DNA is then placed at the top of a square gel and the DNA is separated by electrophoresis in a second dimension, under conditions that resolve molecules by their shape as well as their size. The separated molecules in these Brewer–Fangman two-dimensional gels are then transferred as a Southern blot and probed for a particular genomic region. Fragments that are passively replicated from an origin outside the restriction fragment are seen as an increasingly large series of Y-shaped molecules that form an arc (*blue curve*), beginning at the unreplicated size (1N) and culminating in a nearly fully replicated duplex (2N). Fragments containing a "bubble," which is indicative of initiation of DNA replication within the restriction fragment, migrate as an arc (*red*) that terminates when the bubble reaches one end of the fragment and becomes a Y-shaped molecule of twice the original molecular weight (2N). Branched molecules containing four-armed Holliday junctions form a spike (*purple*) that rises from the point where nearly fully replicated molecules will migrate. (D) A replication fork barrier (*red rectangle*) causes the accumulation of molecules of one size, which can be seen as a "blob" on the arc of replicating molecules (i). Deletion of the *RRM3* gene (*rrm3Δ*) encoding a 5′ to 3′ DNA helicase in budding yeast enhances replication fork stalling at both a centromere (*CEN3*) (ii) and at a tRNA gene (iii). Fork stalling in the absence of Rrm3 also produces some chromosomal breakage (BR). WT, wild type. (Adapted from Ivessa AS, Lenzmeier BA, Bessler JB et al. [2003] *Mol Cell* 12:1525–1536. With permission from Elsevier.)

In some special instances, such as ribosomal DNA replication, the replication fork is forced to go primarily in one direction by the presence of a *replication fork blocking sequence* that, when bound to a specific protein, acts as a directional barrier to fork movement. A well-studied example is the replication fork blocking (RFB) sequence in budding yeast ribosomal DNA that binds a fork-blocking protein, Fob1. This kind of unidirectional impairment of replication is also found at the replication termination sites (Ter) of bacterial DNA.

The most potent barriers are chemical modifications of DNA, exemplified by UV-induced cyclobutane dimers (for example, thymine dimers) that stop replication by the normal DNA polymerase machinery. Treatment of cells with methyl methanesulfonate (MMS) causes base modifications that also impair replication. In the lab, replication fork stalling is frequently induced by treating cells with hydroxyurea (HU)—an inhibitor of ribonucleotide reductase—to deplete deoxynucleotide triphosphate pools and thus limit the rate of DNA synthesis. Alternatively, aphidicolin, an inhibitor of DNA polymerase α, can be used to reduce the rate of synthesis. Both HU and aphidicolin act synergistically at "fragile" DNA sites to create nonrandom chromosome breaks.

The way the replication fork stalls depends on the nature of the block. In HU-treated cells or with some chemical lesions such as MMS, the helicase on the leading strand outruns the DNA polymerases, leaving a region of single-stranded DNA (ssDNA). ssDNA is coated by the single-strand DNA binding protein called SSB in bacteria or archaea or replication protein A (RPA) in eukaryotes. In contrast, replication fork blocking sequences apparently do not create ssDNA regions because the helicase itself is stopped just ahead of the DNA polymerases. This distinction is important in how the cell detects fork stalling; the creation of ssDNA regions at many replication forks can trigger a cascade of replication checkpoint protein kinases that prevent the cell from entering mitosis until the damage is repaired.

1.4 AN INTRODUCTION TO THE HOLLIDAY JUNCTION

At stalled forks, for example when the fork encounters a thymine dimer (**Figure 1.6A**), the replication machinery may dissociate, allowing the ends of the two newly synthesized strands of DNA to unwind from their templates and pair with each other, creating a symmetric four-armed and four-stranded structure known as a *regressed replication fork* (Figure 1.6A). The pairing of the newly copied strands leads also to the re-formation of the template duplex. This results in regression of the fork away from the site that caused replication stalling. The resulting four-armed structure is sometimes referred to as a "chicken foot," because of its appearance in electron micrographs (**Figure 1.6B**).

This four-armed, branched structure is an example of a *Holliday junction* (HJ). This key intermediate in DNA repair by recombination was proposed by Robin Holliday in 1964, in the context of explaining recombination events in meiosis, but it has proven to be at the center of many types of repair events in mitotic cells as well. The HJ is a symmetrical structure when viewed in three dimensions (**Figure 1.7A**). In a flattened two-dimensional representation (**Figure 1.7B**), one sometimes loses sight of the equivalence of each arm, although the key features of the HJ are still evident. First, all possible base pairs can be formed within the four arms of the structure. Second, the crossing point at the intersection of the four strands can move (branch migrate) by the breaking and making of base pairings (**Figure 1.7C**). To move the branch point one step, two base

(A)

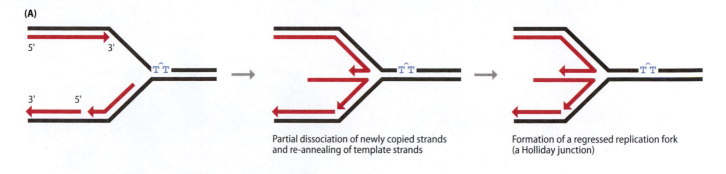

Partial dissociation of newly copied strands and re-annealing of template strands

Formation of a regressed replication fork (a Holliday junction)

(B)

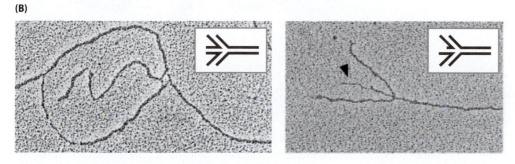

Figure 1.6 Replication fork reversal at a site of DNA damage. (A) Depicted is fork stalling that occurs at the site of a UV-induced thymine photodimer (TT). Fork reversal allows the two newly synthesized strands to pair (*middle*) and the resultant Holliday junction (HJ) to migrate away from the site of damage (*right*). This structure is sometimes referred to as a "chicken foot." (B) Visualization of "chicken foot" DNA molecules by electron microscopy. (B, from Sogo JM, Lopes M & Foiani M [2002] *Science* 297:599–602. With permission from the American Association for the Advancement of Science.)

pairs need to be broken, but they simultaneously can form two new base pairs with a different partner. Consequently, there is no net energy cost to branch migration (**Figure 1.7D**).

At a stalled replication fork, formation and branch migration of the HJ allows fork regression away from the lesion that blocked replication. This movement may allow repair enzymes access to the lesion that blocked replication, so that it can be excised and replaced with an unmodified base.

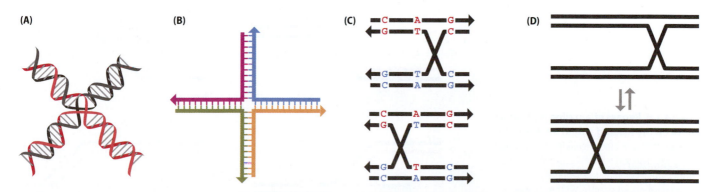

Figure 1.7 Structure of a Holliday junction (HJ). (A) Three-dimensional depiction of an HJ, a structure in which two DNA helices share their double strands at a four-stranded juncture known as a branch point. (B) A flattened diagram of a synthetic HJ formed by the annealing of four oligonucleotides, each depicted in a different color. (C) Holliday junctions can branch migrate by the simultaneous breaking of two base pairs and the formation of two new base pairs, thus displacing the point of strand exchange. (D) Branch migration can move the branch point over hundreds of base pairs and occurs without the net expenditure of energy, as an equivalent number of base pairs are broken and created at each step. (A, from Liu Y & West SC [2004] *Nature Reviews Molecular Cell Biology* 5:937–944. With permission from Macmillan Publishers, Ltd. B, courtesy of David Lilley, University of Dundee.)

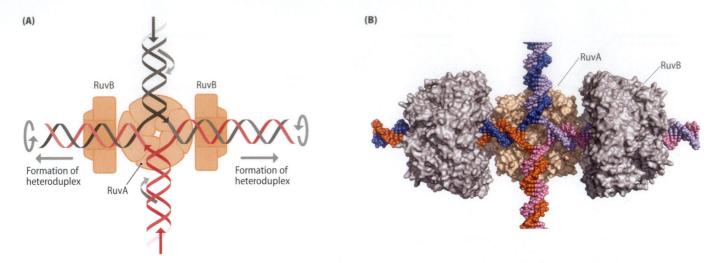

(A)

RuvB   RuvB

Formation of
heteroduplex

RuvA

Formation of
heteroduplex

(B)

RuvA

RuvB

Figure 1.8 Branch migration of HJs can be catalyzed by a pair of oppositely oriented RuvB helicases in *E. coli.* (A) After the HJ-recognition protein RuvA binds to the HJ, two hexameric RuvB complexes form that can pull a pair of strands DNA through the center of the complexes, driving branch migration. (B) Shown is a three-dimensional model of RuvA (*tan*) and RuvB (*gray*) associated with an HJ (an animation can be viewed at http://www.shef.ac.uk/mbb/ruva). (A, adapted from Parsons CA, Stasiak A, Bennett RJ & West SC [1995] *Nature* 374:375–378. With permission from Macmillan Publishers, Ltd. B, courtesy of Peter Artymiuk, University of Sheffield.)

The movement of the HJ by branch migration can occur without the mediation of proteins, but more rapid, and directional, migration can be effected by HJ-binding proteins that associate with a specific DNA helicase. The proteins required for branch migration have so far only been well characterized in bacteria. The best-analyzed pair is the RuvA protein of *Escherichia coli*, which binds to the junction itself, and RuvB, which promotes its migration (**Figure 1.8**). Two RuvB hexamers bind to a pair of opposite arms of the HJ and promote the breaking and forming of bonds as they pull two strands in opposite directions and thus translocate the HJ branch point.

The branch-migration properties of the Holliday junction provide alternative ways to bypass some DNA lesions. **Figure 1.9A** shows the situation where a lesion has blocked leading-strand synthesis. Formation of a regressed fork (**Figure 1.9B**) enables the pairing of the two newly copied strands, in which the 5′-ended product of lagging-strand synthesis extends beyond the point where the new 3′-ended leading strand was blocked. In this reversed fork, this 3′ end can be extended by a DNA polymerase (using the 3′ end as its primer), to the end of protruding

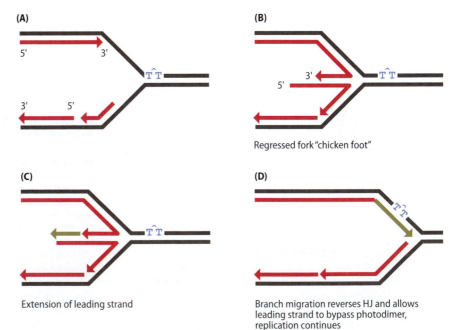

(A)

5′   3′

T̂T

3′   5′

(B)

3′

5′

T̂T

Regressed fork "chicken foot"

(C)

T̂T

Extension of leading strand

(D)

T̂T

Branch migration reverses HJ and allows leading strand to bypass photodimer, replication continues

Figure 1.9 Bypass of a stalled replication block on the leading strand by fork reversal. (A) A photodimer blocks leading-strand synthesis, but lagging-strand replication proceeds a bit further. (B) Replication fork reversal pairs the newly synthesized strands. (C) The recessed 3′ end of the leading strand, now paired again with the end of the newly copied lagging strand, is used as a primer to fill in to the end (*green arrow*). (D) Reversal of the HJ now places newly copied DNA across from the photodimer and replication can proceed.

5′ strand (**Figure 1.9C**). At this point, the HJ can branch migrate in the opposite direction, restoring the original three-stranded replication fork; but now the leading strand has been extended to base pair beyond the site of the lesion (**Figure 1.9D**). Reassembly of the replication fork machine will then allow DNA synthesis to resume, leaving only a lesion on one strand that will either be repaired or will provoke a replication block in the next cell cycle.

1.5 A HOLLIDAY JUNCTION CAN BE RESOLVED BY ENZYMATIC CLEAVAGE

In *E. coli*, the RuvA and RuvB proteins associate with another enzyme, RuvC, which was the first-identified and is still the best-studied example of an HJ *resolvase*, an enzyme that can cleave the four-stranded HJ. RuvC creates symmetrical nicks on two opposite strands of the HJ (**Figure 1.10A**), thus converting the four-stranded structure into two duplexes with single-strand nicks. HJ resolvases can cleave either pair of opposite strands. The two different ways of cleaving the HJ may have important biological consequences. If we look at an HJ joining two sister chromatids, one of which has a BrdU-labeled strand, we can see one consequence of the alternative ways of cleaving the HJ (**Figure 1.10B**). If the HJ is cleaved between strands marked by the arrows labeled "a," then one of the two resulting molecules will have BrdU along one strand while the second

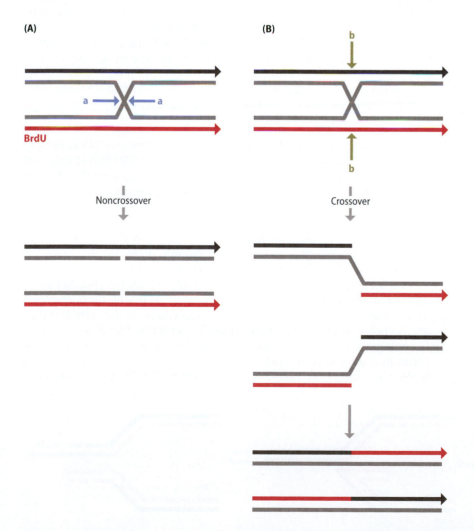

Figure 1.10 HJs can be resolved by cleaving two of the four strands. (A) An HJ is shown in which one of the strands contains BrdU (*red*). Cleavage (*arrows*) of the crossed strands (a) separates the branched structure without a crossover, leaving BrdU only on one chromatid. (B) Cleavage of the noncrossed strands (b) results in two molecules that are only connected by the crossing strands. The resulting molecules have a crossover, each with a nick that can be readily re-ligated. Here, BrdU is found on both sister chromatids.

DNA will have no BrdU label. If, however, the HJ is cleaved by the arrows marked "b," there is a crossover, so that there is a discontinuity in which BrdU label "jumps" from one sister chromatid to the other at the point of the crossover.

As we shall see in more detail later, resolvase activity proves to be essential for the completion of many types of homologous recombination, both in meiotic and somatic cells. We will revisit HJ resolvases and their biochemical properties in more detail in Chapter 14.

1.6 HJ RESOLVASES CAN PROMOTE REPLICATION RESTART BY BREAK-INDUCED REPLICATION

The regressed fork can also be a substrate for an HJ resolvase. In this instance, cleavage of the branched structure leads to the formation of one replicated and intact strand and one broken replicated end of DNA (**Figure 1.11**). This outcome is similar to that envisioned when the replication fork encountered a nick in a template strand (Figure 1.4). It is the starting point for reestablishing the replication fork by homologous recombination. *Break-induced replication* (BIR), also known as recombination-dependent DNA replication, proceeds by a series of steps, as outlined below.

1. *Creation of a recombinogenic end.* The first essential step in BIR is that the broken end needs to be processed so that it can become an active participant in DNA repair. The DNA end must be able to invade into the intact duplex and open it up, so that a new replication fork can be established. This can be accomplished—with the aid of many proteins—if the DNA end can make base pairs with the intact duplex. For this purpose, the blunt end of the broken DNA is resected by a 5′ to 3′ exonuclease, to create an ssDNA tail (**Figure 1.12A** and **1.12B**). This 3′-ended single-stranded region of DNA is the key substrate needed for homologous recombination to proceed.

2. *Formation of the RecA/Rad51 recombination machine.* ssDNA is rapidly bound by single-stranded binding proteins (called SSB in bacteria and archaea, and RPA in eukaryotes). This binding protects ssDNA from being degraded and prevents the formation of secondary structures. SSB/RPA binding enables the important next step in recombinational repair, the formation of a nucleoprotein filament of the recombinase proteins (RecA in bacteria, RadA in archaea, and Rad51 in eukaryotes). With the aid of a number of mediator proteins, RecA/Rad51 is able to displace SSB/RPA and to form a helical filament of the recombinase protein, with three bases of ssDNA per RecA/Rad51 monomer (**Figure 1.12C**). The nucleoprotein filament then engages in a remarkable search for sequences elsewhere in the genome that are homologous to those encased by the recombinase (**Figure 1.12D**). Homologous sequences could be located on a homologous chromosome in a diploid or at some ectopic location, but in the context of replication restart, the homologous sequences that the filament is most

Figure 1.11 HJ resolution of a regressed replication fork. A regressed fork, similar to that shown in Figure 1.9, is cleaved by an HJ resolvase, producing an intact chromatid (*top*) and a one-ended, partly replicated chromatid (*bottom*). Recombination will allow the broken-ended chromatid to re-initiate DNA replication, as shown in Figure 1.12.

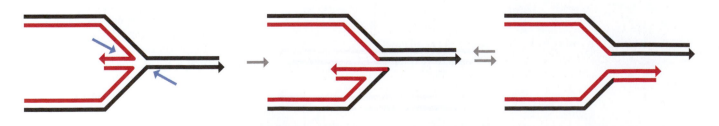

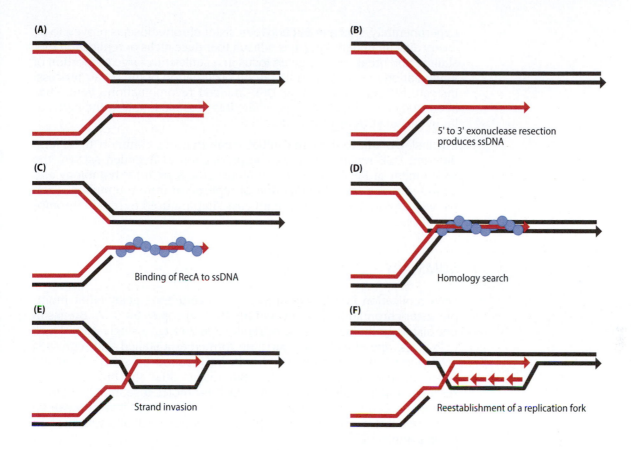

(A)

(B)
5' to 3' exonuclease resection produces ssDNA

(C)
Binding of RecA to ssDNA

(D)
Homology search

(E)
Strand invasion

(F)
Reestablishment of a replication fork

likely to encounter are those of the intact, partially replicated sister chromatid. How, exactly, this search is accomplished will be discussed in detail later. Here we need to know that RecA/Rad51 can somehow promote a base-pair-by-base-pair comparison of the sequences in the filament with those of an intact double helix. When a complementary strand of the helix can make a sufficient number of base pairs with the ssDNA, a strand exchange takes place (**Figure 1.12E**). Strand exchange leads to the formation of base pairs between the invading strand and the creation of a complementary strand of the helix and the formation of a *displacement loop* (a D-loop) of one strand of the original helix.

3. *Conversion of the D-loop into a new, unidirectional replication fork.* The D-loop contains an invading single strand with a 3' end. This end can be used as a primer to allow DNA polymerase to initiate the copying of one strand of the D-loop (Figure 1.12E). However, to re-initiate DNA replication at this broken replication fork, there must also be recruitment of the lagging-strand replication machinery (Polα and primase), so that both strands of the D-loop serve as templates (**Figure 1.12F**). In addition, a DNA helicase must be recruited to allow the replication fork to proceed down the template. This recombination-dependent initiation of both strands of DNA replication results in the reestablishment of the replication fork, which can progress hundreds of kilobase (kb) pairs toward the end of a chromosome or be terminated by encountering a normal, oncoming fork.

BIR appears to be important not only in restoring replication at broken forks, but in maintaining telomere sequences when elongation by the telomerase enzyme is inhibited or absent. BIR thus provides one possible alternative lengthening of telomere (ALT) mechanism that is an essential feature of some immortal cell lines. We will examine BIR in detail in Chapter 8.

Figure 1.12 Resumption of replication at a broken fork by break-induced replication. (A) A one-ended break is created by cleavage of a regressed replication fork (a Holliday junction). (B) The tip of the broken end is resected by a 5' to 3' exonuclease, allowing the RecA recombinase protein to form a filament on single-stranded DNA (ssDNA) (C). The RecA::ssDNA complex engages in a search for homology (D), allowing the ssDNA to pair with a homologous sequence and (E) displace one strand into a displacement loop (D-loop). The D-loop can then be converted into a unidirectional replication fork (F).

Experimentally, studying BIR between sister chromatids has proven to be a very difficult undertaking. It is difficult to induce nicks or replication fork stalling and breakage at a given locus in a sufficiently large proportion of a population of cells to permit careful observation of the repair process. Instead, BIR, like many other DSB-induced recombination events, has been studied in model systems, where it is possible to follow the molecular events of DNA recombination in detail.

But first we will look at the initial steps that are common to several different DSB repair pathways: the production of 3′-ended ssDNA, the recruitment of RecA/Rad51 recombinase, the search for homology and strand invasion, and the initiation of replication from a template. Then we will explore the variety of pathways that are used to repair chromosomal DSBs.

SUMMARY

DNA replication is a fragile process, and vertebrate cells suffer multiple sister chromatid breaks every time the cell copies its DNA. Breaks in one sister chromatid can arise by replication through a single-strand nick or by cleavage of a Holliday junction formed at a stalled and regressed replication fork. The broken fork can be repaired by sister chromatid recombination, especially by BIR. BIR results after the broken end is resected and is coated with RecA, which promotes a search for homology and strand invasion. The D-loop created by strand invasion can be converted into a unidirectional replication fork that will allow replication to be completed.

SUGGESTED READING

Anand RP, Lovett ST & Haber JE (2013) Break-induced DNA replication. *Cold Spring Harb Perspect Biol* doi:pii: cshperspect. a010397v1. 10.1101/cshperspect.a010397.

Haber JE (1999) DNA recombination: the replication connection. *Trends Biochem Sci* 24:271–275.

Kreuzer KN (2000) Recombination-dependent DNA replication in phage T4. *Trends Biochem Sci* 25:165–173.

Llorente B, Smith CE & Symington LS (2008) Break-induced replication: what is it and what is it for? *Cell Cycle* 7:859–864.

Michel B (2000) Replication fork arrest and DNA recombination. *Trends Biochem Sci* 25:173–178.

Petermann E, Helleday T (2010) Pathways of mammalian replication fork restart. *Nat Rev Mol Cell Biol* 11:683-687.

CHAPTER 2

DOUBLE-STRAND BREAK REPAIR PATHWAYS

For the most part, this book will focus on repair of chromosome double-strand breaks (DSBs). Broken chromosomes can be repaired in a variety of ways (**Figure 2.1**). One set of pathways manages to re-join the broken ends without relying on another homologous sequence as a template. There are several related *nonhomologous end-joining* (NHEJ) processes that can re-join blunt DNA ends or ends with 5′ or 3′ overhangs (**Figure 2.1E**). In many cases the re-joinings result in the loss of sequences around the ends and thus these processes are inherently mutagenic. NHEJ mechanisms will be examined in detail in Chapter 15. These mechanisms have enormous importance in human cancers as they account for many of the most potent oncogenic translocations between different chromosome segments.

2.1 SOME DNA REPAIR OCCURS AT SINGLE-STRANDED GAPS

This book focuses on the more accurate processes of homologous recombination in which the broken end or ends rely on a homologous template to restore the sequences that were interrupted or lost by the creation of a DSB. It would be misleading, however, to suggest that all important DNA repair involves DSBs. Repair of single-strand gaps also plays an important role in some circumstances, especially in repairing lesions by *sister chromatid recombination*. Gaps can occur during DNA replication if, for example, a UV-induced photodimer is on the lagging strand such that a new Okazaki fragment is unable to extend to the previous fragment (**Figure 2.2**). One way that such a gap can be repaired is by *template switching* of the leading strand, to get past the lesion. Single-strand gaps are also intermediates in the excision and replacement of oxidized bases during base excision repair (BER), in the removal of cyclobutane dimers and other damaged bases by *nucleotide excision repair* (NER), and during correction of mismatched bases in *mismatch repair* (MMR); but in all of these cases, the excision of the offending bases and some surrounding region is simply filled in by a DNA polymerase. We will examine sister chromatid recombination in Chapter 9.

2.2 REPAIR OF DSBs CAN OCCUR IN SEVERAL WAYS

As just stated, our focus is on DSB repair, mechanisms of which are shown in **Figure 2.3**. The BIR repair pathway reviewed in Chapter 1 represents a special case of homologous recombination when only one DSB

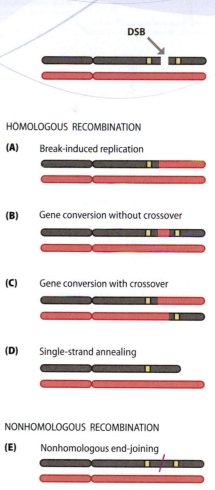

Figure 2.1 Major pathways of DSB repair. Homologous recombination can proceed by break-induced replication (BIR) (A) or by two different gene conversion pathways, synthesis-dependent strand annealing (SDSA) (B) or a double Holliday junction (dHJ) pathway (C). In addition, extensive 5′ to 3′ resection can produce single-strand annealing (SSA) between flanking homologous sequences (D). Alternatively DSBs can be repaired by several pathways of nonhomologous end-joining (NHEJ) (E).

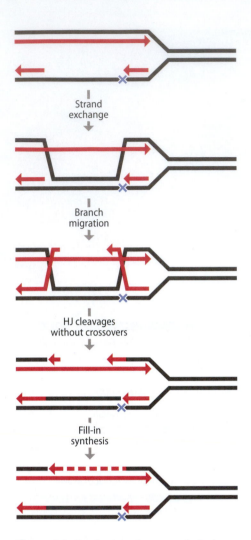

Figure 2.2 Single-strand gap repair during DNA replication, an example of template switching. A lesion on the lagging strand template (*blue* x) will prevent the completion of an Okazaki fragment, leaving a gap on one of the replicating chromatids. Newly synthesized strands are shown in *red*. Strand transfer from the leading strand chromatid's template strand can fill in the gap. This exchange leads to the formation of branched molecules, initially "half-Holliday junctions" that can be converted into full HJs by branch migration. This movement bypasses the lesion on the lagging template strand (although it is still not repaired). When these two Holliday junctions are resolved by cleavage—removing the crossed strands—there is now a gap on the leading strand. The 3′ DNA end flanking the gap can then be used as a primer to fill in the gap, using a repair DNA polymerase.

end is present and shares homology with an intact template. A BIR event between two homologous chromosomes is illustrated in **Figure 2.1A** and in detail in **Figure 2.3A**. BIR between homologous chromosomes yields a nonreciprocal *loss of heterozygosity* (LOH) for all markers further toward the telomere (that is, distal to the point of repair). LOH can expose recessive genetic markers in what was originally a heterozygote and can thus pose a genetic risk for the organism; BIR will be examined in detail in Chapter 8.

But most chromosome breaks arise when both ends of a DSB are homologous to a sister chromatid or a homologous chromosome. In special cases the template may be a short, nearly identical sequence located in an ectopic location, far away on the same chromosome or even on an unrelated chromosome. In these cases, BIR is suppressed in favor of two-ended DSB repair events, in which only a short patch of new DNA needs to be synthesized to restore chromosome integrity (**Figure 2.1B** and **2.1C**). Right around the DSB, gene conversions are also nonreciprocal, with the broken chromosome having a patch of sequences copied from the donor template. If the donor template is identical to the recipient then there is no genetic alteration—the break is perfectly restored—but if there are any nucleotide differences between the donor and recipient, then the DSB repair event results in what is termed a *gene conversion* (**Figure 2.4**). In a diploid such a gene conversion would result in a gene that was heterozygous (+/−) becoming homozygous (+/+ or −/−). For simplicity, and because the molecular process is the same whether or not there is actually a genetic difference between donor and recipient, we will refer to all such short-patch DSB repair events as gene conversions. In many instances, gene conversions have no other chromosomal or genetic alteration, beyond the replacement of sequences at the site of the break. In mitotic cells, these *noncrossover* outcomes (Figure 2.1B) strongly predominate. One mechanism by which these noncrossover gene conversions arise is through a mechanism known as *synthesis-dependent strand annealing* (SDSA) (**Figure 2.3B**).

However, gene conversions can also be accompanied by crossing over, resulting in a reciprocal crossing over between the donor and recipient chromosomes. We already saw examples of such crossovers in the sister chromatid exchanges discussed in Chapter 1, where repair occurs between two sister chromatids. But crossovers can also occur between homologous chromosomes in diploids as illustrated in Figure 2.1. Such reciprocal exchanges are relatively rare in somatic (mitotic) cells, but—depending on how the chromatids segregate in mitosis—when crossovers do occur they may result in a loss of heterozygosity from the point of crossing over to the end of the chromosome (**Figure 2.5**). LOH in this situation differs from LOH in BIR because these gene conversion events give rise to a *reciprocal crossover*, so that one of the mitotic segregants after recombination is homozygous for markers distal on one parental chromosome and the other mitotic segregant is homozygous for markers on the other chromosome (Figure 2.5).

While crossovers associated with DSB repair are infrequent in mitotic cells, they are frequent in meiosis, during recombination between homologous chromosomes. As pointed out above, the SDSA mechanism of patching up a DSB occurs in a way that yields only noncrossover outcomes. Crossovers arise from a second, very important mechanism of DSB repair which proceeds through a recombination intermediate that has not one, but two Holliday junctions. We will refer to this mechanism as the double HJ, or dHJ, mechanism (**Figure 2.3C** and *Box 2.1*).

In later chapters, especially Chapters 6 and 7, we will devote considerable attention to the molecular mechanisms of gene conversions, by both SDSA and dHJ, in mitotic cells. We will also look at variations on

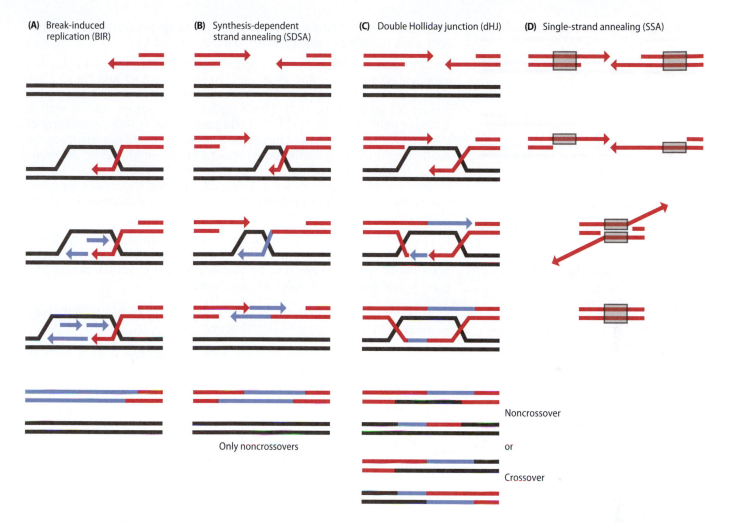

(A) Break-induced replication (BIR)

(B) Synthesis-dependent strand annealing (SDSA)

(C) Double Holliday junction (dHJ)

(D) Single-strand annealing (SSA)

Only noncrossovers

Noncrossover

or

Crossover

Figure 2.3 Molecular mechanisms of DSB repair by homologous recombination. All homologous recombination repair of DSB damage begins with 5′ to 3′ resection of DSB ends, to produce long single-strand DNA (ssDNA). (A) In BIR, the ssDNA can—with the help of recombination proteins (not shown)—invade and pair with a homologous donor template producing a displacement loop. In the absence of a second homologous DSB end, the D-loop is converted to a unidirectional DNA replication fork, using leading- and lagging-strand synthesis to copy the template chromosome to its end. New DNA synthesis is shown with *light blue* lines. (B) SDSA, one of two major pathways of gene conversion, begins similarly to BIR, but the D-loop assembles only one DNA polymerase that uses the 3′ end of the invading strand as a primer to initiate new DNA synthesis. DNA copying proceeds with the movement of the D-loop, displacing the newly copied strand as it moves and leaving the template unaltered. The second end of the DSB, also resected, can pair with the newly copied strand and initiate a second round of copying, to complete DSB repair in the absence of crossing over. (C) An alternative gene conversion mechanism that also copies only a short patch of DNA to repair the DSB involves similar initial strand invasion steps to SDSA and BIR but creates a stable D-loop by "capturing" the second end to base pair with some of the exposed D-loop sequences. Copying of the two template strands results in the formation of a double Holliday junction (dHJ) that can be resolved either to produce noncrossover or crossover outcomes. (D) In the absence of a homologous template on another chromosome, 5′ to 3′ resection may continue until flanking homologous sequences, indicated by *gray* boxes, are exposed. The complementary ssDNA strands can then anneal. The protruding nonhomologous 3′ ends are clipped off and the 3′ ends are used as primers to fill in the gaps, yielding a deletion between the flanking repeats in a process termed single-strand annealing (SSA).

these mechanisms in which it is possible to obtain crossovers without dHJs and noncrossovers without SDSA. We will also spend time examining the elaborate ways in which the basic recombination machinery has been modified to deal with the complex problem of meiotic recombination. Meiotic cells have to cope with dozens or hundreds of DSBs along the chromosomes while mitotic recombination usually occurs when cells suffer relatively few breaks. The large number of meiotic DSBs facilitate the genetic exchanges which are essential for both generating diversity and, even more importantly, assuring the proper segregation

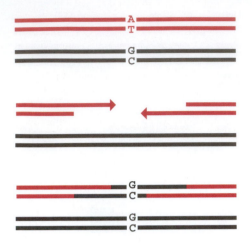

Figure 2.4 Gene conversion. Repair of a DSB by copying a short patch of DNA from a nearly identical donor template can result in the replacement of one allele's sequences by those of a second allele, changing a heterozygous locus to one that is homozygous for one allele.

of chromosomes. The meiotic modifications of gene conversion mechanisms are presented principally in Chapters 12 and 13.

An alternative pathway of homologous recombinational repair is known as single-strand annealing (SSA), shown in **Figure 2.1D** and **Figure 2.3D**. Here extensive 5′ to 3′ resection reveals complementary, homologous sequences flanking the DSB. These repeated sequences can be tens of kilobase pairs from the break and still participate in SSA. Annealing of the ssDNA regions, followed by clipping off the 3′-ended nonhomologous sequences and filling in the gaps, results in potentially large deletions of the regions surrounding the DSB, between the flanking homologous sequences. SSA events between dispersed repeated sequences are associated with a number of human diseases. SSA will be discussed in detail in Chapter 5.

In BIR and in gene conversion, the initial steps of repair (5′ to 3′ resection of the DSB ends and strand-invasion of the single strand into a homologous donor template) seem to be identical or quite similar. So we first need to address several important questions:

1. How are the DSB ends processed to create long ssDNA tails that are needed in homologous recombination?

2. How is the RecA/Rad51 filament formed and how does it work in searching for homologous template sequences?

These questions will occupy our attention in the next two chapters.

Figure 2.5 Loss of heterozygosity resulting from a reciprocal crossover and subsequent segregation of mitotic chromosomes. The case of a recessive loss-of-function rb mutation is shown. A crossover in two of the four chromatids after DNA replication can segregate in two different ways as the pairs of sister chromatids align in metaphase. In one arrangement (*top right*) chromosome segregation produces cells homozygous for the *gray* arm of the chromosome distal to the point of crossover, including the recessive rb mutation. Thus, the two daughter cells are homozygous for one or the other allele (RB or rb). In the other mode of segregation (*bottom right*) there is a change in the linkage of rb relative to markers near the centromere, but the cells remain heterozygous for RB/rb.

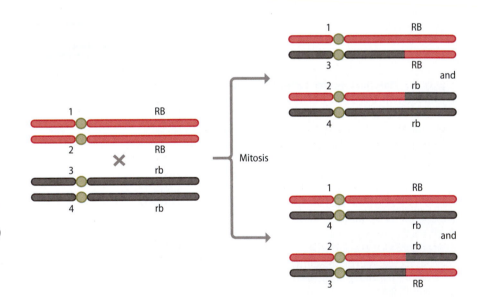

Box 2.1 Naming several mechanisms of DSB repair

The Appendix of this book presents a detailed history of the evolution of molecular models of homologous recombination. The dHJ mechanism was first proposed by Michael Resnick in 1976 but was extended and elaborated by Jack Szostak, Terry Orr-Weaver, Rodney Rothstein, and Frank Stahl in 1983. At the time they called this mechanism "the double-strand break repair mechanism (DSBR)". But as we now know that there are several DSBR mechanisms, it seems difficult to continue to call one of them "the" DSBR mechanism. Hence, I have elected to call it the dHJ model after its most important feature. Resnick also anticipated a second mechanism of DSB repair, now called synthesis-dependent strand annealing (SDSA). A third mechanism, initially called recombination-dependent DNA replication by Ann Skalka in 1974, is now commonly called break-induced replication (BIR).

SUMMARY

While some DNA repair mechanisms involve the repair of single-stranded gaps in DNA, most repair events require the repair of double-strand breaks. Breaks can be repaired by nonhomologous end-joining, but the most accurate ways of repairing such lesions involve homologous recombination. Repair mechanisms can be categorized as either break-induced replications or gene conversions, depending on whether the DSB ends share homology on one or both sides of the break. There are two major types of gene conversion repair mechanisms: synthesis-dependent strand annealing that produces only noncrossover outcomes, and a double Holliday-junction mechanism that can yield both crossover and noncrossover results. In addition, repair can occur by single-strand annealing in instances in which the DSB is flanked by direct repeat homologies.

SUGGESTED READING

Brugmans L, Kanaar R & Essers J (2007) Analysis of DNA double-strand break repair pathways in mice. *Mutat Res* 614:95–108.

Krejci L, Altmannova V, Spirek M, & Zhao X (2012) Homologous recombination and its regulation. *Nucl Acids Res* 40:5795–5818.

Lieber MR (2008) The mechanism of human nonhomologous DNA end joining. *J Biol Chem* 283:1–5.

Maher RL, Branagan AM, & Morrical SW (2011) Coordination of DNA replication and recombination activities in the maintenance of genome stability. *J Cell Biochem* 112:2672–2682.

O'Driscoll M & Jeggo PA (2006) The role of double-strand break repair – insights from human genetics. *Nat Rev Genet* 7:45–54.

Pâques F & Haber JE (1999) Multiple pathways of recombination induced by double-strand breaks in *Saccharomyces cerevisiae. Microbiol Mol Biol Rev* 63:349–404.

Persky NS & Lovett ST (2008) Mechanisms of recombination: lessons from *E. coli. Crit Rev Biochem Mol Biol* 43:1–24.

San Filippo J, Sung P & Klein H (2008) Mechanism of eukaryotic homologous recombination. *Annu Rev Biochem* 77:229–257.

CHAPTER 3

RecA/Rad51 AND THE SEARCH FOR HOMOLOGY

At the heart of nearly all homologous recombination mechanisms is the RecA/Rad51 nucleoprotein filament that carries out a search for homologous sequences and then effects strand exchange. In this chapter we examine the nature of the recombinase filament and how it is able so efficiently to carry out recombination.

3.1 RecA AND Rad51 ARE THE KEY STRAND EXCHANGE PROTEINS

The central strand exchange protein in *E. coli* was identified genetically by John Clark in 1969 as the first recombination-defective mutation (hence it was called RecA). RecA protein was purified by several labs a decade later. In budding yeast a group of ionizing radiation-sensitive mutations were identified by Robert Mortimer's group and called Rad50, Rad51, and so forth, to distinguish them from a set of UV-sensitive complementation groups that began with Rad1. The UV-radiation-sensitive mutations identified most of the genes needed to carry out nucleotide excision repair, while the X-ray-sensitive mutations proved to have roles in DSB repair. Protein sequence homology between RecA and Rad51 was only noted in 1992. In most eukaryotes Rad51 homologs are also known as Rad51 but in fission yeast it is called Rhp51 and in *Drosophila*, SpnA, while homologs in archaea are called RadA. For simplicity, in discussing these proteins below the eukaryotic protein will unusually just be called Rad51. In some eukaryotes there is a second, meiosis-specific Rad51-like protein, Dmc1, which will be discussed in detail in Chapter 13, on meiotic recombination.

RecA and Rad51 carry out the same function in recombination. RecA/Rad51 are proteins of approximately 43 kilodaltons (kD). Their most common feature is an ATPase domain containing Walker A and B motifs that are found in many ATPases (**Figure 3.1**). ATPase activity is stimulated by binding to either single-stranded DNA (ssDNA) or double-stranded DNA (dsDNA). There is a greater stimulation by ssDNA. A mutation in the Walker A box changing a key lysine to alanine abolishes both DNA binding and ATPase activity and creates a functionally inactive RecA/Rad51. However, mutations that change this lysine to arginine, such as the Rad51-K191R mutation in budding yeast or human Rad51-K133R, yield a protein that binds ATP but cannot hydrolyze it. Cells carrying these mutations are recombination and repair defective, but surprisingly, the mutant yeast protein—when overexpressed—partially complements the repair and recombination defects seen in a *rad51* deletion. Hence,

(A)

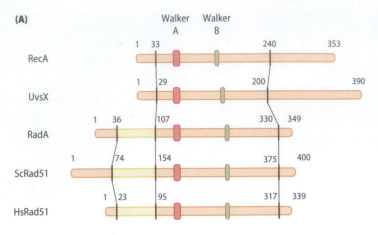

(B)

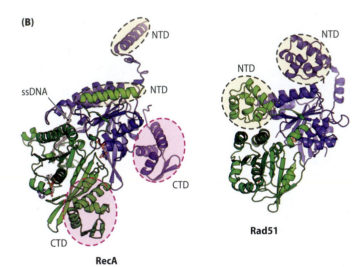

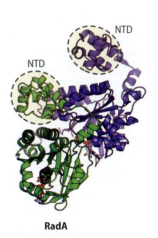

Figure 3.1 Similarities and differences among RecA homologs. (A) Protein sequence alignments of RecA and homologs UvsX (phage T4), RadA (*Pyrococcus furiosus*), *Saccharomyces cerevisiae* Rad51 (ScRad51), and human Rad51 (HsRad51). The conserved core domain with its ATPase active site, including the Walker A (*red*) and Walker B (*green*) motifs are shown. (B) Crystal structures of dimers of RecA, Rad51, and RadA showing the strong similarity of Rad51 and RadA structures in contrast to RecA. The Rad51 structure does not show 80 amino acids that are specific to *S. cerevisiae*. The amino (N)-terminal domains (NTD) of each monomer are circled. The NTDs of RecA and RadA line the upper surface of the filament whereas the carboxy (C)-terminal domains (CTD) of RecA (also circled) line the lower surface. The path of ssDNA is known and shown for RecA. (A, adapted from Lin Z, Kong H, Nei M et al. [2006] *Proc Natl Acad Sci USA* 103:10328–10333. With permission from the National Academy of Sciences. B, courtesy of Phoebe Rice, University of Chicago.)

ATP binding but not ATP hydrolysis is essential for the activity of these recombinases; nevertheless, full activity certainly requires ATP turnover.

Although the ATP binding domain is well preserved, RecA and Rad51 are surprisingly structurally dissimilar and there are significant biochemical differences that may be important during repair *in vivo*. Most notably, RecA contains a large carboxy (C)-terminal domain that is absent in Rad51; conversely, Rad51 proteins (and archaeal RadA) have an N-terminal extension that is absent in the RecA protein (Figure 3.1). Both RecA's C-terminal domain and Rad51's N-terminal domain bind DNA, but it is possible they evolved by convergent evolution. Both RecA and Rad51 can bind both ssDNA and dsDNA and in both cases each monomer binds to three nucleotides of ssDNA or three base pairs of dsDNA. RecA and Rad51 (and RadA) all form right-handed filaments. Viewed by electron microscopy the filaments have at least superficially a similar architecture (**Figure 3.2**), but the pitch of the filament is different.

RecA, Rad51, and RadA filaments can simultaneously bind both single- and double-stranded DNA. The primary binding site preferentially associates with ssDNA, while a secondary site will accommodate dsDNA. When only dsDNA is present *in vitro* it preferentially binds to it in the primary binding site, with one strand occupying the site where ssDNA would lie (and where, after strand exchange had occurred, a new dsDNA molecule would lie). The bound DNA is stretched and underwound, so that the DNA inside a recombinase filament is 1.5 times its normal B-form DNA length. This stretching for ssDNA can clearly be seen in electron micrographs of the recombinase filament comparing the contour length of circular ssDNA

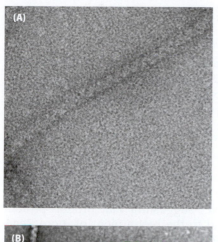

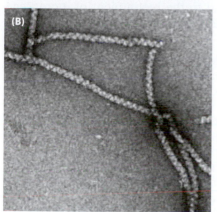

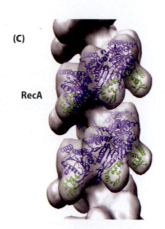

RecA

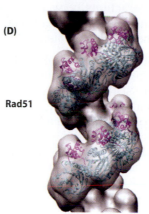

Rad51

Figure 3.2 Electron micrographs of RecA and Rad51 filaments bound to dsDNA and their three-dimensional reconstruction. Electron micrographs of negatively stained filaments of RecA-dsDNA (A) and budding yeast Rad51-dsDNA filaments (B), showing a highly repeated right-handed filament structure. (C) Three-dimensional reconstructions of RecA–DNA filaments. The reconstructions are displayed as *gray* transparent surfaces, with ribbon models for the recombinase shown within each. The RecA core is shown in *purple*, with the C-terminal domain shown in *green*. (D) Three-dimensional reconstructions of Rad51-DNA filaments. The Rad51 core is shown in *blue*, while the N-terminal domain is shown in *magenta*. (A, from Galkin VE, Britt RL, Bane LB et al. [2011] *J Mol Biol* 408:815–824; B, from Galkin VE, Wu Y, Zhang XP et al. [2006] *Structure* 14:983–992; C, D, from Lucarelli D, Wang YA, Galkin VE et al. [2009] *J Mol Biol* 391:269–274. With permission from Elsevier.)

to that of RecA-coated DNA (**Figure 3.3**). This untwisting of the DNA facilitates strand exchange. When dsDNA is added to the recombinase filament, it too is stretched and underwound in a similar fashion.

3.2 FILAMENT ASSEMBLY OF RecA AND Rad51 CAN BE ASSAYED *IN VITRO*

Different assays have been used to determine how RecA or Rad51 polymerizes on ssDNA to form a filament. *In vitro*, the development of rapid kinetic techniques, the use of fluorescence resonance energy transfer (FRET) methods and single-molecule analysis have made it possible to study the assembly of RecA/Rad51 filaments in great detail. These recent analyses confirmed and extended many earlier experiments showing that RecA assembles preferentially in a 5′ to 3′ direction along an ssDNA molecule, but in fact RecA can polymerize from either end (**Box 3.1**). In the presence of ATP, the filament dissociates from its ends (**Inline Figure 3.1**).

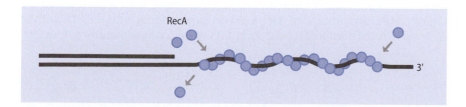

E. coli RecA cannot readily displace the single-strand binding protein SSB from ssDNA without the help of mediator proteins (RecF, RecO, and RecR), but once a dimer of RecA has nucleated on ssDNA it has no

(A)

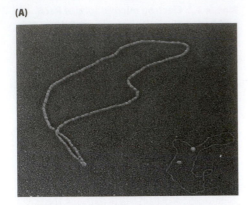

(B)

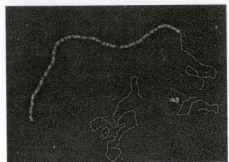

(C)

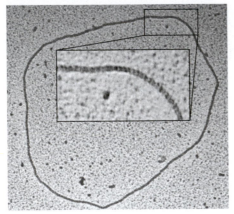

Figure 3.3 RecA proteins bound to single-stranded and double-stranded DNA as viewed by electron microscopy. RecA protein was bound to relaxed nicked circular replicative form of fd phage (A) or to linearized DNA (B). Note the 1.5-fold increase in length of the filament-bound DNA compared to unbound molecules in the same fields. (C) Single-stranded circular mp18 phage DNA was coated with RecA. One segment is enlarged to show the regular spacing of RecA protein along the filament. Again there is a 50% increase in contour length relative to unbound circular ssDNA. (A, B, from Dunn, K, Chrysogelos S & Griffith J [1982] *Cell* 28:757–765; C, from Drees, JC, Lusetti SL, Chitteni-Pattu S et al. [2004] *Mol Cel* 15:789–798. With permission from Elsevier.)

Box 3.1 Recent results depend on earlier pioneers
The recent insights into strand invasion on which we have focused are examples of scientists being able to see so far because they were "standing on the shoulders of giants." A more detailed account of the many individual contributions can be found in the review articles listed at the end of the chapter and in the detailed account in the Appendix.

problem displacing SSB without further requirement for the mediators. RecA assembly is highly cooperative, suggesting that protein–protein interactions between monomers lead to the recruitment of more RecA once the process gets started. RecA does not apparently dissociate from the middle of a filament.

Stephen Kowalczykowski's group has used single-molecule technology to look at fluorescently labeled Rad51 or RecA binding to dsDNA (**Figure 3.4**) or ssDNA (**Figure 3.5**). DNA is tethered at one end to a polystyrene bead and the DNA molecule is stretched by the flow of fluid. An optical trap holding the bead allows the bead and its DNA to be moved into a parallel flowing solution containing GFP-tagged Rad51 (Figure 3.4). After briefly "dipping" the DNA into the Rad51 solution, the bead and its single DNA molecule are returned to the channel with no Rad51 to observe where and how much GFP-Rad51 has been bound. Repeated dipping into the Rad51 solution shows how small clusters of Rad51 first form at several sites along the DNA and then nucleate the formation of a long, nearly continuous filament. This work shows that filament growth begins at several independent nucleation sites where a cluster of 4–5 Rad51 (or RecA) molecules promotes subsequent elongation, at different rates from both ends. The assembly is most efficient using a nonhydrolyzable ATP analog, such as ATP-γS, since ATP hydrolysis can promote filament disassembly.

The assembly of RecA onto SSB-coated ssDNA has also been visualized by Kowalczykowski's lab (Figure 3.5) using similar microscopic analysis of individual tethered DNA molecules in a microfluidic device that allows them to expose the molecule to solutions of two different fluorescently labeled RecA proteins. SSB-bound ssDNA is first dipped into a solution of red-labeled RecA, which forms small clusters along the molecule. Then the molecule is repeatedly immersed in green-labeled RecA and the extent of RecA addition to each end of the RecA cluster is observed after each "dipping". This analysis demonstrated that RecA was added about twice as rapidly to the 3' end of the initial cluster as to the 5' end, but clearly filament growth can occur in either direction.

An analogous experiment to examine Rad51 assembly on ssDNA has not yet been performed but other biochemical experiments suggest that Rad51 filament assembly is less cooperative than RecA assembly. Moreover, Rad51 apparently assembles a filament with a preference for the opposite direction (3' to 5'), based on experiments with dsDNA with 5' or 3' ssDNA tails (**Inline Figure 3.2**) but as with RecA, assembly can occur

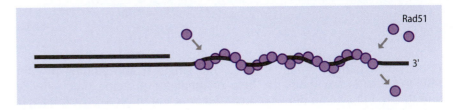

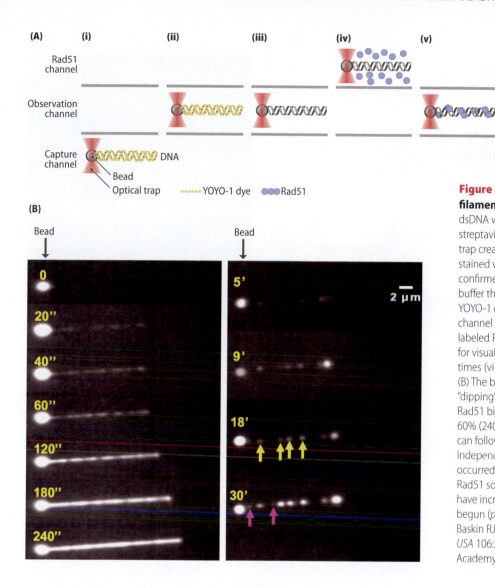

Figure 3.4 Single molecule analysis of Rad51 filament formation. (A) 50 kilobases (kb) phage λ dsDNA with biotin at the terminus is captured on a streptavidin-coated polystyrene bead held in an optical trap created by focusing a laser. (i) The DNA is first stained with YOYO-1 so that its capture on a bead can be confirmed. (ii) The bead can be moved into a stream of buffer that stretches out the DNA and washes away the YOYO-1 dye (iii). (iv) The DNA is then moved into another channel containing a concentration of fluorescently labeled Rad51 and pulled back into the buffer channel for visualization (v); this process is repeated several times (vi–vii) to follow the growth of the Rad51 filament. (B) The binding of Rad51 is shown for different times of "dipping" the DNA into the Rad51 solution (*left*). As more Rad51 binds the length of the DNA is stretched about 60% (240"). Using a lower concentration of Rad51, one can follow the assembly in more detail (*right*). At 18' four independent nucleation events (*yellow arrows*) have occurred. When the same molecule is returned to the Rad51 solution for further incubation, by 30' these foci have increased in intensity while two new ones have begun (*pink arrows*). (Adapted from Hilario J, Amitani I, Baskin RJ & Kowalczykowski SC [2009] *Proc Natl Acad Sci USA* 106:361–368. With permission from the National Academy of Sciences.)

at either end. This direction of assembly correlates well with the observation that resection of DSB ends proceeds 5′ to 3′, leaving dsDNA with an increasingly long, 3′-ended ssDNA tail. For both RecA and Rad51, since the invading 3′ ssDNA end will be used as a primer for new DNA synthesis during repair, filament formation and the strand exchange process should ensure that the 3′ end is assimilated into the strand exchange intermediate.

3.3 X-RAY CRYSTALLOGRAPHY HAS REVEALED A GREAT DEAL ABOUT RecA STRUCTURE AND FUNCTION

The clearest view of the RecA filament has come from directly looking at the conformation of DNA and protein at very high resolution by X-ray crystallography. To create a structure long enough to mimic a filament in solution, Nikola Pavletich's lab used recombinant DNA methods to encode five RecA proteins into a single protein polypeptide, each subunit separated from the next by a 14 amino acid linker sequence (**Figure 3.6**). As expected, the ssDNA molecule is bound by RecA with 3 nucleotide (nt) per RecA monomer. Similar to measurements made previously, the DNA is stretched by about 50% from the length expected for right-handed helical B-form DNA. What was unexpected—and indeed astonishing—is

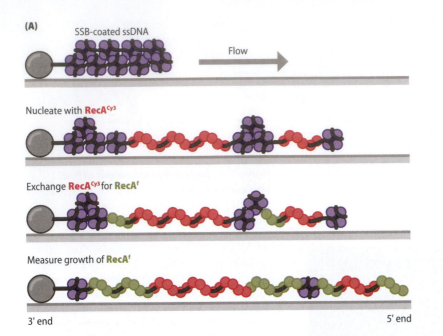

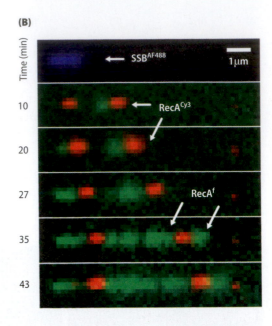

Figure 3.5 Single-molecule analysis of RecA binding to SSB-coated ssDNA. As in Figure 3.4, a ssDNA molecule is tethered to a polystyrene bead. (A) As illustrated schematically, the molecule is first incubated with SSB and then exposed to a channel containing *red* (Cy3) labeled RecA until several small clusters of RecA form on the DNA. Then the molecule is repeatedly exposed to GFP-labeled RecA (*green*, RecA^f) to observe the growth of the RecA filament, which extends about twice as fast in the 5′ to 3′ direction as in the opposite direction. (B) Microscopic images of the actual experiment. (Adapted from Bell JC, Plank JL, Dombrowski CC & Kowalczykowski SC [2012] *Nature* 491:274–278. With permission from Macmillan Publishers, Ltd.)

that the stretching is not uniform. Within the 3-nt unit bound by one RecA monomer, the DNA has nearly the conformation of B-DNA; but between each RecA-bound triplet the strand is stretched from about 3.4 Å per base to a distance of 7.8 Å and a 42° left-handed twist is introduced (Figure 3.6). This means that the 3 nt within one monomer could contact B-form dsDNA without having to distort the dsDNA helix, but making any subsequent adjacent base pairs would require that the dsDNA molecule also be both stretched and underwound. Inspection of the RecA structure showed that two loops within the protein that are not well structured in the RecA monomer become well structured and act as "fingers" to push the region between the three bases apart (**Figure 3.7**).

Figure 3.6 X-ray crystallography of a recombinant RecA molecule with six monomers, bound to ssDNA. (A) The complete structure, shown as a ribbon of the protein backbone with *orange* ssDNA held along the filament axis. (B) The ssDNA (*orange*) is shown in isolation with the protein "fingers" that constrain the three nucleotides bound to each RecA monomer in a B-DNA configuration. In between the sets of three nucleotides, the ssDNA is stretched and underwound relative to B-DNA. (From Chen Z, Yang H & Pavletich NP [2008] *Nature* 453:489–494. With permission from Macmillan Publishers, Ltd.)

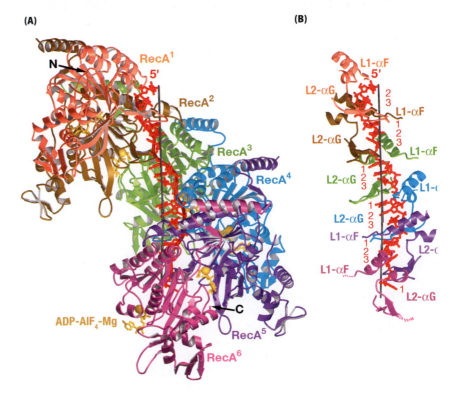

3.4 STRAND EXCHANGE CAN BE STUDIED *IN VITRO*

In the late 1970s the labs of Charles Radding and Robert Lehmann purified RecA and showed that it could catalyze a strand exchange reaction *in vitro*. Specifically, they showed that the incubation of radioactively labeled, supercoiled dsDNA with RecA and complementary ssDNA would create a *displacement loop* (D-loop) in which the displaced strand of the labeled DNA would be single stranded and thus could be retained on a nitrocellulose filter (**Figure 3.8**). The concentration dependence of RecA protein showed that the active form involved a highly cooperative assembly process (to make a filament), as indicated by the "S-shaped" curve (**Figure 3.8B**). They also photographed the D-loops using electron microscopy (**Figure 3.8C**).

The simplest and most widely used assay of the ability of RecA/Rad51 to carry out strand exchange does not precisely replicate the events that occur *in vivo*, where a broken DNA end with a 3′-ended ssDNA tail coated with RecA or Rad51 invades an intact duplex DNA donor (which is naturally negatively supercoiled around nucleosomes in eukaryotes). Instead, a closed circular ssDNA molecule, coated with RecA or Rad51 is confronted with a homologous linearized duplex DNA that is usually ^{32}P-labeled at its 5′ ends (**Figure 3.9**). When strand invasion is initiated, one strand of the dsDNA will begin to pair with the circular ssDNA, displacing the complementary strand. Initially a *joint molecule* (JM) is formed, in which the circular ssDNA is partially base paired with one strand of the duplex. This three-stranded recombination intermediate (a joint molecule) migrates more slowly in gel electrophoresis compared to the labeled dsDNA and is easily visualized. When strand exchange is completed, one strand of the original duplex has been incorporated into what is now a nicked circular dsDNA and the complementary strand is now seen as a ssDNA molecule (Figure 3.9). This assay has been used extensively in analyzing various RecA or Rad51 mutants as well as in determining the roles of various mediator proteins.

Although the strand exchange reactions are similar, it turns out that RecA and Rad51 display opposite *polarity*. RecA on the ssDNA captures the 3′ end of the duplex molecule and progresses in a 5′ to 3′ direction along the ssDNA as shown in **Figure 3.9A**; that is, the sequences being transferred from the donor molecule are assimilated from the 3′ end toward the 5′

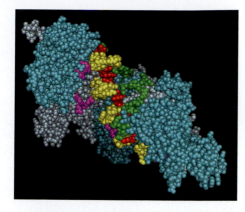

Figure 3.7 A space-filling model of the five RecA subunits and ssDNA. The image shows the two protein "fingers" (*yellow* and *green*) that hold the three nucleotides of ssDNA (*red*) bound to each monomer in B-DNA configuration while underwinding the backbone in between subunits. Adjacent RecA monomers are shown in alternating *gray* and *cyan*. The *magenta* areas are basic residues that will hold dsDNA in a second binding site. The ssDNA is shown in *red*. (Derived by Douglas Bishop, University of Chicago, from data in Chen Z, Yang H & Pavletich NP [2008] *Nature* 453:489–494. With permission from Macmillan Publishers, Ltd.)

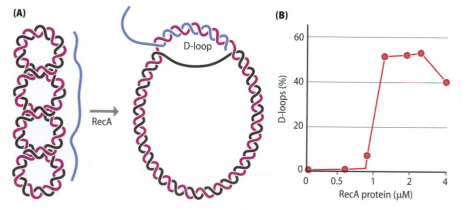

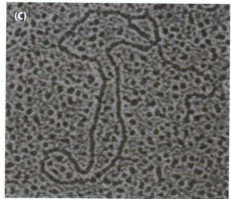

Figure 3.8 The first RecA-mediated strand exchange assay. (A) Negatively supercoiled DNA becomes less wrapped around itself (that is, its "writhe" is reduced) as unwinding of the duplex occurs during the assimilation of ssDNA (*blue*) into the D-loop (an increase in "twist"). (B) The formation of the D-loop means that part of the dsDNA is single-stranded, which causes the DNA to be retained on a nitrocellulose filter. The abrupt increase in the concentration-dependence curve as more RecA is added argues that RecA's binding involves a highly cooperative step (that is, filament formation). (C) An electron micrograph showing a D-loop as illustrated in (A). (B, C, adapted from Shibata T, DasGupta C, Cunningham RP & Radding CM [1979] *Proc Natl Acad Sci USA* 76:1638–1642. With permission from the National Academy of Sciences.)

end. By contrast, Rad51 engages the identical substrates by grabbing the 5′ end of the dsDNA and elongates the strand assimilation in a 5′ to 3′ direction (**Figure 3.9B**). From the vantage point of the ssDNA, the ssDNA is

Figure 3.9 Strand exchange assays for RecA and Rad51. (A) RecA-coated circular ssDNA (SS) is mixed with homologous linear duplex DNA (L). First a joint molecule (JM) is formed and then a nicked circle (NC) and a single-strand (SS) are produced as products, as seen after agarose gel electrophoresis and staining DNA with a green dye. Closed circular (CC) DNA was used as a reference. (B) Linear dsDNA (DS) was labeled with ^{32}P (*) at both 5′ ends and mixed with circular homologous ssDNA coated with Rad51. The products separated by gel electrophoresis and visualized by autoradiography reveal joint molecules (JM) and nicked circular (NC) and linear single-stranded (LSS) products. (A, adapted from Calmann MA & Marinus MG [2004] *Proc Natl Acad Sci USA* 101:14174–14179. With permission from the National Academy of Sciences. B, adapted from Namsaraev E & Berg P [1997] *Mol Cell Biol* 17:5359–5368. With permission from the American Society for Microbiology.)

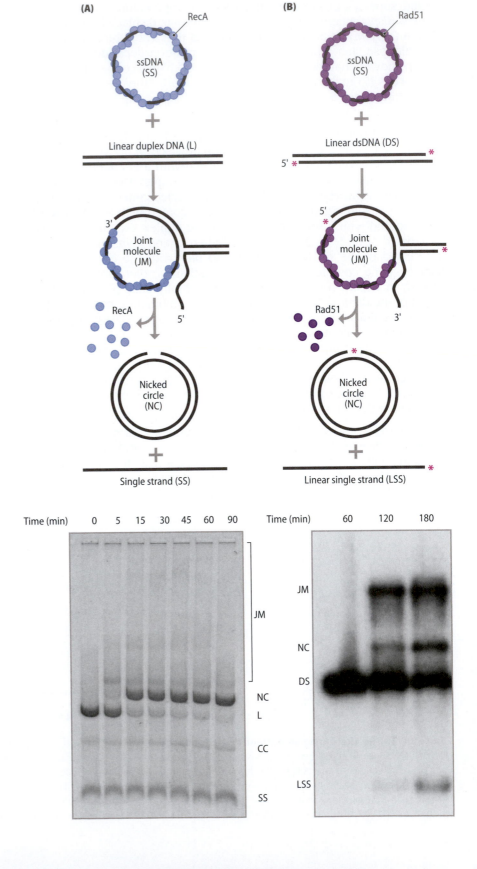

being "taken up" in a 3' to 5' direction. Thus the eukaryotic enzyme seems evolved to polymerize on a 3' protruding ssDNA end and to extend the strand exchange to make longer regions of hybrid DNA.

Another very informative assay employs a gapped circular molecule and homologous linear dsDNA (**Figure 3.10**). RecA will promote strand exchange, thus creating a joint molecule intermediate—a four-strand, branch-migrating Holliday junction. The initial strand invasion occurs with the displacement of a homologous strand, to create what is called

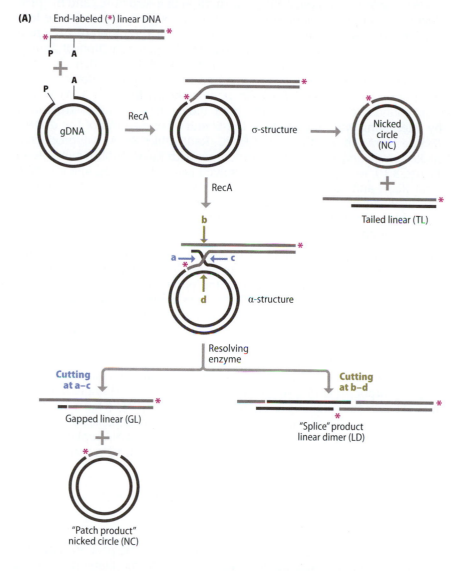

Figure 3.10 Formation of HJ-containing "α-structures" during strand-exchange mediated by RecA. (A) A 5'end-labeled (*) linear DNA and a gapped, homologous circular molecule can be recombined by RecA first into a sigma (σ) structure and then into an alpha (α) structure containing a HJ. The addition of the HJ resolving enzyme RuvC will cleave the HJ to form two novel products by either cutting along the a–c or b–d axes. (B) Southern blot analysis of the radioactively labeled strand from the assay shown in (A). Without the HJ resolvase RuvC, RecA creates joint molecules (α) that progress later to form nicked circles (NC) and gapped linear (GL) exchange outcomes. When RuvC is included in the reaction, a novel linear dimer (LD) rapidly appears after formation of α-structures. (Adapted from Eggleston AK, Mitchell AH & West SC [1997] *Cell* 89:607–617. With permission from Elsevier.)

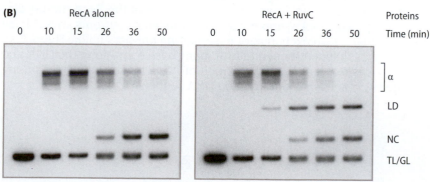

a *sigma-structure* (σ-structure). Then strand invasion continues, generating the so-called *alpha-structure* (α-structure) containing an HJ. The α-structure can be branch migrated around the circle by the continued action of RecA. α-Structures are substrates for HJ cleaving enzymes such as *E. coli*'s RuvC. These enzymes can cut the HJ in two different ways, yielding in one case a nicked circle and a gapped linear molecule, and in the other, a linear dimer (**Figure 3.10B**). Note that if the initial substrates are completely homologous, then RecA, without RuvC, can create a nicked circle and a tailed linear outcome.

Rad51 is able to carry out both the formation of α-structures and the production of the two final products *in vitro*, although it appears to be less efficient than RecA. The differences may reflect the different preferred polarity of filament assembly but may simply be a reflection of finding the most optimal biochemical assay conditions. But again there is a distinct difference in the process compared to RecA. Rad51 (for example, the Rhp51 of *Schizosaccharomyces pombe* shown in **Figure 3.11**) shows a profound preference to create joint molecules (σ- and α-structures) by interacting with homology at the 5′ end of one strand of the dsDNA (**Figure 3.11A**). In the absence of RuvC, Rad51, like RecA, will produce a nicked circle and a tailed-linear outcome. With RuvC, as shown before in Figure 3.10, α-structures can be resolved to yield a nicked linear dimer. That RecA and Rad51 should arrive at the same intermediate in opposite ways again raises a question whether they represent convergently evolved proteins that achieve the same results without having descended from a common ancestor.

RecA and Rad51 will promote strand exchange so long as the substrates are homologous. If the linear DNA contains a long segment of nonhomology, then exchange will progress to the point where homology ceases (**Figure 3.12A**). But RecA and Rad51 are able to bypass smaller regions of nonhomologies so long as there are additional homologous sequences beyond the point of discontinuity (**Figure 3.12B**). *In vitro*, regions of a few hundred base pairs can be assimilated; *in vivo*, probably with the aid of other proteins such as DNA helicases, large heterologous loops can be formed. The strand exchange proteins can also—*in vitro* and *in vivo*—incorporate DNA that has a significant number of single base-pair mismatches, thus creating a region of *heteroduplex DNA* (**Figure 3.12C**). As we will see in later chapters, cells use the *mismatch repair* proteins—the enzymes that correct misincorporation errors during DNA replication—to repair mismatches in heteroduplex DNA that arise during recombination.

3.5 STRAND EXCHANGE TAKES PLACE INSIDE THE RecA FILAMENT

Despite having purified RecA available to study for the past 30 years, we are still not certain exactly what happens inside the filament to allow strand exchange. If we imagine a cross section of the RecA/Rad51 filament, the process looks simple: a single-stranded A confronts an A::T base pair and there is an exchange of base-pairing partners (**Inline Figure 3.3**). But this must happen sequentially at many base pairs along the DNA and the big question is how such regions of homology are located among hundreds of thousands of unrelated DNA sequences.

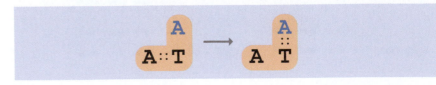

A topological approach to studying strand invasion

A very clever approach taken by Andrzej Stasiak used RecA-coated ssDNA oligonucleotides that could pair with closed circular DNA, so that strand exchange would occur, but the displaced strand (a D-loop) remained part

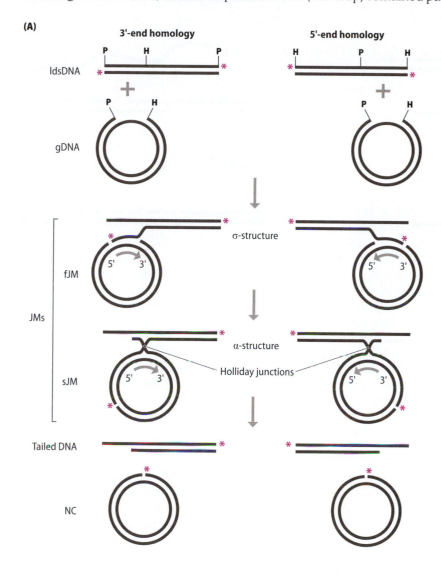

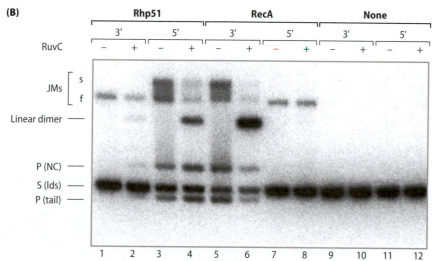

Figure 3.11 *S. pombe* **Rad51 promotes a DNA four-strand exchange reaction** *in vitro* **using opposite polarity from** *E. coli* **RecA.** (A) Illustration of the formation of α-structures beginning with homology between a ssDNA gap (gDNA) and the 3' or 5' end of a radioactively labeled (*) linear dsDNA (lds). In the absence of a HJ resolvase, the products are a tailed linear DNA and a nicked circle (NC). The assimilation of the strand from the dsDNA and direction of migration of the HJ, relative to the ssDNA segment, is shown by an arrow. (B) Gel image of 3' end-labeled (and 3' end-homology) or 5' end-labeled (and 5' end-homology) dsDNA recombining with a gapped circle. The Rhp51 of *S. pombe* shows a strong preference to form joint molecules (JM) with 5' end-homology whereas RecA has an equivalent selectivity for 3' end-homology. Without RuvC, the products (P) are nicked circles (NC) and tailed linear (tail). With RuvC, one sees the appearance of nicked linear dimers. Note that one does not see the initial gapped circle because it is not radioactively labeled. (Adapted from Murayama Y, Kurokawa Y, Mayanagi K & Iwasaki H [2008] *Nature* 451:1018–1021. With permission from Macmillan Publishers, Ltd.)

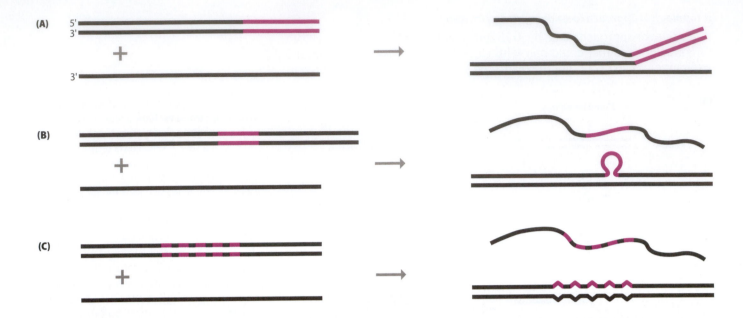

Figure 3.12 Strand exchange and nonhomology. (A) RecA/Rad51 can promote strand exchange up to the point where the sequences become nonhomologous. (B) A nonhomologous sequence of up to a few hundred base pairs can be assimilated into heteroduplex DNA if there is substantial homology on the other side where the recombinase can again effect strand exchange. (C) Strand exchange between generally homologous DNAs containing multiple mismatches results in heteroduplex DNA with single base-pair multiple mismatches that can be corrected by mismatch repair.

of the structure (**Figure 3.13**). After formation of the synaptic structure, the dsDNA was relaxed by topoisomerase I treatment. Then RecA was removed by phenol extraction. The removal of the protein re-introduces supercoiling inherent in the strand exchange structure into what had been a fully relaxed closed circular molecule. The degree of supercoiling reflects the helical turns that had been created by RecA as it incorporated ssDNA into the D-loop. Stasiak concluded that the strand exchange intermediate takes place in a structure in which the DNA is underwound, with 18.6 base pairs per turn compared to normal B-DNA's 10.4, and the DNA is stretched to 1.5 times its normal B-DNA length. A general view of what appears to happen during strand exchange is shown in **Figure 3.14**.

Single molecule analysis of recombination

A recent important advance has been the ability to manipulate single tethered molecules of DNA to make biophysical measurements. The DNA, tethered at one end to a glass slide and with a bead attached to the other end, can be stretched, twisted, and measured with molecular tweezers or magnetic fields. These approaches have shown that the initiation of nucleoprotein filament formation can occur simultaneously at several sites along a ssDNA molecule and that RecA filament assembly at each site is highly cooperative. These approaches have confirmed electron microscopic observations that RecA binding extends the length of a dsDNA filament by roughly 50%. When untethered RecA-coated ssDNA is mixed with a tethered, homologous dsDNA molecule, it is possible to measure changes in length and twist during formation of the synaptic filament. A paper from the Kanaar, Wyman, and Dekker labs, examining single molecules, confirmed measurements made on millions of molecules in a test tube that the rate of strand exchange is rather slow—about 2 nt/s at 22°C. Consistent with the idea that ATP binding but not ATP hydrolysis is essential for recombinase activity, they found that the nonhydrolyzable ATP-γS analog is as effective as ATP in driving strand pairing. However, in the presence of ATP, the reaction continues, allowing RecA to dissociate after completing strand exchange.

From analyzing the length and topology of DNA during strand invasion, these authors came to two other very important conclusions. First, only about 80 bp are actively involved in synapsis at any time while a long

(A)

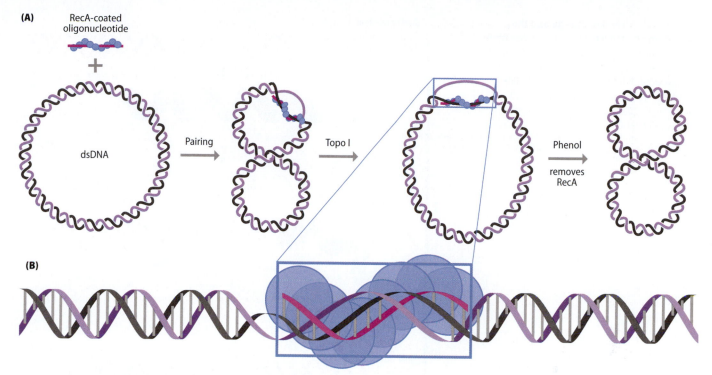

(B)

Figure 3.13 Measurement of the topological state of the RecA-mediated strand invasion intermediate. (A) A strand assimilation assay in which relaxed dsDNA recombines with an oligonucleotide coated with RecA. The formation of the D-loop creates some torsional stress in the dsDNA circle that can be removed by treatment with topoisomerase I (Topo I). When RecA is removed by phenol extraction and the oligonucleotide dissociates, the unwinding that was inherent in the strand invasion structure is redistributed throughout the circular molecule and can be measured as an increase in supercoiling. (B) The proposed DNA structure within RecA-mediated synaptic filaments. All three strands in the pairing region are shown in a coaxial right-handed arrangement with approximately 18.6 bases per turn and that the incoming single strand (*red line*) is base paired with its complement in duplex DNA but the strand to be displaced in the reaction is not base paired but still well structured within the helical RecA–DNA filament. (From Kiianitsa K & Stasiak A [1997] *Proc Natl Acad Sci USA* 94:7837–7840. With permission from the National Academy of Sciences.)

(1 kb) ssDNA molecule is undergoing strand exchange into the duplex (**Figure 3.15**). Second, when a long ssDNA has been incorporated by RecA in the presence of ATP (that is, when RecA protein can dissociate), the displaced strand that was initially part of the dsDNA ends up wrapped around the new duplex (Figure 3.14). This "D-wrap" as opposed to D-loop,

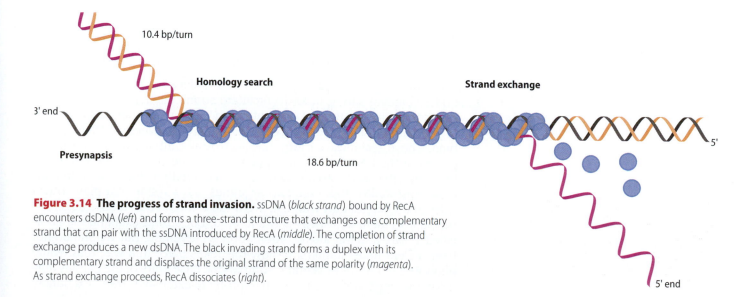

Figure 3.14 The progress of strand invasion. ssDNA (*black strand*) bound by RecA encounters dsDNA (*left*) and forms a three-strand structure that exchanges one complementary strand that can pair with the ssDNA introduced by RecA (*middle*). The completion of strand exchange produces a new dsDNA. The black invading strand forms a duplex with its complementary strand and displaces the original strand of the same polarity (*magenta*). As strand exchange proceeds, RecA dissociates (*right*).

Figure 3.15 Strand exchange and the formation of a D-wrap. Strand exchange occurs as shown in Figure 3.14. As deduced from single-molecule experiments, the D-loop is actually a D-wrap around the strand invasion structure. (Adapted from van der Heijden T, Modesti M, Hage S et al. [2008] *Mol Cell* 30:530–538. With permission from Elsevier.)

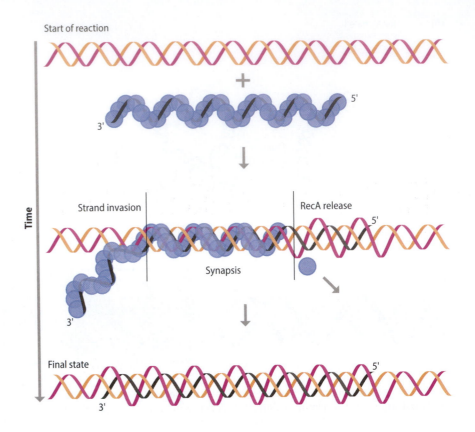

as one normally imagines it, challenges our imagination as to how this strand would be involved in subsequent steps in DSB repair.

Similar studies are just now emerging for Rad51. It should be noted that *in vitro* measurements of Rad51 show that its affinity for ssDNA is not much greater than for dsDNA; indeed the conditions to get an efficient strand exchange reaction *in vitro* proved to be much more stringent than for RecA. This difference makes some of the single-molecule approaches with Rad51 much more tricky to set up and analyze.

The X-ray structure of ssDNA and RecA has enabled the Pavletich lab to model how the strand exchange takes place inside the filament. **Figure 3.16** shows RecA with dsDNA in the primary site, as would occur after strand invasion. **Figure 3.17** imagines ssDNA in the process of base pairing with a complementary strand. The important finding is that this second strand of the new double helix makes little contact with the protein and is held to the bound strand only by Watson–Crick base pairings.

It is apparently necessary to stretch and underwind dsDNA in the secondary binding site in order to facilitate strand exchange, but how this is accomplished is not yet evident. Each RecA monomer has a pair of arginine and lysine residues that lie in the groove where dsDNA should bind. These residues, by forming electrostatic interactions with phosphates in the DNA backbone, appear to provide some of the energy to help open the dsDNA to the same extended length as the ssDNA. In budding yeast Rad51, there are two arginines and a lysine that play critical roles in the interaction with dsDNA; if these three basic amino acids are mutated to alanine, ssDNA can still bind, but strand exchange is abolished *in vitro* and mitotic recombination is prevented. But we still do not have a clear understanding of how this takes place in a way that would allow RecA both to transiently "sample" the base pairs of the dsDNA during homology searching and then to promote extensive strand exchange.

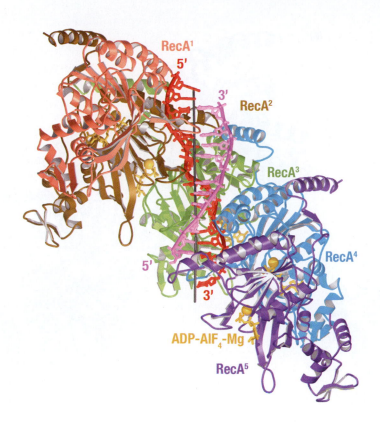

Figure 3.16 DNA configuration of dsDNA that would result from strand exchange. Model of dsDNA within the RecA filament, bound to the primary binding site. (From Chen Z, Yang H & Pavletich NP [2008] *Nature* 453:489–494. With permission from Macmillan Publishers, Ltd.)

The stretching of dsDNA should eliminate the energy of base-pair stacking and weaken base pairing.

A recent theoretical model by Yonatan Savir and Tsvi Tlusty suggests that homology searching is driven not only by the energy of forming proper base pairs but also by the necessary distortion of the dsDNA, in a process they term conformational proofreading. They argue that the searching is effectively done not base pair by base pair but in groups of three—the three base pairs in B-form conformation within each RecA monomer.

3.6 *IN VITRO* ANALYSIS SUGGESTS HOMOLOGY SEARCHING PROCEEDS BY A TETHERED THREE-DIMENSIONAL SEARCH

There is still much to learn about how RecA and Rad51 work. It is remarkable how fast the RecA/Rad51 filament can survey hundreds of thousands of base pairs of dsDNA to locate the relatively tiny patch of homologous DNA with which strand invasion can take place. How is this accomplished? Does the filament land at random and slide around to maximize base pairing and, if so, over what distance does the filament slide and sample dsDNA until it dissociates and tests another (random?) location? Obviously it would be a fatal situation if the filament bound to one chromosome in one orientation and just slid around, as this would doom it to fail half the time, even going all the way around a circular chromosome.

It is now possible to visualize in single-molecule experiments the pairing of a RecA::ssDNA filament with its homologous site on dsDNA (**Figure 3.18**). One of the most striking results from this analysis is that the efficiency of stand invasion and synapsis is dramatically changed depending on whether the dsDNA is randomly coiled up or is pulled into a more linear

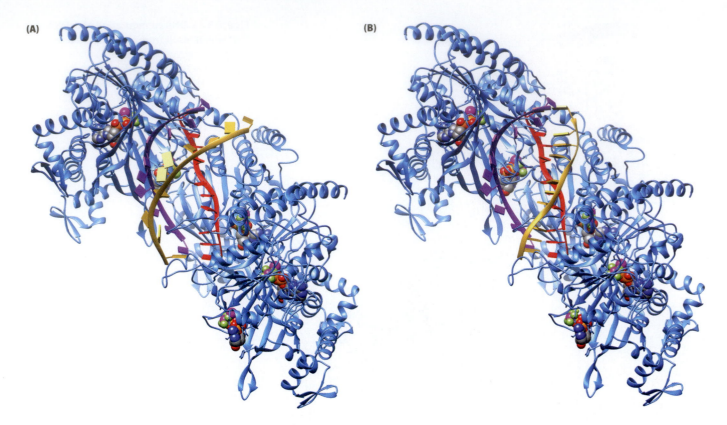

(A)

(B)

Figure 3.17 Possible mechanism of strand exchange modeled into the RecA crystal structure. (A) Initial step in strand exchange showing one strand of dsDNA (*gold*) beginning to pair with ssDNA (*orange*). (B) Post-exchange configuration, with a newly formed dsDNA in the primary binding site and a displaced ssDNA strand (*purple*) in the secondary DNA binding site. The bases are shown as rectangles. In this ribbon diagram, ADP-AlF$_4$-Mg molecules (mimics of the transition state of ATP hydrolysis) are shown as space-filling molecules bound to each of the five RecA units. (Courtesy of Nikola Pavletich, Memorial Sloan-Kettering Cancer Center, based on Chen Z, Yang H & Pavletich NP [2008] *Nature* 453:489–494.)

arrangement. RecA is much less able to locate its homologous target when the DNA is manipulated to be linear, even though this pulling is not enough to create any topological stress that would prevent strand invasion (**Figure 3.18C**). Kowalczykowski has suggested that this result can be explained if efficient homology searching involves a three-dimensional search in which part of the RecA filament engages one dsDNA region that might have a few base pairs homology while another part of the filament can contact other, physically nearby sequences that might have more homology, even though the sequences are distant on the DNA molecule (**Figure 3.18D**). Thus the exploration for homology may not be a series of completely independent collisions of the filament. Kowalczykowski has likened the process to the way an ape can move from tree to tree by holding on to one branch while grabbing another, even branches on another intersecting tree, rather than having to get to the second tree by climbing down one tree and climbing up the next from the ground. When the DNA is stretched out, this kind of searching is lost. We need to apply these ideas to thinking about homology searching *in vivo*, especially as it becomes possible to visualize the conformations of an entire set of chromosomes by using chromosome conformation capture technology.

3.7 THE REQUIREMENTS OF HOMOLOGY SEARCHING AND STRAND INVASION CAN BE ASSAYED *IN VIVO*

Exactly how the encounter between the homology-searching ssDNA RecA/Rad51 filament and dsDNA occurs remains a mystery, but experiments in the past few years have provided considerable insight. *In vivo*, it seems that homology searching must depend initially on random collisions between the nucleoprotein filament and segments of dsDNA. For DSBs arising during DNA replication the search should be greatly facilitated by the close proximity of an intact sister chromatid. In eukaryotes this association is reinforced by the presence of cohesins that hold

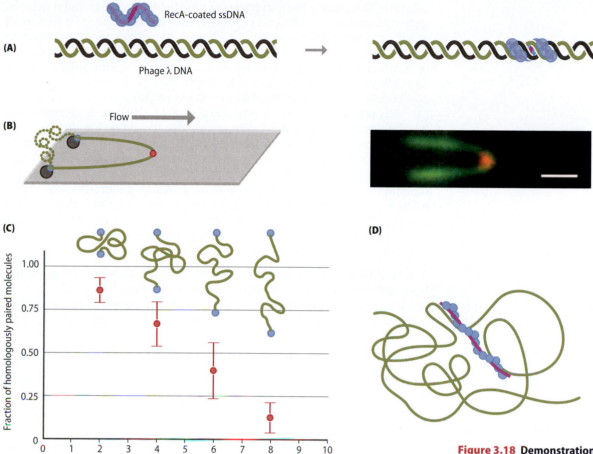

Figure 3.18 Demonstration of three-dimensional homology searching by RecA. (A) Red-labeled RecA bound to an approximately 1-kb ssDNA encounters and strand-exchanges with phage λ DNA under conditions that prevent RecA from dissociating. (B) Left, strand exchange takes place on a DNA molecule (stained *green*) that is tethered at both ends. RecA binding is displayed by flowing buffer to stretch the molecule away from its coiled relaxed form (also shown on *left* as a chain of beads). Right, microscope image of the DNA displayed in flow showing RecA binding. (C) Strand exchange (monitored by a stable red spot on phage DNA) is less efficient as the ends of the dsDNA are drawn further apart before the oligonucleotide covered with RecA is added. (D) These results suggest that homology searching by RecA is more efficient when the DNA is more compact and can be explained if the ability for RecA to guide a small piece of ssDNA to its homologous site involves a three-dimensional search in which one part of RecA explores a neighboring region while another part is still bound, though nonproductively, elsewhere. (A, C, D, adapted from Bell JC, Plank JL, Dombrowski CC & Kowalczykowski SC [2012] *Nature* 491:274–278. With permission from Macmillan Publishers Ltd. B, adapted from Forget AL & Kowalczykowski SC [2012] *Nature* 482:423–427. With permission from Macmillan Publishers Ltd.)

the sisters together; in fact, after a DSB is created, even more cohesins assemble around the sisters, near the break site. The problem becomes much more complex when the DSB must be repaired from a non-sister, homologous chromosome and even more difficult if homology consists of only a few hundred base pairs of template sequences located in an ectopic location, embedded in a different chromosome. Although most DSBs probably arise spontaneously during S phase, they may also appear in G1 or in G2, from ionizing radiation or the consequence of topoisomerase failures. DSBs sometimes arise in G1 by programmed site-specific enzymatic cleavages, as in the case of the rearrangements of the budding yeast *MAT* locus or the mammalian immune system. In these cases, repair of a DSB by homologous recombination will demand the use of a non-sister template. In most eukaryotes there seems to be no special arrangement of homologous chromosomes in the nucleus—a notable exception being the pairing of homologous chromosomes in *Drosophila* somatic cells—so that the search must be able to sample most of the nuclear volume.

If it is a daunting task to find a recombination partner even in haploid budding yeast, where there are *only* 10,000 kb to search; it seems almost impossible to imagine how searching occurs in diploid mammalian cells with 6,000,000 kb of DNA. In mammals it appears that some chromosomes are preferentially located near the nuclear periphery and others more centrally, so that one would expect in some cases at least that recombination between homologous chromosomes would be faster and more efficient than between ectopic sites on nonhomologous chromosomes. In one study, Baker and colleagues used strong positive genetic selection to examine the correction of a surface immunoglobulin gene by gene conversion involving an ectopic donor. Repair required a minimum of 2 kb and was 10,000 times more efficient for a donor on the same chromosome (but more than 1 Mb away) compared to a site on a homologous or nonhomologous chromosome. In another experiment with mammalian cells in which a DSB was induced with the I-SceI endonuclease, Jasin and colleagues again found that allelic events (that is, between identical sites on homologous chromosomes) were much less efficient than intrachromosomal or inter-sister repair events; however, in this case the intrachromosomal donor sequences were part of the same inserted construct and thus only a few kilobases distant from the DSB. Allelic recombination would have the advantage of extensive homology shared on either side of the DSB, whereas ectopic events generally involve only a few kilobases of shared homology. So the differences in allelic versus ectopic efficiency may not only reflect the ease of locating the sequences by the recombinase filament but may have to do with the amount of adjacent homology that can be used to stabilize the strand invasion.

A first consideration is what is the minimum size of homology to initiate recombination? How long does a RecA/Rad51 filament have to be, to carry out recombination? Henry Huang provided a conceptual framework to think about this question by defining a *minimum efficient processing segment* (MEPS)—the smallest length of homologous DNA that will initiate recombination *in vivo*. Based on recombination in *E. coli* between related but diverged (homologous) sequences, in which stretches of perfect homology are interrupted by single base-pair mismatches, Shen and Huang estimated that as short a region as 25–30 bp of perfect homology could weakly initiate genetic exchange. The efficiency of RecA-dependent recombination increased linearly, up to a success rate of 12% when the homology was 400 bp (**Figure 3.19**). As the size of homology (L) increases there are $L - M + 1$ total MEPS (where M = the size of MEPS). So, for $L = 100$ bp, the total number of MEPS = $100 - 30 + 1 = 71$, suggesting that recombination will be 71 times more efficient than homology of 30 bp. At $L = 400$ bp, the total MEPS would be 371, which in this system resulted in about 12% efficiency. Whether the efficiency of recombination would continue to increase linearly with even longer lengths of homology was not determined. We could extrapolate to say that recombination values could reach 100% if the length of homology were about 3200 bp. (However, there could be other factors in the experiment that could limit recombination to a value much lower than 100%.) The idea that searching is based on MEPS suggests that each initial contact between the filament and the target represents a distinct test of finding a suitable partner sequence and that success depends primarily on making an initial homologous alignment.

In budding yeast, Rad51-mediated recombination appears to have a MEPS of between 70 and 100 bp and recombination reaches about 75% of its maximal value by a length of 500 bp (**Figure 3.20**). In one particularly well-studied DSB-initiated intrachromosomal gene conversion event—mating-type gene (*MAT*) switching—recombination is nearly 100% efficient when an ectopic donor locus carries only 230 bp homology on the side of the DSB that appears to make initial contact. Even when the donor locus is placed on a different chromosome, this much homology is sufficient for 70–80% of cells to complete the repair event. So, with Rad51 covering about 230 nt of homology, this filament must be able to search the entire genome of >10,000 kb for a partner! In yeast cells it is possible to induce a specific DSB and monitor how long it takes for Rad51 to associate with a donor on the same chromosome or on another chromosome. The pairing of the DSB with a donor on the same chromosome takes at least 15 min, but the process is substantially slower when the

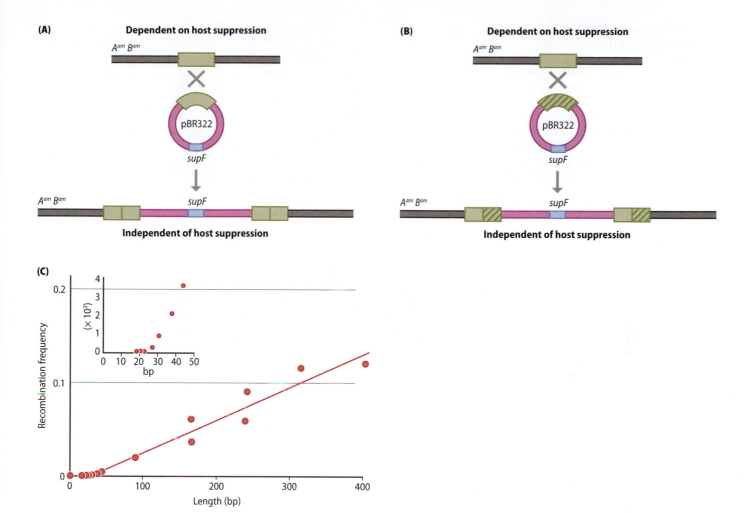

donor is located on another chromosome. It is not especially surprising that intrachromosomal searching is a more rapid process; we all learned that intramolecular chemical reactions are more rapid and concentration independent compared to interchromosomal events. But it is surprising that the same rule applies at the scale of an entire chromosome, where a donor is 200 kb away.

It is not at all clear what happens in the 20–30% cells that fail to complete an interchromosomal DSB repair event. One possibility, now supported by some experiments, is that during the time that a search for homology is proceeding, 5′ to 3′ resection of the DSB ends continues to expose more and more sequences, including short repeated sequences that can also be coated with Rad51 and that could engage in a totally different homology search, to produce recombination events that would create inviable chromosomal rearrangements.

3.8 BROKEN DNA ENDS BECOME MORE MOBILE

We do not yet know if there are significant differences in the ability of a DSB end to find a donor depending on its position on a chromosome arm. A DSB in the middle of a chromosome arm might have more difficulty finding a donor that was located close to the centromere or telomere of another chromosome; this needs to be tested systematically. There is good evidence that budding yeast chromosomes are arranged in a "Rabl

Figure 3.19 Determination of a minimum effective pairing segment (MEPS) using mismatched substrates in *E. coli* based on the largest shared region of homology in the homeologous region. (A) Recombination between 400-bp segments of DNA carried on phage λ and a pBR322 plasmid that also carries the amber suppressor, *supF*. The amber mutants in the A^{am} and B^{am} genes prevent it from propagating unless the host carries an amber suppressor tRNA such as *supF*. Recombination between the plasmid and the phage creates a phage that carries its own *supF* amber suppressor, rendering the cell independent of host suppression. (B) The homologous segment on the plasmid was varied by using evolutionarily diverged sequences (*hatched lines*), so that the longest region of perfect homology between the phage and the plasmid ranged from 19 bp to 400 bp. (C) MEPS is calculated from the point of inflection of the recombination data, seen in the insert. (Adapted from Shen P & Huang HV [1986] *Genetics* 112:441–457. With permission from the Genetics Society of America.)

Figure 3.20 Effect of homology length on recombination between inverted homologous segments in a plasmid.
(A) A plasmid system to assay homology-length requirements for HO-induced DSB repair in *S. cerevisiae*. The *MAT**a**-inc* donor sequence is identical to the HO-cuttable *MAT**a*** sequence except for a single base-pair difference that prevents HO cleavage. In a series of plasmids, the size of homology that this donor shared with the HO cut *MAT**a*** was varied from 24 bp on each side to 1000 bp. (B) Homology dependence of DSB repair in wild type (WT) and several mutant strains (*rad51*, *rad50*, and *rad59*). (C) There is significant recombination with as little as 33 bp in a WT cell, and with such short homology, deletion of *RAD51* surprisingly increases the efficiency of repair (by a Rad51-independent, but Rad50- and Rad59-dependent, mechanism). In the absence of Rad59, Rad51-mediated recombination needs about 70 bp homology. Recombination is nearly eliminated in a *rad50 rad51* double mutant. (Adapted from data in Ira G & Haber JE [2002] *Mol Cell Biol* 22:6384–6392.)

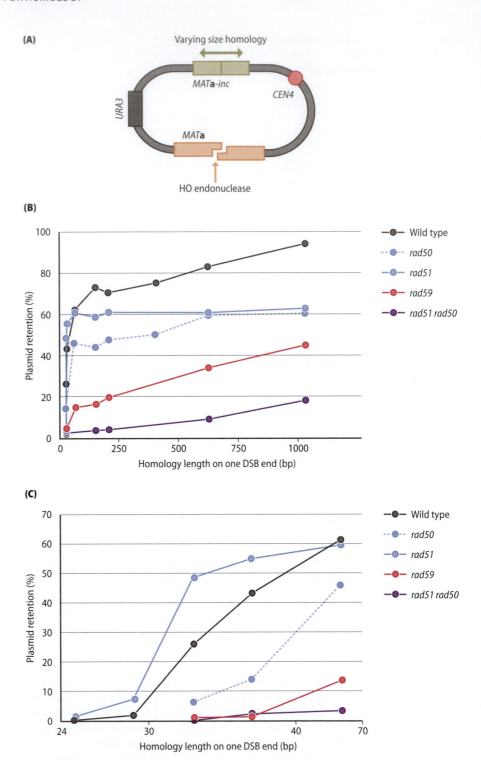

orientation," as first suggested by the nineteenth century cytologist Carl Rabl, with all the centromeres clustered near the spindle pole body and the chromosome arms extended toward the opposite side (where one finds the nucleolus). Some evidence suggests that a DSB is more likely to recombine with a donor at the same distance from the centromere but on another chromosome arm than with a donor at a greater distance from the kinetochores; but this idea has not been thoroughly tested.

Several recent studies in budding yeast suggest that one important factor in broken DNA molecules being able to find their homologous partners is that the broken molecules explore a greater nuclear volume than unbroken molecules (**Figure 3.21**). Studies from the Rodney Rothstein and Susan Gasser labs have monitored the movement of chromosomes in which sites near a DSB created by a site-specific endonuclease are marked by arrays of LacO or TetR sequences to which fluorescent LacI or TetR molecules can bind. The DSB itself can be observed using fluorescent Rad52 protein, which binds at or near DSBs independent of Rad51. Induction of DSBs by site-specific endonucleases or by chemical treatment with drugs such as zeocin causes a significant increase in chromosome movement (an increase in the *radius of confinement*) that may facilitate collisions with possible homologous targets to enable DSB repair to take place. Surprisingly, the wider exploration of the nucleus is prevented in cells lacking Rad51 (**Figure 3.21C**). Further elucidation of the basis of changes in chromosome dynamics will be an important area of study.

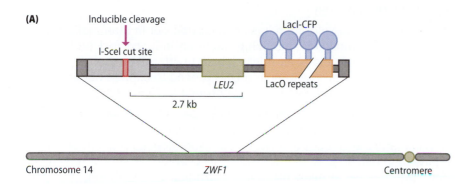

(A)

Figure 3.21 Increased chromosome mobility after induction of a DSB. (A) An I-SceI-induced DSB is formed near a LacI-CFP mark inserted on chromosome 14 at the *ZWF1* locus. (B) Images of chromosome movement taken over a 5 min interval before and after DSB induction by I-SceI cleavage. (C) The broken molecule shows an increased radius of confinement (determined from the plateau value reached at equilibrium), although there is not much difference in the initial slope (indicating rate of diffusion) of the mean squared displacement (MSD) plot, the damaged chromosome explores a much larger fraction of the nuclear volume, as shown in the left plot. Surprisingly, this increased mobility is absent in a strain in which Rad51 was deleted (*right plot*). (Adapted from Dion V, Kalck V, Horigome C et al. [2012] *Nat Cell Biol* 14:502–509. With permission from Macmillan Publishers Ltd.)

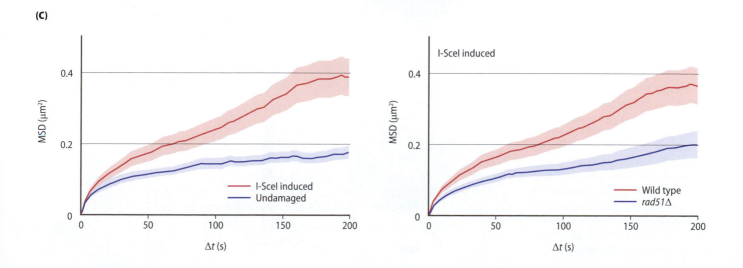

In contrast to the finding that broken chromosome ends explore more of the nuclear volume, other reports suggest that broken ends in budding yeast—at least if not repaired—become tethered to the nuclear periphery through the interaction of the Rad52 protein with an nuclear envelope protein, Mps3. So, it may be that DSB ends provoke a "search mode" that increases chromosome mobility but then go into a peripherally tethered state, where it is known that some components of NHEJ, including the Ku proteins, are present in high concentrations. Alternatively, the ends might be repaired by new telomere addition, as telomerase and other telomere binding proteins are also enriched at the nuclear periphery.

Another issue, still unresolved, is whether the two ends of DSB frequently carry out independent homology searches or if they act in a more coordinated fashion. We will examine this question more in later chapters, but here we will note that several experiments in budding yeast and in mammals, placing tags on either side of an induced DSB, suggest that most of the time the two ends remain together and do not move separately in searching for homology.

Recently a similar exploration has been carried out in mammalian cells, monitoring the mobility of ionizing-radiation-induced foci (IRIF) compared to chromatin that was briefly labeled with a fluorescently tagged nucleotide (Cy3). Chromatin near a DSB proved to explore about a two-fold greater volume than undamaged chromatin (**Figure 3.22**).

In later chapters we will examine in more detail how homology searching occurs *in vivo*, in experimental situations where loading of Rad51 at a DSB end and the kinetics of its search for homologous sequences can be analyzed in real time. Before we leave this chapter, we should remember that strand invasion takes place in the context of chromatin, in supercoiled DNA, where nucleosomes may have to be moved or displaced in order for Rad51 to initiate strand invasion (**Figure 3.23**). To date, none of the recombination assays *in vitro* accurately model the invasion of a resected DSB end into chromatin.

Figure 3.22 Increased mobility of chromatin domains after inducing a DSB in human cells. Chromatin domains were labeled by brief incorporation of Cy3-dUTP to produce fluorescent spots whose mobility could be compared with ionizing radiation induced DSB foci (IRIF) that were marked with 53BP1-GFP. (A) 100 randomly selected trajectories of labeled chromatin domains or 53Bp1 spots during 60 min, plotted as if they all originated from a common point in the two-dimensional plane. The color reflects the time after start of imaging, beginning with *red* and ending at 60 min in *blue*. (B) Mean squared displacement (MSD) plots of the mobility of chromatin domains (*red*) and IRIF (*green*) showing that DSB ends exhibit increased mobility compared with undamaged chromatin spots. (Courtesy of Roland Kanaar, Erasmus Medical Center, and Przemek Krawczyk, Van Leeuwenhoek Centre for Advanced Microscopy, based on Krawczyk PM, Borovski T, Stap J et al. [2012] *J Cell Sci* 125:2127–2133. With permission from The Company of Biologists.)

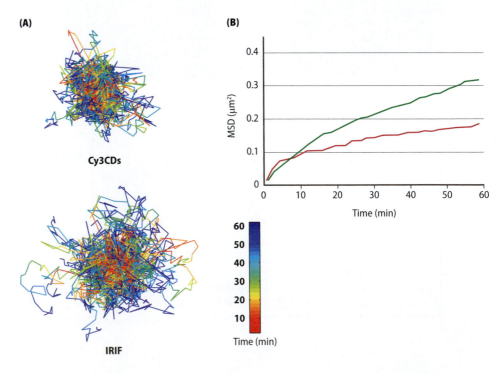

(A)
Nucleosomes

RPA   Rad51

(B)

(C)

3.9 Rad51's PROPERTIES CAN BE ALTERED BY ASSOCIATED PROTEINS

Although Rad51 carries out the essential steps of homology searching and strand exchange, this machine only works *in vivo* with the aid of a number of additional proteins. In the next chapter we will examine the role of several proteins in the construction of the RecA/Rad51 filament. These include the bacterial proteins RecFOR and the eukaryotic proteins Rad52, Rad55, Rad57, and BRCA2. In later chapters, as we examine some specific recombination events in more detail, we will encounter a number of other factors that are either important or absolutely essential for the successful completion of strand invasion (especially in eukaryotes where the DNA is wrapped around nucleosomes) and so that the 3′ end can be used to initiate the next step in the process. Among them are several helicases (BLM/Sgs1, Srs2, and FANCM/Mph1) and chromatin remodeling complexes (Rad54, Rdh54/Tid1, Swi2-Swi5, and RSC).

As noted previously, nearly all DSB repair begins with 5′ to 3′ resection of the end, leading to a 3′-ended ssDNA tail that Rad51 will enable to carry out strand invasion and initiate new DNA synthesis from the 3′ end. However, William Holloman has found that the properties of Rad51 from the corn smut yeast *Ustilago maydis* are altered in the presence of the BRCA2 homolog, Brh2. Normally, if confronted with a dsDNA with either a 5′ or 3′ protruding segment, Rad51 carries out strand exchange with a supercoiled duplex most efficiently on the 3′-ended substrate. But in the presence of Brh2, Rad51 can also use dsDNA with a 5′ ssDNA extension, leading to supercoiled DNA forming a four-stranded structure (**Figure 3.24**). Holloman suggests that the ability of Brh2 and Rad51 to use a 5′ end provides another way to restart DNA replication if stalling and breakage left a 5′-ended tail. So we must add another possible mechanism to repair a stalled and broken replication fork as we outlined in Chapter 1. Here, the 5′ end stabilizes an open structure, allowing the 3′ end to initiate new synthesis, with presumably the 5′ end elongating by Okazaki fragment synthesis. Whether these intriguing *in vitro* results reflect an important pathway *in vivo* is hard to say, but they certainly point to the fact that the Rad51 machinery, when interacting with BRCA2, can display unexpected properties.

3.10 STRAND INVASION CAN APPARENTLY OCCUR WITHOUT Rad51

In budding yeast, some recombination can occur without Rad51. Rad51-independent events can occur either in the absence of any induced DNA damage (spontaneous recombination), where we do not know the nature of the initiating lesion, or after creating a DSB. The idea that there is a Rad51-independent but Rad52-dependent pathway is reinforced by two findings, both of which suggest that this pathway uses much less homology

Figure 3.23 Recombination takes place in the context of chromatin. (A) Rad51 (*blue*) binds to ssDNA, as does RPA (*green*). (B) Rad51-mediated strand invasion may require the movement or displacement of nucleosomes to initiate a search for homology. (C) D-loop formation can be stabilized by RPA binding to the displaced strand while Rad51 dissociates after strand exchange is complete.

Figure 3.24 The *Ustilago* Brh2 homolog of BRCA2 is capable of directing strand invasion of 5'-overhanging ssDNA ends. If an overhanging end is simply clipped off and resected (A) then Rad51 will mediate strand invasion with a 3' end that can act as a primer to initiate new DNA synthesis. (B) At least *in vitro*, without end-processing, Brh2 can aid Rad51 in directly assimilating a 5' ssDNA end into a recombination structure that—after branch migration of the partial Holliday junction—can restart DNA replication from the 3' end. (Adapted from Llorente B & Modesti M [2009] *Mol Cell* 36:539–540. With permission from Elsevier Ltd.)

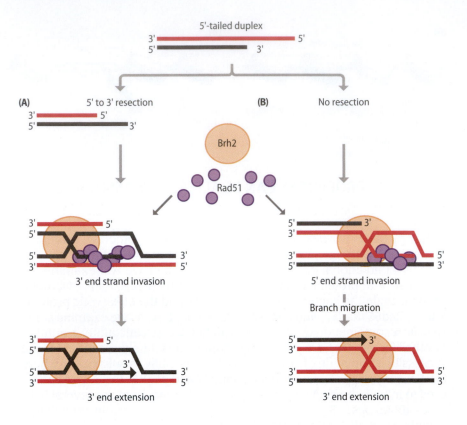

to initiate recombination than does Rad51. Telomere elongation in the absence of telomerase can occur by two pathways, one of which is Rad51 independent. This pathway involves recombination within the irregular TG_{1-3} telomere sequences themselves, suggesting that the extent of perfect homology between recombining molecules must be short. These events still require the Rad52 protein whose properties we will discuss in the next chapter, as well as Rad59 and the Mre11-Rad50-Xrs2 complex.

Rad51-independent recombination is also seen HO-induced in DSB repair using plasmid substrates in which the donor shares very little homology (for example, 33 bp) with a donor in inverted orientation (see Figure 3.19). In fact Rad51 appears to *inhibit* repair with such short regions of homology, because successful repair is much greater when Rad51 is deleted. This observation suggests that Rad51 can bind to ssDNA ends but becomes stymied if there is not sufficient homology to direct repair. As above, Rad51-independent recombination requires Rad52, Rad59, and the MRX complex, but it is not yet evident how these proteins can carry out homology searching, strand invasion, or strand annealing.

Further evidence of Rad51-independent recombination has come from studies of DSBs generated either by site-specific nucleases or at fragile sites within yeast chromosomes. For example in a diploid yeast cell with a DSB on one homolog, repair of an induced DSB normally results in a simple gene conversion event. But in the absence of Rad51, repair still occurs in a surprising 35% of the cells. A detailed analysis of these repair events has revealed that nearly all of the repair events have involved extensive 5' to 3' resection and removal of about 30 kb of DNA proximal to the DSB, apparently exposing an ~6 kb Ty retrotransposon that could recombine with one of many other copies of Ty elements located on different chromosomes (**Figure 3.25**). The resulting recombination events

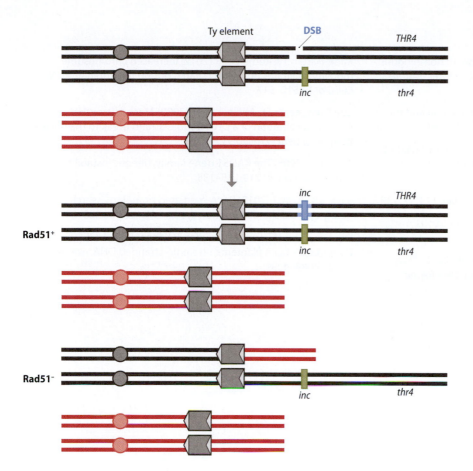

Figure 3.25 Rad51-independent recombination in a diploid. An HO-induced DSB is created on one homologous chromosome; the homologous chromosome has an "*inc*" mutation that cannot be cleaved. When Rad51 is active, nearly all DSBs are repaired by gene conversion, by a short patch of new DNA synthesis (*blue*) and cells usually remain heterozygous for the distal *THR4* gene. Without Rad51, gene conversion is eliminated but Rad51-independent BIR is possible. However, almost always BIR is initiated between dispersed repeated Ty sequences. In this case, cells become hemizygous for *thr4* and all sequences distal to the location of the Ty element. (Adapted from Malkova A, Signon L, Schaefer CB et al. [2001] *Genes Dev* 15:1055–1060. With permission from Cold Spring Harbor Laboratory Press.)

mostly yield nonreciprocal translocations, apparently arising from break-induced replication, with Ty elements at the junctions. There is clearly something special about the Ty regions, because in principle BIR should have been possible between the broken chromosome and its intact homolog, using sites anywhere along the 30-kb region between the DSB and the nearest Ty element; but in the absence of Rad51, this homology is ignored. How recombination is favored at Ty elements in the absence of Rad51 remains to be solved.

SUMMARY

RecA and Rad51 are the agents of homology searching and strand invasion. These recombinases form a right-handed filament on ssDNA in a cooperative process that begins with the nucleation of a small number of monomers. ATP binding, but not ATP hydrolysis, is required to form the filament and to engage in homology searching, but ATP hydrolysis is needed to remove RecA/Rad51. RecA/Rad51 binding stretches both ssDNA and dsDNA about 60%, but the crystal structure of RecA surprisingly revealed that the three nucleotides of ssDNA bound to a monomer retain a B-DNA configuration while the backbone between the monomers is greatly underwound. Homology searching requires a minimum effective pairing segment that is about 30 nt for RecA and about twice as large for Rad51. Single-molecule experiments have suggested that the searching process is facilitated by a three-dimensional search between nearby segments of dsDNA.

SUGGESTED READING

Chen Z, Yang H & Pavletich NP (2008) Mechanism of homologous recombination from the RecA-ssDNA/dsDNA structures. *Nature* 453:489–494.

Cox MM (1999) Recombinational DNA repair in bacteria and the RecA protein. *Prog Nucleic Acid Res Mol Biol* 63:311–366.

Forget AL & Kowalczykowski SC (2012) Single-molecule imaging of DNA pairing by RecA reveals a three-dimensional homology search. *Nature* 482:423–427.

Galletto R & Kowalczykowski SC (2007) RecA. *Curr Biol* 17:R395–397.

Heyer WD (2007) Biochemistry of eukaryotic homologous recombination. *Top Curr Genet* 17:95–133.

Holthausen JT, Wyman C & Kanaar R (2010) Regulation of DNA strand exchange in homologous recombination. *DNA Repair* 9:1264–1272.

Mine-Hattab J & Rothstein R (2012) Increased chromosome mobility facilitates homology search during recombination. *Nat Cell Biol* 14:510–517.

Oza P, Jaspersen SL, Miele A et al. (2009) Mechanisms that regulate localization of a DNA double-strand break to the nuclear periphery. *Genes Dev* 23:912–927.

Thacker J (2005) The RAD51 gene family, genetic instability and cancer. *Cancer Lett* 219:125–135.

Savir Y & Tlusty T (2010) RecA-mediated homology search as a nearly optimal signal detection system. *Mol Cell* 40:388–396.

van Loenhout MT, van der Heijden T, Kanaar R et al. (2009) Dynamics of RecA filaments on single-stranded DNA. *Nucl Acids Res* 37:4089–4099.

CHAPTER 4

PREPARATION OF THE RecA/Rad51 FILAMENT

The seemingly simple steps of creating 3'-ended ssDNA at the DSB end and then loading RecA/Rad51 are both surprisingly complex. Resection—the steps for which are outlined in this chapter—often involves multiple exonucleases or helicase–endonucleases. Recombinase loading requires the interplay of a number of *mediator* proteins. In eukaryotes these include several Rad51 paralogs as well as other mediator proteins, and—depending on the organism—either Rad52 or BRCA2.

4.1 DSB ENDS ARE RESECTED IN A 5' TO 3' DIRECTION

Although the RecA and Rad51 recombinases can bind to both dsDNA and ssDNA, they cannot effect recombination in the absence of ssDNA. This has been shown *in vivo* in budding yeast by demonstrating that repair of a site-specific DSB requires 5' to 3' resection; there is no repair in mutants that lack 5' to 3' resection of the DSB end. A similar requirement for resection appears to prevail in mammalian cells. Most pathways of homologous recombination appear to be dependent on 5' to 3' resection, generating 3'-ended ssDNA tails. Resection can be accomplished by two different enzymatic processes: either a true exonuclease can progressively remove one or more bases at each step or else a combination of a helicase and an endonuclease can remove segments of ssDNA (**Figure 4.1**).

4.2 RESECTION HAS BEEN WELL STUDIED IN *E. COLI*

Genetic studies of phage recombination in *E. coli* by Ray White and Maurice Fox argued that the joints between two parent molecules involved pairing of 3'-ended single-stranded DNA. Subsequent experiments have supported the idea that most recombination in *E. coli* begins with creating 3'-ended tails. The best-studied resection activity in bacteria is the three-protein complex, RecBCD helicase–endonuclease (**Figure 4.2**). If unregulated, RecBCD can degrade a dsDNA end at an *in vitro*-measured rate of 1800 bp/s. (For a frame of reference, consider that in budding yeast resection moves about 1 bp/s *in vivo*.)

RecBCD contains two motors: a 5' to 3' helicase RecD pulls the strand ending 5' at the DSB end while the 3' to 5' RecB helicase pulls on the opposite strand. In its unregulated state, RecBCD rapidly degrades the 3'-ended strand, but in addition, the helicase–nuclease occasionally clips the 5'-ended ssDNA as well, so that both strands are shortened (Figure 4.2).

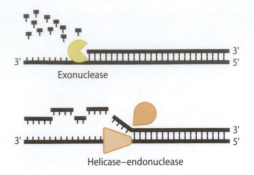

Figure 4.1 5′ to 3′ resection of DSB ends.
Resection can be carried out either by an
exonuclease that generally removes one
nucleotide at a time, or by a combination of a
helicase and an endonuclease that will remove
deoxyoligonucleotides.

Acting in this way RecBCD is less a recombination enzyme than a
DNA degradation machine. RecBCD may exist in bacteria principally
to destroy injected linearized phage DNA. To fight back, the phage λ of
E. coli expresses the Gam protein that inhibits RecBCD. Phage λ also
expresses its own more modest 5′ to 3′ exonuclease, λ exo, to promote its
recombination. But *E. coli* has its own way of modulating RecBCD activity,
through the presence of 8-bp Chi (χ) sequences (CGTGGTGG) that are
scattered along its chromosome. When RecBCD exposes this sequence,
it binds to the enzyme and inhibits 3′ to 5′ degradation, now moving at
about half the rate and generating 3′-ended (and recombinogenic) ends
(Figure 4.2).

The crystal structure of RecBCD provides a clear understanding of its
action (**Figure 4.3**). When it binds to a dsDNA end, the ends are split apart
at a "pin" in RecB and each end is bound by a different helicase. RecD
moves more rapidly on the 5′-ended strand and RecB tracks on the 3′ end
(Figure 4.3). As the DNA is pried open by these helicases the 3′-ended
strand is frequently cleaved by a single endonuclease site in RecC; but
the same site occasionally cleaves the opposite strand. When RecBCD
encounters a *Chi (χ) sequence*, Chi is bound by RecC, somehow causing an alteration in the function of RecD, thus leaving only the slower
5′ to 3′ resection promoted by RecB and the creation of the recombinogenic 3′-ended ssDNA. RecBCD also binds RecA and thus also directly
promotes RecA filament assembly on the 3′-ended ssDNA (Figure 4.3B).

A special case is created by deleting RecD. Now resection is much slower,
since RecD is the "fast" helicase and resection by RecBC—in the absence
of Chi—goes 5′ to 3′, producing 3′-ended recombinogenic ends. But surprisingly it is not the inherent nuclease activity in RecC that does the
cleavage. Instead another 5′ to 3′ exonuclease, RecJ, does the cutting, still
relying on the helicase activity of RecBC.

The Chi sequence appears to have evolved to ensure that RecBCD does
not just trash its own DNA. *E. coli* has to cope with broken replication
forks that can be restarted by BIR, if the DSB end is not just destroyed.
Interestingly, Chi sequences are preferentially oriented in the active
orientation with respect to the moving bidirectional replication fork, suggesting that they are present to allow repair of broken forks. Moreover,
Chi sequences are 5–10 times more frequent than one would expect by
chance, and these sites may enable *E. coli* to cope with the introduction
of linear *E. coli* DNA during conjugation. To promote gene transfer by
recombination during conjugation, RecBCD must be attenuated. So, when
E. coli conjugation introduces a linear segment of DNA, RecBCD begins
its rampant resection, but when it encounters Chi, its 3′ to 5′ exonuclease
is greatly attenuated and the necessary 3′-ended ssDNA is created.

In the absence of RecB or RecC, DSB ends can still be resected in *E. coli*.
Again, this requires RecJ, but now another helicase—RecQ—is needed
to open up the DNA for RecJ. RecQ-RecJ moves much more slowly than
RecBCD, so its presence is usually masked unless there is a *recBCD*
mutant. Moreover, RecQ-RecJ's slow 5′ to 3′ resection is counteracted
by two 3′ to 5′ ssDNA exonucleases known as Exo1, or SbcB (suppressor
of recB mutants), and SbcCD that destroy the 3′-ended ssDNA substrate.
Consequently RecQ-RecJ-mediated recombination is only evident in the
absence of both RecB and SbcB or SbcCD. Exo1 and RecJ can also act
(one at the 3′ end and one at the 5′ end) to enlarge ssDNA gaps that are
also substrates for recombinational repair.

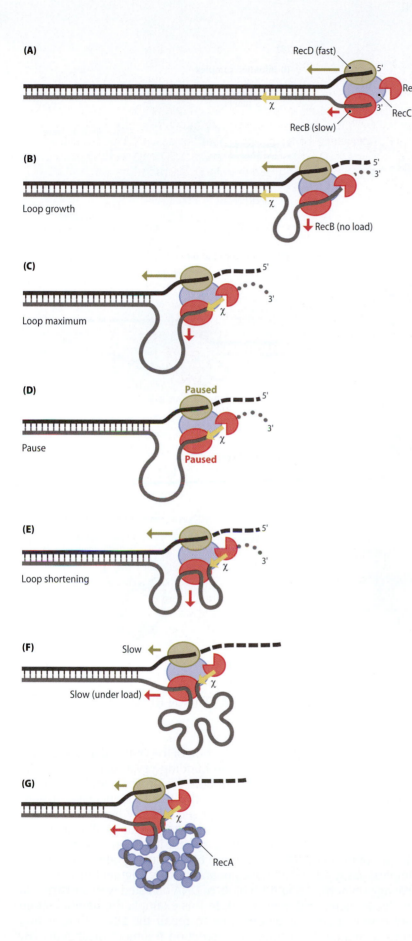

(A)

RecD (fast)

5'

RecB^n

RecC

χ

RecB (slow)

3'

(B)

Loop growth

5'

3'

χ

RecB (no load)

(C)

Loop maximum

5'

3'

χ

(D)

Paused

Pause

Paused

5'

3'

χ

(E)

Loop shortening

5'

3'

χ

(F)

Slow

Slow (under load)

5'

χ

(G)

χ

RecA

Figure 4.2 Degradation and resection of DNA by the RecBCD enzyme complex. Two helicase motors in RecD and RecB unwind DNA and allow a nuclease domain in RecB to remove both the 5'- and 3'-ended strands (A, B). The 5' end is cleaved infrequently. The slower moving RecB accumulates a loop of ssDNA (B). When RecBCD encounters the Chi sequence (χ) it is tightly bound by RecC (C), pausing and slowing RecD's helicase but allowing 5' to 3' resection to continue while the 3' end is held at the Chi site (D, E). Continued resection allows an unwound 3'-ended strand to accumulate and to bind RecA (F, G). (Adapted from Spies M, Amitani I, Baskin RJ & Kowalczykowski SC [2007] *Cell* 131:694–705. With permission from Elsevier.)

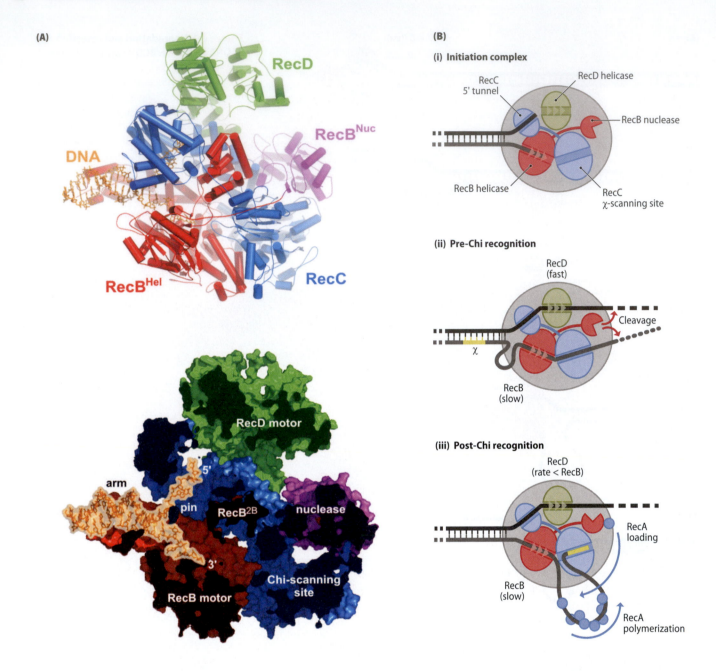

Figure 4.3 Crystal structure analysis of RecBCD. (A) General structure of RecBCD. (B) Proposed action of RecBCD involves binding of two helicase motors on the two strands and splitting apart the duplex by a "pin" in RecC. A single nuclease domain of RecB cleaves both strands, with different sized products. When the Chi site (χ) is encountered, it is bound by RecC, thus forcing the accumulation of an ssDNA loop as more DNA is unwound. The loop is bound by RecA. (Adapted from Dillingham MS & Kowalczykowski SC [2008] *Microbiol Mol Biol Rev* 72:642–671. With permission from the American Society for Microbiology.)

4.3 RESECTION IN BUDDING YEAST CAN BE MONITORED *IN VIVO*

Most of what we understand about resection of the DSB end in eukaryotes comes from studies of specific DSBs in budding yeast; but these results have been extended to experiments in both fission yeast and mammals. As we will mention innumerable times, the HO endonuclease of budding yeast will efficiently cleave a 24-bp sequence. When the HO gene is induced from a galactose-inducible promoter, virtually all cells in the population suffer a DSB within 20 min, so that it is then possible to follow the fate of the cut DNA on Southern blots or using other techniques. The simplest assay follows the disappearance of a fragment produced by HO cleavage on normal Southern blots of DNA digested with a restriction endonuclease (here, *Sty*I) (**Figure 4.4**). In this example the haploid strain lacks an ectopic homologous template to repair the DNA, so resection continues unabated. Over time the restriction fragment disappears. By

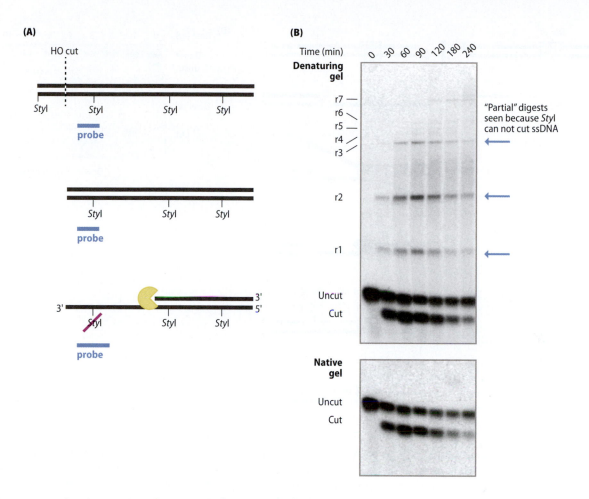

(A)

(B)

Figure 4.4 Physical monitoring of 5′ to 3′ resection *in vivo* in budding yeast. Expression of the HO endonuclease creates a DSB that reduces the size of a *Sty*I restriction fragment homologous to a probe on a Southern blot (A). (B) The cut fragment disappears on a native gel, but a denaturing gel reveals that resection has gone past the *Sty*I site to the right of the DSB, thus producing one strand that will reflect the length of the next intact *Sty*I site. (B, from Mimitou EP & Symington LS [2008] *Nature* 455:770–774. With permission from Macmillan Publishers, Ltd.)

itself this assay only shows that the DNA is being degraded and made heterogeneous, so that it does not migrate as a distinct band. A more revealing assay uses a denaturing gel (usually just performing gel electrophoresis in the presence of 0.1 N NaOH) and strand-specific probes to examine the fate of each strand. Here one sees that the 3′-ended strand remains intact while the 5′-ended strand is removed: evidence that resection is progressing 5′ to 3′. Moreover, as resection proceeds, sites that should be cut by the restriction enzyme become single-stranded and are not cleaved; consequently the 3′-ended strand is now found as a ladder of what appear to be partial digestion products but are in fact evidence of an increasingly long ssDNA end (Figure 4.4). During resection the increasingly long 3′-ended ssDNA is remarkably resistant to any cleavage. Were this not the case the "ladder" on denaturing gels would not appear, since the probe is located very close to the DSB end.

These kinds of physical monitoring of DSB ends reveal that ends are degraded in a 5′ to 3′ direction at a rate of about 4 kb/hr (roughly 1 nt/s). In the absence of any repair, resection inexorably proceeds for at least 24 hr, as monitored by the disappearance of restriction fragments as far as 100 kb from the DSB.

Of course we are most interested in cases where the DSB can be repaired. One simple repair mechanism is *single-strand annealing* (SSA), introduced in Chapter 2. The rate-limiting step in SSA is the generation of complementary single-stranded regions flanking the DSB that can then anneal and create a deletion (**Figure 4.5**). SSA can be monitored on Southern blots after creating a DSB by turning on expression of a site-specific endonuclease. When the repeat located furthest from the DSB is 2 kb from the DSB, repair takes about 1 hr; when the repeat is 25 kb away,

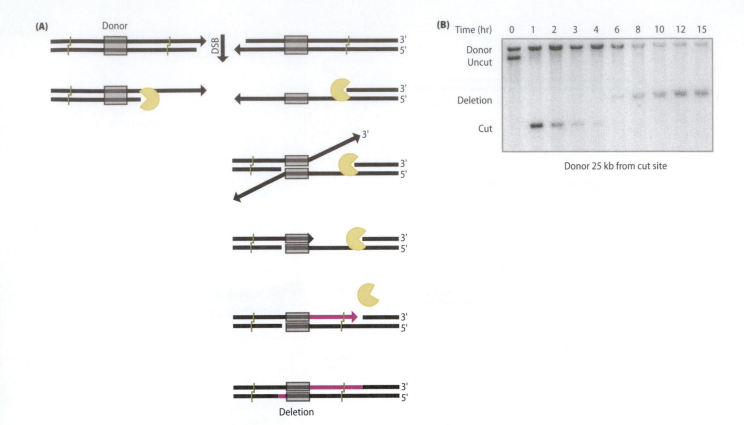

Figure 4.5 Resection of DSB ends during single-strand annealing. (A) A DSB created between two identical flanking sequences (*gray boxes*) is resected 5′ to 3′ on both sides of the break by an exonuclease (or helicase–endonuclease). Exposure of the more distant repeated sequence will determine when SSA can occur. Resection removes some restriction sites (*green lines*). Following annealing and clipping off of nonhomologous 3′-ended tails, the 3′ ends of the annealed structure are used as primers to enable a repair DNA polymerase to initiate new DNA synthesis (*purple arrow*) to fill in the single-strand gaps and to displace the resection machinery. The deletion product can be recognized by a unique restriction fragment carrying the repeat sequences. (B) Southern blot, probed with sequences homologous to the gray boxes. Two restriction fragments are seen initially, one of which is cleaved by HO endonuclease to produce a new restriction fragment (cut). At later times one sees the loss of the donor sequence as a deletion product is formed. When the left-hand repeat is 25 kb from the DSB, resection takes roughly 6 hr to complete SSA. (B, from Vaze M, Pellicioli A, Lee S et al. [2002] *Mol Cell* 10:373–385. With permission from Elsevier.)

SSA occurs only after 6 hr. SSA will be discussed in detail in Chapter 5. Note that the 3′ end must be very stable; a single nick beyond the region of homology would be fatal, but SSA is remarkably efficient even if 50 kb must be resected.

One can also see evidence of 5′ to 3′ resection when a DSB is repaired by gene conversion. Again, we will take up this issue in detail in Chapter 6, but suffice it to say that if one induces a DSB in a situation where there is a distant homologous donor sequence to allow the HO-induced DSB to be "patched up" by recombination, then by isolating DNA at intervals and running them by denaturing gel electrophoresis, one can readily see the transient appearance of the higher-molecular weight bands that were shown in Figure 4.4 (for example, bands r1 and r2). Resection progresses substantially beyond the length needed to initiate efficient gene conversion. These bands disappear as the 3′-single stranded end engages in recombination.

When repair has taken place, resection stops. How resection is terminated is not known in detail, but it seems most likely that the slow-moving resection machinery is overtaken by a more rapidly progressing DNA polymerase that must fill in single-stranded regions after using the 3′ end of the recombination intermediate as a primer (Figure 4.5).

4.4 SEVERAL DIFFERENT PROTEIN COMPLEXES ARE INVOLVED IN RESECTION IN BUDDING YEAST

Resection in yeast is a surprisingly complex process involving at least four exonucleases or helicase–endonucleases. Mutations in the genes encoding these proteins were originally identified as radiation-sensitive (*rad*) or X-ray sensitive (*xrs*) or meiotic recombination-defective (*mre*). The first of these activities is contained in a complex that we will encounter frequently in many guises, the Mre11-Rad50-Xrs2 (MRX) complex. In fission yeast and vertebrates, the Xrs2-like subunit is known as Nbs1 and the complex is called MRN. Deletion of any subunit in exponentially growing budding yeast causes a twofold reduction in resection, as analyzed by the loss of restriction fragments adjacent to the DSB, but this twofold difference masks the complications of cell cycle-specific differences. MRX is required for essentially all end degradation of an HO-induced DSB in the G2 phase of the cell cycle and—even in wild type cells—there is essentially no resection in G1-arrested cells. So the fact that *rad50Δ* only reduces resection by half in cycling cells suggests that there must be an MRX-independent resection process in S-phase cells (**Box 4.1**).

A complication in understanding the action of MRX is that the *in vitro* activity of the Mre11 subunit exhibits both endonuclease activity and 3′ to 5′ resection of dsDNA (that is, the opposite direction of the observed 5′ to 3′ resection *in vivo*). A *mre11-H125N* mutation that removes its *in vitro* nuclease activity surprisingly exhibits nearly wild-type degradation of HO-cut ends. However, Mre11's endonuclease activity is critical in opening DNA hairpin ends and in the removal of the Spo11 protein covalently attached to the 5′ end of a DSB made in meiosis. Rad50 is reported to have helicase activity, which may account for the observation that, when other exonucleases are deleted, MRX acts to cut off sequentially several approximately 100-bp segments from the DSB end in a 5′ to 3′ direction. In this situation it seems that MRX acts as a helicase–endonuclease.

MRX/MRN are associated with another protein, known as Sae2 in budding yeast or CtIP in mammals. Sae2 also exhibits endonuclease activity *in vitro*, removing a 5′ ended flap from a splayed-ended substrate. Deletion of *SAE2*, like *rad50Δ* or *mre11Δ*, reduces resection about twofold and a double mutant, *rad50Δ sae2Δ*, is no more severely blocked than either single deletion; but why both activities are needed is not yet clear. Interestingly, *sae2Δ* has a less severe effect in G2-arrested cells compared to MRX deletions, which almost completely block resection. Sae2 is, like

Box 4.1 A word on genetic nomenclature

Unfortunately, there is no uniform genetic nomenclature used to describe genes and mutations in different organisms. If we imagine an evolutionarily conserved gene, *ECG1*, then—in the highly unlikely chance that the gene had the same name in all these organisms—it would be described as *ecg1⁺* in E. coli, *ECG1* in budding yeast, *ecg1⁺* in fission yeast, *ecg-1⁺* in Drosophila, *ecg-1⁺* in worms, and *ECG1* in vertebrates. The protein encoded by this gene would be Ecg in bacteria, Ecg1 (and sometimes Ecg1p) in yeasts, ECG-1 in flies, ECG-1 in worms, but ECG1 in mammals. Knockout mutations are often given by adding a Δ to the end in budding yeast (*ecg1Δ*) and more often *Δecg1* in fission yeast. In mammals the mutant (in a diploid) would be *ECG1^{Δ/Δ}*.

Things become even more complicated when one begins to discuss specific alleles. The simplest situation is when the precise mutation is known so that a mutation in budding yeast is identified as *ecg1-L45E*, but in worms and flies it would be *ecg-1^{L45E}* and *ecg-1(L45E)*, respectively. Mammalian alleles are superscripted: *ECG1^{L45E}*. Often, though, the allele indication is just the number that a mutation appeared in a screen: *ecg1-108*, with no

indication of the specific defect. The real problem is that hardly ever is the evolutionarily conserved gene named the same way in different organisms. RecA is the key recombination protein in bacteria, while it is called RadA in archaea, and just about anything you can imagine in eukaryotes: Rad51 in budding yeast and Rhp51 (rad homolog protein 51) in fission yeast; in flies it is SPN-A ("spin A" for strange historical reasons and for the fact that drosophilists adore very strange names for genes—"bride of sevenless" being one example); happily it is *rad-51* in worms and *RAD51* in humans. Over time there is a tendency to unify names by appending the two-letter genus–species abbreviation to a common protein name: ScRad51, SpRad51, DmRad51, CeRad51, and HsRad51 for example.

In this book, the genes and proteins will often be identified by their names in *Saccharomyces cerevisiae*, where the most extensive genetics and biochemistry has been carried out; but sometimes the gene and protein names used in each organism will be used.

MRX, needed in meiosis to remove the Spo11 protein from meiotic DSBs, but in this case it seems clear that it is Mre11 that does the actual cutting. However, no physical interaction between Sae2 and MRX has been found. Mammalian CtIP interacts directly with MRN. Inhibition of CtIP in mammals causes an apparently far more severe defect as resection is almost completely blocked. However, the phenotypes of MRN deletions in vertebrates are difficult to assess, because null mutations are lethal.

Recently a closer examination of resection defects of MRX mutants in budding yeast suggests that the reduction in resection in logarithmically growing cells is only evident within a few kilobases of the DSB; as resection proceeds, the rates of degradation 10 kb or more from the break are comparable in wild type and *mre11Δ* or *rad50Δ* mutant cells, although there is a delay in the onset of degradation in the mutants. These data suggest that MRX acts close to the DSB and may prepare the ends for the action of other exonucleases (possibly by creating the 100-bp resections mentioned above). Resection further from the DSB end depends on two other nucleases. One major activity resides in a 5′ to 3′ helicase–endonuclease consisting of the Sgs1–Top3–Rmi1 helicase complex and the endonuclease Dna2 (**Figure 4.6**). Dna2 has a RecB family nuclease domain, but curiously, although it also has helicase activity, this function is not essential in resection. Knocking out Sgs1 or other components of this complex reduces the resection rate to about 1 kb/hr. This remaining resection is carried out by the exonuclease Exo1, which also plays an important role in mismatch repair, where it chews away the region containing a mismatched base. Knocking out Exo1 by itself reduces resection somewhat, but ablation of Exo1 and either Sgs1, or Rmi1, or Dna2 blocks nearly all resection of an HO-induced DSB, except that MRX still chops off a few 100-nt segments very close to the DSB end (**Figure 4.7**).

Together these data suggest that MRX-Sae2 initiates resection at the DSB end but then responsibility for resection is handed over to these other two activities. MRX apparently does not move processively down the DNA and it is possible that the resection takes place with MRX remaining associated with the DSB end. MRX also may help in loading the other exonucleases. In another guise, MRX and MRN—without Sae2/CtIP—tether

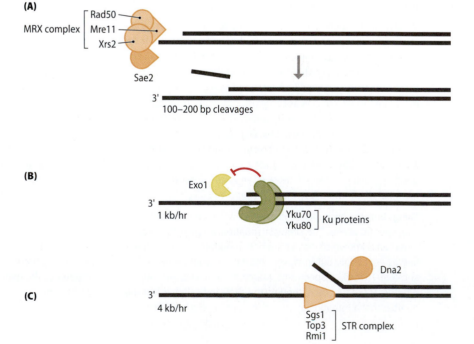

Figure 4.6 Resection nucleases in budding yeast. MRX and Sae2 proteins initiate resection close to the DSB end (A), while Exo1 (B) and the STR-Dna2 complex (C) carry out resection further from the DSB. Exo1's activity is impaired by the binding of the Ku proteins (Yku70–Yku80), which also have affinity for the resected 5′ end structure. MRX also facilitates both Dna2 and Exo1 loading and antagonizes Ku protein loading. (Adapted from Shim EY, Chung WH, Nicolette ML et al. [2010] *EMBO J* 29:3370–3380. With permission from Macmillan Publishers, Ltd.)

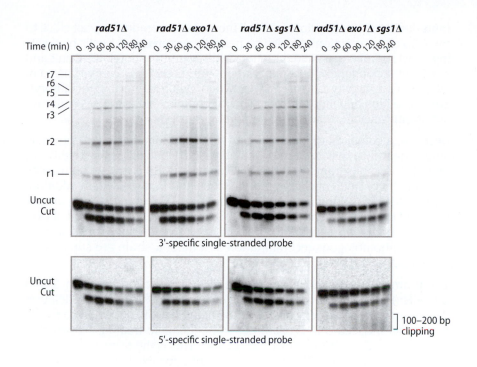

3'-specific single-stranded probe

5'-specific single-stranded probe

Figure 4.7 Exo1 and Sgs1 control two different 5′ to 3′ resection activities. An HO endonuclease cleavage was initiated in a *rad51Δ* strain which cannot repair the DSB by homologous recombination. Deleting Rad51 does not affect the wild-type rate of resection as assayed by Southern blot of denaturing gel electrophoresis (see Figure 4.4). Deletion of Exo1 has only a small effect on resection whereas deleting Sgs1 results in the persistence of the cut fragment because resection is slowed and there is less accumulation of the "partial digest" bands that arise from extensive resection. An *exo1Δ sgs1Δ* double mutant has virtually no resection although the initial 100–200 bp cleavages carried out by MRX and Sae2 are still visible. (From Mimitou EP & Symington LS [2008] *Nature* 455:770–774. With permission from Macmillan Publishers, Ltd.)

the ATM DNA damage checkpoint kinase (Tel1 in *Saccharomyces*) to the DSB end.

Resection is strongly regulated during the cell cycle (**Figure 4.8**). In G1-arrested yeast cells, there is virtually no resection and as a consequence HO-cleaved ends are predominantly rejoined by DNA ligase 4-Xrcc4 and Ku-dependent end-joining processes. The lack of 5′ to 3′ resection in G1-arrested cells reflects the absence of Cdk1 cell cycle kinase activity. The same strong inhibition of resection and increase in end joining can be accomplished in G2-arrested cells by inactivating Cdk1. At least one target of Cdk1 appears to be Sae2, as creation of a phosphomimetic

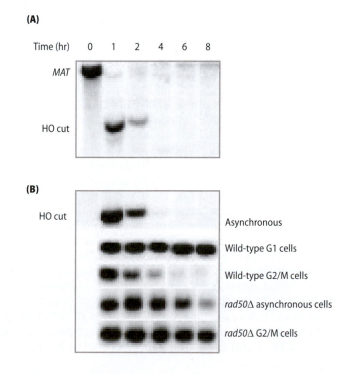

Figure 4.8 Cell cycle effects on 5′ to 3′ resection in budding yeast. (A) An HO-induced DSB at *MAT* generates a smaller restriction fragment on a Southern blot. This fragment vanishes as it is resected and becomes diffuse. (B) The HO-cut fragment is not degraded in G1-arrested cells but shows normal resection in G2-arrested cells. Deletion of Rad50 reduces resection in cycling cells but eliminates resection in G2-arrested cells. (From Ira G, Pellicioli A, Balijja A et al. [2004] *Nature* 431:1011–1017. With permission from Macmillan Publishers, Ltd.)

mutation Sae2-S367D allows a small increase in resection at least close to the DSB end. Resection can also be partially restored in G1 cells by deleting yeast's Yku70 or Yku80 proteins. Cdk1 also phosphorylates Dna2 and a mutant lacking these phosphorylation sites is severely compromised in its resection activity *in vivo*. In fact, when Dna2 cannot be phosphorylated, resection depends on Exo1. The Ku complex binds to DSB ends and apparently excludes Exo1 from degrading the ends.

Recently, a combination of *in vivo* and *in vitro* data have added considerable insight into how the Sgs1/Dna2 and Exo1 nucleases are regulated. The MRX complex rapidly binds to HO-induced DSB ends and stimulates recruitment of Dna2 nuclease, but this does not require Sae2. In addition, MRX assists Exo1 binding to the DSB, both by a direct interaction and by blocking the recruitment of Ku proteins to the DSB ends. As noted above, Ku protein binding appears to prevent Exo1 from acting; consequently, defects in homologous recombination in a *rad50Δ* strain are suppressed by *yku70Δ*, so long as Exo1 is present.

There are also very significant differences in resection between S and G2 cells (Figure 4.8). In asynchronous cells, a *rad50Δ* or *mre11Δ* deletion causes only a twofold defect in resection near the DSB; but when degradation of HO-induced DSB is examined in nocodazole-treated, G2-arrested *rad50Δ* cells, there is virtually no resection. This is the expected outcome if MRX-Sae2 is required to prepare the ends for Exo1 and Sgs1/Dna2; but if this is so, then in S-phase cells there must be still an alternative, as-yet-unknown way to allow Exo1 and Sgs1/Dna2 to act on DSBs. In fact, since there is no resection in G1-arrested cells, even wild-type cells, and no resection in G2-arrested cells lacking MRX, we are left with the conclusion that resection in S phase must be largely MRX independent. Finally it is now clear that four serine residues in Exo1 are phosphorylated by yeast's Chk2 kinase, Rad53, resulting in a reduction in activity. Since budding yeast's DNA damage checkpoint response is triggered by even a single DSB, it suggests that Exo1 may normally be phosphorylated under the circumstances that are routinely analyzed. Clearly there is still work to be done concerning the regulation of resection in budding yeast.

4.5 BLM PROTEIN IS IMPORTANT IN TWO EXONUCLEASE COMPLEXES FOR RESECTION IN MAMMALIAN CELLS

Biochemical evidence suggests that the Bloom's syndrome BLM protein, the human homolog of yeast Sgs1, and Exo1 act together rather than in separate complexes, and the Sae2 homolog, CtIP, is also of central importance. There is both a BLM–RPA–Dna2–MRN complex and a BLM–RPA–Exo1–MRN activity whose activities have been examined *in vitro* (**Figure 4.9**). The ssDNA binding protein complex RPA is required to enforce the 5′ to 3′ polarity of Dna2. *In vivo*, resection is monitored by the accumulation of RPA at sites of DSB damage, viewed as foci by indirect immunofluorescent staining with an anti-RPA antibody (**Figure 4.10**). siRNA depletion of either BLM or Exo1 reduces the rate of resection at DSB ends created by X-irradiation (**Figure 4.10A**) or by treating cells with the topoisomerase I inhibitor camptothecin in S-phase cells. A double knockdown of both BLM and Exo1 has a more severe effect, as might be expected if BLM is used in both the Dna2 and Exo1 pathways, remembering also that siRNA knockdowns are not necessarily equivalent to null mutants. As noted above, knocking down the Sae2 homolog, CtIP, has a more severe effect than the double knockdown of BLM and Exo1 (**Figure 4.10B**); this could reflect the different efficiencies of siRNA inactivation or could indicate that CtIP and perhaps the MRN complex must act prior to the actions of BLM and Exo1, reminiscent of what happens in budding

(A)

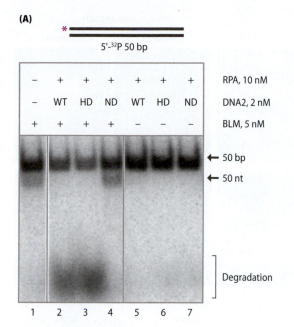

(B)

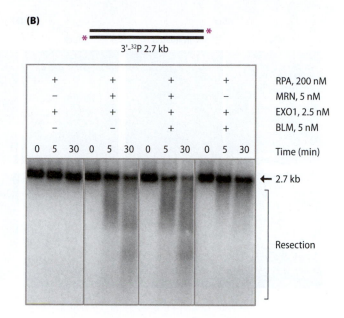

Figure 4.9 *In vitro* assays of human 5′ to 3′ resection protein complexes. A BLM helicase, DNA2, and RPA promote 5′to 3′ degradation. (A) 5′-end-labeled 50-bp double-stranded deoxyoligonucleotide can be unwound into a 50-nt ssDNA by BLM helicase (lane 1), but is instead degraded when RPA and DNA2 are added (lane 2). As the label is rapidly lost as rapidly migrating oligonucleotides, resection must start at the 5′ end. A helicase-dead (HD) mutant of DNA2 is still able to execute resection (lane 3) while a nuclease-dead (ND) mutation is not (lane 4); but BLM can still unwind the dsDNA, as seen by the appearance of a 50 nt single-strand product. Both activities depend on BLM helicase. Resection is stimulated by inclusion of the MRN complex (not shown). (B) Resection is also carried out by a complex containing BLM, EXO1, MRN, and RPA. Here, longer 3′-end-labeled DNA is assayed and resection is seen as a smear extending down from the intact fragment in which the label is retained on the shortening molecules. Significant resection is seen without BLM and a small amount of resection is seen without MRN. (From Nimonkar AV, Genschel J, Kinoshita E et al. [2011] *Genes Dev* 25:350–362. With permission from Cold Spring Harbor Laboratory Press.)

yeast. As noted before, evaluating MRN is difficult because the complex is essential.

The control of resection in vertebrate cells is strikingly similar to what has been seen in budding yeast. For example there is a similar lack of resection activity in G1 mammalian cells as in yeast, as inferred by the lack of appearance of the RPA foci revealed by indirect fluorescence microscopy. Furthermore, *Xenopus* cell extracts taken in S phase and in mitosis reveal significant differences in resection activity. Most notably, all resection in the M-phase extracts becomes dependent on MRN-CtIP, just as resection depended on the MRX proteins in G2-arrested yeast cells. Moreover, resection depends on Cdk1-mediated phosphorylation of CtIP just as yeast needs phosphorylation of Sae2.

4.6 RESECTION MUST PASS THROUGH CHROMATIN

The control of resection must also depend on chromatin remodeling activities, since resection has to be able to get past nucleosomes (**Inline Figure 4.1**). It isn't yet evident how the different exonucleases or

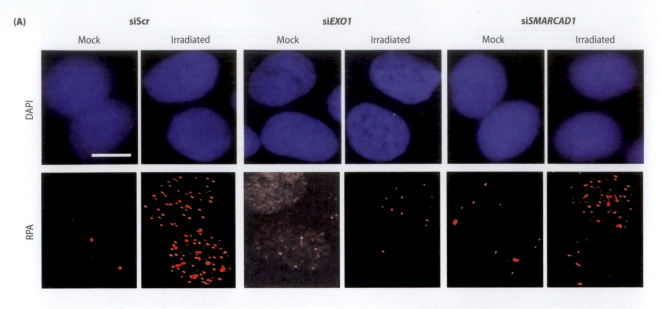

(B)

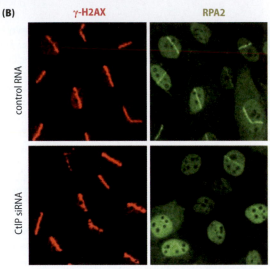

Figure 4.10 Appearance of RPA foci mark sites of resection in mammalian cells. (A) X-irradiation of human cells reveals the assembly of RPA foci in cells expressing a control siRNA (siScr) but these are reduced when either *EXO1* or *SMARCAD1* are knocked down. *SMARCAD1* is the mammalian homolog of budding yeast Fun30, which is also implicated in controlling 5′ to 3′ resection. Nuclei are stained with DAPI. (B) DNA damage is inflicted on human cells by moving a laser beam across the cell in a straight line. Damage is monitored by the appearance of a phosphorylated histone protein, γ-H2AX. RPA, as seen by antibody staining for RPA2, is clearly visible above the general background across the laser stripe, but is absent when siRNA is used to deplete CtIP. (A, from Costelloe T, Louge R, Tomimatsu N et al. [2012] *Nature* 489:581–584. With permission from Macmillan Publishers, Ltd. B, from Sartori AA, Lukas C, Coates J et al. [2007] *Nature* 450:509–514. With permission from Macmillan Publishers, Ltd.)

helicase–endonucleases can carry out their degradation of one strand of a duplex wrapped around a nucleosome. We do know from yeast studies that as resection proceeds, histones H2A and H2B are displaced, at a rate of about 4 kb/hr, as we would expect from the rate of resection. Surprisingly, chromatin immunoprecipitation of histone H3 or a damage-modified form of histone H4 (phosphorylated at the S1 residue) suggests that the H3–H4 dimer may not be fully removed from resected regions. How H3–H4 "hemisomes" might coexist with RPA and Rad51 is an unsolved question.

In budding yeast, deletion of the Arp8 subunit of the Ino80 chromatin remodeling complex causes a twofold reduction in the rate of resection near the DSB ends, but deleting other Ino80-related subunits has no effect. No role has yet been found for the SWR-C, Swi2, and RSC chromatin remodelers; however, another ATP-dependent Swi2-related chromatin remodeler, Fun30, has proven to have a very significant effect on the resection of DSB ends. When *FUN30* is deleted, the rate of resection is reduced to about one-quarter of the normal rate, both close to the DSB and further away (**Figure 4.11**). Moreover, in *fun30Δ sgs1Δ* and *fun30Δ exo1Δ* double mutants, resection is slower than in the single mutants.

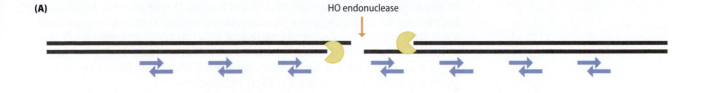

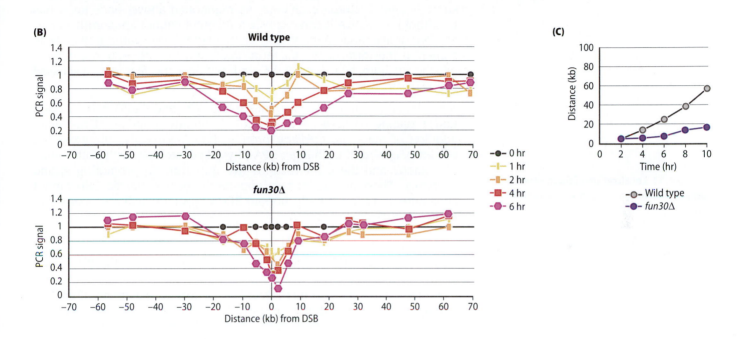

These data suggest that Fun30 may act ahead of both the Sgs1-Top3-Rmi1-Dna2 and Exo1 resection machinery to make the DNA accessible for resection. The mammalian homolog of Fun30 is *SMARCAD1*, and siRNA depletion of *SMARCAD1* dramatically reduces the formation of RPA foci after X-irradiation, suggesting that there is less ssDNA generated without SMARCAD1 (Figure 4.10A).

4.7 PREPARING ssDNA FOR THE LOADING OF RecA/Rad51 REQUIRES ssDNA BINDING PROTEINS

ssDNA can readily form secondary structures by base pairing. Such structures can be extensive in the case of regions containing inverted repeated DNA or in regions such as those encoding tRNAs. The ability of RecA/Rad51 to form a continuous filament in such regions is impaired, but the problem is overcome by the prior binding of ssDNA binding proteins SSB or RPA, which apparently are undeterred by such secondary structures. Thus, both *in vitro* and *in vivo*, the strand exchange reaction is facilitated by these ssDNA-binding proteins.

Bacterial (and eukaryotic mitochondrial) SSB is a 19 kD protein that forms a homotetramer that has been reported to bind between 30 and 65 nt of ssDNA *in vitro*. Eukaryotic RPA (sometimes called RP-A) comprises a heterotrimer of 70, 32, and 14 kD proteins. In yeast these are called Rfa1, Rfa2, and Rfa3, respectively. Recent crystallographic analysis has shown that the eukaryotic complex contains four DNA-binding OB folds, analogous to the single OB fold found in each of the four SSB subunits. RPA appears to protect about 30 nt of ssDNA. ssDNA binding proteins are essential (that is, yeast cells are inviable if the gene encoding one

Figure 4.11 Deletion of yeast *FUN30* retards 5′ to 3′ resection. (A) Quantitative PCR is used to follow resection of an HO endonuclease induced DSB, using many primer pairs across the adjacent regions. The signal decreases to 50% of the initial signal as resection leaves one strand that can be PCR amplified. Decreases below 50% may reflect late loss of the 3′ ended strand. (B) Progressive loss of signal occurs as one strand of DNA is removed by resection. This process is greatly retarded in the absence of the chromatin remodeling protein Fun30. (C) The change in the rate of resection is plotted on the rate. (B, C, from data in Eapen V V, Sugawara N, Tsabar M et al. [2012] *Mol Cell Biol* 32:4727–4740. With permission from the American Society for Microbiology.)

of these proteins is deleted). RPA plays a central role in DNA replication by binding to ssDNA regions that are unwound as replication proceeds. That RPA plays an apparently different role during recombination was revealed by a budding yeast mutation *rfa1-t11*, an L45E mutation in the large subunit, which is unable to carry out DSB repair either by SSA or gene conversion but is unaffected in replication.

It is surprising that, *in vitro*, adding SSB to ssDNA before adding RecA strongly *inhibits* strand exchange. As mentioned above, once RecA has a foothold on ssDNA it can displace SSB, but it cannot apparently nucleate four or five monomers on SSB-coated DNA *in vivo* without help. Help comes in the form of the bacterial RecF and RecOR protein complex, which can both help load RecA and stabilize the filament, discussed below in more detail.

A similar situation prevails with eukaryotic RPA and Rad51, where again Rad51 loading *in vitro* is impaired if RPA is already bound to ssDNA. Here, addition of the yeast *mediator* protein, Rad52, strongly stimulates the efficiency of strand exchange. RPA physically interacts with both Rad51 and Rad52 and these interactions are important in promoting strand exchange. These interactions are apparently species specific; SSB cannot substitute for RPA. Similarly, human BRCA2 protein facilitates loading of Rad51 on RPA-coated ssDNA.

4.8 RPA AND Rad51 ASSEMBLY CAN BE ASSAYED *IN VIVO*

One method to demonstrate that Rad51 has formed a filament after DNA damage is through the use of indirect immunofluorescence using an anti-Rad51 antibody in fixed cells. The Rad51 antibody is visualized by a second, fluorescently modified antibody that recognizes the first antibody. Using this approach, after exposing cells to ionizing radiation, one can examine how rapidly Rad51 assembles into aggregates (foci) that presumably represent Rad51 filaments on ssDNA (**Figure 4.12A**).

Figure 4.12 Localization of fluorescently tagged recombination proteins after DNA damage. (A) Co-localization of Rad52-GFP with indirect immunofluorescence against Rad51 protein in mammalian cells after ionizing radiation. Co-localization is indicated by the yellow color and by white arrows. (B) Co-localization of RPA (Rfa1-CFP) and Rad52-YFP foci in budding yeast after ionizing radiation. Yeast cells are shown in a differential interference contrast (DIC) image. (A, from van Veelen LR, Essers J, van de Rakt MW et al. [2005] *Mutation Res* 574:34–49. With permission from Elsevier. B, from Lisby M, Barlow JH, Burgess RC & Rothstein R [2004] *Cell* 118:699–713. With permission from Elsevier.)

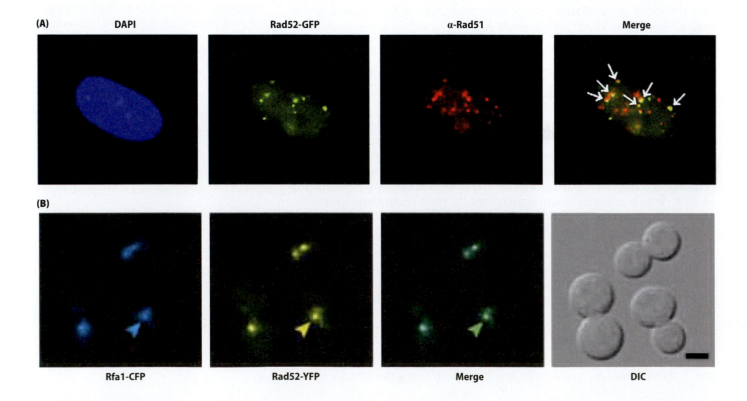

(A) DAPI Rad52-GFP α-Rad51 Merge

(B) Rfa1-CFP Rad52-YFP Merge DIC

With living cells, it is possible to use genetic fusions between various recombination proteins and green fluorescent protein (e.g. Rad52-GFP) as shown in Figure 4.12A or fusions with fluorescent proteins with other colors (for example Rfa1-CFP or Rad52-YFP, as shown in **Figure 4.12B**). Following treatment of cells with ionizing radiation, one can observe the formation of foci whose persistence presumably reflects the association of RPA with ssDNA. In live cells, the use of GFP-tagged proteins allows one to determine the kinetics of focus formation. Using many different fluorescently tagged proteins, Rodney Rothstein's lab showed that there was a clear order to the arrival of different proteins at the sites of DSBs, with the DNA end-binding proteins Ku70-Ku80 and the MRN complex present before RPA, which in turn was present before Rad52 and Rad51. Especially in situations where DSB repair is blocked, RPA can be detected by immunofluorescence or with a fluorescently tagged protein as foci that presumably contain long regions of ssDNA. These foci are frequently used to show the sites of DNA damage, to which other proteins may co-localize (Figure 4.12B).

The assembly of RPA and Rad51 onto the ends of a DSB can also be measured by *chromatin immunoprecipitation* (ChIP). DNA and proteins are cross-linked in living cells by formaldehyde and then chromatin is sonicated into small fragments that can then be immunoprecipitated by an antibody, in this case one against Rad51 or RPA. After reversing the cross-linking, one can assay for the enrichment of particular DNA sequences by PCR. The assembly of both RPA and Rad51 at a site adjacent to a DSB created by HO endonuclease is shown in **Figure 4.13**. Induction of the DSB is essentially complete 20 min after inducing expression of HO, as seen in the Southern blot (Figure 4.13A). By 20 min RPA has already begun to assemble at the *MAT* locus (Figure 4.13B). However, Rad51 doesn't arrive in abundance for another 10 min. Interestingly, RPA does not go away as Rad51 accumulates; this might indicate that Rad51 filaments are not continuous and that some RPA remains, or it might reflect a more dynamic situation in which Rad51 assembles and disassembles so that RPA is displaced and then binds again.

(A)   **Southern blot of DNA fragments**

(B)   **ChIP at *MAT***

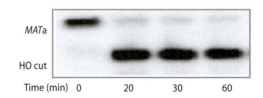

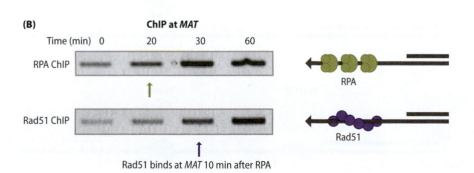

Figure 4.13 Detection of RPA and Rad51 binding by chromatin immunoprecipitation (ChIP) after HO induction of a DSB in budding yeast at the *MAT* locus. (A) A Southern blot of restriction endonuclease digested DNA shows HO cleavage of *MAT***a** is complete in 20 min (note that the *MAT*α product [not shown] does not appear until 1.5–2 hr). (B) Binding of RPA to the HO-cut end is seen by 20 min whereas there is about a 10 min delay in the binding of Rad51. (Adapted from Wang X & Haber JE [2004] *PLoS Biol* 2:104–111.)

4.9 CREATING A RECOMBINASE FILAMENT IN BACTERIA NEEDS THE PARTICIPATION OF SEVERAL MEDIATORS

An efficient strand exchange reaction cannot be carried out *in vivo* without the participation of a number of mediator proteins. In *E. coli*, there are two distinct pathways of recombination, a RecBCD-dependent pathway and RecQ- and RecJ-dependent pathway that also requires the RecF, RecO, and RecR proteins. RecFOR collectively act as mediators of RecA loading. When RecBCD is functional, deletion of RecF, RecO, or RecR has almost no effect on the efficiency of recombination during conjugation, but single-strand gap repair is markedly reduced. Remarkably, RecBCD not only acts as a Chi-regulated resection machine, it also directly participates in loading RecA onto ssDNA. This RecA-loading activity involves its interaction with the RecB subunit and is inhibited by RecD in the absence of Chi sequences. When RecBCD encounters Chi, alterations in the association of RecD within the complex relieve the inhibition of RecA loading. Thus, in cells lacking RecD, there are two consequences. First, the absence of the 5' to 3' helicase means that DNA is only acted on by RecB's slower 3' to 5' helicase. This results in degradation of the 5'-ended strand—by RecJ nuclease—and generation of 3'-ended tails at the end of the DNA molecule, even without encountering Chi. Second, without RecD's interference, RecA can be loaded on these new ssDNA regions, by RecFOR. Consequently, recombination occurs close to the DSB end.

Without RecBCD, when ssDNA is generated by RecQ/RecJ, RecFOR becomes essential for displacement of SSB and RecA loading. The three proteins form both a RecFOR and a RecOR complex. The FOR complex is needed to load RecA specifically at the junction of dsDNA and a ssDNA tail or at the edge of a ssDNA gap, whereas RecOR is sufficient to initiate RecA loading and displacement of SSB in the middle of a ssDNA region (**Figure 4.14**). *In vitro*, RecO has single-strand annealing activity and RecOR help stabilize RecA on ssDNA. Filament formation can be "capped" by the RecX protein, but this action is antagonized by RecF, so that RecF, too, promotes RecA filament growth. Another protein, DinI, binds to and stabilizes the RecA filament.

It is possible to create hybrid resection and RecA loading activities in *E. coli* by using RecF pathway components to complement the missing activities in RecBCD. A mutation of RecB (*recB-D108A*) that has normal helicase activity but impaired nuclease activity also lacks RecA loading capacity. However, recombinational activity *in vivo* can be reestablished if RecJ substitutes for nuclease deficiency and RecFOR takes up the loading of RecA onto ssDNA. RecQ isn't needed because RecB^{D108A}CD still has normal helicase function.

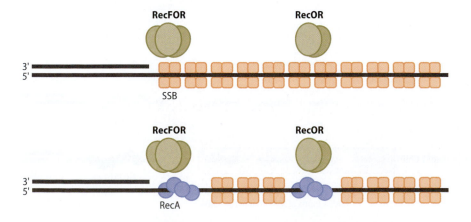

Figure 4.14 Roles of bacterial RecFOR proteins in loading of RecA. RecF, RecO, and RecR combine to load RecA onto ssDNA coated with SSB (*orange*). RecO and RecR are able to load RecA (*blue*) by themselves but RecF acts specifically with RecOR to facilitate loading at the junction of a resected end. (Adapted from Lusetti SL, Hobbs MD, Stohl EA et al. [2006] *Mol Cell* 21:41–50. With permission from Elsevier.)

RecFOR does not have clear eukaryotic homologs, but recent evidence does suggest that RecF has a similar globular domain of the eukaryotic Rad50 protein, which is part of the MRX/MRN complex involved in DNA end binding and processing. RecO appears to be functionally analogous to yeast Rad52 in promoting single-strand annealing.

4.10 MEDIATORS IN EUKARYOTES ARE SURPRISINGLY UNRELATED TO THOSE IN BACTERIA

Although several mediator proteins are evolutionarily well conserved, their importance (and even existence) in different organisms is quite different; hence it is most useful to introduce the core set of mediators in terms of their well-studied roles in budding yeast and then introduce the evolutionary variations on the theme that appear in other fungi, worms, flies, plants, and vertebrates.

4.11 Rad52 AND Rad55–Rad57 ARE THE PRINCIPAL MEDIATORS IN *SACCHAROMYCES CEREVISIAE*

In budding yeast, there are three key mediator proteins, Rad52, Rad55, and Rad57 (**Figure 4.15A**), all identified by the ionizing radiation sensitivity of null mutations. Of these, Rad52 is the most critical protein, as almost all homologous recombination events are eliminated in its absence. Indeed, there are well-documented Rad51-independent processes including single-strand annealing (SSA) and one of the two "survivor" pathways for maintaining telomeres in the absence of telomerase; but these Rad51-independent events all depend on Rad52.

Without Rad52 there is no Rad51 filament formation (**Figure 4.15B**). Rad52, or perhaps a Rad52-Rad51 dimer, can facilitate the displacement of RPA from ssDNA. Rad51 interacts with the C-terminus of Rad52. Although a C-terminal Rad52 truncation is still able to bind ssDNA, it is highly defective in filament formation. However, overexpression of Rad51 will bypass this truncation mutation. Whether there is residual binding to an undefined alternative site on Rad52 or whether the second mediator complex, Rad55–Rad57, can bypass Rad52's lack of interaction with Rad51 has not been established.

(A)

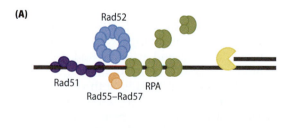

(B)

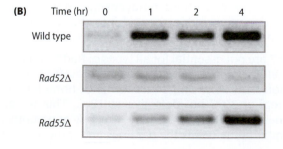

Figure 4.15 Roles of Rad52 and the Rad51 paralogs, Rad55 and Rad57, in recruitment of Rad51 to an HO-induced DSB. (A) The mediators Rad52 and Rad55–Rad57 facilitate the displacement of RPA and the loading of Rad51 onto the resected DSB end. (B) A Rad51 chromatin immunoprecipitation (ChIP) experiment analogous to that in Figure 4.13 shows that Rad51 loading requires Rad52 and is much slower in the absence of Rad55. (Adapted from Sugawara N, Wang X & Haber JE [2003] *Mol Cell* 12:209–219. With permission from Elsevier.)

An *in vivo* experiment that emphasizes the species-specificity of these interactions was carried out by Todd Milne and David Weaver, who showed that expressing the *Kluyveromyces lactis* Rad52 (Kl-Rad52) protein in *S. cerevisiae* could partially complement a *rad52Δ* in *S. cerevisiae*, but surprisingly expressing Kl-Rad52 made wild-type *RAD52* cells sensitive to DNA damage. This result suggests that Kl-Rad52 acts as a *dominant-negative* protein, possibly by sequestering some component from properly performing its role in repair. A similar dominant negativity is found when truncation mutations of Sc-Rad52 are expressed, again suggesting that some process is poisoned even though the entire wild-type recombination machinery is present. Overexpressing Sc-Rad51 prevents the dominant-negative interactions of both Kl-Rad52 and the truncations of Sc-Rad52. Some of the poisoned interactions must be with other recombination proteins because co-expressing both Kl-Rad52 and Kl-Rad51 did not complement a *rad51Δ rad52Δ* strain of *S. cerevisiae*. In a similar vein, expressing *S. cerevisiae* Rad51 in mouse cells is also dominant-negative.

Rad52 is a roughly 50 kD protein with a strong Rad51 interaction domain at its C-terminus. Its N-terminal half has *in vitro* strand annealing activity that is similar to the full-length protein. Crystallographic studies have revealed that the N-terminal portion of human Rad52 forms an 11-membered ring that binds ssDNA and presumably "presents" the ssDNA to a complementary Rad52-bound strand (see Figure 5.7). Rad52 associates with RPA and facilitates the replacement of RPA by Rad51 to form Rad51-ssDNA filament.

As we will see in more detail in later chapters, strand-annealing activity is critical both in SSA and in later steps of Rad51-mediated gene conversion—the capture of the second end of homology by annealing to a D-loop. Although Rad52 is well conserved from yeasts to humans, it is surprisingly absent both in *Drosophila* and *Caenorhabditis*. There are no obvious bacterial homologs of Rad52, but recently a bacteriophage ul36 that infects *Lactococcus lactis* was shown to have a remarkably similar protein that forms an 11-member ring and promotes strand annealing activity and *E. coli* RecO seems to perform similar functions. Phylogenetic analysis suggests that there are bacterial proteins that share some domains of this protein.

Rad51 paralogs: Rad55 and Rad57

Budding yeast Rad55 and Rad57 form a heterodimer. Both proteins are about the size of Rad51 and they share sufficient homology, primarily in the Walker ATP binding domains, to be called Rad51 *paralogs*. Mutations of key ATPase residues in Rad55 are recombination defective, while the same mutations in Rad57 have no effect. However, neither Rad55 nor Rad57 has demonstrable strand exchange activity. These proteins function to facilitate Rad51 loading and without Rad55–Rad57, loading of Rad51 onto an HO-induced DSB is delayed. In their absence Rad51 filament assembly at an HO-induced DSB is both delayed and much less robust (Figure 4.15B) and recombination is absent, a result that suggests that the "quality" of the Rad51 filament without Rad55–Rad57 is defective—perhaps there are only short, discontinuous patches of Rad51. A recent observation by Wolf-Dietrich Heyer showed that a small number of Rad55–Rad57 dimers are incorporated into the Rad51 filament, and stabilize the filament against its dissociation by the Srs2 helicase.

Rad55 and Rad57 are not essential for all types of Rad51-mediated recombination, even with complete gene deletions. For *rad55Δ* or *rad57Δ*, both ionizing radiation sensitivity and some forms of homologous recombination are normal at 30°C but defective at 18°C. This result has been interpreted to mean that Rad51 assembly, which is apparently more difficult at lower temperature, needs the paralogs under these conditions, but can assemble sufficiently well at higher temperatures. So far this idea has not been tested by ChIP. Moreover, the low-temperature defects of *rad55Δ* and *rad57Δ* can be suppressed either by overexpressing Rad51 or by creating mutations in Rad51 (such as I345T) that exhibit stronger binding to both ssDNA and dsDNA. Exactly how Rad55–Rad57 act alongside Rad52 has not been established. Interestingly, although overexpressing Rad51 will suppress a C-terminal mutation of Rad52, the Rad51-I345T mutation will not. In one respect the absence of Rad55–Rad57 appears more severe than the loss of Rad51; spontaneous sister chromatid exchange (assayed by an unequal crossing-over assay that will be described later) is reduced more in *rad57Δ* at low temperature than *rad51Δ*. Lorraine Symington has suggested this might mean that Rad55–Rad57 play a key role in recombination events resulting from ssDNA gaps rather than from DSBs.

Additional mediators: the PCSS complex

Four other proteins apparently play mediator roles in budding yeast. Deletions of these proteins were first identified as DNA damage sensitive and then were shown to have distinctive genetic interactions in cells defective for the helicase complex Sgs1–Top3–Rmi1; specifically, they were **s**uppressors of the **h**ydroxy**u**rea sensitivity of *sgs1Δ* (SHU). The SHU complex consists of four proteins, Psy3, Csm2, Shu1, and Shu2, that physically interact; this complex is now referred to as PCSS. The phenotypes of individual deletions and double mutants indicate that they function in a common pathway, but deletions of Csm2 and Psy3 show more severe defects than the absence of Shu1 and Shu2. Furthermore, *csm2Δ* and *psy3Δ* also have stronger defects in meiotic recombination. The Shu mutants also suppress the slow growth of *top3Δ*, as does the "slow growth suppressor" *sgs1Δ*. Recent structural studies of PCSS show that the proteins form two separate dimers, Psy3–Csm2 and Shu1–Shu2. The crystal structure of the Psy3–Csm2 dimer revealed that these proteins share significant structural similarity with Rad51's ATPase domain and can bind dsDNA nonspecifically. A mutation that eliminates Csm2's DNA binding causes cells to be sensitive to the DNA damaging agent, MMS.

The most specific indication of a specific role for these proteins comes from treating S-phase cells with MMS, which results in Rad51-dependent accumulation of branched DNA intermediates that reflect alterations at stalled replication forks. As shown in Figure 1.6, stalled replication forks can regress into Holliday junctions. These X-shaped intermediates can be seen as a "spike" in two-dimensional Brewer–Fangman gels (**Figure 4.16**).

Figure 4.16 Evidence that PCSS proteins play a role in recombination. Branched X-shaped molecules accumulate in two-dimensional gel electrophoresis in the absence of Sgs1 when cells are treated with MMS for 3 hr. This spike is markedly reduced in the absence of Rad51 or in the absence of the four proteins of the PCSS complex. (Adapted from Mankouri HW, Ngo HP & Hickson ID [2007] *Mol Biol Cell* 18:4062–4073. With permission from Elsevier.)

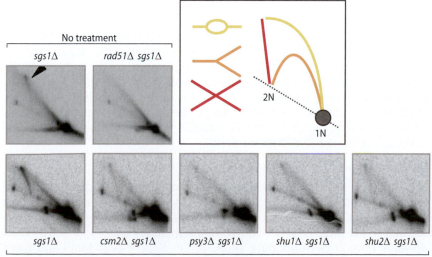

No treatment

sgs1Δ   *rad51Δ sgs1Δ*

sgs1Δ   *csm2Δ sgs1Δ*   *psy3Δ sgs1Δ*   *shu1Δ sgs1Δ*   *shu2Δ sgs1Δ*

0.033% MMS

These intermediates are Rad51 dependent but also PCSS dependent. It is possible that the PCSS complex functions specifically during S phase in the formation of recombinant structures at replication forks, because deleting these genes has little impact on gene conversion. In another assay, HO-induced BIR was reduced about 50% by a Shu2 deletion, much less than a four- to eightfold reduction in *rad55Δ* or *rad57Δ*; however, the residual BIR in *rad55Δ* or *rad57Δ* was reduced significantly further by *shu2Δ*. Importantly, overexpressing Rad51 will suppress the effects of *shu2Δ* or *shu2Δrad57Δ* on BIR. These data suggest the PCSS complex acts similarly but independently of Rad55-Rad57.

Rad59, sharing homology with Rad52, is not a mediator but plays an important role

Rad59 was identified because of the difference between *rad52Δ* and *rad51Δ* in spontaneous recombination between two different alleles (heteroalleles) located in inverted repeats. In this assay, gene conversion can produce prototrophic recombinants. Whereas *rad52Δ* reduced these events > 100-fold, *rad51Δ* reduced them only tenfold. A *rad51Δ rad59Δ* double mutant, however, was nearly—but still not quite—as deficient as *rad52Δ*, implying that there may be still another Rad52-independent pathway. Rad59 shares homology with the N-terminus of Rad52 (the part involved in strand annealing) and indeed, *in vitro*, Rad59 also promotes strand annealing. Rad59 interacts not only with itself but also with Rad52. The precise role Rad59 plays in any HR event is not yet evident. Rad59 will not complement Rad52 even when overexpressed. As expected from the way it was identified, Rad59 plays an important role in Rad51-independent recombination events, namely, SSA and Rad51-independent BIR, including one pathway of telomere maintenance in the absence of telomerase. In SSA, Rad59 is essential when the annealing segments are short (about 200 bp) but its importance diminishes greatly when longer homologous segments are available.

4.12 Rad51 FILAMENT ASSEMBLY IN FISSION YEAST INVOLVES EVEN MORE PROTEINS

Fission yeast has two homologs of Rad52, known as Rad22, and Rti1 (or Rad22B) (**Table 4.1**). A *rad22* deletion is strongly recombination defective, but for some time this conclusion was obscured by the fact that *rad22Δ* strains rapidly acquire mutations in a helicase, Fbh1, that somehow suppresses *rad22Δ*'s lethality. Fbh1 is not found in budding yeast, though it is related to another helicase shared by budding and fission yeast, Srs2. Based on observations about the ability of Srs2 in budding yeast to dismantle Rad51 filaments, it has been suggested that deleting Fbh1 will allow Rad51 to form filaments in the absence of Rad22. A *rad22Δ* strain is strongly defective in both SSA and gene conversion, as studied in a duplication containing *ade6* heteroalleles flanking a *URA4* gene (**Figure 4.17**). Spontaneous Ade6+ recombinants can be either Ura4− deletions (most likely SSA events that are Rad51 independent) or else Ade6+ Ura4+ gene conversions that require *S. pombe*'s Rad51 (Rhp51) and its paralogs. All of these events are Rad22 dependent.

Rad55/Rad57 versus Swi5/Sfr1

Fission yeast differs from budding yeast in having two distinct sets of mediator proteins that affect different kinds of mitotic recombination events. One pair are the *S. pombe* homologs of Rad55 and Rad57 (known as Rhp55 and Rhp57). A second complex includes Swi5 and Sfr1, which do have homologs in budding yeast, but in *S. cerevisiae* Mei5 and Sae3 apparently function only in meiosis. In UV-irradiated cells both pairs of

Table 4.1 DSB repair proteins

Saccharomyces cerevisiae	*Schizosaccharomyces pombe*	*Drosophila melanogaster*	*Caenorhabditis elegans*	*Arabidopsis thaliana*	*Homo sapiens*
Mre11 Rad50 Xrs2	Rad32/Mre11 Rad50 Nbs1	MRE11 RAD50 NBS	MRE-11 RAD-50 no homolog reported	MRE11 RAD50 NBS1	MRE11 RAD50 NBS1/NIBRIN
Sae2 Com1	Ctp1	no homolog reported	COM-1	SAE2	CTIP
Rad51	Rhp51/Rad51	SPN-A	RAD-51	RAD51A, B, C, D	RAD51
Rad52	Rad22/Rad52	no homolog reported	no homolog reported	RAD52-1	RAD52
Rad54	Rhp54	OKRA	RAD-54	RAD54L	RAD54
Rdh54	Rdh54	OKRA	no homolog reported	no homolog reported	RAD54B
Rad55 Rad57	Rhp55 Rhp57 Rdl1 Rpl1	SPN-B SPN-D CG2412 CG6318	RFS-1 only homolog reported	RAD51B RAD51C RAD51D XRCC2 XRCC3	RAD51B RAD51C RAD51D XRCC2 XRCC3
Mei5	Sfr1	no homolog reported	no homolog reported	no homolog reported	SFR1
Sae3	Swi5	CG14104	no homolog reported	no homolog reported	SWI1
Rad59	no homolog reported	no homolog reported	no homolog reported	no homolog reported	no homolog reported
Rdh54/Tid1	no homolog reported	no homolog reported	no homolog reported	no homolog reported	no homolog reported
no homolog reported	no homolog reported	BRCA2	BRC-2	BRCA2	BRCA2
Exo1	Exo1	TOSCA	EXO-1	EXO1	EXO1
Sgs1	Rqh1	MUS309/DmBLM	HIM-6	BLM/RECQl4	BLM
Dna2	Dna2	CG2990	DNA-2	DNA2	DNA2
Top3	Top3	TOP3α	TOP-3	TOP3	TOP3α
Rmi1	Rmi1	no homolog reported	no homolog reported	RMI1	RMI1 RMI2
Mus81	Mus81	MUS81	MUS-81	MUS81	MUS81
Mms4	Eme1	MMS4	no homolog reported	EME1	EME1 EME2
Gen1	no homolog reported	GEN1	GEN-1	GEN1	GEN1
Mph1	Fml1	FANCM	FANCM	FANCM	FANCM
Srs2	Srs2	no homolog reported	no homolog reported	no homolog reported	no homolog reported
no homolog reported	no homolog reported	no homolog reported	RTEL-1	no homolog reported	RTEL
no homolog reported	Fbh1	no homolog reported	no homolog reported	no homolog reported	FBH1

The table gives a noncomprehensive list of recombination proteins important in homologous recombination.

When several names are currently in use for the same protein, both are given, separated by a /. Recently many *S. pombe* genes have been renamed to match *S. cerevisiae* names. The proteins are sometimes distinguished by giving them their organism's two-letter prefix, for example SpRad51 or DmRad54.

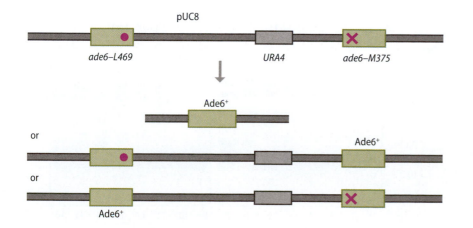

Figure 4.17 Genetic assay for SSA and gene conversion in *S. pombe*. Noncrossover recombination between repeats of the *ade6* gene carrying different alleles can yield Ade⁺ recombinants by noncrossover gene conversion (retaining *URA4*) or by deletion (Ura⁻), nearly all of which are by SSA. (Adapted from Gysler-Junker A, Bodi Z & Kohli J [1991] *Genetics* 128:495–504. With permission from the Genetics Society of America.)

mediators play important roles. The number of UV-induced Rhp51 foci is partially reduced in *swi5Δ* or in *rhp57Δ* mutants and completely abolished in a *swi5Δ rhp57Δ* double mutant. *In vitro*, both Swi5–Sfr1 and Rhp55–Rhp57 stimulate Rhp51's strand exchange activity (neither complex has strand exchange activity on its own). Whether Swi5–Sfr1 are mediators in the sense of facilitating Rhp51 loading or whether they activate Rhp51 in some other way is not yet evident. *In vitro*, addition of RPA after Rhp51 and Swi5–Sfr1 have been incubated with ssDNA still impairs recombination, whereas the late addition of RPA has much less effect on Rhp51 in the presence of Rhp55–Rhp57. Similarly, adding Rad22 to RPA-coated DNA is insufficient to stimulate much strand exchange without Swi5–Sfr1. An example of the stimulation of strand exchange by Swi5–Sfr1 can be seen in **Figure 4.18**, where this complex stimulates completion of the four-strand *in vitro* strand exchange reaction shown in Figure 3.11. This result again supports the idea that the Swi5–Sfr1 duo somehow activates Rhp51. Further evidence of a distinct difference in function has been suggested from *in vivo* studies of the outcomes of recombination induced by the expression of the HO endonuclease. Hiroshi Iwasaki's group has reported that both complexes appear to affect noncrossover outcomes of repair (presumably by SDSA); in contrast Swi5–Sfr1 have no effect on crossovers, whereas Rhp55–Rhp57 are required. However, a recent study found that *rhp57Δ* caused an increase in crossovers in the same assay system. The contradiction has not yet been resolved.

It should be noted that Swi5 joins a growing list of proteins that have more than one function. Independent of Sfr1, it plays a distinctly different role in combination with the Swi2 protein, both of which are named for complementation groups of *swi* mutants that impair *S. pombe* mating-type gene switching, which will be discussed in detail in Chapter 8.

The PCSS complex was also isolated in *S. pombe*, based on an analogous screen looking for suppressors of the HU sensitivity of the Sgs1 homolog Rqh1. In this way, Sws1, the homolog of Shu2, was identified. An epitope-tagged version of Sws1 proved to co-precipitate two other proteins, Rlp1 and Rdl1 that are apparently the homologs of *S. cerevisiae* Shu1 and Psy3, respectively. Rlp1 and Rdl1 also have weak homology to the mammalian paralogs Rad51D and Xrcc2. The Sws1 of *S. pombe* has a human homolog (Hs-Sws1). Where the PCSS proteins act during homologous recombination in fission yeast is not yet known, but deletion of these proteins reduces the number and intensity of Rad52–YFP foci in both budding and fission yeast. Paul Russell has suggested that these proteins might act upstream of Rad52, though deletions have only a modest (twofold) effect on recombination or the formation of Rad52–YFP foci.

Figure 4.18 Stimulation of the four-strand exchange reaction by *S. pombe* Swi5-Sfr1. An autoradiogram monitors the *in vitro* four-strand recombination reaction between a gapped circular dsDNA and linear dsDNA substrates (S [lds]) (see Figure 3.10). Addition of Swi5-Sfr1 stimulates the yield of joint molecules (JM) and products (P) consisting of nicked circles (NC) and gapped linear molecules (GL). However Swi5-Sfr1 do not increase the kinetics of strand exchange or branch migration compared to the reaction lacking the complex. (From Murayama Y, Kurokawa Y, Mayanagi K & Iwasaki H [2008] *Nature* 451:1018–1021. With permission from Macmillan Publishers, Ltd.)

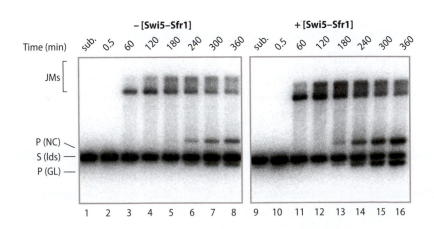

4.13 MEDIATORS OF RECOMBINASE FILAMENT ASSEMBLY HAVE BEEN IDENTIFIED IN VERTEBRATE CELLS

The core machinery of recombination has been conserved from yeasts to vertebrates. There has been intensive investigation of three organisms: chicken DT40 cells and both mouse and human cells. In all organisms, Rad51 is required for efficient DSB repair—in fact in all three organisms the absence of Rad51 is lethal. The lethality appears to be caused by the failure of cells to repair chromosome breaks arising during replication, as illustrated in Figure 1.1. The fact that Rad51 is not essential in yeast can be explained by the fact that its genome is so tiny compared to a vertebrate; but when breaks occur they require Rad51 for repair. In the absence of Rad51 in DT40 one can count perhaps a dozen chromatid breaks (that is, where one of two sister chromatids has a visible chromosome break that would require Rad51 for repair). In budding yeast there is a replication-associated DSB in about 1 in 10 cells. This translates to 1 DSB per 1.5×10^8 bp in yeast, which is not so different from about 12 DSBs in 6×10^9 bp in diploid vertebrate cells (1 DSB in 5×10^8 bp). This estimate of the number of events that require Rad51 every cell cycle is supported by the observation of multiple spontaneous sister chromatid exchanges in mammalian cells, as shown in Figure 1.2. As noted before, in the absence of knowing what proportion of lesions are repaired by a repair event accompanied by crossing over, we don't know the total number of lesions arising every replication cycle, or what fraction of them are DSBs.

The essential nature of Rad51 in mammalian cells has made it difficult to analyze its role in great detail *in vivo*, but biochemical experiments have shown clearly that human Rad51 forms a filament on ssDNA essentially indistinguishable from that seen with yeast Rad51 proteins. In all eukaryotes, filament formation depends on Rad51 paralogs as well as other mediators.

Vertebrate Rad51 paralogs

All metazoans have Rad51 paralogs. Vertebrates have five identified paralogs, called Rad51B, Rad51C, Rad51D and—just to be difficult—Xrcc2 and Xrcc3. The first three were named because of their protein sequence homology with Rad51, the last two were identified as the proteins encoded by hamster genes whose mutation caused X-ray sensitivity. Biochemical studies and immunoprecipitations using antibodies against the different proteins revealed that the five proteins form two stable subcomplexes, one containing Rad51B, Rad51C, Rad51D, and Xrcc2 (BCDX2) and the other comprising of Rad51C and Xrcc3 (CX3).

In chicken DT40 cells, each of the paralogs has been deleted without affecting viability. In the absence of any of the five paralogs, Rad51 fails to form radiation-induced foci and homologous recombination is severely reduced (**Figure 4.19**). However, as with yeast cells, overexpression of Rad51 will compensate for the lack of one or more of the paralogs. In DT40, the role of the paralogs must be partially overlapping with other mediators such as Rad52, because an Xrcc3$^{-/-}$ Rad52$^{-/-}$ cell line is inviable, similar to Rad51$^{-/-}$.

All five paralogs are essential in mice although, surprisingly, certain cell lines lacking at least some of the paralogs are viable. Recent experiments by Simon Powell studying human cell lines suggest that the BCXD2 and CX3 complexes play different role in the early steps leading to recombination. In response to ionizing radiation, cells with siRNA depletions of BCDX2 subunits are deficient in forming Rad51 filaments, but the mediator BRCA2 can still be seen to assemble at DSBs (**Figure 4.20**). In

Figure 4.19 Role of Rad51 paralogs in the formation of Rad51 foci after γ-irradiation of chicken DT40 cells. Rad51 foci are visualized by indirect immunofluorescence (A). In the absence of Rad51C (B), Rad51D (C), Xrcc2 (D), or Xrcc3 (E), foci fail to appear. Focus formation is restored when the paralog mutations are complemented by expressing the genes corresponding to the human or mouse homologs of these proteins (F–I). (From Takata M, Sasaki MS, Tachiiri S et al. [2001] *Mol Cell Biol* 21:2858–2866. With permission from the American Society for Microbiology.)

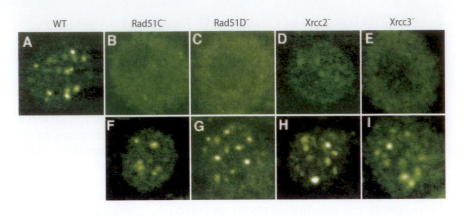

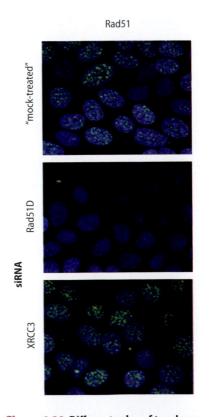

Figure 4.20 Different roles of two human Rad51 paralog complexes in Rad51 loading at sites of DNA damage. The effect of siRNA reducing the abundance of Rad51D and Xrcc3 was assessed by examining Rad51 focus formation after ionizing radiation. (From Chun J, Buechelmaier ES & Powell SN [2013] *Mol Cell Biol* 33:387–395. With permission from the American Society for Microbiology.)

contrast, depletions of CX3 do not block Rad51 assembly but still impair homologous recombination. Depletions of members of either complex are synthetically lethal in the absence of Rad52 and knocking down either complex does not enhance the defects already present in a BRCA2 mutant.

Rad52 and BRCA2

It initially appeared that mammalian Rad52 stimulates Rad51-mediated strand exchange reactions *in vitro* in much the same way as it does in yeast. However, the first studies were done using just Rad51, circular ssDNA, and homologous dsDNA (see Figure 3.8). When RPA is added to coat the ssDNA, although the yeast Rad52 can still facilitate Rad51 filament formation and the displacement of RPA, the human enzyme cannot. Moreover, deleting Rad52 in DT40 or mouse cells has almost no impact on cell viability or radiation resistance or gene targeting, whereas in *S. cerevisiae*, a *rad52* deletion is the most severe of any of the ionizing radiation sensitive mutations. Finally, in some well-studied model organisms—*Drosophila* and *Caenorhabditis*—Rad52 is absent from the genome sequence. The conservation of (some of) Rad52's biochemical function but the absence of a severe phenotype (and its total absence in some organisms) can be best explained by the discovery of another protein—BRCA2—that is not found in *S. cerevisiae* or *S. pombe* and that has apparently taken over the key mediator roles that Rad52 plays in yeast.

BRCA2 was first identified by apparently dominant mutations that greatly increase the likelihood of early-onset, familially inherited breast and ovarian cancer. In fact, the mutation is recessive and cancer arises in tissues that exhibit loss of heterozygosity. Deletion of the entire BRCA2 gene is lethal in mammalian cells, though viable in chicken DT40 cells. In DT40, the absence of BRCA2 prevents formation of Rad51 foci after ionizing radiation.

In addition to its essential nature, a problem in studying BRCA2 in mammals is the huge size of the protein—3418 amino acids in humans. Only recently has it been possible to purify the complete protein. A notable feature of BRCA2 is the presence of eight 39-amino-acid BRC repeats in the middle of the protein (**Figure 4.21**). These segments bind Rad51 protein and are critically important for its function. The structure of a C-terminal part of the protein has been solved, revealing two helix–turn–helix (HTH) "towers" and three OB folds that are characteristic of proteins that bind single-stranded DNA (similar to those found in each RPA subunit and in telomere-binding proteins). The HTH regions appear to bind dsDNA. In addition, there is another single Rad51 binding site at the C-terminus that is regulated by cell cycle dependent kinase activity. This terminal Rad51 binding site appears to be critical. In mammals, C-terminal truncations are viable but confer a radiation-sensitive and recombination-deficient

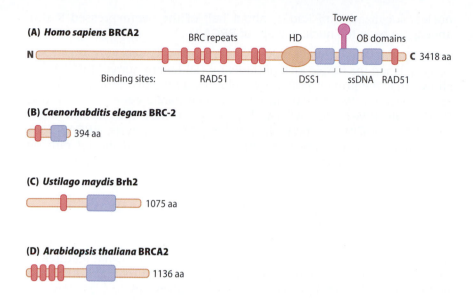

(A) Homo sapiens BRCA2

(B) Caenorhabditis elegans BRC-2
394 aa

(C) Ustilago maydis Brh2
1075 aa

(D) Arabidopsis thaliana BRCA2
1136 aa

Figure 4.21 Structures of mammalian BRCA2 and its homologs in model organisms. (A) Mammalian BRCA2 is a huge protein of 3418 amino acids (aa). It contains a ssDNA binding domain consisting of three OB-fold domains and a "tower" as well as a binding site for the Dss1 protein that plays an important role in its function. In addition there are nine RAD51 binding sites, eight of which are in a set of BRC domains, and a single RAD51 binding site is located at the C-terminus. Very small homologs of BRCA2 are found in *C. elegans* (B), the fungus *Ustilago maydis* (C), and in *Arabidopsis thaliana* (D), where a single OB fold and a single BRC domain are sufficient to carry out the mediator functions of BRCA2. (A, adapted from Venkitaraman AR [2009] *Annu Rev Pathol* 4:461–487. With permission from Annual Reviews. B–D, adapted from Martin JS, Winkelmann N, Petalcorin MI et al. [2005] *Mol Cell Biol* 25:3127–3139. With permission from the American Society for Microbiology.)

phenotype. Without BRCA2's C-terminus, Rad51 is not recruited to sites of DNA damage even though only one of nine possible Rad51 binding sites is ablated (**Figure 4.22**).

The purification of the portion of BRCA2 that has been crystallized depends on the presence of another protein, Dss1, which is tightly associated with BRCA2 in vertebrates but also in worms and in the yeast *Ustilago*. Curiously, although budding yeast lacks BRCA2, it retains Dss1, known as Sem1, which proves to be a subunit of the regulatory 19S proteasome. Expression of the worm Dss-1 protein complements the proteasome defects of *sem1Δ*, suggesting either that the protein has distinct dual functions or that the proteasome is somehow involved in homologous recombination. Indeed, in budding yeast a *sem1Δ* strain does not show any distinctive sensitivity to DNA-damaging agents. However, proteasome subunits are apparently recruited to the site of an HO-induced DSB. One idea that has been advanced is that the proteasome, or some of its subunits, participate in chromatin remodeling at DSB sites, as has also been suggested by Steve Reed for transcriptional activation.

Because of the difficulty in studying such a huge protein, many of the experiments concerning BRCA2 have been carried out with a viable truncation mutation that retains Rad51 binding at the BRC repeats but lacks the HTH towers and the more C-terminal Rad51 binding domain. This mutation is apparently recombination defective, but some activity must be present to rescue the viability. Mammalian BRCA2 with smaller numbers of BRC repeats also appears to be recombination deficient. Importantly, a number of hereditary Brca2 mutations are found in BRC repeats. Alternatively, the overexpression of some or all of the BRC repeats has been used to create a dominant-negative phenotype. But, at least in some eukaryotes, there is no need for eight BRC repeats—a single repeat suffices in the Brca2 homologs of *Caenorhabditis elegans* or *Ustilago maydis*, as discussed below.

One conjecture is that BRCA2 escorts perhaps as many as nine Rad51 monomers to the site of DNA damage and then facilitates their loading onto ssDNA. This escort function may entail transporting Rad51 from the cytoplasm into the nucleus, as most Rad51 is outside the nucleus in BRCA2 mutants. Interestingly, BRCA2 deficiency in recombination assays in mouse cells can be partially suppressed by overexpression of Rad51 protein. Consistent with the idea that BRCA2 functions in facilitating

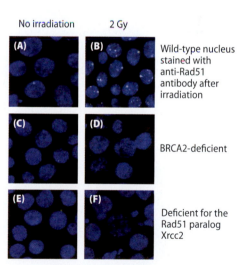

No irradiation   2 Gy

Wild-type nucleus stained with anti-Rad51 antibody after irradiation

BRCA2-deficient

Deficient for the Rad51 paralog Xrcc2

Figure 4.22 BRCA2 and the Rad51 paralog Xrcc2 are required for Rad52 focus formation in DT40 cells. Indirect immunofluorescent staining with anti-Rad51 antibody shows the formation of Rad51 foci after X-irradiation in wild-type cells (B). Foci are not formed when cells lack BRCA2 (D) or XRCC2 (F). Cells without irradiation are shown in panels A, C, and E. (From Hatanaka A, Yamazoe M, Sale JE et al. [2005] *Mol Cell Biol* 25:1124–1134. With permission from the American Society for Microbiology.)

nuclear localization of Rad51, about half of the overexpressed Rad51 appears to be in the nucleus without BRCA2.

Recently the labs of Wolf Heyer and Stephen Kowalczykowski have accomplished the heroic task of purifying full-length human BRCA2 and shown that the protein promotes Rad51 loading onto ssDNA in preference to dsDNA. BRCA2 can do what human Rad52 cannot: it can load Rad51 onto RPA-coated ssDNA, displacing the RPA. BRCA2 also stabilizes RAD51-ssDNA filaments by blocking ATP hydrolysis. BRCA2 does not anneal ssDNA complexed with RPA, implying it does not directly function in repair processes that involve single-strand annealing.

4.14 HOMOLOGS OF BRCA2 ARE FOUND IN MANY EUKARYOTES, INCLUDING A YEAST

It is useful to look at nonvertebrate metazoans to learn what features of the Rad51, Rad52, and BRCA2 machinery have been retained. Although neither *Drosophila melanogaster* nor *Caenorhabditis elegans* has an evident homolog of Rad52, both have homologs of Brca2.

Mutation or RNAi inhibition of Brca2 in both worms and in flies appears to severely impair meiotic recombination and Rad51 recruitment to foci after ionizing radiation. Compared to the more than 3400 amino acids of human BRCA2, the *Caenorhabditis* BRC-2 protein is composed of an astonishingly small 394 residues, with a single BRC repeat and a single OB fold (Figure 4.21). Point mutations in the one Rad51 binding domain of BRC-2 abolish its ability to promote efficient homologous recombination.

The simplicity of the BRC-2 protein calls into question models that rely on the multiple BRC repeats to "line up" Rad51 for filament formation. However studies of the full-length mammalian BRCA2 reveal that the BRC repeats play distinct roles. BRC1, -2, -3, and -4, bind to free Rad51 with high affinity, whereas the second group of BRC5, -6, -7, and -8, bind to free Rad51 with low affinity but bind to the RAD51-ssDNA filament with high affinity. Brc1–4 reduce the ATPase activity of Rad51 and also limit Rad51 binding to dsDNA. Thus, both types of BRC repeats bind to and stabilize the Rad51 nucleoprotein filament on ssDNA.

Equally curious is that although neither *Saccharomyces* nor *Schizosaccharomyces* have a BRCA2 homolog, another yeast, the corn smut *Ustilago maydis*, has both Rad52 and a clear homolog of Brca2, called Brh2 (Figure 4.21). In this case, as in vertebrates, the absence of Rad52 has a relatively minor effect on radiation sensitivity or DSB repair, whereas the Brh2 homolog of Brca2 has a very important role. As with worms, the 1075-amino-acid *Ustilago* protein has a single BRC domain and one OB fold. Overexpression of Brh2 can suppress the defects in repair caused by deletion of *Ustilago*'s single Rad51 paralog, Rec2. More striking is the finding that the OB-fold domain of Brh2 can be replaced by completely different ssDNA binding domain; a Brh2::RPA70 fusion protein also suppresses *rec2Δ* and is in fact hyper-recombinogenic in spontaneous recombination between alleles in homologous chromosomes in a diploid. This surprising finding is strengthened by the demonstration that fusion proteins of mammalian BRC repeats fused to either RPA70 or to the ssDNA binding domain of Rad52 will improve homologous recombination in a Brca2 mutant cell line.

The small size of both the purified *Caenorhabditis* and *Ustilago* proteins has made it possible to characterize their biochemical properties, revealing a number of important features. The worm protein, BRC-2, stimulates Rad51-mediated D-loop formation. This activity depends on using *C. elegans* Rad51, as there is no stimulation of D-loop formation for either *E. coli* RecA or human Rad51. D-loop formation also depends on the integrity of its single BRC domain and is eliminated with single amino acid substitutions. Moreover, a wild-type peptide containing the BRC domain is insufficient to promote D-loop formation; there must be both an association between Rad51 and the BRC domain, and the binding of BRC-2 to ssDNA.

The association of the worm BRC-2 with Rad51 and ssDNA causes a reduction in the rate of Rad51's ATP hydrolysis. Simon Boulton suggests that this reduction may hold Rad51 in an ATP-bound form that will stabilize the filament and would be more active in promoting strand exchange and D-loop formation; however, a deeper understanding of this point will await structural studies of the effect of binding BRC-2 to Rad51.

Another key observation is that purified BRC-2 accelerates annealing of complementary single strands, similar to Rad52 from yeast or human cells. A comparison of the phenotypes of worms that are siRNA depleted for either Rad51 or BRC-2 suggests that BRC-2 plays an important role in a Rad51 independent process that Boulton's lab believes is single-strand annealing. This would mean that BRC-2 facilitates strand annealing in the same way that Rad52 acts in budding yeast and explains how worms can carry out this process without Rad52.

However, there may be some interesting differences between these two homologs. With worm's BRC-2, pre-incubation with Rad51 and ssDNA prior to adding the dsDNA substrate was needed to promote efficient D-loop formation. This result suggests that a BRC-2:Rad51:ssDNA intermediate is important for promoting strand exchange. In contrast, the *Ustilago* protein is most efficient in promoting strand exchange when Rad51 and ssDNA are mixed with Brh2 that is already associated with the dsDNA substrate. Whether either of these assays reflects what happens *in vivo* is hard to know. Another apparent difference between the worm and *Ustilago* homologs is that Brh2, at high concentration, can promote weak D-loop formation between an oligonucleotide and a supercoiled dsDNA substrate.

A detailed study of how BRCA2-like proteins facilitate Rad51 filament formation has come from the Holloman and Pavletich lab's study of *Ustilago* Brh2. They show that Brh2 works at substoichiometric concentrations to promote Rad51 loading. This occurs preferentially at a dsDNA/ssDNA junction of molecules with the 5′ strand resected, so that the Rad51 filament elongates in the 5′ to 3′ direction. Brh2 appears to remain at the ds/ssDNA junction, suggesting that it is required to displace RPA to load the first few Rad51 monomers but that subsequent elongation of Rad51 itself might be able to displace RPA without Brh2. The results are reminiscent of studies of RecA where RecFOR acts at the ds/ssDNA junction in apparently a similar manner.

4.15 MANY QUESTIONS REMAIN UNANSWERED

How resection, SSB/RPA loading and displacement, and RecA/Rad51 filament formation are coordinated *in vivo* is not well understood. But there are other important questions we need to consider. How much resection is really necessary for efficient repair of a DSB? How are the long 3′ ssDNA ends protected? Is there a sufficient supply of RPA and Rad51 to form long, continuous filaments? And how does the filament then search the genome for homologous sequences to accomplish homologous recombination and repair?

SUMMARY

RecA and Rad51 must assemble into a filament on ssDNA before they can carry out their central role in recombination and repair. The first key step in creating the recombinase filament is the process of 5′ to 3′ resection to create 3′-ended ssDNA on which the recombinases can assemble. Resection can be carried out by exonucleases or by helicase–endonuclease complexes. In *E. coli*, the major resection activity is performed by the RecBCD complex, whose activity is attenuated and altered by encountering a Chi site. In eukaryotes, at least four nuclease complexes have been implicated. The MRX/MRN complex appears to be needed to start resection (although resection in cycling cells is only reduced and not eliminated in the absence of these proteins). The Sae2/CtIP protein plays a major role, but the activities of Exo1 and the Sgs1/BLM complex, in some cases using Dna2 as the nuclease, are also important. The second key step is filament assembly. ssDNA is first bound by SSB/RPA, which are then displaced by mediators of filament assembly. In bacteria the RecFOR proteins are important in this process, while in eukaryotes assembly requires a set of Rad51 paralogs and either Rad52 or BRCA2. There is a surprising diversity of mediators in different eukaryotes.

SUGGESTED READING

Carreira A & Kowalczykowski SC (2011) Two classes of BRC repeats in BRCA2 promote RAD51 nucleoprotein filament function by distinct mechanisms. *Proc Natl Acad Sci USA* 108:10448–10453.

Jensen RB, Carreira A & Kowalczykowski SC (2010) Purified human BRCA2 stimulates RAD51-mediated recombination. *Nature* 467:678–683.

Kojic M, Zhou Q, Fan J & Holloman WK (2011) Mutational analysis of Brh2 reveals requirements for compensating mediator functions. *Mol Microbiol* 79:180–191.

Krejci L, Van Komen S & Sehorn MG (2003) Rad51 recombinase and recombination mediators. *J Biol Chem* 278:42729–42732.

Liu J, Ehmsen KT, Heyer WD & Morrical SW (2011) Presynaptic filament dynamics in homologous recombination and DNA repair. *Crit Rev Biochem Mol Biol* 46:240–270.

Mimitou EP & Symington LS (2009) DNA end resection: many nucleases make light work. *DNA Repair* 8:983–995.

Petalcorin MI, Galkin VE, Yu X et al. (2007) Stabilization of RAD-51-DNA filaments via an interaction domain in *Caenorhabditis elegans* BRCA2. *Proc Natl Acad Sci USA* 104:8299–8304.

Saeki H, Siaud N, Christ N et al. (2006) Suppression of the DNA repair defects of BRCA2-deficient cells with heterologous protein fusions. *Proc Natl Acad Sci USA* 103:8768–8773.

Singleton MR, Dillingham MS, Gaudier M et al. (2004) Crystal structure of RecBCD enzyme reveals a machine for processing DNA breaks. *Nature* 432:187–193.

Spies M, Amitani I, Baskin RJ & Kowalczykowski SC (2007) RecBCD enzyme switches lead motor subunits in response to chi recognition. *Cell* 131:694–705.

Sugawara N, Wang X & Haber JE (2003) *In vivo* roles of Rad52, Rad54, and Rad55 proteins in Rad51-mediated recombination. *Mol Cell* 12:209–219.

Wang X & Haber JE (2004) Role of *Saccharomyces* single-stranded DNA-binding protein RPA in the strand invasion step of double-strand break repair. *PLoS Biol* 2:104–111.

Wolner B, van Komen S, Sung P & Peterson CL (2003) Recruitment of the recombinational repair machinery to a DNA double-strand break in yeast. *Mol Cell* 12:221–232.

Yang H, Li Q, Fan J et al. (2005) The BRCA2 homologue Brh2 nucleates RAD51 filament formation at a dsDNA-ssDNA junction. *Nature* 433:653–657.

CHAPTER 5
SINGLE-STRAND ANNEALING

In Chapters 3 and 4 we saw that DSBs are resected in a 5′ to 3′ direction to produce long ssDNA tails. One consequence of this resection is the possibility of single-strand annealing (SSA) between homologous sequences that flank the DSB, thus creating deletions. Resection can continue unabated for a very long time, so that sequences even 25 or 50 kb apart can be joined by this mechanism, creating large deletions within a chromosome. SSA differs from other homologous recombination events in that it occurs efficiently without Rad51. Here we examine SSA in detail.

5.1 5′ TO 3′ RESECTION PROMOTES SINGLE-STRAND ANNEALING

5′ to 3′ resection is an apparently obligatory prelude to homologous recombination to repair a DSB. However, whereas the initiation of resection is well-regulated (as we saw in Chapter 3), its termination is not. At least in budding yeast, once resection begins, it will continue inexorably at a constant rate of 4 kb/hr for at least 24 hr. Continuing resection can be seen by measuring the progressive loss of double-stranded DNA restriction fragments far from the site of an induced DSB (see Figure 4.4). Resection will go through budding yeast's tiny centromere, and recent evidence suggests that resection will invade the large, heterochromatic centromere of fission yeast. Single-strand annealing may therefore be an unselected consequence of having launched resection with no limitation on its progression, what Stephen J. Gould termed a "spandrel." A spandrel is defined as "the class of forms and spaces that arise as necessary byproducts of another decision in design, and not as adaptations for direct utility in themselves." Here, that idea has been extended to include a process rather than a form or a space. In yeasts, the near absence of dispersed repeated sequences in DNA means that SSA could occur only rarely, except in the tandem repeats of ribosomal DNA (rDNA), but in mammals, the widespread presence of many types of non-tandem repeats (LINEs, SINEs, and other retrotransposon sequences) throughout the genome may result in SSA having a far more important role in DSB repair than one would find in unicellular eukaryotes.

A commonly used strategy for studying SSA in yeast involves the creation of an artificially duplicated region with a selectable marker in between, for example by the integration of a pBR322 plasmid carrying the *URA3* gene (as the genetic marker) and a region homologous to a chosen site on the host chromosome (**Figure 5.1**). The plasmid also carries the

Figure 5.1 A simple single-strand annealing (SSA) assay induced by a DSB. A pBR322 plasmid containing an HO cleavage site (*orange*) and the *URA3* selectable marker can be integrated by homologous recombination at the *LEU2* locus in a *ura3* strain by selecting for Ura⁺ transformants. In the absence of *HO* gene expression, the integrated sequences can be lost by a "pop-out" event that reverses the integration step. Induction of HO endonuclease creates a DSB that is resected in both directions by 5′ to 3′ exonucleases. Once single-stranded DNA regions containing complementary DNA sequences are exposed the DSB can be repaired by SSA, causing a deletion that removes the *URA3* locus. The annealing step is followed by clipping off the 3′-ended nonhomologous tails and filling in of the gaps to complete the process.

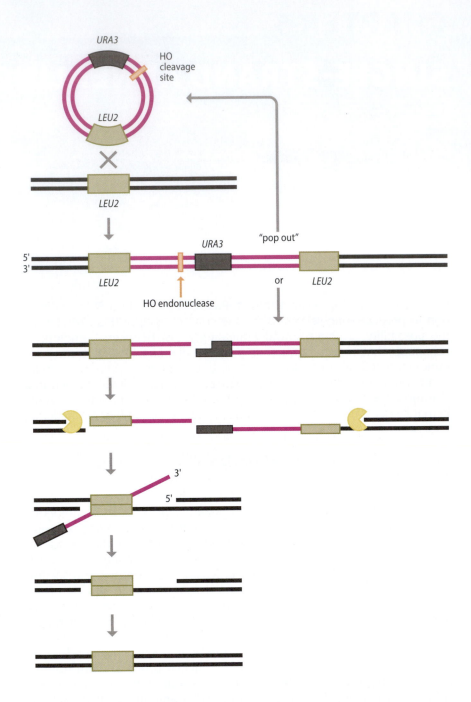

recognition site for a site-specific endonuclease such as the HO endonuclease. Integration of the plasmid creates a duplication. In the absence of HO cleavage to create a DSB, reversal of the integration event (a "pop out") can cause the loss of the *URA3* marker and the other integrated sequences (Figure 5.1). But when HO cuts this region, repair occurs by SSA, again resulting in the loss of *URA3* and all the integrated sequences.

Such events can be selected by plating cells on media containing 5-fluoroorotic acid (5-FOA), which selects against the presence of the *URA3* gene. Although the loss of the integrated sequences can occur by such a "pop out," in fact most of the losses of *URA3* and other sequences appear to result from SSA.

SSA events can also be analyzed in the absence of an induced DSB. One frequently used approach involves a tandem gene duplication containing

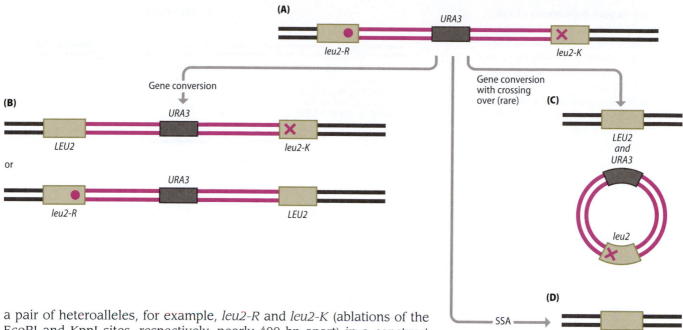

a pair of heteroalleles, for example, *leu2-R* and *leu2-K* (ablations of the EcoRI and KpnI sites, respectively, nearly 400 bp apart) in a construct that also carries a nutritional or a drug-resistance marker (for example, *URA3*) in between (**Figure 5.2**). Then it is possible to select for spontaneous Leu2$^+$ recombination events (that is, without intentionally creating a DSB). These recombinants can either be Ura$^+$, where gene conversion has converted one of the flanking alleles to *LEU2* and retained the *URA3* gene, or else Ura$^-$, where SSA joined the segments of the *leu2* alleles and created a deletion. Much less often, deletions can arise by a true pop-out, resulting from crossing over between the *leu2* alleles. In this case the reciprocal recombination product is a circular DNA carrying the *URA3* marker and a *leu2* mutant sequence. If the circular segment carries an origin of DNA replication, this reciprocal product should replicate and be retained; but in fact such outcomes are rare: most deletions appear to arise by SSA.

Using tandem duplications such as the one shown in Figure 5.2, it is possible to analyze the genetic requirements of SSA versus gene conversion. For example, whereas gene conversion events require Rad51, most deletion events (that is, those arising by SSA) do not. Other genetic requirements will be discussed below.

Figure 5.2 An SSA assay based on recovering prototrophic recombinants. Integration of a circular plasmid containing *URA3* and a *leu2-K* allele by homologous recombination into a *leu2-R* strain creates a tandem duplication of *leu2* genes carrying the two alleles (A). (B) Selection of spontaneous Leu$^+$ recombinants can occur by gene conversion in the absence of crossing over, converting one or the other allele to *LEU2*, leaving cells still Ura$^+$. *LEU2* cells can also arise by gene conversion when the event is accompanied by crossing-over (C), but these events are rare. If the "popped-out" recombination product contains an origin of replication, the Leu$^+$ cells can retain the circular product; otherwise the cells will be Ura$^-$. Most Leu$^+$ Ura$^-$ cells arise from SSA (D).

5.2 SSA IN BUDDING YEAST CAN BE STUDIED AFTER INDUCING A SITE-SPECIFIC DSB

The efficiency, kinetics, and molecular intermediates of SSA have been best studied in budding yeast by using enzymatically induced DSBs that are flanked by artificially created repeated sequences as shown in Figure 5.1. Such DSBs have been introduced by inducing expression of either of two different "meganucleases": the HO endonuclease, which recognizes a degenerate 24-bp sequence and has only a single normal cleavage site in the yeast genome; or I-SceI, which recognizes an 18-bp cleavage site but has no natural site in yeast chromosomes. SSA is an efficient process; in the absence of any competing ways to repair the DSB (that is, no ectopic donor sequences homologous to sequences around the DSB), nearly 100% of the cells will use SSA to repair a break surrounded by ≥ 400-bp identical sequences (**Figure 5.3**). SSA can be detected with flanking regions of only 30 bp.

Figure 5.3 Length dependence of SSA.
A simple DSB-induced SSA assay in which one of the flanking regions can be varied in length to assess the efficiency of SSA. When the extent of shared homology is approximately 400 bp, viability is nearly 100%; however, as the length of shared homology decreases, the recovery of cells that have completed SSA repair diminishes. At a minimum SSA requires approximately 30 bp. Especially with short flanking homologies, SSA requires the Rad59 protein. (From data in Sugawara N, Ira G & Haber JE [2000] *Mol Cell Biol* 20:5300–5309. With permission from the American Society for Microbiology.)

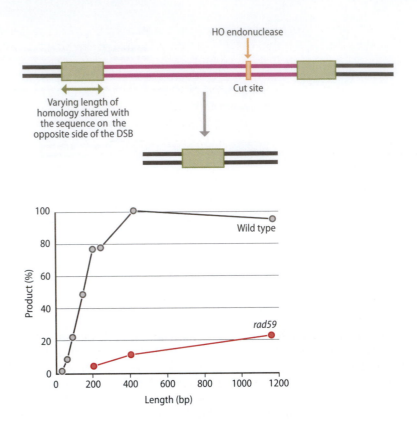

Even if a homologous donor sequence is available to repair a DSB by gene conversion, SSA is a remarkably successful competitive process. In one experiment, the *MAT* locus was artificially surrounded with a pair of 1-kb *URA3* genes and also introduced a *LEU2* gene (**Figure 5.4**). *MAT*

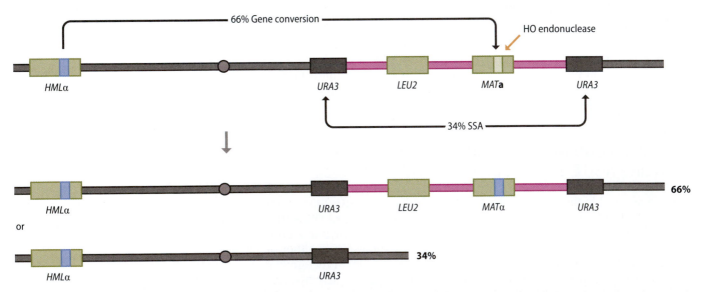

Figure 5.4 Competition between intrachromosomal gene conversion and SSA. HO endonuclease catalyzes mating-type (*MAT*) gene switching by a DSB-induced gene conversion that replaces *MAT***a** sequences with *MAT*α. In the normal chromosomal arrangement, where *MAT* is not surrounded by flanking homologous sequences such as the *URA3* genes shown here, switching to *MAT*α is nearly 100% efficient. However, when *MAT* is surrounded by flanking *URA3* genes, one-third of the repairs of the DSB are by SSA, leading to the deletion of *MAT* and the other internal sequences between the repeats. These data show that 5′ to 3′ resection goes far beyond the short regions of homology required for the gene conversion event and that SSA competes with gene conversion even in a case where the DSB is part of a genetically programmed chromosomal rearrangement. (From data in Wu X, Wu C & Haber JE [1997] *Genetics* 147:399–407. With permission from the Genetics Society of America.)

switching is known as an exceptionally efficient intrachromosomal gene conversion event, but when induced in the presence of flanking repeats, 34% of the outcomes proved to be deletions of *MAT*, *LEU2*, the plasmid sequences, and one copy of *URA3*. When *MAT* and the surrounding *URA3* and *LEU2* genes were integrated on a different chromosome, so that repair by gene conversion became an interchromosomal process, the proportion of SSA events increased to about 68%. These results demonstrate that resection continues far beyond the borders of the homology shared by the sequences that are needed to repair the DSB by gene conversion (that is, the *MAT* sequences) and show that SSA is a disturbingly efficient competitor to gene conversion, which repairs the DSB without a chromosomal rearrangement such as the deletion arising by SSA.

The apparent rate-limiting step in SSA appears to be the rate of resection to expose complementary sequences that can then anneal. When a DSB was created in the middle of the *LEU2* gene (by inserting a synthetic HO cleavage site), repair could occur by SSA if a cloned 1-kb "*U2*" segment was placed distal to the DSB site (**Figure 5.5**). When *U2* was 0.7 kb away from the break, repair could be completed within an hour. When it was located about 5 kb distant, repair was visible in about 2 hr, but when *U2* was located 25 kb away, repair occurred after about 6 hr, consistent with a rate of resection of roughly 4 kb/hr. SSA has been shown to occur even when one repeat is 50 kb from the DSB site.

Figure 5.5 The kinetics of SSA are determined by the distance to the furthest homology from the DSB. (A) HO cleavage is induced within the *LEU2* gene. A second "*U2*" sequence is inserted at different distances away from the DSB in strains YMV80, YMV45, and YMV86. Repair of the DSB occurs by SSA. (B) Southern blot analysis shows that the deletion product appears in 30 min when the sites are very close together, in about 120 min when they are 5 kb apart, and in 6 hr when they are 25 kb distant. (A, from data in Vaze M, Pellicioli A, Lee S et al. [2002] *Mol Cell* 10:373–385. With permission from Elsevier. B, from Jain S, Sugawara N, Lydeard J et al. [2009] *Genes Dev* 23:291–303. With permission from Cold Spring Harbor Laboratory Press.)

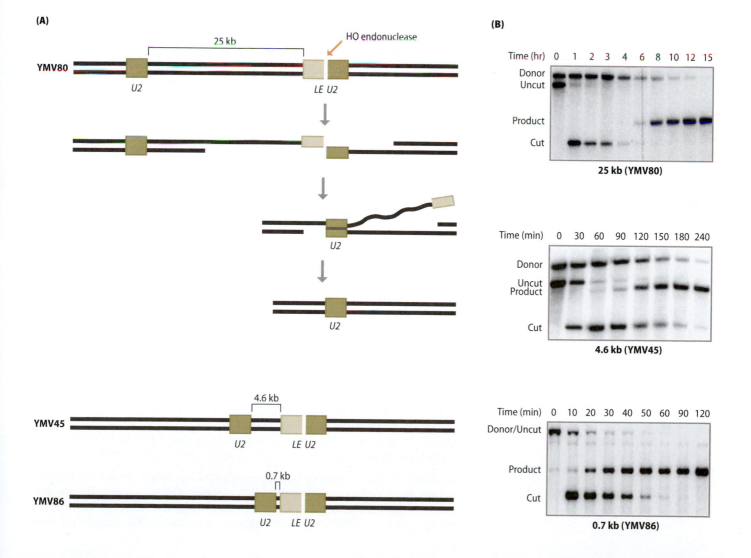

5.3 SSA IS Rad51 INDEPENDENT

Like all other homologous recombination processes, SSA requires Rad52, but in sharp contrast to gene conversion or break-induced replication (BIR), SSA does not depend on Rad51. SSA also occurs independently of the Rad51 paralogs, Rad55–Rad57, or the Rad54 protein. In fact, the level of SSA increases in some assays when Rad51 is absent, likely because competing repair pathways are eliminated. SSA is impaired when Rad51 is overexpressed, suggesting that the Rad51 filament prevents strand annealing.

Strand annealing is catalyzed by the Rad52 protein. However, even Rad52 may be dispensable when the length of the homologous segments is large; for example, when an HO-induced DSB is created within 9-kb tandem repeats of the rDNA locus, SSA becomes Rad52 independent. It is possible that the Rad52 independence of the rDNA deletions could have something to do with other features of the nucleolar compartment; however, Brad Ozenberger and Shirleen Roeder looked at SSA when an HO cut site was inserted in the middle of a tandem array of 18 copies of the 2-kb *CUP1* cluster and found that here, too, repair was possible without Rad52. Many of the repair events had lost two to five copies of the *CUP1* repeats, but losses of more than one repeat were rare in a Rad52⁺ cell. These data suggest that annealing in the absence of Rad52 requires segments of ssDNA that are much longer—on the order of several kilobases—than the few hundred base pairs that are efficiently annealed in the presence of Rad52.

The Rad59 protein, which has some homology to Rad52 and has been shown to have strand-annealing activity *in vitro*, also plays an important role in SSA, but it does not substitute for Rad52. The importance of Rad59 is strongly dependent on the length of the sequences to be annealed (Figure 5.3). With flanking repeats of only 205 bp, Rad59 proved to be almost as important as Rad52, yielding only 2% the wild-type level of SSA. But with 1200-bp repeats, deleting Rad59 only reduced SSA to 20% whereas *rad52Δ* yielded essentially no deletions. Whether Rad52-independent SSA of long homologous regions has a strong dependence on Rad59 has not been tested. In any case, it is clear that Rad59 will not substitute for Rad52 but plays an important secondary role.

5.4 SSA DEPENDS ON Rad52's STRAND ANNEALING ACTIVITY

Rad52 is a multifunctional protein that both facilitates the loading of Rad51 onto ssDNA and promotes SSA in the absence of Rad51. The N-terminal "business end" of the protein has been crystallized. Electron micrographs had suggested both yeast and human Rad52 form heptameric rings, but two X-ray crystal structures found 11-member rings. It is not known why there is a difference in the number of subunits analyzed under these different conditions or if one of them is preferentially active. ssDNA binds in a groove around the ring, with a pronounced 4-nt periodicity (that is, every fourth base is more exposed to hydroxy-radical modification), consistent with each monomer binding 4 nt (**Figure 5.6**). ssDNA would then be held in an extended conformation with which a second strand—either free or bound to Rad52—would permit annealing. The groove in which ssDNA sits is too small to accommodate the annealed dsDNA.

Recently, Rad52's activity has been studied by single molecule experiments in which the complementary ssDNA strands have fluorescent markers that can be used to look at annealing by fluorescence resonance energy transfer (FRET). Eli Rothenberg and Taekjip Ha suggested that

(A)   **(B)**

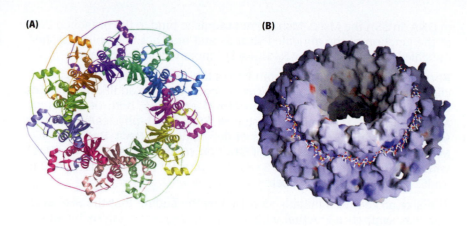

Figure 5.6 Structure of the human Rad52 protein. Ribbon structure (A) and space-filling model (B) of the 11-subunit Rad52 protein ring, showing a groove in which ssDNA can bind. Rad52 facilitates the loading of Rad51 onto ssDNA and promotes Rad51-independent SSA. (From Singleton MR, Wentzell LM, Liu Y et al. [2002] *Proc Natl Acad Sci USA* 99:13492–13497. With permission from the National Academy of Sciences.)

Rad52-mediated annealing begins with each ssDNA wrapped around multimers of Rad52. Random collisions facilitate an initial contact between complementary segments within the Rad52 complexes, which then remain in contact as they slide by each other to locate a region of strong base pairing (**Figure 5.7**). Initial pairing is followed by dissociation of at least one of the ssDNA regions from a multimer, so that more base pairs can zipper up (there is no room for dsDNA to form in the groove holding the ssDNA on Rad52). This same process is then extended to adjacent Rad52 multimers, allowing the annealed region to grow. As we will see later, Rad52's strand-annealing activity, independent of Rad51, also plays a crucial role in gene conversion.

5.5 THE REMOVAL OF NONHOMOLOGOUS TAILS IS REQUIRED FOR THE COMPLETION OF SSA

Once complementary sequences are revealed by resection and they anneal, the next step is the removal of the protruding 3′-ended single-stranded tails; without this excision, it would not be possible to fill in the gaps adjacent to the annealed region (**Figure 5.8A**). Clipping off the tails depends on the Rad1–Rad10 endonuclease, which was initially identified as a key nuclease in nucleotide excision repair (NER), where it cuts on the 5′ side of a photodimer. Subsequently it was found that the Msh2–Msh3 proteins that recognize insertions/deletions in mismatch repair (MMR) were also important in nonhomologous tail cleavage. No other NER or MMR proteins appear to have any role in the processing of the 3′ tails during SSA. Whereas Rad1–Rad10 are always required, no matter what the length of the annealed region, Msh2–Msh3 are much more important when the annealed region is short than when it is longer. For example deleting *MSH3* nearly eliminated SSA between 205-bp repeats, whereas the *msh3Δ* deletion allowed 80% of the wild-type level of SSA with 1200-bp repeats. How Msh2–Msh3 act is not certain. In mismatch repair the Msh2–Msh3 dimer recognizes a small loop of unpaired bases

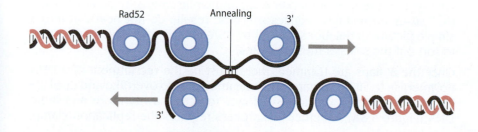

Rad52   Annealing   3′

3′

Figure 5.7 A model showing how Rad52 rings carrying complementary ssDNA strands can promote annealing. Multiple Rad52 rings as shown in Figure 5.6 bind to ssDNA. Once a short region of the complementary ssDNA strands anneals, as shown in the center, the movement of two sets of rings in opposite directions can promote extensive strand annealing. (Adapted from Rothenberg E, Grimme JM, Spies M & Ha T [2008] *Proc Natl Acad Sci USA* 105:20274–20279. With permission from the National Academy of Sciences.)

(A)

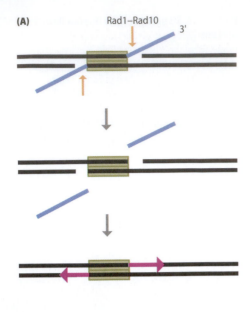

(B)

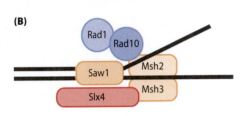

Figure 5.8 Clipping of nonhomologous tails to complete SSA. (A) After Rad52 has promoted strand annealing, the 3′-ended nonhomologous tails must be removed. Rad1–Rad10 cleavage (*arrows*) of 3′ flap structures permits the new 3′ ends to serve as primers for filling in the ssDNA gaps and completion of SSA. (B) Proteins required for 3′ flap excision. (A, adapted from Fishman-Lobell J & Haber JE [1992] *Science* 258:480–484. With permission from the American Association for the Advancement of Science. B, adapted from Li F, Dong J, Pan X et al. [2008] *Mol Cell* 30:325–335, and from Flott S, Alabert C, Toh GW et al. [2007] *Mol Cell Biol* 27:6433–6445. With permission from Elsevier and the American Society for Microbiology respectively.)

in DNA. In SSA the Msh2–Msh3 dimer seems to bind to the branched DNA at the junction of the annealed strands and may act to stabilize a short annealed segment to allow Rad1–Rad10 more efficient cleavage.

Recently, however, our understanding of the roles of these factors in tail-clipping has been complicated by the discovery of two additional proteins that have key roles in the process, Saw1 and Slx4, both of which are required for the tail-processing step of SSA. Saw1 physically interacts with virtually all the components of flap removal—Rad1–Rad10, Msh2–Msh3, and Rad52—and deletion of Saw1 prevents the association of Rad1 with SSA intermediates. Thus Saw1 may directly recruit Rad1–Rad10 to Rad52-coated SSA intermediates (**Figure 5.8B**).

The Slx4 protein was initially identified by the finding that the Slx4 deletion was synthetically lethal with *sgs1Δ*. Subsequent analysis by Steven Brill's laboratory showed that Slx4 formed a complex with Slx1, and that the complex has nuclease activity able to cleave model substrates that mimic replication intermediates, removing 5′-ended flaps. The Slx1 endonuclease activity is stimulated 500-fold by Slx4. Although it is still not at all clear what Slx1–Slx4 does that makes it essential in the absence of Sgs1, the big surprise was the discovery that Slx4, but not Slx1, has an essential role in SSA, in the tail-clipping step. Based on its apparent role in the Slx1–Slx4 complex, where it apparently directs the nuclease to branched DNA structures, Slx4 could be imagined to help direct the Rad1–Rad10 nuclease to the precise site for tail-clipping; indeed Slx4 has been shown to physically interact with Rad1.

Unlike Msh2–Msh3, Slx4 is required for SSA whether the length of the annealed region is 1 kb or 205 bp. Another unexpected finding was that Slx4 is phosphorylated rapidly after induction of DNA damage, even a single DSB. Phosphorylation depends on the Mec1 and/or Tel1 checkpoint kinases. Slx4 mutants lacking phosphorylation sites are inactive in promoting the clipping of intermediates in SSA, and a phosphomimetic mutation (for example, replacing the threonine that can be phosphorylated with a glutamic acid that always has a negative charge that mimics the phosphorylated state) partially restored cleavage function. At present, Slx4 is the only protein directly involved in DSB repair that must be phosphorylated in order to function. Why such a modification is necessary is difficult to understand, unless the modification is tied to preventing Slx4 from functioning with Slx1 in the presence of DNA damage.

We will revisit the role of Slx4 in later chapters where studies of Slx4 homologs in worms, flies, and mammals have shown that it plays a central role in controlling crossing over in meiosis. A large number of biochemical and molecular biological studies have suggested that Slx4 acts as a scaffold or "toolbelt" on which a number of different nuclease complexes are assembled. Indeed, in mammals it seems that the Slx4 homolog, BTBX12, associates not only with the mammalian homologs of Slx1 and Rad1–Rad10 but also with the Mus81 nuclease.

As noted earlier, the only naturally occurring set of tandemly repeated genes in budding yeast are the ribosomal DNA (rDNA). Deletions of either Slx4 or Saw1 alter the stability of rDNA, as measured by loss of a marker gene inserted into the rDNA array. However, *rad1Δ* does not differ from wild type in the rate of marker loss, and while *saw1Δ* increases marker loss, *slx4Δ* surprisingly reduces it! These results do not neatly fit into a simple picture of deletions caused by SSA. Clearly, more work is needed to sort out the roles of these proteins.

Once the 3′ flaps are trimmed, there must be the recruitment of a DNA polymerase to fill in the gaps and, one imagines, to overtake and displace the still-degrading exonuclease slowly resecting away more and more DNA (Figure 5.8A). Whether this process requires the replication clamp,

PCNA, has not been directly tested, and there is no evidence at this point which DNA polymerase acts to complete the deletion.

5.6 SSA CAN OCCUR BETWEEN MISMATCHED SEQUENCES

In both gene conversion and SSA, the efficiency of recombination is greatly reduced when the strand-invading or annealing sequences have mismatched base pairs. Such *homeologous recombination* events become less and less efficient as the level of divergence increases. In one study of SSA between diverged DNA sequences, researchers examined repair when a 205-bp repeat differed at seven sites, six of which were base-pair substitutions and one was a 1-bp insertion/deletion. The mismatches were clustered in the first 150 bp, so locally the sequences diverged by about 5% and over the whole fragment by 3.5%. HO-induced SSA was reduced sixfold compared to perfectly matched flanking repeats. The discouragement of SSA is under the surveillance of the mismatch repair proteins Msh2 and Msh6; deletions of Msh6 restored SSA to the level seen with identical repeats (**Figure 5.9**). Because Msh2 also works with Msh3 in

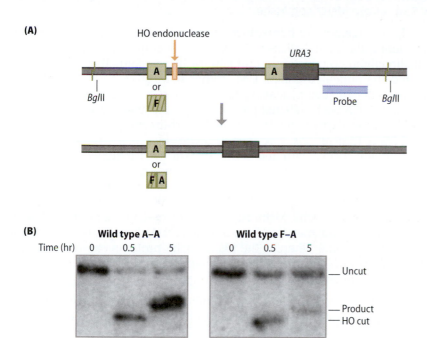

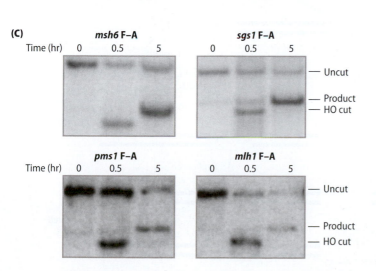

Figure 5.9 Heteroduplex rejection in SSA by Sgs1 and Msh6 but not Pms1-Mlh1.
(A) A DSB is flanked by the first 205 bp of the *URA3* gene. Two situations are compared: when the flanking sequences are 100% identical to the sequence on the right (designated A); and when the sequence on the left differs by 3.5% from A (designated F). Repair of the DSB occurs by SSA. (B) A Southern blot showing repair of the DSB by SSA. Thirty minutes after inducing expression of HO endonuclease, the *Bgl*II restriction fragment, detected by the indicated probe, is cleaved in most cells. After 5 hr, SSA product formation is seen as a new band on the Southern blot. In the F–A strain, SSA is reduced to about 15% of the wild-type (A–A) level. (C) SSA in the F–A strain is restored to near wild-type levels by deleting *SGS1* or *MSH6* but is unaffected by deleting *PMS1* or *MLH1*. (B, C, from Sugawara N, Goldfarb T, Studamire B et al. [2004] *Proc Natl Acad Sci USA* 101:9315–9320. With permission from the National Academy of Sciences.)

the clipping of nonhomologous 3′ tails, the role of Msh2 in mismatch surveillance during SSA was demonstrated using a separation-of-function mutation that had lost mismatch recognition but preserved the capacity to promote clipping of the protruding tails. Unexpectedly, the low level of SSA in homeologous SSA was not suppressed by deleting any of the MutL homologs (Mlh1, Mlh2, Mlh3, or Pms1) that are as important as the Msh2 in normal replication-dependent mismatch repair. The absence of a role for Mlh1 or Pms1 was surprising in view of the well-established finding that spontaneous homeologous gene conversion events are suppressed both by Msh2–Msh6 and by Mlh1–Pms1 (although the Mlh–Pms1 proteins consistently have a smaller effect).

Nevertheless, once strand-annealing occurs, the standard mismatch repair machinery (involving both Msh and Mlh proteins) acts to correct mismatches. Consequently, among the reduced level of deletions formed by SSA in the homeologous case for wild-type cells, mismatch correction was efficient, so that few colonies contained the sectored colonies that would be expected if there were no MMR (**Figure 5.10**). Thus, while Pms1 and Mlh1 are not needed to reject homeologous strand annealing, these proteins are still required for mismatch correction of the duplexes formed by SSA, as are Msh2 and Msh6.

The discouragement of homeologous recombination is also suppressed by deleting the Sgs1 helicase (**Figure 5.9C**), though it is not known if it acts directly in conjunction with Msh2-Msh6. It seems that Sgs1 acts to unwind the annealed duplex DNA rather than provoke the excision of the mismatched region, leaving the ssDNA regions intact to continue searching for a well-matched partner. This conclusion was drawn from a competition experiment in which the arrangement of sequences is A–F–DSB–A (that is, the F sequence differs by 3.5%) (**Figure 5.11**). SSA preferentially forms with the more distant well-matched A segment (Figure 5.11), but the total efficiency does not go down compared to an A–A–DSB–A control. In a *sgs1Δ* strain, deletions are again efficient between the F and A segments flanking the DSB. These results suggest a scenario in which Msh2–Msh6 recognize mismatches and interact with Sgs1 to unwind mismatched heteroduplex DNA during strand invasion or strand annealing whereas the Msh2–Msh6 proteins can alternatively interact with Mlh1–Pms1 and Exo1 when the duplex is "sealed up" as part of a SSA-mediated deletion.

Sgs1 has also been shown to play a similar role in discouraging spontaneous homeologous gene conversion. Deletion of another helicase, Srs2,

Figure 5.10 Mismatch repair of heteroduplex DNA formed by SSA. If SSA occurs between strands that carry base-pair differences, then the resulting structure will contain mismatches in the heteroduplex DNA. If mismatch repair takes place then the cell will give rise to a homogeneous colony with either one or the other genotype. If, however, Mlh1 or Pms1 is absent, then the heteroduplex DNA will not be mismatch corrected and each DNA strand will be the template for DNA replication, after which each of the two daughter cells will give rise to a sectored colony in which all the descendants from each of the two daughter cells of the annealed DNA will have one or the other genotype.

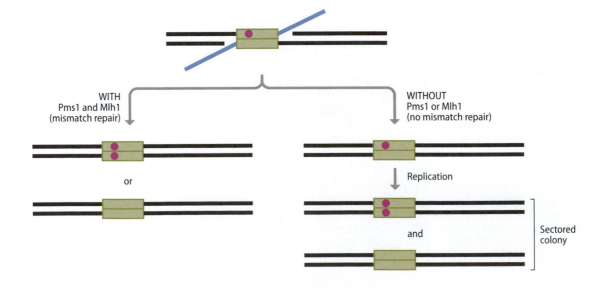

(A)

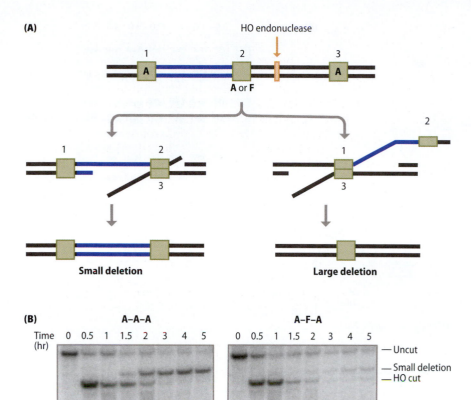

Small deletion   **Large deletion**

(B)

Figure 5.11 Competition between SSA outcomes. (A) A DSB in which there are two equivalent 205-bp homologous regions on one side can generate two different sized deletions by SSA events, as seen in the Southern blots (B). If the middle homology contains seven mismatches (3.5% mismatch in sequence F), then forming the short deletion is impaired compared to the perfectly matched case (A–A–A). However, in the A–F–A case, the amount of large deletion (which results from annealing perfectly homologous strands) increases. This result indicates that the attempt to use sequence F was discouraged but the annealing strands were not destroyed; rather they were apparently unwound and able to find the more distant homology. (Adapted from Sugawara N, Goldfarb T, Studamire B et al. [2004] *Proc Natl Acad Sci USA* 101:9315–9320. With permission from the National Academy of Sciences.)

had no effect. A third helicase that might be involved, Mph1, has not been tested. Thus, unlike what happens in mismatch repair during replication, where mismatched DNA is nicked and the offending region is excised by Exo1 and other exonucleases and then re-copied, it seems that the initial rejection of homeologous sequences occurs by helicase-driven unwinding of the DNA. Consequently, after a homeologous sequence has been "rejected" at the point of annealing, the sequences remain intact and able to carry out SSA with another, more distant and homologous repeat.

5.7 THE BEHAVIOR OF DSB ENDS CAN BE EXPLORED USING SSA

In mammalian cells, chromosomes occupy different domains of the nucleus, and even in budding yeast, one imagines that the two ends of a DSB are much more likely to interact with each other than with DSB ends on a different chromosome. To explore whether DSB ends are more likely to remain within a chromosomal territory, a yeast strain was created in which there were two possible HO-induced SSA events (**Figure 5.12A**). The integration of a *URA3*-containing plasmid carrying a 1-kb central region of the *HIS4* gene ("IS") results in the arrangement "HIS"–*URA3*–HO-cut-site–"IS4". In a similar fashion, on a different chromosome, a plasmid carrying *URA3* and the "IS" part of *HIS4* was integrated into the *ura3-52* allele which has a retroviral insertion that renders it Ura⁻. These two constructs were arranged in such a way that, after induction of the DSBs by expressing HO endonuclease, cells could employ SSA to form either a pair of deletions (annealing in *cis*, to produce a His⁺ Ura⁻ outcome) or a pair of reciprocal translocations (annealing in *trans*, to yield a His⁻ Ura⁺ result). In each case there are annealings

(A)

(B) Time (hr)   0   0.5   1.0   1.0   2.0   3.0   4.0   5.0

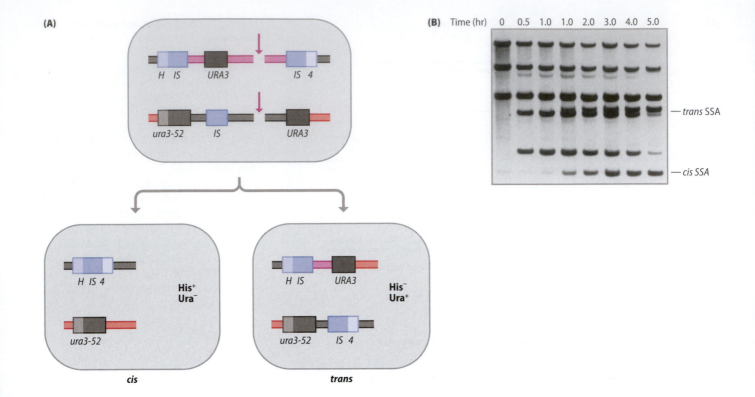

— *trans* SSA

— *cis* SSA

His⁺
Ura⁻

His⁻
Ura⁺

cis

trans

Figure 5.12 SSA does not show territoriality.
(A) DSBs on two different chromosomes are flanked by equivalently sized alternative pairs of homologous sequences ("*IS*" and *URA3*) that can be used to create two SSA events in *cis* or an alternative set in *trans*. (B) A Southern blot showing the kinetics of SSA. At 0.5 hr one sees the appearance of three new restriction fragments created by HO endonuclease cleavage, homologous to an "*IS*" probe. At 2 hr two additional bands characteristic of the *cis* and *trans* SSA products are visible, in equal amounts. The two outcomes appear with similar kinetics and at roughly equal frequencies. (Adapted from Haber JE & Leung WY [1996] *Proc Natl Acad Sci USA* 93:13949–13954. With permission from the National Academy of Sciences.)

between two "*IS*" segments and annealings between *ura3-52* and a *URA3* gene. The striking result was that the two translocations were as frequent as the pair of internal deletions (**Figure 5.12B**). These data argue that DSB ends—at least when they are subject to the extensive resection needed to carry out SSA—are just as likely to engage ssDNA from another chromosome as from the same chromosome. This result suggests that there is no "territoriality" to chromosomes that would make it more likely that one single-stranded end of a DSB would locate homologous ssDNA on the same chromosome.

However, this freedom of ends to interact independently may only occur when they are well separated from each other and have undergone extensive resection. If one makes a single DSB on each chromosome and asks how often they will be joined in *cis* or in *trans*, not by SSA but by NHEJ, it turns out that nearly all are intrachromosomal joinings. Similarly, as discussed below, the two ends of one DSB are more efficient in finding a homologous partner to repair by gene conversion than are two ends coming from two different DSBs. So chromosome ends from a single DSB may initially be held together, possibly by the Mre11-Rad50-Xrs2 (MRX) protein complex that has been shown to facilitate NHEJ in budding yeast, but this association is apparently lost once the ends are extensively resected.

5.8 SSA CAN BE SEEN IN OTHER ORGANISMS

A great deal of the early work on SSA was done by Ed Maryon and Dana Carroll using *Xenopus* extracts. They were the first to show that SSA involved the creation of intermediates with protruding 3'-ended nonhomologous tails. Even earlier work from Nat Sternberg's lab showed that SSA events could be assayed after transformation into mammalian cells.

In bacteria, SSA can occur in cells lacking RecBCD and mutated for the *sbcA* (suppressor of RecBCD mutation), which allows expression of a dormant phage-lambda-like bacteriophage. The prophage expresses a 5′ to 3′ exonuclease (RecE) and a strand annealing protein, RecT. When linearized DNA is created by expressing a restriction enzyme, SSA results if there are homologous sequences flanking the DSB.

Relatively little is known about the proteins involved in SSA in other eukaryotic organisms, but in fission yeast, *Drosophila*, *Caenorhabditis*, and mammals, SSA events appear to be Rad51 independent, as it is in budding yeast. In *S. pombe*, SSA depends on the Rad52 homolog, Rad22, and on the homologs of Rad1–Rad10. In the absence of normal telomere maintenance, the erosion of *S. pombe* telomeres and adjacent sub-telomeric regions can result in the creation of circular chromosomes. These fusions arise by SSA that requires the *S. pombe* homologs of Rad52, Rad1–Rad10, the helicase Srs2, and the ssDNA binding protein complex RPA. In worms, SSA depends on the BRCA2 homolog, CeBRC-2 (worms do not have a Rad52 homolog). In mouse, ablation of Rad52 reduces SSA, as does a BRCA1 mutant.

Recently, a number of SSA assays have been used in *Drosophila*, where DSBs can be induced either with P-element excision or with expression of the I-SceI endonuclease. SSA in *Drosophila*, as in yeast, accounts for a surprisingly high proportion of DSB repair events even when SSA is in competition with gene conversion, with about half of the I-SceI-induced events being SSA and half gene conversion (**Figure 5.13**). Interestingly, the recovery of deletions declines as a function of distance between the repeats, up to 2.4 kb. In yeast, the time to accomplish SSA, but not the overall efficiency, varies proportionally with length. It is possible that as resection proceeds along a fly chromosome, additional repeated sequences are exposed and these engage in other, inviable attempts at DSB repair.

In *Drosophila*, I-SceI-induced SSA between sequences that are diverged by as little as 0.5% is strongly inhibited by the Sgs1 homolog, BLM, so that homozygous BLM$^{-/-}$ flies show a much higher level of homeologous SSA. However, mutation of the Msh2 homolog, *spel1*, does not seem to affect either the efficiency of SSA between short repeats or homeologous SSA. With 3-kb repeats, the Rad1 homolog *mei-9* does not seem to be required to clip off nonhomologous tails, but another study, using ~150-bp repeats, showed that *mei-9* activity was required.

There is great interest in the role of SSA in mammals. A number of human diseases have been shown to result from deletions between dispersed repeated sequences such as Alu sequences. In some cases, researchers have suggested that intrachromosomal deletions between different Alu sequences could arise by unequal sister chromatid exchange, but it is at least as likely that these result from SSA. Maria Jasin's laboratory has studied the competition between gene conversion and SSA using an *I-SceI*-induced DSB where there is a nearby sequence homologous

Figure 5.13 SSA occurs in *Drosophila*.
An *I-SceI*-induced DSB in the *red* target sequence can be repaired by interchromosomal gene conversion (GC) (using the *green*, nearly identical sequences, that cannot be cut by I-SceI as the template) or by SSA. The two events occur at roughly equal frequencies. Some repair also occurs by nonhomologous end-joining (NHEJ; not shown). The two chromosomes in the diploid were followed by linkage to the Sb$^+$ gene. (Adapted from Rong YS & Golic KG [2003] *Genetics* 165:1831–1842. With permission from the Genetics Society of America.)

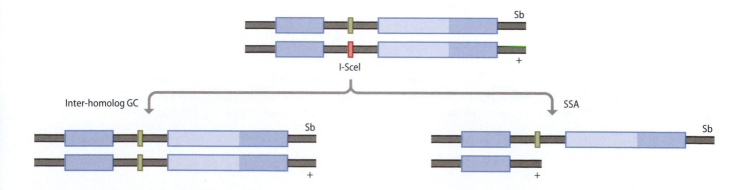

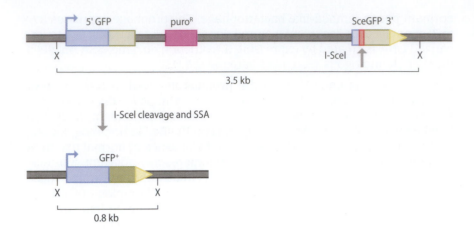

Figure 5.14 SSA occurs in a model assay system in mammalian cells. A construct containing the 5′ two-thirds of a GFP gene is separated from the 3′ two-thirds of the gene, also containing an I-SceI cleavage site. A DSB promotes SSA, leading to a GFP⁺ Puro⁻ recombinant. (Adapted from Jasin M [2002] *Oncogene* 21:8981–8993. With permission from Macmillan Publishers, Ltd.)

Figure 5.15 Interchromosomal SSA also occurs in mammalian cells. A mouse cell line contains inserts into the *pim* and *Rb* genes on chromosomes 17 and 14, respectively. Each carries an 18-bp I-SceI cleavage site flanked by part of the puromycin-resistance (puro) and by a Alu sequence, as well as by the 5′and 3′ segments of the G418-resistance *neo* gene, split in the middle of an intron. I-SceI cleavages on the two chromosomes can be repaired by two SSA events that create a pair of reciprocal translocations (17;14 and 14;17). One SSA event joins two 300-bp Alu sequences (*green* boxes) and the second anneals overlapping homologies of the 5′ and 3′ segments of the puro gene (*pink* boxes). The SSA event joining Alu sequences also creates a functional intron allowing mRNA splicing to occur between splice donor (SD) and splice acceptor (SA) sequences of the *neo* gene, resulting in G418 resistance. The efficiency of the Alu × Alu SSA event is reduced when the two sequences are divergent. (Adapted from Weinstock DM, Richardson CA, Elliott B & Jasin M [2006] *DNA Repair* 5:1065–1074. With permission from Elsevier.)

to both sides of the DSB (**Figure 5.14**). Consistent with results from yeast, SSA competes strongly with gene conversion among the outcomes. SSA is unaffected when cells express a dominant-negative Rad51 mutant protein, whereas gene conversion is reduced. Rad52 deficiency reduces SSA but has no effect on gene conversion, consistent with the idea that the BRCA2 protein has taken over many of Rad52's roles in mammals. That Rad52 reduces SSA suggests that BRCA2 may not substitute for Rad52 in this respect; in fact, mutations of BRCA2 cause an increase in SSA while gene conversion goes down. SSA is, however, reduced by BRCA1 and BARD1 mutants (as is gene conversion), but to date, how BRCA1/BARD1 operate in homologous recombination is poorly understood.

In a different assay, Jasin's lab has examined translocations in which two DSBs were created on different chromosomes (**Figure 5.15**). The DSBs were arranged so that two different SSA events could form reciprocal translocations, generating a selectable phenotype. One pair of sequences consisted of identical 300-bp Alu sequences. A similar outcome could be generated by NHEJ if the adjacent homologies were not used. In wild-type cells, 84% of the junctions in such translocations arose by SSA, but when the Alu sequences were 20% divergent, the efficiency of SSA dropped to only 7%, with the rest resulting from NHEJ. Thus homeology strongly suppresses SSA. The opposite reciprocal product still formed by SSA between other, identical sequences more than 90% of the time. At present, the roles of MMR genes or the BLM helicase have not been explored in mammals.

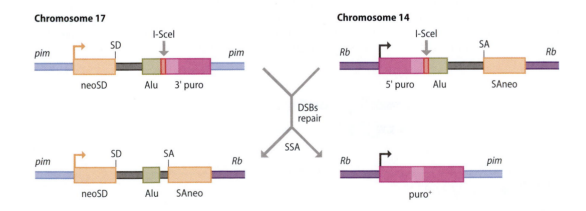

SUMMARY

DSB ends can be extensively resected and thus can promote SSA even when the flanking sequences are tens of kilobases distant from the break. SSA is Rad51 independent but Rad52 dependent. Clipping of the 3'-ended nonhomologous tails requires Rad1–Rad10 and a number of additional proteins, including the Slx4 scaffold. SSA is impaired by a relatively low level of mismatches between the flanking sequences. Heteroduplex rejection requires the BLM helicase Sgs1 and the Msh2–Msh6 proteins but, surprisingly, not the Pms1–Mlh1 proteins that normally work together in mismatch repair.

SUGGESTED READING

Fishman-Lobell J, Rudin N & Haber JE (1992) Two alternative pathways of double-strand break repair that are kinetically separable and independently modulated. *Mol Cell Biol* 12:1292–1303.

Flott S, Alabert C, Toh GW et al. (2007) Phosphorylation of Slx4 by Mec1 and Tel1 regulates the single-strand annealing mode of DNA repair in budding yeast. *Mol Cell Biol* 27:6433–6445.

Gould SJ (1997) The exaptive excellence of spandrels as a term and prototype. *Proc Natl Acad Sci USA* 94:10750–10755.

Paques F & Haber JE (1999) Multiple pathways of recombination induced by double-strand breaks in *Saccharomyces cerevisiae*. *Microbiol Mol Biol Rev* 63:49–404.

Rong YS & Golic KG (2003) The homologous chromosome is an effective template for the repair of mitotic DNA double-strand breaks in *Drosophila*. *Genetics* 165:1831–1842.

Silberstein Z, Tzfati Y & Cohen A (1995) Primary products of break-induced recombination by *Escherichia coli* RecE pathway. *J Bacteriol* 177:1692–1698.

Singleton MR, Wentzell LM, Liu Y et al. (2002) Structure of the single-strand annealing domain of human RAD52 protein. *Proc Natl Acad Sci USA* 99:13492–13497.

Weinstock DM, Nakanishi K, Helgadottir HR & Jasin M (2006a) Assaying double-strand break repair pathway choice in mammalian cells using a targeted endonuclease or the RAG recombinase. *Methods Enzymol* 409:524–540.

CHAPTER 6
GENE CONVERSION

Gene conversion involves the nonreciprocal transfer of genetic information between homologous sequences. Gene conversions may be accompanied by reciprocal crossing over. Unfortunately much of the literature on homologous recombination contrasts "gene conversions" (defined as events without crossing over) with "crossovers." This distinction is misleading as it suggests that crossovers are not accompanied by gene conversions. Simple crossovers can be generated by site-specific enzymes (see Chapter 11) but most crossovers that will interest us arise from mechanisms of DSB repair that can give rise to both crossover and noncrossover outcomes and generally involve an asymmetric transfer of sequences from a donor to a recipient.

6.1 GENE CONVERSIONS WERE INITIALLY DEFINED FROM ABERRANT SEGREGATION OF ALLELES IN MEIOSIS

Nearly all of the analysis of gene conversions has come from situations in which recombination is initiated by a DSB, although much of the early analysis was carried out in the absence of any knowledge that the initiating lesions were indeed DSBs. The first indications of gene conversion came from aberrant ratios of meiotic products observed in fungi, where all four meiotic products could be recovered from a single meiosis. Accurate transmission of genetic information should mean that a diploid heterozygous for a marker, **A/a**, should yield 2**A** and 2**a** segregants, but at a surprisingly high frequency—often greater than 1%—meiotic tetrads sometimes contain 3**A** and 1**a** or 1**A** and 3**a** segregants, where there is clearly an asymmetric inheritance of information. In addition, in meiosis involving two different *auxotrophic* recessive alleles (heteroalleles) within a gene (**a-1/a-2**), recombination could produce wild type, *prototrophic*, recombinants. These recombinants could arise by a simple reciprocal crossing over between the two alleles, to produce a wild type and a double-mutant (**a-1 a-2**) recombinant (**Figure 6.1A**). In fact the great majority of such events are again produced by nonreciprocal genetic transfer—gene conversion (**Figure 6.1B**). Gene conversions between heteroalleles can also be studied in mitotic cells, although they occur much less often. But, as in meiotic cells, most prototrophic recombinants arise from nonreciprocal transfers of sequences rather than by simple crossovers.

Figure 6.1 Intragenic recombination between two different mutant alleles can produce a wild-type allele. Alleles on two different chromosomes can recombine to generate a wild-type (+ +) allele either by a reciprocal exchange (A) or by gene conversion (B).

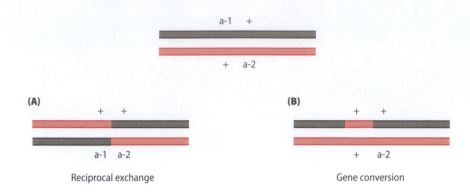

6.2 ANALOGOUS GENE CONVERSIONS ARISE IN MITOTIC CELLS

For the moment we will focus our attention on gene conversion events studied in somatic cells, without the complications of a large number of meiosis-specific attributes. Again, it is in budding yeast that we have garnered the most detailed understanding, but analogous information has emerged from many other organisms. A great deal of our understanding of this process has come from the genetic analysis of recombination between pairs of different auxotrophic alleles (heteroalleles) in genes encoding enzymes involved in amino acid or nucleotide metabolism, where gene conversion can yield prototrophic recombinants. A deeper molecular picture has come from studies of recombination involving linearized DNA transformed into cells as well as from DSB-induced repair events using site-specific endonucleases.

Consider a diploid yeast carrying two different alleles of *leu2* gene in budding yeast (*leu2-K/leu2-R*) and heterozygous for a recessive *his4* allele, located further away from the centromere on the same chromosome arm (**Figure 6.2**). Initially the cells are auxotrophic for leucine (Leu⁻); that is, their growth requires that nutritional medium be supplemented with leucine. The diploids are initially prototrophic for histidine (His⁺). However,

Figure 6.2 Heteroallelic recombination in an auxotrophic budding yeast diploid. (A) A diploid heterozygous for *leu2-K* and *leu2-R* can produce a prototrophic (Leu⁺) recombinant. (B) Spontaneous Leu⁺ papillae of patches of cells replica-plated from rich medium to medium lacking leucine. (C) Leu⁺ papillae in cells that were UV-irradiated on rich medium and then replica-plated to medium lacking leucine. The rate of UV-induced Leu⁺ recombination is much higher than the spontaneous value. (D) Leu⁺ papillae of patches of diploid cells that were induced to go through meiosis and sporulation and then replica-plated to leucine dropout medium. The rate of prototroph formation is 100–1000 times higher after meiosis; hence without diluting the number of cells plated one sees a lawn of Leu⁺ cells instead of individual papillae. (B–D, courtesy of Qiuqin Wu, Brandeis University.)

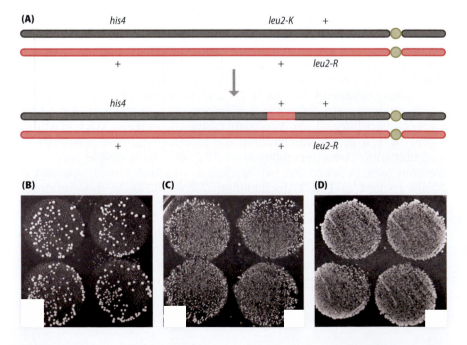

one can readily obtain Leu$^+$ recombinants. The spontaneous rate of protototroph formation is about 1 in 10^7 cells—high enough that several Leu$^+$ papillae emerge and grow from the background of cells replica-plated onto medium lacking leucine (**Figure 6.2B**). As noted above, the simplest way one could obtain such Leu$^+$ recombinants would be by a crossover between the two alleles, producing a diploid of genotype *LEU2/leu2-K,R*, but in fact almost none of the recombinants—even when there is an associated crossing over—arise by such an event. Instead, these protophic diploids are either *LEU2/leu2-R* or *leu2-K/LEU2*, in which one or the other allele has undergone gene conversion.

6.3 MOST GENE CONVERSION ARISES FROM MISMATCH CORRECTION OF HETERODUPLEX DNA

In the molecular models outlined in Chapters 1 and 2, the key intermediate that will lead to gene conversion is the formation of heteroduplex DNA (hDNA), after strand invasion or strand annealing (**Figure 6.3A**). This heteroduplex can be repaired by mismatch repair enzymes either to its original genotype (a restoration) or it can be gene converted, in this case yielding a *LEU2* recombinant. In the dHJ model, correction in favor of the template strand will be a restoration of genetic information, whereas correction of the heteroduplex in favor of the invading strand genotype will lead to a gene conversion.

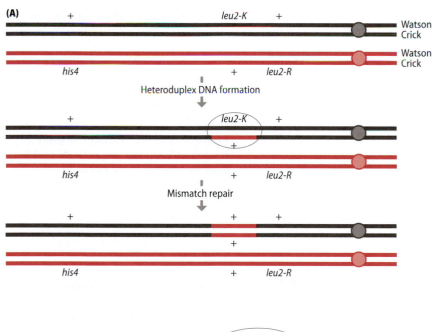

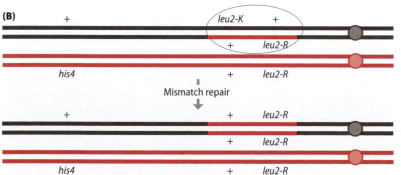

Figure 6.3 Gene conversion can occur by DNA strand transfer to form heteroduplex DNA, followed by mismatch correction. (A) If the transferred strand creates heteroduplex DNA (*circled*) covering only one of the two alleles, mismatch correction can produce a Leu$^+$ recombinant. (B) If strand transfer creates heteroduplex DNA covering both alleles, it is most likely that mismatch repair will correct both sites coordinately so that there will be a gene conversion (from *leu2-K* to *leu2-R*) but the cells will remain Leu$^-$.

If hDNA covers both alleles, the most likely outcomes of mismatch repair are either a complete *restoration* back to the original genotype or a *co-conversion*, where one allele is replaced by the other; so hDNA covering both alleles will not produce a wild-type recombinant unless the mismatches are independently corrected (**Figure 6.3B**). Thus the strand invasion events that are most productive in producing gene conversions are usually those which terminate between the two markers. The further apart the heteroalleles, the higher the chance that heteroduplex will cover only one of the alleles. Thus, heteroalleles that are about 200 bp apart will yield Leu⁺ recombinants about half as often as those that are 400 bp apart.

In the decades before easy DNA sequencing, a set of alleles within a gene could be ordered by making pairwise crosses to create a series of heteroallelic diploids and measuring their rates of recombination. One could order the alleles along a linear map just as one would create a genetic map of different genes by their frequencies of meiotic recombination. Because spontaneous rates are low, such intragenic mapping was often done by stimulating a significantly higher level of recombination by UV- or X-irradiating the cells (Figure 6.2C). Pairwise crosses could also be analyzed by inducing meiosis, to stimulate intragenic recombination, which again occurs at rates 100–1000 times as often as they arise spontaneously in mitotic cells (Figure 6.2D). An example is shown in **Figure 6.4**. These observations support the idea that the greater the distance between the alleles, the higher the probability that the recombination intermediate will terminate between the two alleles and that a wild-type recombinant will arise.

(A)

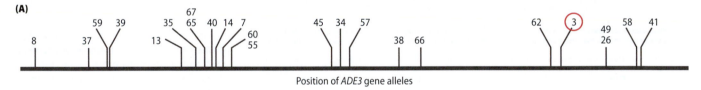

Position of *ADE3* gene alleles

(B)

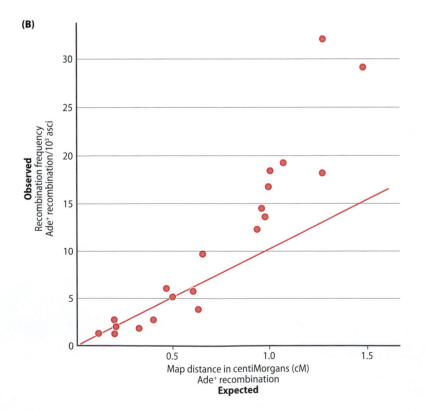

Figure 6.4 The rate of intragenic gene conversion is roughly proportional to the distance separating the alleles. A haploid strain (A) carrying the *ade3-3* allele (*circled*) was crossed to a series of strains of opposite mating type carrying different *ade3* alleles and the frequencies of prototrophic gene convertants arising during meiosis were then used to create an intragenic map. Similar maps can be made by measuring mitotic recombination, but the rates of conversion are orders of magnitude lower. (B) Comparison of the observed rate of Ade⁺ recombinants versus the expected rates assuming that recombination is strictly proportional to genetic map distance and extrapolated from recombination events involving nearby alleles. Deviations of the experimental observations from the extrapolated, expected rate are seen at longer distances. These deviations result from the fact that the rate of recombination is underestimated when alleles are close to each other, as a result of co-conversion of both alleles, which will not produce an Ade⁺ recombinant. (Adapted from Jones EW [1972] *Genetics* 70:233–250. With permission from the Genetics Society of America.)

6.4 GENE CONVERSIONS CAN BE ACCOMPANIED BY CROSSING OVER

Although Leu⁺ recombinants do not arise by a simple crossing over between the two alleles, about 5–10% of these recombinants in budding yeast do show evidence of a crossover associated with gene conversion. In the example shown in **Figure 6.5**, a crossover accompanying gene conversion at *LEU2* can produce cells that are homozygous for the distal *his4* marker. Such *loss of heterozygosity* (LOH) will occur only half of the time that there is an exchange of chromosome arms, assuming that the crossover occurs in G2 cells, and depending on which chromatids segregate with each other in mitosis. Because we can only see visible evidence of a crossover in half the cases, as explained above, we can conclude that about 10–20% of the spontaneous mitotic Leu2⁺ gene conversion events are accompanied by crossing over. For future reference, the frequency of crossing over for the same alleles undergoing meiotic recombination is closer to 50%. Note that if gene conversion and an associated exchange occur in G1 cells, the linkage between genetic markers will be changed, but the cells will still be heterozygous for the flanking marker. However,

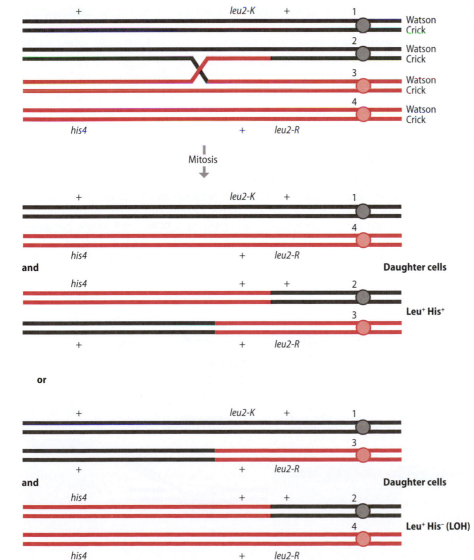

Figure 6.5 Gene conversion arising in G2 cells can give rise to loss of heterozygosity. Gene conversion to generate a Leu⁺ recombinant can be accompanied by crossing over in the G2 phase of the cell cycle. A heteroduplex-containing strand exchange intermediate associated with a Holliday junction is mismatch corrected to create a *LEU2* (Leu⁺) allele. At mitosis, the segregation of sister centromeres is random. In one of two configurations the segregation results in loss of heterozygosity (LOH) of distal markers (here *his4/his4*).

most spontaneous gene conversions likely depend on chromosome breaks or nicks arising during S phase.

6.5 GENE CONVERSION CAN BE ASSAYED IN HAPLOID YEAST

Gene conversions can also be studied in haploids, by creating the opportunity for *ectopic* gene conversion, where the recombining alleles are located in different places on the same chromosome or even on different chromosomes. For example, in a yeast strain carrying *leu2-K* at its normal location, one could integrate a cloned copy of *leu2-R* carried on a *URA3*-marked plasmid at the *leu2* locus, creating a haploid with the tandem duplication *leu2-K–pBR322–URA3–leu2-R* (**Figure 6.6**). Again one can select for Leu+ recombinants. Some of these Leu+ recombinants are also Ura+, and result from gene conversions without crossing over. The remaining recombinants are Ura− and have lost one copy of the *leu2* sequences as well as the plasmid sequences and the selectable marker, as expected for a crossover outcome. Because the "popped-out" plasmid that will carry the *URA3* gene has no origin of replication, it cannot be maintained in the growing colony. But, in fact, even if one inserts an origin of replication into the plasmid, only a minority of the apparent crossovers in this tandem arrangement of heteroalleles arise from gene conversion accompanied by exchange; most of the Leu+ Ura− recombinants came from single-strand annealing (SSA) (Figure 6.6). These data reinforce the conclusion that the majority of mitotic gene conversions occur without an accompanying crossover. By selecting for Ura+ Leu+ cells, one can be assured that the events arose by gene conversion.

SSA can also be avoided by constructing a pair of alleles in inverted orientation, so that prototrophs should arise by intrachromosomal gene conversion (**Figure 6.7**). In this case, crossovers would be seen as an inversion of sequences between the two *leu2* segments and can be detected by analyzing the DNA using Southern blots or PCR rather than phenotypically.

In the constructs illustrated in Figure 6.6, gene conversions could also arise by recombination between sequences on sister chromatids. For the most part, since mitotic gene conversions are generally not accompanied by crossing over, one cannot tell the difference between an intrachromosomal and an intersister event. However, crossovers do occur and if they accompany gene conversions between alleles on sister

Figure 6.6 Intragenic gene conversion between *leu2* heteroalleles. Heteroalleles of *leu2* in a tandem duplication can be assessed in haploids by the appearance of *LEU2* recombinants. As noted in Chapter 5, although reciprocal crossing over can accompany gene conversion to "pop out" the plasmid sequences that originally created the duplication, most of the deletions arise by SSA.

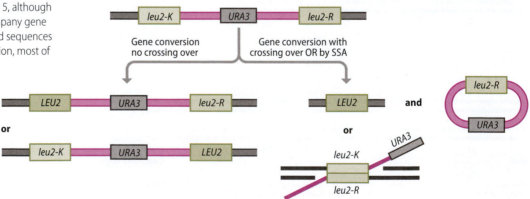

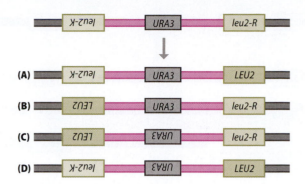

Figure 6.7 Gene conversions between inverted repeated heteroalleles. In this inverted arrangement SSA will not occur. (A, B) Either *leu2-K* or *leu2-R* can be gene converted to Leu⁺. (C, D) *LEU2* recombinants accompanied by crossing over can be seen by the inversion of sequences between the two *leu2* segments.

chromatids, one can observe not only deletions but also duplications of the region between repeats, depending on which allele was converted to wild type (**Figure 6.8**). Sister chromatid exchange will be discussed in detail in Chapter 9.

Ectopic gene conversion can also be observed when the second *leu2* allele is inserted by transformation either further away on the same chromosome (making SSA less likely) or on a different chromosome. In the intrachromosomal case (**Figure 6.9**), most outcomes are without crossing over, but if a crossover accompanies the gene conversion, a deletion can be recovered, so long as there are no essential genes in the intervening interval. Here, too, intersister gene conversions can sometimes be accompanied with the formation of deletions or duplications. In the interchromosomal case (**Figure 6.10**), if the two sequences are oriented in the same direction relative to the centromere, a crossover associated with *LEU2* gene conversion will produce a reciprocal translocation that is readily detected by Southern blot or PCR analysis.

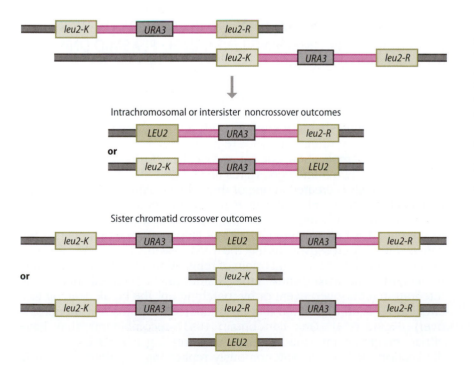

Figure 6.8 Unequal sister chromatid exchange can also produce prototrophs from heteroalleles in a tandem duplication. A tandem duplication of *leu2* genes carrying different alleles (created by plasmid integration) can give rise to Leu⁺ prototrophs either by intrachromosomal or interchromatid recombination. Crossovers associated with gene conversion in unequal sister chromatid recombination will produce either a duplication of the intervening sequences (*URA3*) or a reciprocal outcome lacking these sequences.

Figure 6.9 Intrachromosomal ectopic gene conversion can occur with crossing over to produce a chromosomal deletion, or without exchange. A cloned copy of *leu2-K* and an adjacent selectable G418-resistance marker (*KAN*) are inserted at an ectopic location distal to *leu2-R* allele. Ectopic recombination accompanied by crossing over will produce a deletion.

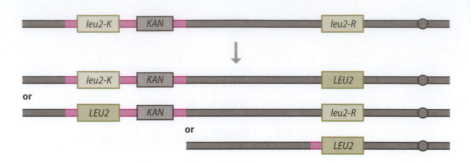

Figure 6.10 Interchromosomal ectopic gene conversion between heteroalleles on different chromosomes can result in a reciprocal translocation. A cloned copy of *leu2-K* and an adjacent selectable G418-resistance marker (*KAN*) are inserted at an ectopic location relative to the *leu2-R* allele. Ectopic recombination accompanied by crossing over will produce a reciprocal translocation.

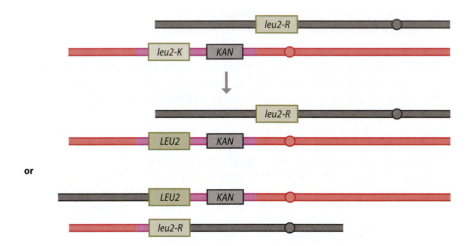

6.6 THE MOLECULAR BASIS OF GENE CONVERSION WAS DEDUCED BY RECOMBINING LINEARIZED PLASMID DNA WITH A CHROMOSOME

The pioneering studies from the Jack Szostak and Rodney Rothstein labs examined a different type of gene conversion: the integration of linear DNA whose ends shared homology with a chromosomal location. If a plasmid carries two different cloned regions of yeast DNA (*LEU2* and *HIS3*) and no origin of replication, one can recover transformants—integrants— at either of the two homologous sites (*leu2-1* and *his3-X*) (**Figure 6.11**). If, however, a DSB is created in one of them, by cleaving with a restriction endonuclease that only cuts once in the plasmid, the fragment will integrate with a very high degree of specificity at the locus homologous to the cut DNA ends (**Figure 6.12**). Thus DSBs are recombinogenic and can result in a crossover during the repair event. If the transforming plasmid contains a functional origin of replication as well as a selectable marker, then the broken ends of the plasmid can be repaired either by an integration (crossover) event or by "patching up" the break, leading to a circular, replicating plasmid and an unaltered chromosome (noncross-over) (**Figure 6.13**). One can obtain His3+ recombinants that have either integrated into the chromosome, creating a *HIS3–URA3–his3-X* duplication, or have an autonomously replicating, repaired *HIS3 LEU2* plasmid and with the *his3-X* copy still on the chromosome. Note that for

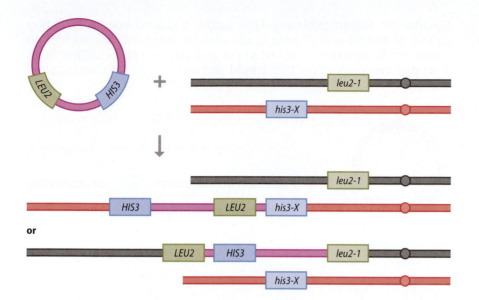

Figure 6.11 Integration of a nonreplicating plasmid by homologous recombination. In a plasmid carrying two regions of homology to different sites in the genome (*LEU2* or *HIS3*), integration by homologous recombination can occur at either site. (Adapted from Orr-Weaver T & Szostak JW [1981] *Proc Natl Acad Sci USA* 78:6354–6358. With permission from the National Academy of Sciences.)

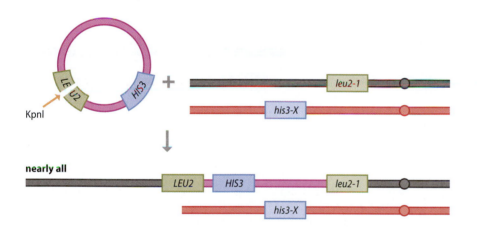

Figure 6.12 DSBs direct plasmid integration. A DSB created by a restriction endonuclease within one of two homologous regions shared by a plasmid with the genome strongly directs integration of a nonreplicating plasmid to that region of homology. (Adapted from Orr-Weaver T & Szostak JW [1983] *Proc Natl Acad Sci USA* 80:4417–4421, and from Orr-Weaver T, Szostak JW & Rothstein RJ [1981] *Proc Natl Acad Sci USA* 78:6354–6358. With permission from the National Academy of Sciences.)

the cut *HIS3* gene to be repaired the site of the DSB must be different from the site of the *his3-X* mutation; otherwise the mutant sequence would be used to patch up the DSB and cells would not become His3⁺. In most experiments, there are significantly more noncrossover outcomes than crossovers.

The fact that one can recover a functional *HIS3* gene argues strongly that the repair event is accurate, without the addition or deletion of any base pairs, but one could not formally tell if any new DNA synthesis was required or if there could be some very precise end joining, aided by the homologous template. A decisive answer came from experiments in which the plasmid sequences were cut by restriction enzymes at two sites within the *HIS3* gene (**Figure 6.14**) so that recovery of a functional *HIS3* copy required copying the sequences missing in the gapped, linearized molecule. Again, for this to work the *his3-X* mutation must lie outside the gapped region. Consequently His3⁺ transformants, both those associated with crossovers and those without a crossover, had a restored *HIS3* gene and an unaltered *his3-X* template.

Because the linearized plasmid DNA carried a second selectable marker (*LEU2*), Rothstein and Szostak could also demonstrate that some DSB repair events would not only fill in the gap but also apparently enlarge it, so that the repair of the DSB would also incorporate into the recipient

Figure 6.13 DSB repair outcomes.
In a plasmid that is able to replicate because of an ARS sequence, a DSB within homology can be repaired by using intact sequences on a chromosome as a donor. DSB repair can occur either without crossing over (leaving an intact, replicating plasmid) or with crossing over (resulting in the integration of the plasmid at the homologous locus). (Adapted from Orr-Weaver T & Szostak JW [1983] *Proc Natl Acad Sci USA* 80:4417–4421, and from Orr-Weaver T, Szostak JW & Rothstein RJ [1981] *Proc Natl Acad Sci USA* 78:6354–6358. With permission from the National Academy of Sciences.)

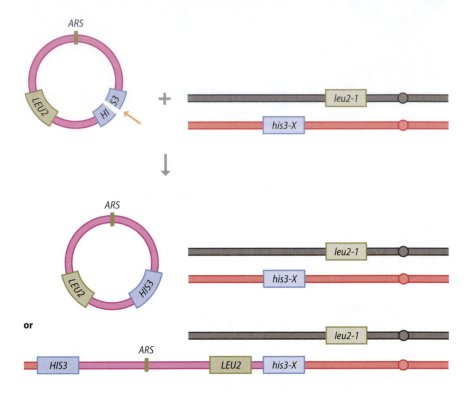

Figure 6.14 Demonstration of gap repair, restoring the full-sized *HIS3* gene.
A linearized plasmid that had been cut at two sites within the *HIS3* gene to remove a central segment was transformed into a *his3-X* strain. DSB repair copied all the intervening sequences from the chromosomal template to produce a His+ transformant. Some repair events (noncrossovers) resulted in a *HIS3*-containing replicating plasmid and an unaltered chromosomal locus. Other repair events (crossovers) resulted in the integration of the plasmid to create a tandem duplication. (Adapted from Orr-Weaver T & Szostak JW [1983] *Proc Natl Acad Sci USA* 80:4417–4421, and from Orr-Weaver T, Szostak JW & Rothstein RJ [1981] *Proc Natl Acad Sci USA* 78:6354–6358. With permission from the National Academy of Sciences.)

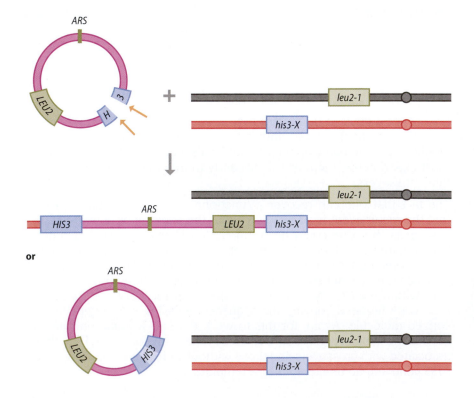

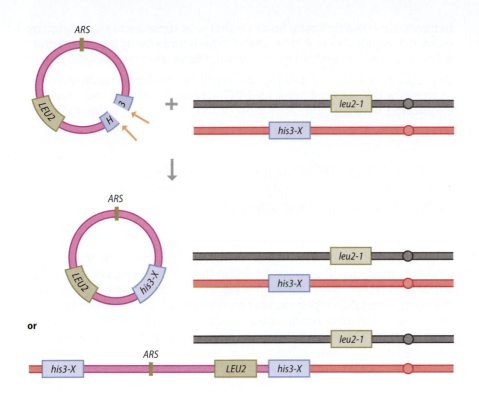

Figure 6.15 Gap repair sometimes results in co-conversion of an adjacent allele. If transformants are selected to be Leu⁺, it is possible to see that some gap repair events incorporate not only the missing sequences but also an adjacent *his3-X* mutation, so that the transformants are Leu⁺ but remain His⁻. (Adapted from Orr-Weaver T & Szostak JW [1983] *Proc Natl Acad Sci USA* 80:4417–4421, and from Orr-Weaver T, Szostak JW & Rothstein RJ [1981] *Proc Natl Acad Sci USA* 78:6354–6358. With permission from the National Academy of Sciences.)

sequences the *his3-X* mutation (**Figure 6.15**). Thus the transformed cells would become Leu2⁺ but still His3⁻. This *co-conversion* event could arise if the gap were enlarged by exonucleases once the fragment entered the nucleus, so that the gap repair now covered the *his3-X* site (**Figure 6.16A**). Alternatively, co-conversions could result from the formation a region of

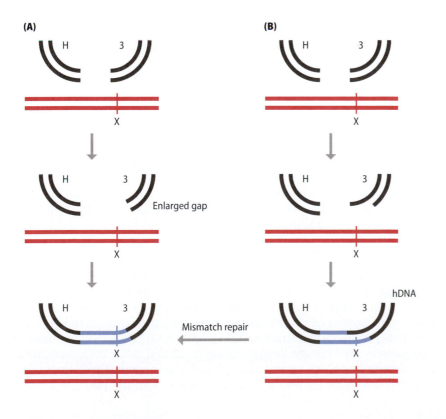

Figure 6.16 Co-conversion of the DSB and an adjacent site. Co-conversion can occur either by widening a gap so that the adjacent marker becomes part of the sequences that must be copied (shown in *blue*) to repair the DSB (A) or by the formation of heteroduplex DNA during repair (B). Heteroduplex DNA can then be mismatch corrected to produce the co-conversion event.

heteroduplex DNA by strand invasion that was subsequently corrected by mismatch repair (**Figure 6.16B**). These experiments using linearized plasmids led the way in understanding how DSBs would promote DSB repair.

Gap repair is capable of "picking up" very large regions of sequence. For example, a transformed linearized DNA molecule whose two ends are engineered to share homology with distant regions of a chromosomal template can acquire at least 40 kb of sequences.

6.7 A QUESTION OF SEMANTICS: CAN THERE BE GENE CONVERSIONS IF THE DONOR AND RECIPIENT CHROMOSOMES ARE IDENTICAL?

Classically, gene conversions were defined as nonreciprocal exchanges of genetic information between homologs and it is precisely those kinds of events that have allowed us to define the molecular mechanisms that allow DSB repair, both with and without an accompanying crossover. However, some DSB repair events can occur in a region that has no polymorphisms; but DSB repair occurs using exactly the same molecular processes—except mismatch repair—that are employed when there are one or more nucleotide differences. In some cases we know there must have been an asymmetric transfer of genetic information—for example when there is a gap that must be filled in by copying sequences from the donor—but in some cases a simple DSB is repaired without any alteration in sequence and we must infer what happened. At a biochemical level we can isotopically label "old" and "new" DNA to demonstrate that the repair event is asymmetric. So, to use the fewest terms with (hopefully) the widest meaning, here and later I will invoke "gene conversion" when DSB repair involves a short patch of new DNA synthesis, copied from a homologous donor template, in contrast to BIR (where there is extensive synthesis). I will distinguish, when possible, gene conversions that occur by SDSA from those arising by the dHJ mechanism.

6.8 HETERODUPLEX CORRECTION OFTEN DEFINES GENE CONVERSION TRACT LENGTHS

If hDNA contains several mismatches, they will usually all be repaired in the same direction—either all as conversions or all as restorations—because very few conversion tracts show an interrupted pattern where a marker in between two converted sites has been restored, although this certainly is seen in a minority of cases. Strongly biased repair of mismatches will occur if the cell preferentially recognizes one strand of the heteroduplex as the preferred template against which to repair the mismatches. For example, the MMR machinery may recognize the end of the invading strand as a signal to repair that strand in favor of the unbroken template strand.

Not all mismatches are equally subject to correction; in most organisms, C–C mismatches evade correction by the MMR machinery. Similarly, small self-annealed palindromes are largely invisible to the MMR system. These palindromes can be created by inserting a small inverted repeat into a gene so that when one strand carries the insertion (as in heteroduplex DNA) it will form a small hairpin (**Figure 6.17**). Thus C–C mismatches and palindromes remain largely uncorrected and, after DNA replication, the Watson and Crick strands will give rise to genetically distinct progeny that will grow up into a sectored colony, with the two halves each

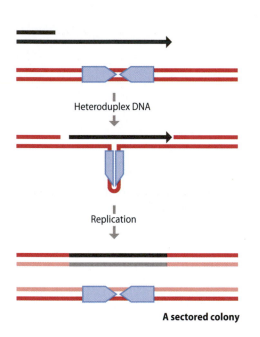

Heteroduplex DNA

Replication

A sectored colony

Figure 6.17 Some heteroduplex DNA sequences escape mismatch repair. hDNA containing a small palindromic sequence, such as a pair of 6-bp inverted sequences separated by 2–3 bp, forms a hairpin that is not effectively repaired by the mismatch repair system. If hDNA is not repaired, then a single cell will replicate and produce genotypically different daughter cells. As the cells continue to replicate they will produce a sectored colony. Among single base-pair mismatches, C–C mismatches also fail to be effectively repaired. (Adapted from Nag DK & Petes TD [1993] *Mol Cell Biol* 13:2324–2331. With permission from the American Society for Microbiology.)

expressing one of the two alleles. This strategy provides a way to ask if a marker at a certain distance from the DSB has been included in heterodu-plex. However in the presence of other, well-repaired mismatches, the situation changes: a poorly repaired mismatch will often be co-repaired along with a well-repaired nearby mismatch. Heteroduplex length can be measured by using strains defective in mismatch repair, as discussed further below.

6.9 WHERE DOES HETERODUPLEX DNA FORM DURING DSB REPAIR?

The most obvious source of hDNA is that created by strand invasion of ssDNA that is produced by resection of the break. Where hDNA forms is different in the two main models of gene conversion that we are considering.

In the dHJ model, the enlargement of the D-loop by the elongating DNA polymerase, starting at the 3′ end of the invading strand, will increase the length of hDNA that can form between the D-loop and the second, resected end of the DSB (**Figure 6.18**). In dHJ events, the heteroduplex on opposite sides of the DSB would be expected to lie on two different participating chromatids. The length of the hDNA can be extended by additional strand exchange mediated by RecA/Rad51 (Figure 6.18 panel ii) or by the initiation of new synthesis from the first 3′ end (Figure 6.18 panel iii). Finally hDNA can be extended by branch migration of a HJ (Figure 6.18 panel iv).

But other actors can change the location of hDNA. After the gaps in the dHJ strand-invasion structure are filled in by DNA polymerase, the dHJ can be "dissolved" by the combined action of a helicase and a topoisomerase

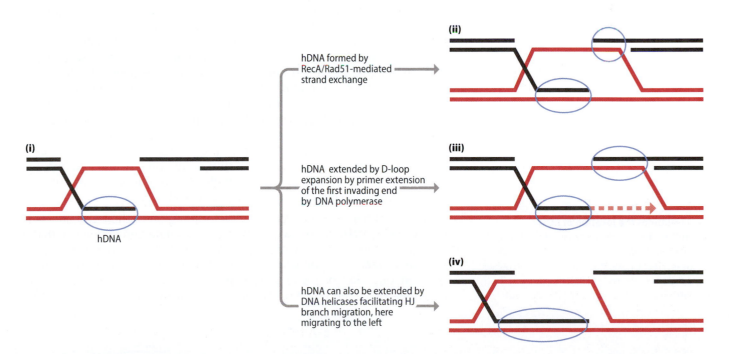

Figure 6.18 Extension of hDNA. Strand invasion creates a region of hDNA (i). The independent strand invasion of the second end mediated by RecA/Rad51 of a DSB creates a second region of hDNA (ii). Alternatively, the action of DNA polymerase extending the 3′ end of the first invading strand can extend a D-loop until a second heteroduplex region forms by strand annealing (iii). In addition, hDNA can be extended by branch migration of a HJ, here shown as the leftward extension of the original strand invasion hDNA (iv).

Figure 6.19 Alternative ways to recover a noncrossover outcome from a dHJ intermediate. Noncrossover resolution of a dHJ by HJ resolvases yields a different outcome from that obtained by unwinding of the dHJ (dissolution). These two outcomes are genetically identical with respect to the arrangement of genetic markers flanking the region where recombination is occurring, but they differ in the locations of heteroduplex DNA. (Adapted from Hastings PJ [1988] *Bioessays* 9:61–64. With permission from John Wiley & Sons, Inc.)

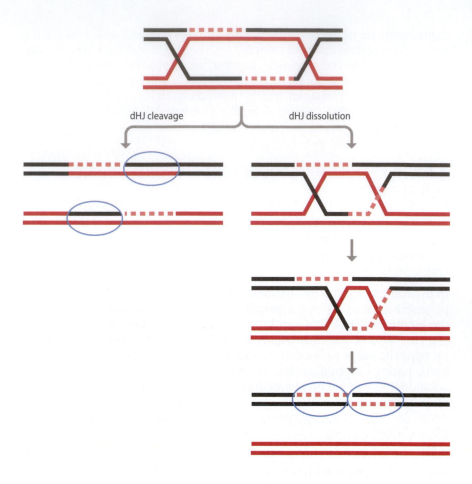

Figure 6.20 Branch migration of a dHJ also changes the location of hDNA. Branch migration is energetically neutral because as many base pairs are broken as will be formed to slide an HJ one base. Movement of the dHJ outside the region where recombination occurred will create two adjacent regions of hDNA (*circled*).

(**Figure 6.19**). After dHJ dissolution, the two heteroduplex regions can end up in only one of the participating chromosomes, in a noncrossover configuration (Figure 6.19). dHJs have been shown to be dissolved in this way by the combined action of the mammalian BLM helicase coupled with TopIIIα and by similar complexes in other organisms. Another way that all the hDNA can end up in one of the two recombining sequences involves branch migration of the dHJ (**Figure 6.20**). Yet another way we need to consider is that strand invasion intermediates can be cleaved by other endonucleases, most notably the Mus81–Eme1 endonuclease or the Rad1–Rad10. Here, these enzymes are thought to be able to attack a "half Holliday junction" (**Figure 6.21**), thus converting what could become a dHJ into an intermediate with a single HJ that can of course branch migrate and again change the location of hDNA.

In SDSA, primer extension can copy a long region of the template. This newly copied DNA then unwinds from the template and anneals with the resected second end of the break. Thus the initial region of heteroduplex will have been unwound so that hDNA may be predominantly found only on one side of the repaired locus (**Figure 6.22**). The majority of noncrossover gene conversions in yeast and in somatic cells of other

dHJ branch migration

organisms, as well as in a significant fraction of those in meiosis, exhibit one-sided gene conversion tracts.

6.10 GENE CONVERSION TRACT LENGTHS HAVE BEEN MEASURED IN BUDDING YEAST

Conversion tract lengths have been studied both by transforming broken or gapped plasmids sharing homology with a chromosomal donor or by using an inducible site-specific endonuclease to create a chromosomal DSB. For example, Jac Nickoloff's lab used a system in which an HO endonuclease cleavage site, present as a small insertion in the *ura3* gene, could be repaired by an allelic *ura3* gene with a XhoI 4-bp insertion 332 nucleotides away (X764) (**Figure 6.23**). About one-third of the repair events were short-patch gene conversions yielding *URA3* cells, whereas the remainder had longer gene conversion tracts that co-converted the X764 mutation leaving the cells Ura3⁻. Thus most mitotic gene conversion tracts extend at least 333 bp to one side of the DSB. A more detailed analysis has been carried out by introducing additional heterologies between the donor and recipient sequences surrounding the HO cleavage site, including a single base pair change in the HO cleavage sequence of the donor, to prevent HO cutting. With five heterologies within 100 bp of the cleavage site and others every ~100 bp, about 10–15% of the events converted only the single base pair of the mutant cleavage site itself, but the average conversion tract was about 700 bp long and nearly all were continuous (that is, there were few cases where a series of gene conversions were interrupted by unchanged site).

A detailed analysis of gene conversion by transformation has been carried out by Sue Jinks-Robertson using a restriction enzyme-cleaved plasmid with a gapped, linearized *HIS3* gene next to an intact *URA3* gene, transformed into a host carrying a 3′-truncated *his3* gene in which there are multiple single nucleotide polymorphisms (SNPs) scattered on either side of the DSB, with a total sequence divergence of about 2% (**Figure 6.24**). His⁺ transformants require gap repair and Ura⁺ His⁺ colonies could either have a tandem duplication of a functional and a truncated *HIS3* gene (an integration, Figure 6.24B) or have an autonomously replicating, repaired plasmid and the unaltered donor (Figure 6.24C). To facilitate measurement of the extent of heteroduplex DNA (hDNA) formation, the strain was defective in mismatch repair, deleted for *MLH1*.

We first consider the noncrossover (NCO) events illustrated in Figure 6.24C. It is evident that the great majority of the NCO events have hDNA only on one side of the DSB. On average the hDNA tract lengths are only a few hundred base pairs (but they cannot be more than about 300 bp because of the size of the *HIS3* gene). These results are consistent with the idea that most gene conversions that occur without an accompanying crossover arise from SDSA, which would predict that there should be hDNA only on the side of the repair event (Figure 6.22). Consistent with this interpretation, when there is a one-sided conversion tract all of the hDNA is found in the recipient (namely, the repaired plasmid). If repair involved a dHJ-containing intermediate, cleaved by HJ resolvases, there should be a patch of hDNA on both the donor and the recipient.

The small proportion of NCO events in which there is hDNA on both sides of the DSB could result from the dissolution of a dHJ by a helicase and topoisomerase or by branch migration or by cleavage of the strand invasion structure (see Figures 6.19, 6.20, and 6.21). Depending on how this HJ migrated across the region, one could get NCO outcomes with hDNA on both sides of the DSB. There are a small number of events where

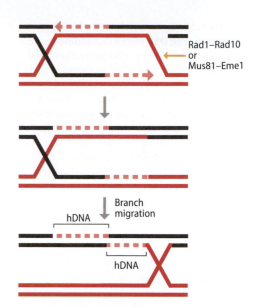

Figure 6.21 Nicking of a D-loop can produce one mobile HJ. Nicking of one strand of the D-loop (a half HJ) creates a single HJ that can branch migrate to place all hDNA on one side of an eventual crossover. Both the Rad1-Rad10 and the Mus81-Eme1 endonucleases have been implicated in cleaving such structures *in vitro*. (Adapted from Osman F, Dixon J, Doe CL & Whitby MC [2003] *Mol Cell* 12:761–774. With permission from Elsevier.)

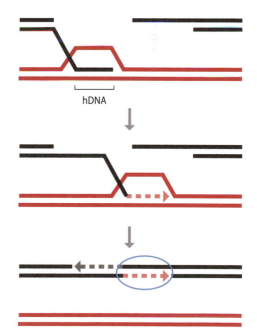

Figure 6.22 SDSA creates hDNA on one side of a DSB. The elongation and displacement of the invading strand during DSB repair removes the initial hDNA on the left side of the DSB and creates hDNA on the other side. Second-strand synthesis (*leftward*) restores the original genotype to the left side of the DSB.

Figure 6.23 Co-conversion of a flanking marker during DSB repair in budding yeast. An HO-induced DSB in a *URA3* gene disrupted by the HO cleavage site (and hence initially Ura⁻) will often be repaired to give a Ura⁺ gene conversion. However, in some instances, repair can result in the co-conversion of an adjacent mutation in the donor sequences to yield a Ura⁻ outcome. (From Weng YS, Whelden J, Gunn L & Nickoloff JA [1996] *Curr Genet* 29:335–343. With permission from Springer Science and Business Media.)

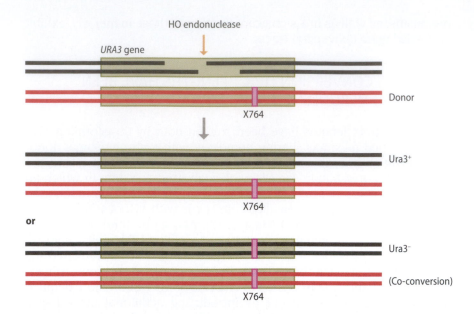

there is a full gene conversion of the sequences (shown in blue in Figure 6.24B and C). Such events could arise in several ways. First, there might be some residual mismatch repair, not dependent on Mlh1. Second, the small gap that was created in the plasmid might be enlarged, for example by losing part of 3'-ended ssDNA tail created by resection, so that a larger region was gap-repaired. Gap-enlargement might be a consequence of introducing the DNA into yeast by transformation, where the DNA was initially "naked" and not wrapped in chromatin. If this is the case, the events preferentially seem to occur on one side of the DSB.

About 15% of the noncrossover transformation events appeared to involve repair that did not leave any heteroduplex DNA, as if the repair involved very short regions of annealed DNA that did not include a heterology; but in some of these experiments the first SNP was only 20 bp from the end of the gap. It is an intriguing possibility that the "capture" of the second end in a DSB repair event might have much shorter homology requirements than what is needed to promote the initial strand invasion in SDSA. These may also represent cases where there was a *restoration* of hDNA back to the genotype of the recipient, in spite of the absence of Mlh1.

6.11 GENE CONVERSIONS ASSOCIATED WITH CROSSING OVER MAY OCCUR BY ALTERNATIVE MECHANISMS

In contrast to the one-sided events in noncrossover gap repair, when the *HIS3* gene conversion was accompanied by a crossing over, most of the events displayed more extensive hDNA on either side of the DSB (Figure 6.24B). This result would be expected from the dHJ mechanism, where fully ligated dHJs are cleaved in opposite planes (**Figure 6.25**). However, a closer inspection of the data suggests either that there is a highly biased way in which dHJs are resolved or that the crossovers stem from an intermediate with a single HJ. If there was a dHJ, then the pair of HJs should be able to be resolved into a crossover (CO) in two different ways. The junction on the left can be cleaved to give a NCO or a CO and—if we want to end up with a CO—the junction on the right can be conversely cleaved to give a NCO or a CO. If the sequence of cleavage is CO (left) NCO (right), then the hDNA should be on the inside of the region repaired by gap-repair (that is, closer to the plasmid sequences) (Figure 6.25), but if the resolution leads to a crossover downstream of the gap-repaired region

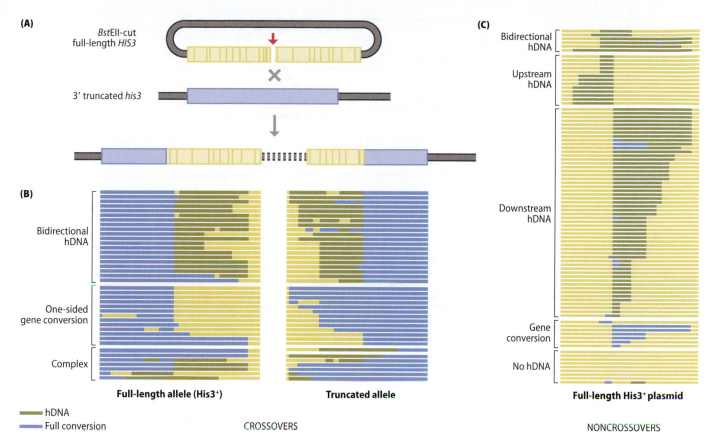

Figure 6.24 Gap-repair outcomes between a gapped, replicating plasmid and its chromosomal homologous site. (A) A plasmid carrying a full-length *HIS3* gene with an 8-bp deletion was cleaved by *Bss*HII restriction endonuclease that marks the deletion. This gapped plasmid was then transformed into a mismatch repair deficient (*mlh1Δ*) yeast strain carrying a small 3' truncation of *his3*, with which the DSB ends of the gapped plasmid can recombine. The two *HIS3* sequences differ at 2% of the sites. (B) Crossovers (CO), leading to the integration of the plasmid, were recovered and analyzed by DNA sequencing to determine the parental identity of each polymorphism (*yellow* or *blue*) and whether the transformant contained unrepaired heteroduplex (*green*) that will allow a single His+ transformant to contain both alleles. Each integrant is shown as a horizontal line, with hDNA (*green*) and full gene conversion (*blue*). A simple crossover would integrate the plasmid so that there would be *blue* sequences followed by *yellow* on the left *HIS3* copy and *yellow* followed by *blue* on the right *his3* copy. Most crossovers had a tract of hDNA (*green*) adjacent to the gap on both sides of the repair event, as expected for a dHJ-containing intermediate. A smaller number showed a single gene conversion tract and some were more complex. (C) Noncrossovers (NCO) were recovered as autonomously replicating plasmids that were converted to HIS3. Among NCOs, a small proportion had hDNA (*green*) on either side of the gap while most had hDNA only on one side. A small proportion exhibited gene conversion that could represent cases either where mismatch repair was complete even in the *mlh1Δ* mutant, or where there had been enlargements of the gap prior to repair. (Adapted from Mitchel K, Zhang H, Welz-Voegele C & Jinks-Robertson S [2010] *Mol Cell* 38:211–222. With permission from Elsevier.)

(that is, NCO [left] and CO [right]) then the hDNA should be on the out-side of the gap repair segment in the two *his3* regions. But only the first pattern was seen. This unanticipated result, which could be seen in the integration of a gapped plasmid but would not be evident in exchanges between two linear molecules, is hard to square with the idea of resolving a fully covalent dHJ intermediate in mitotic gap repair.

6.12 THERE ARE ALTERNATIVE WAYS TO GENERATE CROSSOVERS

Studies over the past few years have made it clear that crossovers can be generated without forming fully ligated dHJs. Crossover outcomes will be obtained if there is enzymatic nicking of the initial strand

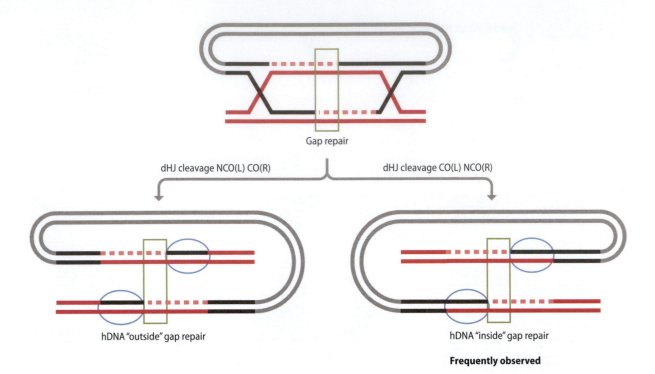

Gap repair

dHJ cleavage NCO(L) CO(R)   dHJ cleavage CO(L) NCO(R)

hDNA "outside" gap repair   hDNA "inside" gap repair

Frequently observed

Figure 6.25 Possible positions of hDNA in the two *his3* genes after integrative transformation via gap repair in a strain lacking mismatch repair. Repair leading to a crossover is assumed to proceed by a dHJ mechanism. Random resolution of each of the two dHJs (one resolved as a NCO and the other resolved as a CO) should produce two types of integration events in which the hDNA is either on the outside, flanking the regions of new DNA synthesis (*dotted lines*), or on the inside of the repaired region. In fact, nearly all the integrants show the "inside" pattern. (Adapted from Mitchel K, Zhang H, Welz-Voegele C & Jinks-Robertson S [2010] *Mol Cell* 38:211–222. With permission from Elsevier.)

invasion structure, as shown in Figure 6.21. This alternative mechanism, first suggested by Matthew Whitby, has been invoked to explain the importance in some gene conversion studies of the Mus81 endonuclease, which exhibits very poor cleavage of fully ligated dHJ but can readily cleave nicked, or incomplete, HJs. We will look at HJ resolvases in more detail in Chapter 15. The result of cleaving the two half HJs in the initial structure will result always in a crossover (**Figure 6.26**). Whether the strand invasion occurs initially on left or the right side of the DSB, during the integration of the *his3* sequences, the hDNA will be found on the insides of the gap repaired region (Figure 6.26), consistent with the experimental results.

These data revealed other surprises. A significant proportion of crossovers apparently have hDNA only on one side of the DSB. One-sided events are not predicted by the canonical dHJ model or even in the nicked version of the model just described, but can be explained in at least two ways. First, it is possible that there is branch migration of one or both Holliday junctions after hDNA formation and before their resolution to give a crossover (Figure 6.20). This idea has strong support from the molecular analysis of meiotic recombination, as we shall see in Chapter 14. Alternatively, it is possible that SDSA can be accompanied by crossing over if the migrating D-loop containing the elongating new DNA synthesis is captured by the second end of the DSB before the nascent strand is completely dissociated (**Figure 6.27**). These events create a dHJ far from the site of the DSB, with essentially all of the hDNA on one strand.

6.13 RECOMBINATION BETWEEN SEQUENCES OF LIMITED HOMOLOGY LENGTH CONSTRAINS MEASURING hDNA

In the *HIS3* transformation experiment discussed above, the estimates of hDNA length in the mismatch repair-deficient strain proved to be similar to the lengths of full gene conversion tracts that were obtained by

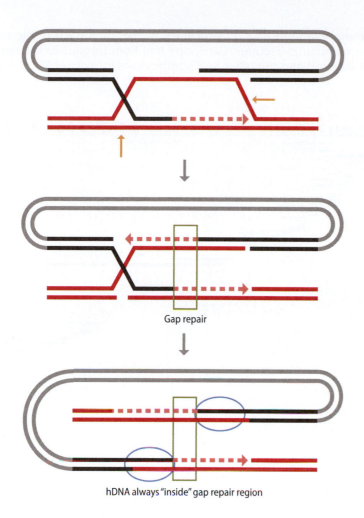

Gap repair

hDNA always "inside" gap repair region

Figure 6.26 Cutting of half HJs will produce a crossover in which hDNA will lie inside the site of gap repair. The initial stage of dHJ repair contains two partial, or nicked, HJs. Despite their different appearances when drawn as flat objects, they are equivalent except for the orientation of the nick. Cleavage of both nicked HJs will always result in a crossover accompanying DSB repair. (Adapted from Osman F, Dixon J, Doe CL & Whitby MC [2003] *Mol Cell* 12:761–774. With permission from Elsevier.)

looking at gene conversions in a wild-type strain. This suggests that the outcomes were not strongly perturbed by removing the mismatch repair system; however, this remains a point of contention in the literature. The mismatch repair system also plays an important role in discouraging recombination between *homeologous* sequences, at the step of strand invasion (see Section 6.15), so some outcomes may be biased by the absence of the MMR system.

Another limitation in many studies of mitotic gene conversion is that they involve transformation or else ectopic recombination in which the donor and recipient share only a few kilobases of homology. Thus hDNA cannot extend very far before it passes the point where the sequences are unrelated. To avoid this complication, the Petes lab examined spontaneous mitotic crossovers in a diploid that was heterozygous for the *can1–100* mutation located near the end of a chromosome arm. Mitotic crossing over will (depending on the segregation of chromatids at mitosis) produce *can1-100/can1–100* cells resistant to canavanine. However, to assure that the events being examined were reciprocal crossovers and not some other type of repair, such as BIR, on the opposite homologous chromosome the strain also carried a *SUP4-o* allele in place of *CAN1* sequence. *SUP4-o* is a tRNA suppressor of the *can1–100* ochre mutation. *SUP4-o* also suppresses another nonsense mutation, *ade2-1* (**Figure 6.28A**). So the initial diploid is canavanine-sensitive (Can^S) and Ade2$^+$. Ade2$^+$ colonies are white, while *ade2-1* mutant colonies are red. The diploid carried 34 *single nucleotide polymorphisms* (SNPs) in the 120 kb between *can1-100* and the

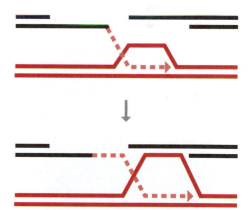

Figure 6.27 Crossovers can occur if an SDSA intermediate is captured into a dHJ structure. The moving D-loop in SDSA repair could be captured by the complementary strand at the second DSB end and be converted to a dHJ structure. The position of the dHJ will depend on where along the resected DNA end the annealing takes place. (Adapted from Pâques F & Haber JE [1999] *Microbiol Mol Biol Rev* 63:349–404, and from Ferguson DO & Holloman WK [1996] *Proc Natl Acad Sci USA* 93:5419–5424. With permission from the American Society for Microbiology and from the National Academy of Sciences, respectively.)

centromere. If there is a reciprocal exchange, then each of the reciprocal recombinants becomes Can^R, one half homozygous to the complete *CAN1* deletion and the other half homozygous for *can1-100*. In addition,

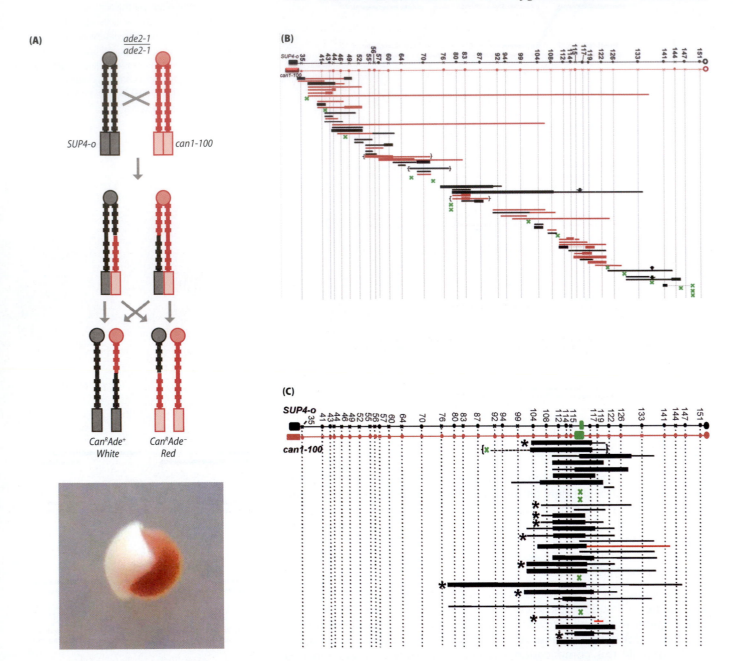

Figure 6.28 A budding yeast assay to examine loss of heterozygosity arising from reciprocal crossing over. (A) A diploid strain homozygous for *ade2-1* and heterozygous for *SUP4-o* and *can1-100* inserted at allelic positions near the end of one chromosome will produce a fully Can^R red/white sectored colony (*below*) if there is a reciprocal crossover between the distal markers and the centromere. The diploid is heterozygous for 34 SNPs along the chromosome arms. (B) Spontaneous gene conversion events. Positions of gene conversion tracts, which are accompanied by reciprocal crossing over, to generate the Can^R red/white sectored colonies. Lines indicate the extent of gene conversion and the color of the line indicates which sequences have been introduced. Thicker *black* lines show that both sister chromatids display evidence of a gene conversion event, suggesting that a DSB was created prior to DNA replication. There

are cases where no adjacent marker was gene converted or, in some instances, where the crossover was separate from the gene conversion tract (*green*). (C) Reciprocal crossovers in a strain carrying GAA repeats that cause localized chromosome breaks during DNA replication. The events all overlap the site where GAA repeats (*green*) are inserted and nearly all of them reflect gene conversions in which the break appears to have occurred on the chromosome with the longer, more fragile GAA tract. A much larger proportion of events involve gene conversion on both sister chromatids, indicative of a DSB forming before the chromatids had fully replicated. (A, B, adapted from Lee PS, Greenwell PW, Dominska M et al. [2009] *PLoS Genet* 5:e1000410. C, adapted from Tang W, Dominska M, Greenwell PW et al. [2011] *PLoS Genet* 7:e1001270.)

the sector lacking *SUP4-o* becomes red because the *ade2-1* allele is not suppressed. Thus the interesting events are red/white sectored colonies that grow in the presence of canavanine.

Among spontaneous mitotic crossovers (**Figure 6.28B**), the great majority showed gene conversion of at least one marker and the mean conversion tract length among those with at least one site converted was 11.7 kb. This is clearly quite different from the conversion tract lengths of much less than 1 kb seen when gene conversions between ectopic sequences of a few kilobases were analyzed. Here, we do not know whether the lesion that caused the crossing over was a DSB, but similar results have been obtained both after inducing damage with ionizing radiation and by the insertion of GAA repeats, which cause replication fork stalling and chromosome breakage (**Figure 6.28C**); so it appears that the rules apply to DSB-induced events. When one chromosome contains a much longer GAA tract than the other, the events are strongly biased toward that chromosome being the recipient (that is, the one that suffered the DSB); by contrast the spontaneous events in Figure 6.28B show that both homologs were equally likely to be used as the template. Moreover, in this case, the great majority of the events exhibit evidence of gene conversion on *both* sister chromatids, but the changes are different on the two. These findings argue that the DSB caused by the GAA repeats occurs at the time of replication so that both sisters end up broken, after which they repair independently. Further analysis suggests that one repair event involves an encounter with the homologous chromosome and is accompanied by a crossover while the second repair event may well use the initially repaired sister as the template.

6.14 GENE CONVERSION TRACT LENGTHS ARE VERY DIFFERENT IN MITOTIC AND MEIOTIC YEAST CELLS

A survey of the literature suggests that gene conversion tracts in mitotic cells are much longer than in meiotic ones, but there have been few direct comparisons, as the manner of initiating recombination in meiosis and vegetative cells is usually quite different. One experiment that attempted to overcome this problem expressed the site-specific HO endonuclease under the control of a meiosis-specific promoter so that the same DSB could be studied at the same position in the same locus as in mitotic cells. The DSB was created in a *leu2* gene next to which a 1.7 kb *ADE1* gene was inserted. *ADE1* lies about 1 kb to the right of the DSB (**Figure 6.29**). Repair by gene conversion used a homologous chromosome carrying *LEU2* without the insertion as the donor. In mitotic cells 48% of the DSB repair events exhibited co-conversion of the *ADE1* gene (that is, the cells lost the insertion). In contrast, among meiotic cells only 6% of the cells showed such co-conversion. The most likely factor that would account for this large difference is that 5′ to 3′ resection of the DSB ends is considerably more extensive in mitotic than meiotic cells. It is, however, possible that there might be differences in the repair of heterologies included in hDNA, if, for example, there were less bias in repairing them in favor of the strand without the insertion.

Another comparison of gene conversion tract lengths comes from the Petes lab's analysis of crossing over between the *can1* gene and its centromere. As noted above, the mean mitotic conversion tract length was 11.7 kb. For the same diploid, passaged through meiosis, a much smaller proportion of crossovers had a detectable gene conversion (that is, the repair often occurred without a visible change in any of the SNPs) and, when at least one marker was gene converted, the mean length was only 4.7 kb, far shorter than that seen in mitotic cells.

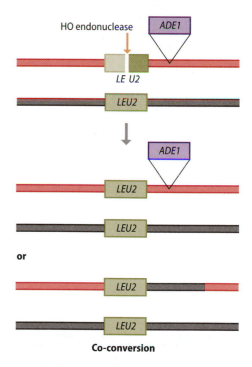

Figure 6.29 A test of co-conversion of a flanking marker. An HO-induced DSB within the *leu2* gene can be repaired in a diploid from an allelic *LEU2* locus. If the break is enlarged, either as a single strand by resection or by creating a gap when both strands are removed, an adjacent heterology (the insertion of *ADE1* more than 1 kb from the DSB) can be lost by co-conversion. Note that such events cannot occur by ectopic recombination, where repair would be limited to the length of the duplicated sequences. (From data in Malkova A, Ross L, Dawson D et al. [1996] *Genetics* 143:741–754. With permission from the Genetics Society of America.)

6.15 A COMPLICATION: HETEROLOGIES INTRODUCE UNCERTAINTY

The length of gene conversion tracts will obviously depend on having enough well-spaced markers at a distance from the DSB to monitor their conversion, but not so many as to perturb the outcome. Recall that sequences that diverge significantly from each other are termed homeologous rather than homologous. The level of divergence that makes a sequence homeologous is somewhat arbitrary; the term is often applied to sequences derived from different species or to rapidly diverging repeated sequences in the same organism, which may differ by as little as a few percent or by as much as 25%. Recombination between homeologous sequences is less successful than between identical sequences.

One way in which heterologies cause a reduction in repair is by *heteroduplex rejection*, in which heteroduplex created during strand invasion is unwound and removed. In budding yeast, plasmid gap repair between sequences that are 2% mismatched causes a one-third reduction in repair. Similarly, a 3% difference in 205-bp flanking regions reduces SSA to only 15% of the wild-type level. In both of these cases, repair with the homeologous sequences is restored to near wild-type levels by mutations in the mismatch repair genes Msh2 and Msh6. In these situations it appears that these MMR proteins act in concert with the Sgs1 (BLM) helicase to unwind the annealed regions, as opposed to promoting single-strand cleavage and excision of the sequences. The consequence of heteroduplex rejection is that too many mismatches may cause the rejection of long heteroduplex regions and favor recovery of only short conversion tracts.

As noted above it is possible that MMR could attack heteroduplex regions by promoting nicking and excision of a region containing a mismatch. This seems to occur during yeast meiosis, where having as few as one mismatched site per kilobase in a 9-kb region proved to be sufficient to provoke a second wave of recombination events. If hDNA includes two mismatches, separated by roughly 1 kb, two independent mismatch repair complexes can form. If these complexes initiate repair by excising sequences on one strand (prior to filling-in the region with the "correct" sequence) and if they initiate this process at two sites on opposite strands, resection on two different strands can lead to the formation of a DSB (**Figure 6.30**). These MMR-generated DSBs will promote a second recombination event, potentially obscuring our ability to monitor where the initial event happened and over what distance.

It should be possible to avoid these complications by monitoring heteroduplex formation in a MMR mutant; however, the meiotic studies that have used this approach led Kolodner's lab to conclude that the MMR proteins may play an important active role in limiting the length of heteroduplex itself. Therefore, eliminating Msh2 or another MMR protein will lead to the recovery of apparently longer heteroduplex regions. One way this might happen is that heteroduplex further from the DSB could be partially unwound, shortening the heteroduplex tract. Alternatively, the MMR proteins might bind to mismatches in a way that blocks further heteroduplex extension. The idea that MMR proteins could limit heteroduplex formation was also seen in Jinks-Robertson's study of mitotic gene conversion tracts within 350-bp segments with 94% identity. The lengths of conversion tracts appear to be 50% longer when Msh2 or Msh3 was absent. However, there was much less effect of deleting Mlh1 in the *HIS3* integration experiment described above in Figure 6.24.

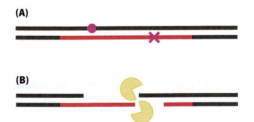

Figure 6.30 Multiple mismatches in hDNA can provoke a second round of recombination. A secondary DSB can arise during recombination between multiply mismatched substrates if hDNA contains two mismatches separated so that they are independently corrected (A), and the exonucleases that carry out resection during mismatch repair converge to create a DSB (B). (Adapted from Borts RH & Haber JE [1987] *Science* 237:1459–1465. With permission from the American Association for the Advancement of Science.)

6.16 THERE IS COMPETITION AMONG POSSIBLE GENE CONVERSION DONORS

Martin Kupiec devised a clever competition experiment in budding yeast to ask if the sequences closest to the 3′ ends of the DSB are more likely to locate a homologous template than sequences further from the ends. An HO-induced DSB was created in the middle of a *ura3* gene, which itself was flanked by the 5′ and 3′ parts of the *lys2* gene (**Figure 6.31**). Repair of the DSB in a diploid could occur by one of two mechanisms. First, gene conversion could recruit an ectopic *URA3* gene as the donor. This donor contained the same inserted HO cleavage site except it had a single extra base-pair mutation that prevented HO cutting. A second way to repair the DSB involved bypassing all the *ura3* homology close to the DSB and using an ectopic 4.9 kb *LYS2* gene as the donor. When the HO-cut *ura3* gene was 1.2 kb, only about 10% of the repair events used the ectopic *ura3* donor as the repair template, the remaining 90% ignored the 0.6 kb of homology adjacent to the DSB on each side and used instead the larger homologies of the *LYS2* gene. When the size of *ura3* surrounding the DSB was increased to 5.6 kb, repair using the *ura3* allele increased to 60%, but still 40% of the time the homologous sequences nearest to the DSB were ignored even though they presumably had to become single-stranded sooner than the flanking *lys2* sequences and even though gene conversion with *LYS2* would require the additional step of clipping off the nonhomologous *ura3* ssDNA before new DNA synthesis could start. Recently Grzegorz Ira's lab has shown that mutations that greatly reduce the rate of 5′ to 3′ resection (for example, *sgs1Δ exo1Δ*) dramatically increase the use of the nearer homologous sequences.

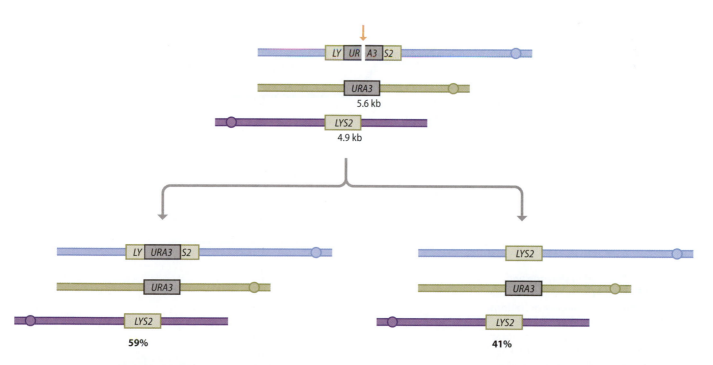

Figure 6.31 Sequences close to a DSB are not necessarily used to repair a DSB if there are other competing homologous sequences further from the DSB site. An HO-induced DSB in a cloned *LY::URA3::S2* sequence in a haploid strain can be repaired by ectopic gene conversion either by recombining with *URA3* sequences, using the homology immediately adjacent to the DSB, or by recombining with a similarly sized *LYS2* locus, whose homologous sequences are more than 1000 bp away on either side from the DSB. Surprisingly the two outcomes are equivalent. (From Inbar O & Kupiec M [1999] *Mol Cell Biol* 19:4134–4142. With permission from the American Society for Microbiology.)

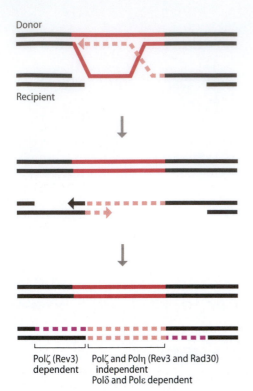

Figure 6.32 Mutations arise at high frequency in the vicinity of a DSB. In ssDNA regions that need to be filled in after DSB repair (*purple dashed lines*), there is a high level of error-prone DNA synthesis by DNA Polζ (Rev3). In a region where both strands are newly synthesized (*pink dashed lines*) the rate of mutation is also elevated but is Polζ independent. (From data in Holbeck SL & Strathern JN [1997] *Genetics* 147:1017–1024; Yang Y, Sterling J, Storici F et al. [2008] *PLoS genetics* 4:e1000264; and Hicks WM, Kim M & Haber JE [2010] *Science* 329:82–85.)

6.17 GENE CONVERSION IS MUTAGENIC: EVIDENCE FOR REDUCED DNA POLYMERASE PROCESSIVITY

Although gene conversion without crossing over should be the most conservative mode of repairing a DSB, it is not without risks. In budding yeast, both in the resected regions that are filled in at the end of DSB repair and in the region where both DNA strands are newly synthesized, the rate of mutation increases ~1000 times relative to the spontaneous rate. (Still, mutations only arise about once in 100,000 repair events.) In ssDNA regions adjacent to the DSB, the rate of mutation is strongly affected by eliminating DNA polymerase zeta (Polζ), although deletion of Polζ does not affect the efficiency of repair (**Figure 6.32**). However, in the region where both strands are newly synthesized, the mutations are Polζ independent but apparently involve both DNA polymerases Polδ and Polε. Most of these mutations are single base pair changes, but about 40% have a distinctive "signature" compared to spontaneous events. The unusual events can all be explained by an increase in template switching and template slippage during repair. First, there are many examples of −1 frameshifts, virtually all of which arise in homonucleotide runs (for example, a CCCC sequence is miscopied as CCC). Second, there are complex mutations that both mutate a base and add or subtract a base; these can be explained by dissociation of the DNA polymerase and self-annealing of the partially replicated strand to quasi-palindromic sequences, leading to the copying of a few bases before the structure apparently again dissociates and realigns with the original template (**Figure 6.33**). Finally, there are remarkable template switchings that jump to another chromosome, where there was (fortuitously) a 72%-identical copy of the *URA3* reporter gene used in these experiments (**Figure 6.34**). Here the newly synthesizing strand apparently dissociated from its template and

Figure 6.33 Replication slippage and transient formation of quasi-palindrome sequence can lead to complex mutations. The partly copied, replicating strand must dissociate from its initial template and form an incompletely self-complementary cruciform structure. Extension of the elongating end "perfects" the palindrome but introduces multiple changes into the new sequence that then resumes copying the original template. (Adapted from Hicks WM, Kim M & Haber JE [2010] *Science* 329:82–85. With permission from the American Association for the Advancement of Science.)

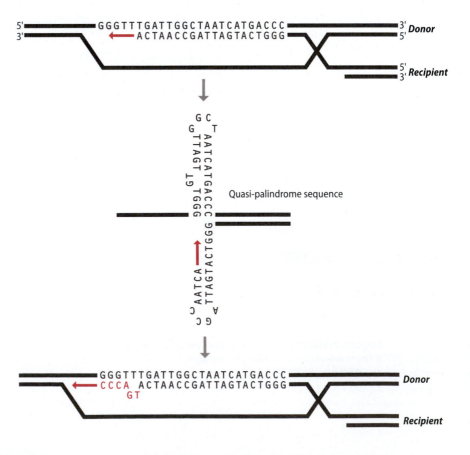

invaded the homeologous *ura3* sequences on another chromosome, at regions of microhomology (2–11 bp), but then had to again dissociate to complete DSB repair at the initial donor. Remarkably, all of these instabilities are eliminated in a proofreading-defective mutation of DNA Polδ that lacks 3′ to 5′ exonuclease activity. In this mutant the number of single base-pair substitutions increases several fold, but this increase does not account for the failure to see the template switches. This result suggests that the wild-type Polδ enzyme is responsible for these template jumps. An *in vitro* biochemical study by Peter Burgers has demonstrated that a proofreading-defective Polδ is actually more processive than the wild type. Whereas a wild-type enzyme that polymerizes from the end of a short primer is blocked from further polymerization by encountering an oligonucleotide paired to the template strand, the proofreading-defective enzyme can displace the blocking oligonucleotide and copy the template strand to the end. DNA Polε must also play some role in new synthesis, possibly copying the second strand, because a similar proofreading mutation also causes a change in the spectrum of mutations, but here the "signature" mutations are still present and the most notable change is the appearance of +1 frameshifts, again in homonucleotide runs. This high rate of mutation could be important in explaining how tumor cells can acquire so many mutations in a small number of cell divisions. It is important to note that the mutation rate is only elevated in the locus undergoing DSB repair and not at unlinked genes.

6.18 GENE CONVERSION FREQUENCIES ARE INFLUENCED BY CHROMOSOMAL POSITION

The arrangement of recombining partners is strongly influenced by chromosomal arrangement and position. First, in diploids intrachromosomal recombination is significantly more efficient than interchromosomal recombination (**Figure 6.35**). A diploid budding yeast with *leu2-K* and *leu2-R* mutations in *allelic* positions yield spontaneous Leu+ recombinants at a rate of 1.2×10^{-7}. If the diploid carried the same two alleles in ectopic locations on the same chromosome (in *cis*), while the homolog had a *leu2-K,R* double mutant that could not yield Leu+ recombinants, the rate was nearly fourfold higher. If the two recombining alleles were in *trans*, then the rate was fivefold lower. In haploid strains one can see again that intrachromosomal recombination rates are usually much higher than interchromosomal events (here the interchromosomal events are between homologous sequences inserted into generally nonhomologous chromosomes). A similar rule applies when the initiating lesion is a DSB (**Figure 6.36**). It is important to recognize that these generalizations may not explain the rate of recombination between two sequences if one or both of them are strongly constrained, either by being near telomeres (and tethered to the nuclear envelope) or near centromeres or when the sequences are heterochromatic.

6.19 GENE CONVERSION AND GENE CONVERSION TRACTS HAVE BEEN DEFINED IN OTHER MODEL ORGANISMS

Experiments carried out in a variety of model systems have reinforced our understanding of gene conversion mechanisms. Below are examples drawn from different organisms.

S. pombe

The HO endonuclease system has been adapted to work in fission yeast, by placing the HO gene under control of the thiamine-repressible *nmt1*

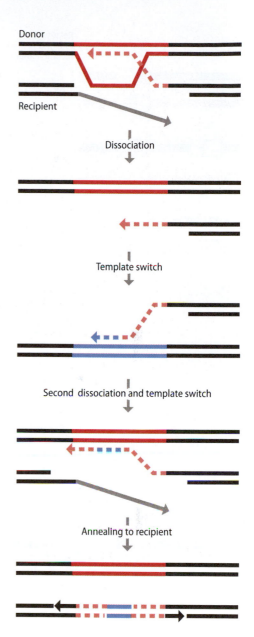

Figure 6.34 Interchromosomal template switches involving homeologous sequences. Dissociation of the partly copied strand from the donor template can result in strand invasion or DNA polymerase-mediated switching to a homeologous template on another chromosome. Completion of the repair event demands that there be a second dissociation and encounter between homeologous strands of the extended, newly copied strand and the donor. The junctions at the entry and exit points contain microhomologies of 2–11 bp. A nonhomologous tail removed during repair is shown in *gray*. (Adapted from Hicks WM, Kim M & Haber JE [2010] *Science* 329:82–85. With permission from the American Association for the Advancement of Science.)

Relative
rate of Leu⁺
recombinants

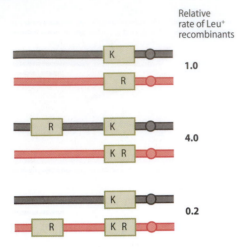

1.0

4.0

0.2

Figure 6.35 Competition between inter- and intrachromosomal donors during spontaneous recombination. The rate at which Leu⁺ spontaneous recombinants arise in diploids carrying *leu2-K* and *leu2-R* alleles is compared with cases in which only an ectopic intrachromosomal or interchromosomal recombination can produce Leu⁺ cells. A cloned *leu2-R* segment is inserted 22 kb away on one or the other homologous chromosome and can recombine with *leu2-K* at its normal location. In these instances the second allelic copy of *leu2* carries the double mutant, *leu2-k,R*. (From data in Lichten M & Haber JE [1989] *Genetics* 123:261–268. With permission from the Genetics Society of America.)

promoter. Induction is delayed by many hours so it has not been possible to analyze the recombination events kinetically, as in budding yeast, but one can characterize the types of recombination events induced by a site-specific DSB. Tim Humphrey created a DSB in a truncated version of chromosome III that is disomic with a normal chromosome III and maintained because the two alleles of *ade6* intragenically complement each other, so that cells are Ade⁺ (**Figure 6.37**). The DSB can be repaired by gene conversion, although apparently less efficiently than in budding yeast because about 20% of cells suffer loss of the broken mini-chromosome. The effects of deleting several recombination proteins, identified by their *S. cerevisiae* names, are shown (**Table 6.1**). As in budding yeast, Rad51 and its paralog Rad55 are needed for probably all gene conversion, although the level of remaining repair events with the same phenotype in their absence is significant. Eliminating Ku70, which is required for most NHEJ events, has little effect by itself, but Ku70 clearly plays a role in a Rad51-independent pathway that creates large deletions which remove both the HO cleavage site and part of the adjacent G418-resistance gene events; consequently the *rad51Δ ku70Δ* double mutant eliminates nearly all repair.

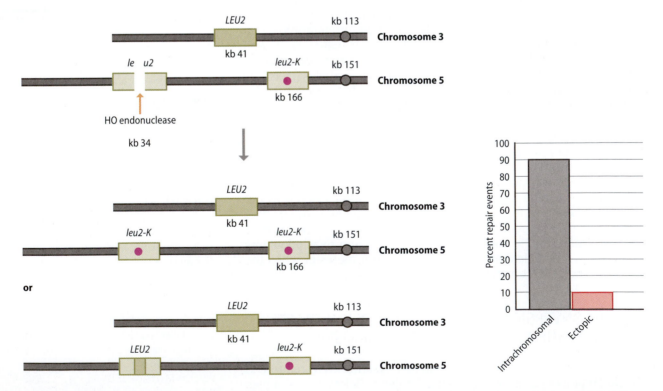

Figure 6.36 Competition between inter- and intrachromosomal donors in DSB repair. An HO-induced DSB in a cloned copy of the budding yeast *LEU2* can be repaired either by *LEU2* sequences on a different chromosome or by an intrachromosomal donor carrying *leu2-K*, which is altered at the same site where the HO cleavage site is inserted.

The locations of the sequences are given as distances from the left telomere. Intrachromosomal recombination is strongly favored over an interchromosomal repair. (Adapted from Li J, Coic E, Lee K et al. [2012] *PLoS Genet* 8:e1002630.)

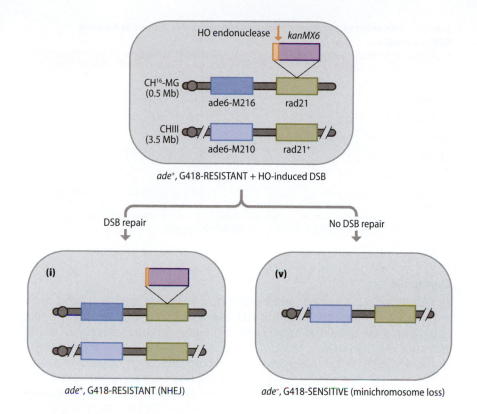

ade⁺, G418-RESISTANT + HO-induced DSB

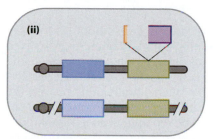

(i) *ade*⁺, G418-RESISTANT (NHEJ)

(v) *ade*⁻, G418-SENSITIVE (minichromosome loss)

(ii) *ade*⁺, G418-SENSITIVE (large NHEJ deletions)

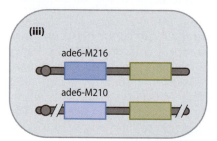

(iii) *ade*⁺, G418-SENSITIVE (gene conversion)

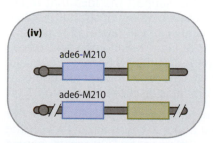

(iv) *ade*⁻, G418-RESISTANT (gene conversion) with co-conversion of *ade6*

Figure 6.37 HO-induced gene conversions in *S. pombe*. A fission yeast strain carrying a minichromosome composed of part of the sequences near the centromere of chromosome III (Ch¹⁶) into which the nonhomologous HO cleavage site (from budding yeast *MAT***a**) and the adjacent *kanMX6* sequences are inserted. The cell maintains this nonessential chromosome by intergenic complementation of two *ade6* alleles (so the cells are Ade⁺). An HO-induced DSB can be repaired in several ways. NHEJ will often ablate the HO cut site but not affect the adjacent *kanMX6* gene (G418ᴿ) (i), but some NHEJ events will result in large deletions removing both the cut site and kanMX6 activity (ii). Short gene conversion tracts will also result in Ade⁺ G418ˢ cells with the loss of the insert (iii). Gene conversion without crossing over is shown but crossovers between Ch III and Ch¹⁶, producing a reciprocal chromosomal rearrangement, can also be demonstrated (iv). Long-tract gene conversions will co-convert *ade6*-M210 in place of *ade6*-M216. Finally, lack of repair of the DSB will result in chromosome loss (v). (Adapted from Prudden J, Evans JS, Hussey SP et al. [2003] *EMBO J* 22:1419–1430. With permission from Macmillan Publishers, Ltd.)

Table 6.1 Role of homologous recombination and end-joining genes in repairing a DSB in *S. pombe*

S. pombe homolog	% gene conversion
Wild type	63.7
rad51Δ	2.8
rad55Δ	5.0
rad50Δ	25.6
ku70Δ	62.5
rad51Δ ku70Δ	0.2

The effect of deleting several different recombination genes on the DSB repair assay shown in Figure 6.37. Genes are identified by their homology to *S. cerevisiae* genes. In wild-type cells nearly all Ade+ G418^S cells arise by gene conversion. The Ade+ G418^S cells remaining when Rad51 or Rad55 are eliminated are most likely large NHEJ deletions (Figure 6.37, panel ii) and are eliminated when the Ku70 gene is ablated.

Drosophila

In fruit flies DSB-induced gene conversion has been studied both by the expression of either the I-SceI or HO endonucleases, but much work has also been done by taking advantage of the DSB left by the excision of the transposable P element. Most of these studies have been carried out by inducing DSBs in the male germ line, where there is no meiotic recombination. The site-specific endonucleases can be expressed under the control of a heat shock-inducible promoter. Similarly, P-element excision can be induced by inducible expression of the P-element transposase gene that is present in a nontransposable construct.

P-element excision leaves 17-nt 3′ overhanging ends about 6 kb apart. These breaks can be repaired by end joining but most often they are fixed by homologous recombination using a sister chromatid in which the P element has not been excised (**Figure 6.38**). Using a sister chromatid as a template, repair events will usually be "silent" and simply restore the P element to its original state, though interesting exceptions are discussed below. P-element-induced DSBs can also be repaired from a homologous chromosome or a donor located ectopically so that gene conversions can be recognized by the transfer of DNA marks. Of course DSBs can also be repaired efficiently by SSA if there happen to be homologous sequences flanking the excision point.

An important experiment showing that repair can occur between sister chromatids made use of a set of 375-bp repeats placed inside the P element (Figure 6.38). P-element excision occurs in male meiosis, where normal meiotic crossing over is absent. Excision apparently occurs after chromosome replication so that the P element is carved out of one chromatid but leaves the sister chromatid as a donor to repair the DSB. The repair events are recovered in offspring so there isn't any way to look at the sister chromatid to be sure what occurred. What was found were P elements that had either more repeats or fewer repeats than the original number. This result was interpreted to support the idea that repair involved an SDSA mechanism in which the annealing of strands resulted in expansions or contractions of the number of the repeated regions. A version of this experiment was carried out later in budding yeast, where it was possible to recover both the donor and the recipient. The results strongly supported the idea that SDSA was the predominant repair mechanism in mitotic yeast cells. This experiment is discussed in Section 7.11.

William Engels has examined P-element repair events in diploids with five to eight polymorphisms encompassing about 2 kb on either side of the P element. In both locations, loss of the P element by DSB repair showed gene conversion tracts of about 1400 bp. Most conversion tracts were

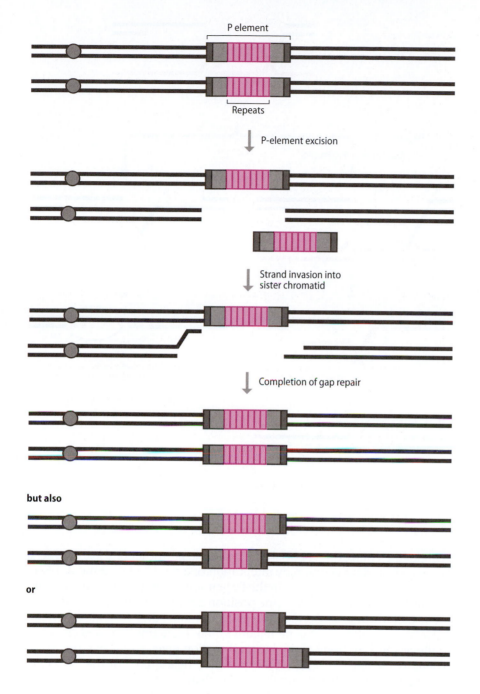

Figure 6.38 Repair of a P-element excision in *Drosophila* provides evidence of an SDSA mechanism. A modified P element containing an array of seven tandemly repeated 375-bp 5S genes is excised from one sister chromatid and the resulting gap is repaired by homologous recombination. Frequently the number of repeats found after repair (presumably in the recipient locus) are either fewer or more than in the template. An analogous experiment in budding yeast, but involving ectopic recombination, showed more directly that all the rearranged copies were indeed in the recipient, with the donor remaining unaltered. These data support an SDSA mechanism in which the donor is unchanged. (Adapted from Pâques F, Bucheton B & Wegnez M [1996] *Genetics* 142:459–470. With permission from the Genetics Society of America.)

bidirectional and nearly all were continuous. Engels also compared the gene conversion tracts created by P-element excision with those created by expression of the HO endonuclease in flies. The results are essentially superimposable, suggesting that there are no special features of either creating the DSB by excision or by using a site-specific endonuclease. The predominance of bidirectional gene conversion tracts suggests that SDSA may not be the predominant mechanisms of DSB repair in these cells. Alternatively, strand invasion may be rapidly accompanied by mismatch correction of heterologies before the primer-extended invading strand is displaced from the template, so that the gene conversion tract covers both sides of the DSB (**Figure 6.39**).

Kent Golic and his colleagues have perfected the use of the site-specific *I-SceI* endonuclease in *Drosophila* to carry out experiments analogous to those that have been performed in yeasts and mammals. The endonuclease

Figure 6.39 Rapid mismatch correction of hDNA can create bidirectional gene conversion tracts. Mismatch repair may act to correct the invading strand before it has a chance to dissociate from the template during SDSA repair (A). Correction almost always uses the unbroken (donor) strand as the template to correct the heterology in the invading strand. As repair proceeds there will be gene conversion on both sides of the DSB, in contrast to what occurs if there is no rapid mismatch correction at the strand invasion step (B) (see also Figure 6.22). (Adapted from Haber JE, Ray BL, Kolb JM & White CI [1993] *Proc Natl Acad Sci USA* 90:3363–3367. With permission from the National Academy of Sciences.)

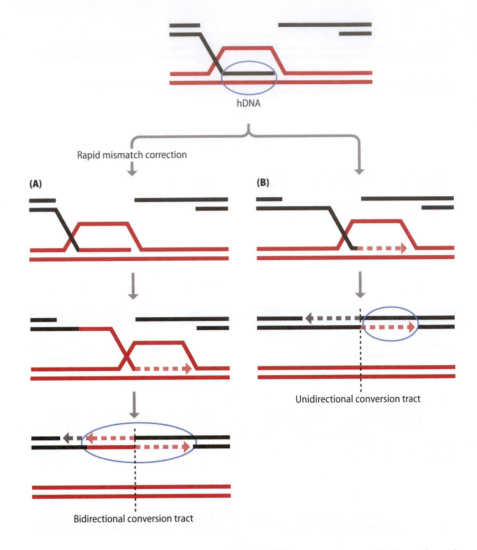

is under the control of a heat-shock-inducible promoter and is introduced at a quasi-random position by integration of a modified P element. I-SceI has no natural cleavage sites in the fly genome, so it will only cut an artificial 18-bp site integrated at some position along a chromosome. Another modified P element, inserted elsewhere, contains an I-SceI cleavage site adjacent to a white (w^+) gene (**Figure 6.40A**). Induction of I-SceI cleavage in the absence of homologous recombination gives rise to various sized deletions created by NHEJ, some of which delete part of the w^+ gene. I-SceI cleavage is robust enough to see cleavage in DNA extracted from flies and sufficient to create alterations by small deletions of the I-SceI site in about 25% of progeny. If such a mutant deleted for the cut site is crossed with a fly with the original (I-SceI-cuttable) construct, it is easy to demonstrate that I-SceI will produce a very high level of gene conversions, repairing the I-SceI cleavage site with the allelic copy, introducing precisely the same alteration of the cut site in the donor chromosome (**Figure 6.40B**).

Although an efficiency of more than 60% repair by allelic gene conversion would be expected in yeast, it is surprising to find such a high level in flies, with their much larger genome. Indeed, in mammals such allelic repair of I-SceI cleavage is an inefficient process. Most likely the high rate of success of allelic repair in flies is that their somatic chromosomes are paired throughout most of the cell cycle, so they are nearly as well lined up as would be sister chromatids. In the most elaborate construct studied by Rong and Golic (**Figure 6.41**), repair could occur by NHEJ, by

(A)

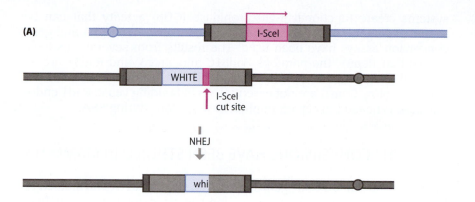

(B)

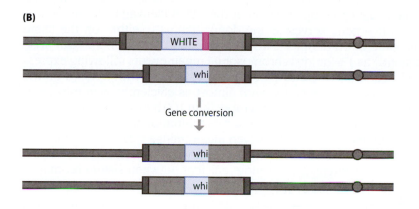

Figure 6.40 I-SceI-based DSB repair assays in *Drosophila*. (A) An induced DSB can sometimes be repaired by NHEJ, leaving a gap around the original I-SceI cleavage site and producing a fly lacking a functional *white* gene. (B) This new deletion allele could then be crossed with a fly carrying the original locus, allowing the demonstration that interchromosomal DSB repair could occur, by gene-converting the cut locus to the *w* deletion allele. (From Rong YS & Golic KG [2003] *Genetics* 165:1831–1842. With permission from the Genetics Society of America.)

SSA of homologous sequences flanking the DSB, or by allelic gene conversion. Intrachromosomal SSA and interchromosomal gene conversion each accounted for more than 40% of the repair events, with the remaining 12% split between imperfect NHEJ and cases where the target was either never cut or was cut and accurately rejoined. The use of I-SceI to promote accurate gene targeting in flies will be discussed in Chapter 10.

6.20 HOMOLOGOUS RECOMBINATION CAN BE ANALYZED IN PLANTS

DSB repair processes in several plant systems have proven to be very similar to what has been seen in other model systems. Both spontaneous and I-SceI-induced recombination have been studied. The main assay

Figure 6.41 A DSB assay system on a *Drosophila*. Induction of a DSB in an engineered P-element system shows that intrachromosomal gene conversion is as frequent as intrachromosomal SSA events. The diploid (left) carries an h+- and Sb+-marked chromosome that suffers an I-SceI-induced DSB. The top chromosome has a mutated cleavage site that can participate in interchromosomal DSB repair. The outcomes are analyzed among male progeny that inherit the h+ and Sb+ markers. Outcomes include interchromosomal gene conversions (i), as well as several intrachromosomal repair pathways, including end joining to recreate the cleavage site (ii), NHEJ deleting the cleavage site (and thus scored as PCR– for sequences around the cleavage site) (iii), and SSA (iv). (Adapted from Rong YS & Golic KG [2003] *Genetics* 165:1831–1842. With permission from the Genetics Society of America.)

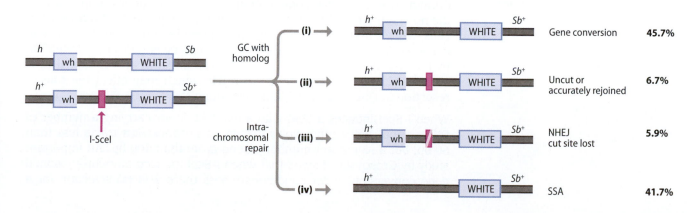

(A)

SSA assay

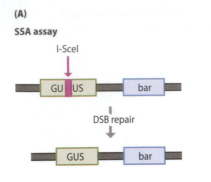

Gene conversion assay

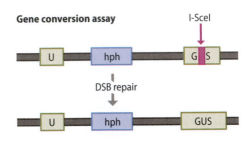

(B)

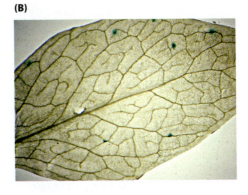

Figure 6.42 Homologous recombination assays in plants. (A) Induction of I-SceI can yield GUS (β-glucuronidase) recombinants either by SSA (*top*) or gene conversion (*bottom*). (B) Example of *blue*-staining GUS-positive spots on the leaf of *Arabidopsis*. Each dot represents an independent recombination event. (A, from Roth N, Klimesch J, Dukowic-Schulze S et al. [2012] *Plant J* 72:781–790. With permission from John Wiley & Sons, Inc. B, from Dubest S, Gallego ME & White CI [2002] *EMBO Rep* 3:1049–1054. With permission from Macmillan Publishers, Ltd.)

systems create functional β-glucuronidase (GUS) activity that can be scored as bright blue patches on a leaf (**Figure 6.42**). Both SSA and gene conversion assays have been used. The results from several labs have shown that Rad51, the paralogs Rad51C and Xrcc3, and Rad54 are all required for gene conversion but not SSA. By contrast, Mre11 and the Sae2 homolog, Com1, are not essential. As in budding yeast, Rad1 endonuclease is needed to clip off nonhomologous tails during SSA.

6.21 GENE CONVERSIONS HAVE BEEN STUDIED IN MAMMALS

In mouse and human cells, gene conversion events have been studied by transient expression of the I-SceI endonuclease, employing a variety of constructs to analyze intrachromosomal and interchromosomal recombination, both allelic and ectopic (**Figures 6.43** and **6.44**). For example, a *neo* gene is disrupted by insertion of the I-SceI cleavage site and a promoterless *neo* gene serves as the donor template, either on the same chromosome or on another chromosome. *Neo*+ recombinants can arise as frequently as 1% for intrachromosomal repair events following expression of I-SceI. But, whereas in budding yeast a DSB with a donor on the same chromosome can be repaired almost as efficiently as in intrachromosomal event, there is a factor of more than 100 between I-SceI-induced intrachromosomal and interhomolog recombination using similarly sized substrates in mammalian cells. Given the immense size of the vertebrate genome and the lack of somatic pairing of homologs, it is not surprising that homology searching is far more efficient between sister chromatids or nearby, along the same chromosome.

One important problem in studying DSB-induced events in mammalian cells is that it has been very difficult to achieve stable, tightly regulated expression of the endonuclease. More recently, better control over the activity of I-SceI has been accomplished by fusing the enzyme to a glucocorticoid receptor protein that retains I-SceI in the cytoplasm until tamoxifen is added, after which there is a rapid import of I-SceI into the nucleus. As measured by the initial appearance of the histone H2AX phosphorylation (γ-H2AX) chromatin modification around a DSB, some cells experience a cleavage of a chromosomal site within minutes, but the production of recombinants continues to increase for more than 24 hr, suggesting that the events are not well synchronized.

Maria Jasin's lab has come up with another approach to make regulated DSBs, by using the RAG1 and RAG2 proteins that create the hairpin-ended DSBs that are used in the generation of immunoglobulin gene rearrangements in B and T cells. These hairpins are cleaved by an endonuclease and are usually then end joined to create fusions between different immunoglobulin gene segments. But the hairpins can also be cleaved to promote homologous recombination. This system can be compared with the now-standard I-SceI assay (Figure 6.43). I-SceI yields about 2% GFP+ cells, while RAG1/RAG2 induction produces about 0.5% successful gene conversion. The levels of gene conversion can be enhanced up to tenfold by preventing NHEJ repair of the ends. The effect of preventing NHEJ is most significant in the absence of Ku70, which may affect the rate of resection at DSB ends as well as preventing NHEJ.

When I-SceI induces a DSB with a plasmid donor carrying a number of SNPs, about half of the repair events have conversion tracts less than 60 bp, but a significant number extend more than 200 bp. An important study by Grosovsky showed that when I-SceI induced interchromosomal gene conversion under conditions where there was no selection for a

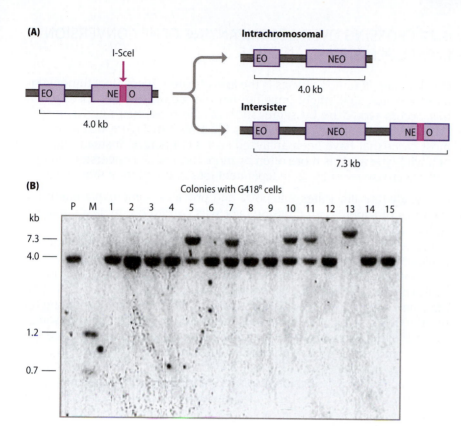

(A)

Intrachromosomal

4.0 kb

Intersister

7.3 kb

(B)

Colonies with G418ᴿ cells

Figure 6.43 **Gene conversion assay in mammalian cells using the I-SceI endonuclease.** (A) An I-SceI site within the *neo* gene can be repaired by intrachromosomal gene conversion using an incomplete, nearby segment of the *neo* gene as a template to produce *neo*⁺ (G418ᴿ) cells. Two types of outcomes are shown, an intrachromosomal gene conversion, yielding a 4.0-kb restriction fragment and an intersister recombination event that produces a 7.3-kb restriction fragment containing a duplication of the region. (B) Cells derived from a single diploid were analyzed by Southern blot, so that the products of both sister chromatids could be recovered. Lane 1 and similar lanes show intrachromosomal gene conversions. Lanes 5, 7, 10, and 11 show the expected pattern for a long-tract gene conversion (LTGC) in which the sister chromatid sequence is unaltered. Lane 13 may be an example of a aberrant event as described below in Figure 6.47. P indicates the parental configuration prior to I-SceI cleavage, and M is a lane containing size markers. (Adapted from Johnson RD & Jasin M [2000] *EMBO J* 19:3398–3407. With permission from Macmillan Publishers Ltd.)

particular outcome, gene conversion tracts were ≥7 kb in more than half the cases. If there is a constraint, such as the selection of a functional gene conversion product that must have created a particular arrangement of alleles, gene conversion tracts may be compelled to be much shorter.

Figure 6.44 **Gene conversion to produce GFP⁺ cells after I-SceI induction in mammalian cells.** (A) An I-SceI cut within a GFP gene can be repaired to produce GFP⁺ cells. (B) Induction of I-SceI in a liquid culture (*right*) can be assayed by fluorescence-activated cell sorting to identify GFP⁺ cells—shown as an arc of cells with strong GFP fluorescence—compared to an uninduced control (*left*). (From Puget N, Knowlton M & Scully R [2005] *DNA Repair* 4:149–161. With permission from Elsevier.)

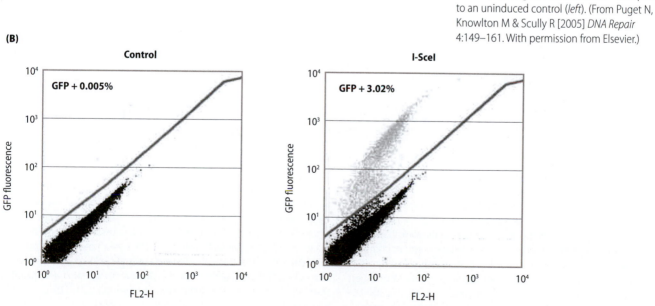

6.22 CROSSING OVER ACCOMPANYING GENE CONVERSION IS RARE IN MAMMALIAN CELLS

One key step in tumorigenesis is the loss of heterozygosity of tumor suppressor genes. One might imagine that the majority of instances when this occurs would be by segregation of chromosomes after a mitotic crossing over, analogous to what is illustrated in Figure 6.28A; but in most cases that have been analyzed such LOH is rare. Instead, the loss of a wild-type allele is more often by more local gene conversion events, without crossover or by an independent loss of function of that allele.

One study that illustrates this point and provides some explanation for the basis of suppressing crossing over was carried out by Jasin and her colleagues. A hybrid mouse cell line with many SNPs along its pairs of homologous chromosomes was modified so that two different, inactive copies of the *neo* gene were inserted at the same site in the middle of one chromosome arm; one copy (*Pneo*) had a frameshift mutation early in the ORF while the second contained an I-SceI cleavage site (**Figure 6.45A**). Induction of I-SceI resulted in *neo*+ colonies that resulted by patching up the DSB with sequences from *Pneo*. As this was an interchromosomal event, the frequency of repair was quite low—about 5×10^{-5} compared to intrachromosomal events that can occur at a rate of 10^{-2} or more. Analysis of *neo*+ recombinants revealed that only about 2% of them exhibited the LOH indicative of a crossover accompanying the gene conversion (**Figure 6.45B**). In most instances these clones showed that I-SceI had cut both

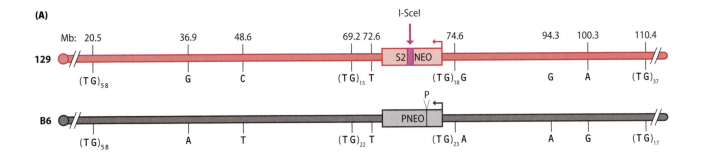

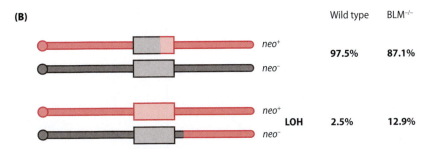

Figure 6.45 Loss-of-heterozygosity assay in mouse cells using I-SceI. (A) An *S2neo* gene interrupted by an I-SceI cleavage site (*red arrow*) is inserted into a mouse chromosome derived from strain 129 and a defective *Pneo* gene, containing the insertion of a frameshift mutation creating a restriction site near the 5′ end of the gene, is present at the allelic position in a chromosome derived from strain B6. Single nucleotide polymorphisms and microsatellite differences are noted along the chromosome, with distances in Mb. (B) Neo+ recombinants were genotyped. The majority show only a short patch of gene conversion without crossing over, leading to colonies still heterozygous for markers along the chromosome. The extent of the gene conversion tract varies in different isolates. Some cells show LOH for markers distal to point of gene conversion. For this to occur, the sister chromatid must also have been involved in an interchromosomal recombination event, leading to a crossover from which the LOH segregant was generated during mitosis. In the example shown, there is a simple crossover, but in many instances the sister chromatid also had evidence of a "patch" to repair the I-SceI cut in the sister chromatid. The frequency of LOH is greatly increased in cells lacking BLM helicase that is used to dissolve dHJs. (From LaRocque JR, Stark JM, Oh J et al. [2011] *Proc Natl Acad Sci USA* 108:11971–11976. With permission from the National Academy of Sciences.)

sister chromatids and each was repaired by a different event. This result is very reminiscent of the events seen in yeast when there is a block to replication (Figure 6.28C). A strong indication of the basis of suppressing crossovers was revealed by inactivating the BLM helicase in these cells. We saw earlier (Figure 6.19) that BLM helicase had been implicated in "dissolving" dHJs into noncrossover outcomes. In this experiment, when BLM was inactivated, there was a marked increase in the proportion of neo^+ cells with an accompanying crossover (Figure 6.45B). So at least a fraction of the repair events are likely to have proceeded through a dHJ intermediate.

6.23 SOME DSB REPAIR EVENTS BEGIN BY HOMOLOGOUS RECOMBINATION BUT TERMINATE WITH A NONHOMOLOGOUS END

In both *Drosophila* and in mammalian cells there is compelling evidence that events that start out with a DSB end finding its homologous donor can sometimes terminate by just sticking ends back together in an imprecise fashion.

A modification of the P-element sister chromatid repair event discussed in Section 6.19 was developed by Jeff Sekelsky's lab by placing the *white* gene in the middle of the P element (**Figure 6.46**). But inside the *white* gene, within an intron, was the insertion of a *copia* transposable element, which has 276-bp *long terminal repeat* (LTR) sequences at its ends. The *copia* insertion reduced the expression of the *white* gene so that flies have an "apricot" eye color. The allele is called w^a. When the P element was excised, complete repair of the element produced apricot-eye offspring that were indistinguishable from siblings that had never excised the P element. But there were several other interesting outcomes. The first yielded a red-eyed fly that is characteristic of a normal w^+ gene and which arose here because during the repair process most of the *copia* element was removed, leaving only one of the LTRs within the intron. This deletion allows full w^+ expression. For this to occur, the gene conversion process had to include either a template jump, from one LTR to the other or an SSA event between partially replicated ends. The second interesting group was flies that had completely lost the *white* gene function. In these instances repair synthesis started at one end of the DSB but copying was aborted and the ends were joined by NHEJ, creating a deletion that removed the w^a gene. Hence in some cases, gene conversion was properly initiated but fell victim to an NHEJ event to end it.

A similar story has emerged in mammalian cells. Jasin's lab set up an I-SceI system to look at interchromosomal recombination induced by I-SceI, similar to the interhomolog assay system illustrated in Figure 6.44, except that one end of the DSB shared only 84-bp homology with the donor and a "perfect" noncrossover gene conversion could not produce a Neo$^+$ recombinant (**Figure 6.47**). The assay was designed to look for crossovers, but no such chromosomal exchanges were found. Yet Neo$^+$ cells were readily recovered. When the recombinants were analyzed, it became clear that repair had begun properly with a homologous recombination event—either SDSA or BIR—in the 384 bp of shared homology between the left end of the DSB and the donor, but they all terminated at a variable distance beyond the 3' end of the *neo* gene into the adjacent HPRT sequences. The copied sequences were joined at or close to the second DSB end by what must have been a nonhomologous end-joining. So, at least under these restrictive conditions, repair involves both homologous recombination and NHEJ.

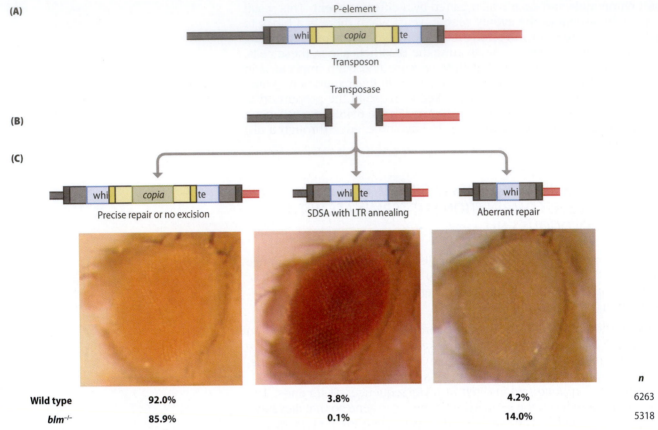

(A)

(B)

(C)

	Precise repair or no excision	SDSA with LTR annealing	Aberrant repair	*n*
Wild type	92.0%	3.8%	4.2%	6263
blm	85.9%	0.1%	14.0%	5318

Figure 6.46 Aberrant gene conversion events during repair of a P-element excision in *Drosophila*. (A) The Pw^a allele is a P element carrying a *white* gene that was interrupted in an intron by the insertion of a *copia* transposable element with 276-bp long terminal repeats (LTRs) (*orange*). (B) When transposase is expressed to induce excision of the P element, it can be repaired by homologous recombination with its sister chromatid (not shown). (C) Three outcomes were recovered, distinguished by their eye phenotypes. Some repair events were apricot-eyed (*left*), indistinguishable from siblings that had not suffered a DSB. Others were red-eyed (namely, w^+) that resulted from gene conversion in which there was a template jump that deleted most of the *copia* element and one of the flanking LTRs (*middle*). A third class lacked the *white* gene entirely but in many cases had evidence of the initiation of copying of one or both ends of the *white* sequences, but copying ceased prematurely and the incompletely replicated segments joined together by NHEJ (*right*). The frequency of abortive events was greatly increased in the absence of the *BLM* helicase gene. (Adapted from Adams MD, McVey M & Sekelsky JJ [2003] *Science* 299:265–267. With permission from the American Association for the Advancement of Science.)

Figure 6.47 Coupled homologous and nonhomologous repair of a DSB in mammals. (A) An I-SceI induced DSB in a *neo* gene lacking its 3′ end was inserted in mouse Chromosome 17, while a 5′-truncated neo gene was inserted in Chromosome 14. The DSB ends share 384- and 84-bp homology with the donor. A simple gene conversion event to repair the DSB would not produce a *neo*⁺ recombinant, but a crossover associated with DSB repair could conceivably produce a *neo*⁺ outcome (B), but no such events were found. All *neo*⁺ recombinants had copied the entire 3′ neo sequence as well as part of the adjacent HPRT region and then joined them to the other end of the DSB (C). (Adapted from Richardson C & Jasin M [2000] *Mol Cell Biol* 20:9068–9075. With permission from the American Society for Microbiology.)

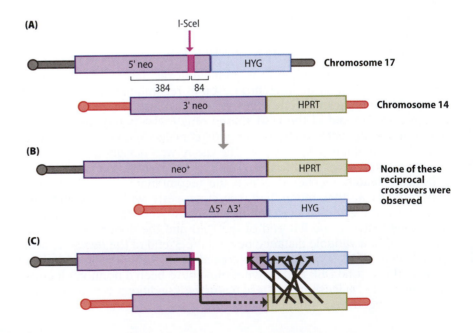

SUMMARY

Gene conversion repair of a DSB involves the nonreciprocal replacement of sequences copied from a donor locus. Gene conversion is sometimes accompanied by crossing over, though much less often in mitotic cells than in meiosis. Much of what we know about DSB repair and gene conversion came from gene targeting experiments that showed that a DSB directed repair to a homologous locus and involved accurate copying of donor sequences to fill in a break or gap. Gap enlargement or mismatch repair of hDNA can result in co-conversion of markers 1 kb or more distant from the break. The use of site-specific HO and I-SceI endonucleases has made it possible to examine repair events in detail. Gene conversion requires that the ends of DSBs be resected to allow Rad51 to form on ssDNA; consequently strand invasion produces heteroduplex DNA that may be mismatch corrected. The majority of mitotic DSB repair appears to proceed by an SDSA mechanism, without exchange. Support for this idea comes from the analysis of gap repair in which repeated sequences are copied and where variations in copy number are found only in the recipient locus. Analysis of gap-repair events suggests that crossovers may occur from intermediates in which branched intermediates are unligated dHJs, preferentially resolved in one outcome. DSB repair is surprisingly mutagenic, often producing complex mutations resulting from template switching. DSB repair has been studied in similar fashion in fission yeast, *Drosophila*, and mammalian cells, again using site-specific endonucleases.

SUGGESTED READING

Gloor GB, Nassif NA, Johnson-Schlitz DM et al. (1991) Targeted gene replacement in *Drosophila* via P element-induced gap repair. *Science* 253:1110–1117.

Hicks WM, Kim M & Haber JE (2010) Increased mutagenesis and unique mutation signature associated with mitotic gene conversion. *Science* 329:82–85.

Mitchel K, Zhang H, Welz-Voegele C & Jinks-Robertson S (2010) Molecular structures of crossover and noncrossover intermediates during gap repair in yeast: implications for recombination. *Mol Cell* 38:211–222.

Nik-Zainal S, Alexandrov LB, Wedge DC et al. (2012) Mutational processes molding the genomes of 21 breast cancers. *Cell* 149:979–993.

Orr-Weaver TL & Szostak JW (1983) Yeast recombination: the association between double-strand gap repair and crossing-over. *Proc Natl Acad Sci USA* 80:4417–4421.

Orr-Weaver TL, Szostak JW & Rothstein RJ (1981) Yeast transformation: a model system for the study of recombination. *Proc Natl Acad Sci USA* 78:6354–6358.

Pâques F & Haber JE (1999) Multiple pathways of recombination induced by double-strand breaks in *Saccharomyces cerevisiae*. *Microbiol Mol Biol Rev* 63:349–404.

Resnick MA (1976) The repair of double-strand breaks in DNA; a model involving recombination. *J Theor Biol* 59:97–106.

Rong YS & Golic KG (2003) The homologous chromosome is an effective template for the repair of mitotic DNA double-strand breaks in *Drosophila*. *Genetics* 165:1831–1842.

Singh SK, Roy S, Choudhury SR & Sengupta DN (2010) DNA repair and recombination in higher plants: insights from comparative genomics of *Arabidopsis* and rice. *BMC Genomics* 11:443.

Szostak JW, Orr WT, Rothstein RJ & Stahl FW (1983) The double-strand-break repair model for recombination. *Cell* 33:25–35.

Weinstock DM, Nakanishi K, Helgadottir HR & Jasin M (2006) Assaying double-strand break repair pathway choice in mammalian cells using a targeted endonuclease or the RAG recombinase. *Methods Enzymol* 409:524–540.

CHAPTER 7

"IN VIVO BIOCHEMISTRY": RECOMBINATION IN YEAST

Much of our understanding about the molecular events in DSB repair has come from physical monitoring of recombination intermediates in both budding and fission yeasts, where it is possible to induce synchronous and efficient DSB repair processes with site-specific endonucleases. This approach has not only allowed us to confirm that the behavior of recombination proteins *in vivo* is similar to what had been worked out *in vitro*, but also to examine steps in DSB repair that have not yet been mimicked *in vitro*.

7.1 BUDDING YEAST *MAT* SWITCHING ALLOWS US TO DESCRIBE THE MOLECULAR EVENTS DURING A GENE CONVERSION EVENT

Haploid yeast have two alternative mating types, **a** and α, which are encoded by the *MAT***a** and *MAT*α genes. Cells of opposite mating type can conjugate to form *MAT***a**/*MAT*α diploids. The HO endonuclease gene is only expressed in haploid cells; its expression leads to the formation of a DSB within the *MAT* locus. Repair of this DSB uses one of two distantly located donors that encode opposite mating type sequences. Consequently there is a homothallic conversion of one mating type to the other, thus allowing budding yeast to go from a haploid state (either *MAT***a** or *MAT*α) to a diploid state (*MAT***a**/*MAT*α), whereupon the HO gene is turned off. The HO endonuclease is under tight genetic control that restricts its normal expression to a small portion of the cell cycle as cells progress from G1 to S. The HO protein is rapidly degraded.

*MAT***a** and *MAT*α differ in their Y regions, which are entirely different DNA sequences (**Figure 7.1**). The replacement of 650 bp of Y**a** with 700 bp of Yα is clearly not a simple mutational event; rather the new Y region is copied from one of two unexpressed donors, *HML*α and *HMR***a**, which actually have intact copies of mating-type genes but are kept transcriptionally silent (and immune to HO endonuclease cleavage) by a Sir2 histone deacetylase-mediated silencing system. Thus when a DSB is made at *MAT*, the original Y sequences are removed and the Y**a** or Yα sequences are copied from one of the two distant donors into *MAT*.

Because the DSB is located such that one side is immediately flanked by the Y sequences that are not homologous to the donor with opposite mating-type sequences, the initial strand invasion and initiation of new DNA synthesis takes place in the Z1 and Z2 regions. *MAT* shares only 327 bp homology with *HML* and even less with *HMR* (230 bp), yet this is sufficient

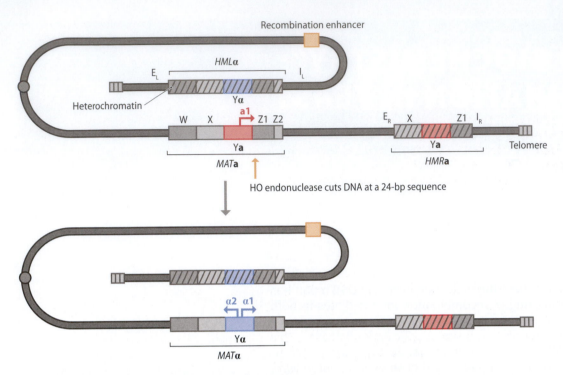

Figure 7.1 *MAT*a **switches to** *MAT*α **in** *S. cerevisiae.* A DSB induced by HO endonuclease at *MAT*a preferentially recombines with *HML*α to replace the Ya (*red*) sequences encoding the *MAT*a1 gene with the Yα (*blue*) sequences encoding *MAT*α1 and *MAT*α2 that determine the cell's mating type. *HML*α and the alternative *HMR*a donor are both transcriptionally silent and heterochromatic (indicated by *hatched lines*). *HML* is located 100 kb from the centromere (*gray circle*) and close to the left end of chromosome III, while *MAT* is 100 kb to the right of the centromere and *HMR* is another 100 kb away, near the right telomere. The preference of *MAT*a for *HML* is enforced by the recombination enhancer (RE). *MAT*α cells preferentially recombine with *HMR*a, as do *MAT*a cells in which the RE is deleted. (Adapted from Haber JE [2012] *Genetics* 191:33–64. With permission from the Genetics Society of America.)

to promote highly efficient recombination. The Z region can be truncated to about 100 bp without drastically affecting the efficiency of DSB repair.

7.2 *MAT* SWITCHING CAN BE PHYSICALLY MONITORED ON SOUTHERN BLOTS

The development of a galactose-inducible HO endonuclease gene has made it possible to inflict the same DSB on essentially all the cells of the population. This has made it possible to follow—in real time using Southern blots—the kinetics of gene conversion, replacing *MAT*a by *MAT*α or vice versa (**Figure 7.2**). Within 20 min after HO induction, essentially all the cells have experienced a DSB, seen by the appearance of a smaller restriction fragment hybridizing to a probe that recognizes sequences just distal to *MAT*. Only after more than an hour does one begin to see the appearance of a new, larger fragment that represents *MAT*α (because Yα lacks a *Sty*I site present in Ya). There are a number of slow steps during the hour before a product is detected (**Figure 7.3**). These are discussed below.

7.3 THE LOADING OF Rad51 ON ssDNA CAN BE VISUALIZED BY CHROMATIN IMMUNOPRECIPITATION

To examine the loading of Rad51 onto the DSB at *MAT*, one can use chromatin immunoprecipitation (ChIP) in which cells are treated with formaldehyde to reversibly cross-link proteins to each other and to DNA.

(A)

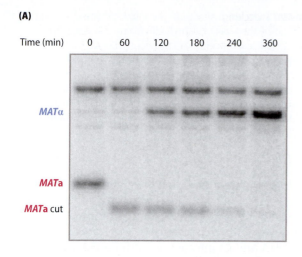

(B)

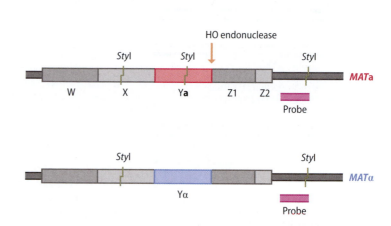

Figure 7.2 Southern blot analysis of the kinetics of *MATa* switching. (A) HO endonuclease was induced for 1 hr. DNA was digested with *StyI* and analyzed by Southern blot after gel electrophoresis. (B) The presence of a *StyI* site in Y**a** (*red*) sequences reveals a *MATa*-specific restriction fragment homologous to the probe that is further truncated by HO cleavage. The probe also weakly hybridizes to the adjacent, larger *StyI* fragment (*top of blot*). Repair of the DSB by recombining with HMLα results in replacement of Y**a** by Yα (*blue*) sequences results in a new, larger restriction fragment homologous to the probe, which is strongly visible after another hour. *MATa* switches predominantly to *MATα* using *HMLα* as the donor, but a small fraction recombines with *HMRa* so the abundance of *MATa* slightly increases. (Adapted from Sugawara N & Haber JE [2006] *Methods Enzymol* 408:416–429. With permission from the authors.)

The cross-linked chromatin is then sheared and immunoprecipitated by an antibody—in this case, anti-Rad51. The cross-links in the enriched, immunopurified chromatin fragments are then reversed by heating and the purified DNA can be analyzed by quantitative PCR or by genomewide methods such as microarrays or deep sequencing. By this technique one can see how rapidly Rad51 becomes associated with the DSB end. As shown in **Figure 7.4**, one can readily see evidence of Rad51 binding to the DSB end about 10 min after the DSB was detected.

Is the delay in Rad51 because the resection process is slow to start or because its loading is preceded by RPA binding, as might be suggested from *in vitro* experiments? To address this question one can do a second ChIP experiment, but using an anti-RPA antibody. The result is surprisingly clear: RPA associates with the DSB end as rapidly as the break is evident (Figure 7.4). This result is consistent with *in vitro* experiments showing that RPA loading apparently is important in facilitating Rad51 loading.

Rad51 loading depends on resection. In G1-arrested cells, where Cdk1 is inactive, or in cycling cells when both Sgs1 and Exo1 are deleted, there is almost no resection of an HO-induced DSB; consequently, there is very little Rad51 (or RPA) loading. We also know that Rad52, Rad55/Rad57, and Rad54 are all required for *MAT* switching, but with ChIP we can gain further insight into their specific roles. Without Rad52, Rad51 fails to assemble onto the DSB (shown in Figure 4.15). Without Rad55 or Rad57, there is some loading, though it is slow and the ChIP signal is reduced. Subsequent experiments show that this loading is insufficient to promote strand exchange. In contrast, deleting Rad54 does not impair Rad51 loading. This result supports a variety of genetic experiments that argue that Rad54 acts at a later step than Rad51, Rad52, Rad55, or Rad57. For example, the double deletion of Sgs1 and Srs2 is lethal, apparently because some sort of aberrant recombination structure is formed spontaneously during replication; but *sgs1Δ srs2Δ* is rescued in combination with *rad51Δ*, *rad52Δ*, *rad55Δ*, or *rad57Δ* that prevent the formation of the lethal intermediate, but not by the later-acting *rad54Δ*.

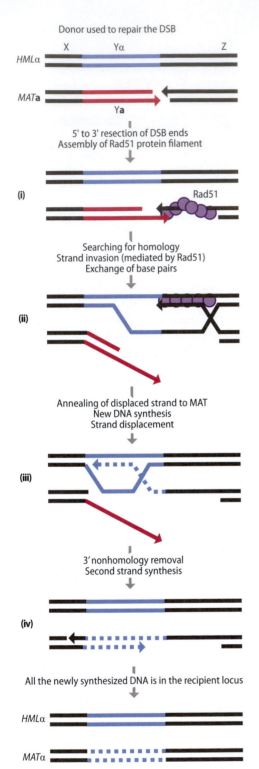

Donor used to repair the DSB

5' to 3' resection of DSB ends
Assembly of Rad51 protein filament

(i)

Searching for homology
Strand invasion (mediated by Rad51)
Exchange of base pairs

(ii)

Annealing of displaced strand to MAT
New DNA synthesis
Strand displacement

(iii)

3' nonhomology removal
Second strand synthesis

(iv)

All the newly synthesized DNA is in the recipient locus

Figure 7.3 Sequential steps in *MAT* switching. After a DSB is created at the Y**a**-Z junction in *MAT***a**, the DSB ends are resected and are then coated with a Rad51 filament (i). A search for homology results in an encounter with *HML*α and strand invasion to form a D-loop (ii). The 3' end of the invading strand is extended by a DNA polymerase and the newly copied strand is displaced from the donor locus (iii). When the strand has been extended to the *HML* X region, it can pair with the single-stranded X sequences. The protruding nonhomologous Y**a** sequence (*red*) is removed by an endonuclease and a second strand is copied by primer extension to complete *MAT* switching (iv). (Adapted from Haber JE [2012] *Genetics* 191:33–64. With permission from the Genetics Society of America.)

ChIP has also made it possible to understand the different roles of the RPA complex in DSB repair. The Rfa1-L45E mutation in the largest subunit of RPA prevents DSB repair, although it is not defective in normal replication. This mutant has no defect in the assembly of the Rad51 filament (that is, RPA's binding to ssDNA and its interaction with Rad52 are apparently normal), but there is no association between Rad51 and the donor locus. One possibility is that the formation of a D-loop requires RPA binding to the displaced strand and this step appears to be compromised. If the displaced strand is not properly sequestered by RPA binding, strand exchange should be reversible, with the invading strand being evicted by re-formation of the donor duplex.

ChIP experiments also confirmed a biochemical estimate of how much Rad51 there is in the cell. In a strain where *HML* and *HMR* are deleted, a DSB cannot be efficiently repaired and 5' to 3' resection progresses at about 4 kb/hr. Examination of regions further from the DSB reveals that the Rad51 ChIP signal diminishes beyond 5 kb of ssDNA on either side of the DSB. Knowing that each Rad51 can bind 3 nt of ssDNA, a total of 10 kb ssDNA implies there are about 3300 Rad51 molecules in the cell even after creation of DNA damage. This result is consistent with biochemical estimates of only about 3500 molecules in a budding yeast haploid cell. If Rad51 is overexpressed, then much longer lengths of ssDNA can be bound by Rad51. There is much more RPA in the cell, so that even when resection proceeds beyond 10 kb and most Rad51 is already recruited, RPA continues to bind the newly exposed ssDNA.

As noted in Chapter 3, *in vitro* measurements of Rad51 show that its affinity for ssDNA is not much greater than for dsDNA, and its subunit cooperativity in assembling a filament is much lower than for RecA; nevertheless, when ssDNA is generated, Rad51 rapidly accumulates at that site. Recent experiments suggest that three Swi1/Snf2-related chromatin remodeling proteins, Tid1/Rdh54, Rad54, and Uls1, play an important role in displacing nonspecifically bound Rad51 from random sites around the genome, freeing it up to bind to ssDNA. A severe problem is seen only when all three proteins are absent.

7.4 THE ENCOUNTER OF THE Rad51 FILAMENT WITH THE DONOR LOCUS CAN ALSO BE MONITORED BY CHROMATIN IMMUNOPRECIPITATION

ChIP can also be used to find out how long it takes for the Rad51::ssDNA filament at *MAT* to form a stable association with the donor sequence, *HML*. Indeed, using anti-Rad51 antibody, this synapsis can be detected about 15–20 min after Rad51 has bound to *MAT* (**Figure 7.5**). This assay therefore reveals the time for a successful intrachromosomal homology search to be accomplished. As noted before, a *rad55*Δ mutant exhibits a delayed and reduced assembly of Rad51 onto a *MAT* DSB, but it is

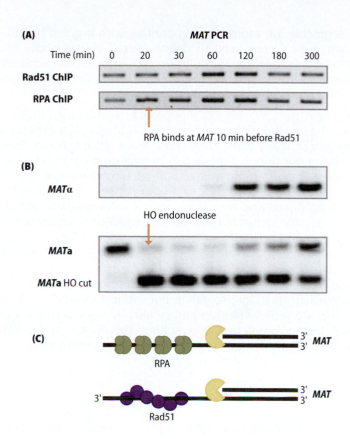

Figure 7.4 Rad51 and RPA recruitment at a DSB end during *MAT* switching. (A) Chromatin immunoprecipitation using antibodies against Rad51 or the largest subunit of RPA (Rfa1) show that RPA is detected at the HO-cut end very soon after HO cleavage whereas Rad51 binding is about 10 min slower. (B) The Southern blot, probed with a *MAT*-specific probe (see Figure 7.2), is shown for reference. (C) Illustration of RPA and Rad51 binding to the DSB end. (Adapted from Wang X & Haber JE [2004] *PLoS Biol* 2:104–111.)

insufficient to enable Rad51 to associate with the donor. Rad51 association with *HML* still occurs in the absence of Rad54.

It is also possible to examine a situation in which the donor is located on a different chromosome (an *ectopic* donor). ChIP experiments indicate that an intrachromosomal search is significantly faster than an interchromosomal search, so that even for sequences 200 kb away on the opposite sides of centromere, their interaction is faster than between a donor sequence on a different chromosome.

A powerful alternative to the ChIP strategy would be to use CCC (chromosome conformation capture) methods to follow the time it takes for a donor and recipient region to come together. In this approach DNA is cross-linked to other regions of DNA with which it happens to be in contact. After restriction enzyme cleavage, there will be two short DNA regions tethered together by the cross-link. The ends of these molecules can be ligated and analyzed by PCR or DNA sequencing to show which

Figure 7.5 Association of Rad51 with donor sequences after homology searching and strand invasion. (A) In a similar experiment to Figure 7.4, formaldehyde cross-linked proteins were immunoprecipitated and analyzed for Rad51 binding to *MAT* and to *HML*α, using pairs of PCR primers specific for *MAT* or for *HML*. (B) An illustration of Rad51 binding to ssDNA at the HO-cut *MAT* locus and after strand invasion into *HML*α. (Adapted from Sugawara N, Wang X & Haber JE [2003] *Mol Cell* 12:209–219. With permission from Elsevier.)

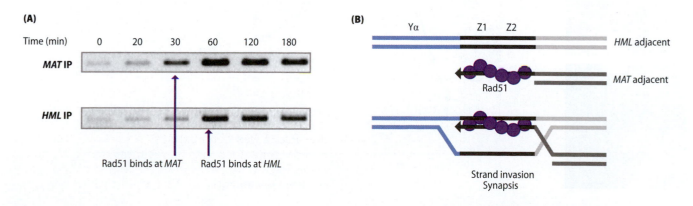

genomic segments are more often in contact with a given region. This method, which has been used to define overall chromosome folding, should also be able to follow the collisions between dsDNA adjacent to a Rad51-coated ssDNA region and the places that the filament explores as it searches for homology.

A limitation of the ChIP (or CCC) approach to monitor chromosome regions coming together for recombination is that one measures the average occupancy of many cells. Thus one cannot tell if synapsis between the DSB end and a donor sequence is initially transient and only later becomes more stable. An alternative approach to visualizing the search for homology is to use chromosomes with fluorescently labeled "tags" adjacent to the donor and the recipient in a recombination event. It is possible to integrate arrays of LacO or TetO sequences adjacent to these sites and to express LacI-GFP and TetR-GFP (or other color combinations) to follow the positions of these chromosome regions before, during, and after recombination is induced in living cells. An example of this approach is shown in **Figure 7.6**, where one can see that prior to HO induction the two loci are usually not close together, but within 15 min after HO cleavage these loci are seen to pair up and to remain associated for a period of time sufficient to complete *MAT* switching. This approach should allow one to understand the dynamics of homology searching in more detail.

Figure 7.6 Single-cell analysis of homology searching and recombination. (A) *MAT*a and *HML*α are marked by approximately 250 copies of the TetO or LacO binding sites, respectively, to which TetR-GFP or LacI-GFP will bind. These sites will come together when *HML* and *MAT* pair so that only one *green* spot will be resolved. (B) Examples of a cell having two distinct GFP foci (that is, *HML* and *MAT* are not paired) or a single focus (when they are paired). (C) There is an increase in cells with a single focus soon after HO induction is complete (*red arrow*). (Adapted from Bressan DA, Vazquez J & Haber JE [2004] *J Cell Biol* 164:361–371. With permission from the Rockefeller University Press.)

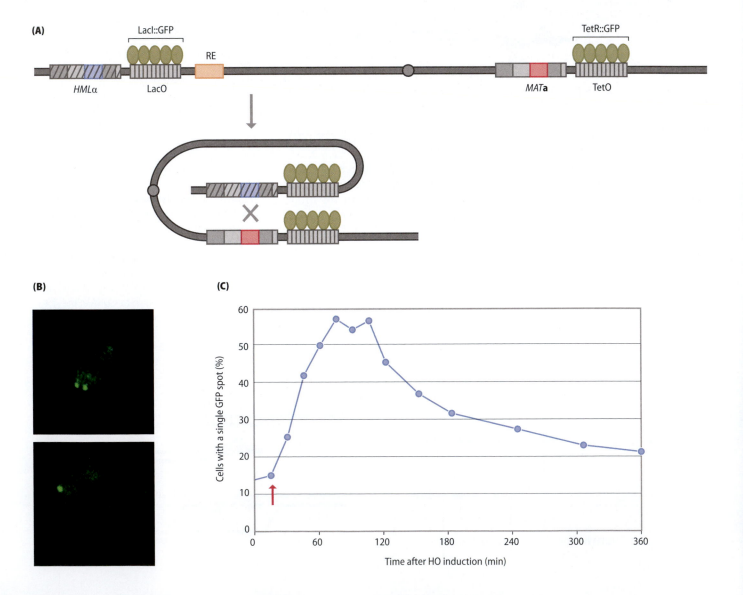

7.5 A PCR ASSAY CAN BE USED TO DETECT THE BEGINNING OF NEW DNA SYNTHESIS

The 3′ end of the invading strand, in conjunction with PCNA and a DNA polymerase, initiates new DNA synthesis, copying one strand of the donor template. We can detect the beginning of new DNA synthesis by a PCR assay in which one primer is in the donor's Yα sequences and the other is located distal to the Z sequences shared by *MAT* and *HML* (**Figure 7.7**). Initially the two primers are 200 kb apart; only when new DNA synthesis has extended the 3′ end of the invading strand by about 50 bp will there be a covalent DNA molecule that can be amplified. Hence it is possible to detect the time of new DNA synthesis relative to the time of strand invasion (by ChIP) and relative to the completion of the event, which can be monitored by a second pair of PCR primers that test if the Yα sequences have been ligated to *MAT*-proximal sequences. By this method we find that new DNA synthesis occurs only 15–20 min after the first detection of synapsis (Rad51 association with *HML*). Presumably during this long time, the DNA replication machinery must be assembled.

With this primer extension assay, it is possible to ask which DNA polymerases or accessory proteins are required for the initiation of new DNA synthesis. Most of the components of DNA replication are essential, so it is necessary to use conditional mutations (high-temperature-sensitive [ts] or cold-sensitive mutations, for example) to examine the role of these proteins in DSB repair. In this way, using a cold-sensitive allele of PCNA, it is possible to show that PCNA is required for *MAT* switching. Similarly, a ts mutation of the Dpb11 replication factor has no defect in completing *MAT* switching at its permissive temperature of 25°C but is blocked in repair at 37°C (**Figure 7.8**). Without Dpb11, there is no primer extension. In contrast, HO-induced gene conversion does not require the Mcm proteins or the GINS complex or Cdc45 that make up the "CMG" helicase that is essential for normal DNA replication. In the experiment shown in Figure 7.8, cells were arrested by blocking them prior to DNA replication, using a drug-sensitive mutation of the Cdc7 kinase. Switching is normal in wild-type cells; hence cells can switch either before or after DNA replication.

As expected, SDSA does not require lagging-strand DNA polymerase α or its associated primase. However, in the course of doing those experiments it became evident that there were unforeseen complications. When

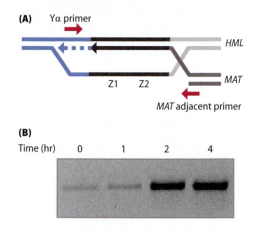

(A)

(B)

Time (hr)   0   1   2   4

Figure 7.7 Detection of new DNA synthesis by PCR. (A) New DNA synthesis (primer extension) from the 3′ end of the invading strand can be detected using one PCR primer specific for *MAT*-distal sequences and a second primer that binds to *HML*-Yα (*red arrows*). Initially the two primers are 200 kb apart, but the synthesis of about 50 nt from the 3′ end of the invading strand creates a covalent template to which both primers can anneal. (B) PCR data showing the time of appearance of primer extension. (A, adapted from White CI & Haber JE [1990] *EMBO J* 9:663–673. With permission from Macmillan Publishers, Ltd. B, from Sugawara N, Wang X & Haber JE [2003] *Mol Cell* 12:209–219. With permission from Elsevier.)

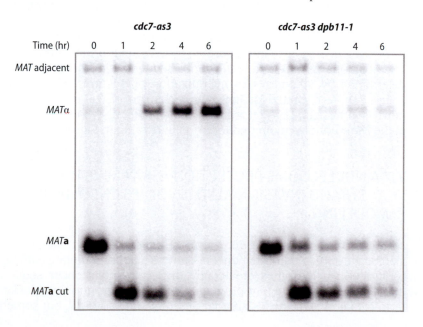

Figure 7.8 *MAT* switching is independent of the Cdc7 kinase and independent of S phase but depends on Dpb11. Yeast cells were arrested prior to S phase by chemically inhibiting the analog-sensitive Cdc7-as3 kinase that is required to initiate normal DNA replication. Cells were incubated at 37°C before HO induction. *MAT* switching is normal under these conditions (*left*). By contrast, if similarly treated cells also carry the temperature-sensitive *dpb11-1* mutation, *MAT* switching is blocked at the restrictive temperature (*right*). (From Hicks WM, Yamaguchi M & Haber JE [2011] *Proc Natl Acad Sci USA* 108:3108–3115. With permission from the National Academy of Sciences.)

such an experiment was first carried out, cells were simply arrested by raising a ts Polα mutant to its restrictive temperature. By this treatment, cells arrested in S phase and when *MAT* switching was then induced it appeared that cells could not repair the DSB without Polα. However, when the experiment was repeated in nocodazole-arrested cells, blocked *after* S phase and prior to mitosis, there was no requirement for Polα or primase. These experiments suggest that when DNA replication was blocked in S phase by inactivating Polα, other components needed both for S phase and for *MAT* switching were somehow sequestered. Hence it is necessary to assay DNA replication factors under conditions where cells are not interrupted in the middle of S phase.

A similar analysis of temperature-sensitive mutations of DNA Polδ and DNA Polε led to the conclusion that they must act redundantly, as inactivation of either one slows and slightly reduces—but doesn't eliminate—*MAT* switching. There is no essential role for either of the translesion DNA polymerases DNA Polη or Polζ. These conclusions were supported by the analysis of mutations arising during *MAT* switching, as mentioned in Chapter 6.

7.6 HEAVY ISOTOPE LABELING CAN BE USED TO SHOW THAT NEW DNA SYNTHESIS IS "CONSERVATIVE"

It is likely that the newly copied strand is displaced from the template, leaving the donor intact. This process is clearly different from semi-conservative DNA replication where the newly made strand is base paired to the template. This step must involve a helicase to unwind the new DNA strand and allow re-formation of the donor duplex (Figure 7.3). It is quite likely that the synthesis of new DNA occurs in the context of a small, moving D-loop, as first suggested by *in vitro* studies of phage DNA replication by Tim Formosa and Bruce Alberts.

Direct evidence that all the newly synthesized DNA ends up at the recipient locus came from using the same "heavy/light" isotope density transfer experiment used by Meselson and Stahl to show that *E. coli* DNA replication was semi-conservative. Cells were grown for several generations in ^{13}C and ^{15}N heavy-isotope-containing medium and then arrested by treating with nocodazole after DNA replication, prior to mitosis so that only repair DNA synthesis would be detected. Cells were transferred to normal "light" (^{12}C and ^{14}N) medium and *MAT* switching was induced. After several hours to allow repair to be complete, DNA was isolated, cut by restriction enzymes and separated by CsCl equilibrium centrifugation. The positions of *MAT* and *HMR* donor sequences in the gradient were determined by Southern blot hybridization of fractions of the gradient (**Figure 7.9**). As expected for an SDSA mechanism of *MAT* switching, all the newly synthesized DNA ended up in the *MAT* locus while the sequences at the *HMR* donor remain unaltered, as illustrated in Figure 7.3.

7.7 A NUCLEOSOME PROTECTION ASSAY REVEALS THAT STRAND INVASION ALSO INVOLVES CHROMATIN REMODELING

MAT switching may represent the most difficult case for gene conversion because the Rad51 filament encounters a donor template with highly positioned nucleosomes created by the *cis*-acting silencer sequences and the Sir2, Sir3, and Sir4 proteins. Nucleosome positioning blocks HO from cleaving near the Y–Z border at either *HML* or *HMR*, but somehow

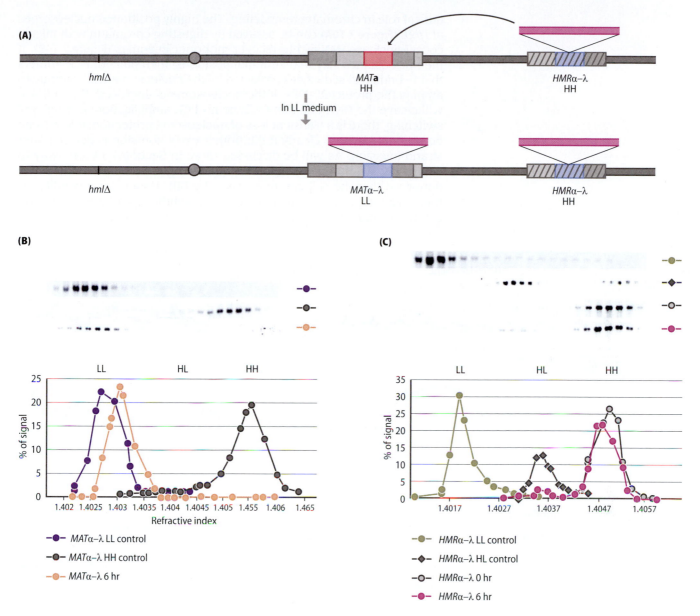

Figure 7.9 *MAT* switching exhibits conservative DNA replication.
(A) A budding yeast strain was modified by deleting *HML* and inserting a 5-kb segment from phage λ (*magenta*) into *HMRα* to create a large enough fragment for density gradient analysis. Cells were grown in ^{13}C- and ^{15}N-containing medium so that all DNA was "heavy" (HH) and then arrested in nocodazole so that there would be no general DNA replication. Cells were then placed in "light" (LL) medium containing ^{12}C and ^{14}N. After induction of HO endonuclease, when recombination was complete, DNA was isolated and cleaved by restriction enzymes and the fragments separated on a CsCl equilibrium density gradient. (B) DNA was fractionated by density from LL to HH and fractions were analyzed by a Southern blot probed with a *MAT*-specific probe. The switched *MATα–λ* migrated as nearly LL DNA. The difference between LL and the position of the *MAT* sequences is explained by the fact that the restriction fragment contained sequences outside the region replaced during switching. (C) A similar blot was probed with an *HMR*-specific probe and also one for an unrelated region (*HIS4*). During switching *HMRα–λ* remained HH. Thus all the newly synthesized DNA is found at the *MAT* locus while the donor is unaltered. (B, C, from Ira G, Satory D & Haber JE [2006] *Mol Cell Biol* 26:9424–9429. With permission from the American Society for Microbiology.)

the strand invasion machinery must arrive and deal with these same nucleosomes.

We best understand the role of the Swi/Snf homolog Rad54 protein in *MAT* switching. As mentioned earlier, strand invasion appears to occur in *rad54Δ* cells, but the next step—primer extension—is blocked. In this case we know that even if the donor is not silenced, Rad54 is still needed for the initiation of new synthesis. Nevertheless, it seems that Rad54 plays a very critical role in chromatin remodeling. The highly positioned nucleosomes at *HML* (**Figure 7.10A**) can be assayed by digesting chromatin with micrococcal nuclease (MNase) to recover mononucleosomes (**Figure 7.10B**). If the nucleosomes stay in their initial positions, then pairs of PCR primers that hybridize to sequences protected by the nucleosome will continue to amplify the protected DNA. If the nucleosome is displaced, then MNase will cleave the DNA and there will be no PCR amplification. During *MAT* switching, there is a transient loss of nucleosome protection, but this can be seen much more clearly if the primer extension step is delayed after strand invasion. As will be discussed more in Section 7.11, one way to delay new synthesis is to delete the homology on the other side of the donor, so that the cell can only repair by BIR. Under these conditions, there is a long delay in initiating new DNA synthesis and one can see an extensive change in nucleosome protection (**Figure 7.10C**).

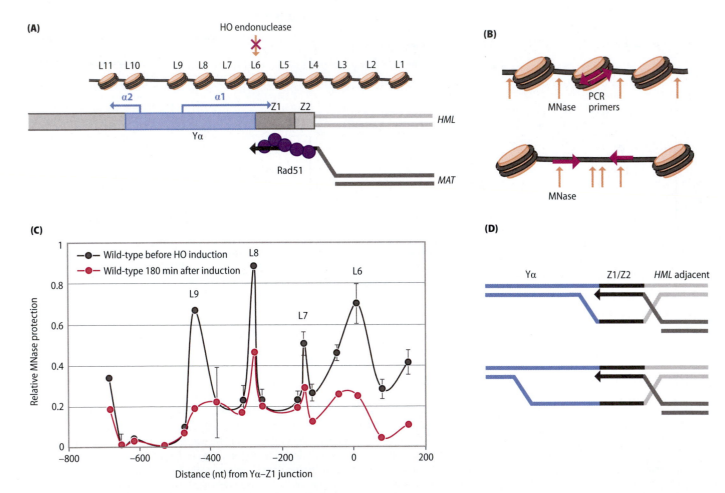

Figure 7.10 Alterations in *HML* heterochromatin during Rad51-mediated strand invasion. (A) Strand invasion by Rad51 and the 327 nt *MAT* Z sequences into the heterochromatic *HML* locus results in a loss of nucleosome protection as assayed by PCR in which each pair of primers (*magenta* arrow) both lie within the sequences normally protected by a highly positioned nucleosome at *HML* (B). If nucleosome positioning is altered, micrococcal nuclease (MNase) will digest the sequences that would be amplified by the PCR primers. (C) To see the loss of nucleosome protection, it was necessary to delete the homologous sequences shared with *MAT* on the left side of the Yα region at *HML*, so that there was a delay after strand invasion before the recruitment of both leading- and lagging-strand DNA polymerase to repair by BIR. Under these conditions, DNA synthesis is delayed for more than 3 hr (T180) and it was possible to see that there was a significant loss of nucleosome protection not only for the L6 and L5 nucleosomes that would be involved directly in strand exchange (*red lines*) but also several nucleosomes to the left (L9, L8, and L7). These data suggest that strand invasion may not simply form a D-loop over the homologous sequences involved in strand invasion (D, top), but may be accompanied by an unwinding of DNA to create an enlarged D-loop (D, bottom). (A, adapted from Weiss K & Simpson RT [1998] *Mol Cell Biol* 18:5392–5403. With permission from the American Society for Microbiology. B, C, adapted from Hicks WM, Yamaguchi M & Haber JE [2011] *Proc Natl Acad Sci USA* 108:3108–3115. With permission from the National Academy of Sciences.)

A striking feature of the change in nucleosome protection seen in this assay is that the loss of protection extends several nucleosomes to the left of the 327-bp region where strand invasion per se takes place. This is an unexpected result as it suggests that the D-loop has been extended well beyond the place where strand exchange would force opening up the DNA (**Figure 7.10D**). How this region is opened up is not yet understood, but additional evidence for the open structure is that RPA binds to the apparently open region. It might also be possible to examine the formation and extent of D-loop formation by using DNA sequence mapping techniques originally developed to look at open RNA polymerase complexes, using chloroacetic acid or $KMnO_4$ to map ssDNA regions or bisulfite treatment to convert C bases in a single-stranded region to U. Such approaches have been used to examine ssDNA in regions where RNA transcripts remain tightly associated with their template.

In contrast to the remodeled chromatin structure in wild-type cells, there is no such alteration in the absence of Rad54 (**Figure 7.11A**). The absence of nucleosome remodeling is striking in view of the fact that strand invasion (as measured by ChIP) is not impaired (**Figure 7.11B**). There is also no

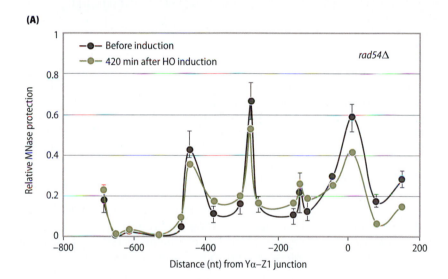

(A)

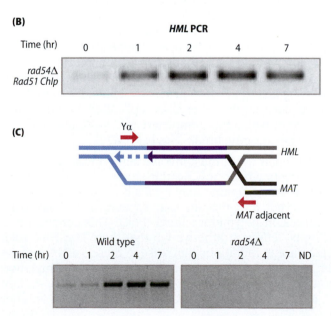

Figure 7.11 Lack of nucleosome disruption without Rad54. (A) An experiment similar to that shown in Figure 7.10 was performed in *rad54Δ* strain. Without Rad54 there is little change in nucleosome protection, even though a ChIP assay similar to that in Figure 7.4 showed that Rad51 is associated with *HMLα* in the absence of Rad54 (B). (C) Despite the Rad51 association with *HMLα*, a primer extension assay showed there was no initiation of new DNA synthesis. (A, adapted from Hicks WM, Yamaguchi M & Haber JE [2011] *Proc Natl Acad Sci USA* 108:3108–3115. With permission from the National Academy of Sciences. B, C, adapted from Sugawara N, Wang X & Haber JE [2003] *Mol Cell* 12:209–219. With permission from Elsevier.)

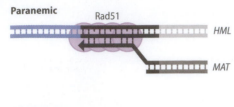

Paranemic

Plectonemic

Figure 7.12 Strand invasion can be paranemic or plectonemic. In the absence of Rad54, strand exchange may be limited to a paranemic (side-by-side) association of the invading and donor strands, so that only some of the possible base pairs can be formed. When Rad54 is functional there may be a complete, plectonemic (interwound) joint molecule in which the invading strand is fully base paired with its complementary strand and a single-stranded D-loop is created.

primer extension (**Figure 7.11C**). These data suggest that without Rad54, strand invasion may result only in a weak *paranemic* (side-by-side, making only some base pairs) joint rather than *plectonemic* joint (fully interwound with a displaced complementary strand) (**Figure 7.12**). Rad54 may also be needed to clear Rad51 off the 3′ end of the invaded strand to allow the assembly of PCNA and DNA polymerase.

Other chromatin remodelers may also play supporting roles. Deleting subunits of the RSC2 complex has been shown to cause changes in the nucleosome positions immediately adjacent to the DSB at *MAT*, as judged by a reduction in HO cleavage, although there is no highly ordered nucleosome arrangement at *MAT* to begin with. In addition, ChIP experiments have shown the RSC protein Sth1 is recruited rapidly to the site of a DSB. The RSC1 complex also appears to play a later, but currently undefined, role in DSB repair. Finally the Snf5 protein, part of the SWI/SNF chromatin remodeling complex, has also been reported to have a decisive role in early steps of DSB repair, but because *snf5Δ* cells are unable to support galactose induction and are impaired in the expression of the normal *HO* gene, this claim needs to be reinvestigated.

7.8 THE CAPTURE OF THE SECOND HOMOLOGOUS END DURING GENE CONVERSION CAN ALSO BE STUDIED

As DNA synthesis progresses, the extruded ssDNA strand becomes long enough to allow annealing to the opposite, resected DSB end. This step requires Rad52's strand annealing activity. The most direct evidence for the role of Rad52 at this step has come from studies of meiotic recombination, where the initial strand exchange can be carried out by Dmc1 instead of Rad51 and does not need Rad52 as a Rad51 loading mediator. But Rad52 is still needed for second-end capture and formation of dHJs. We assume Rad52 is equivalently needed for its second-end annealing function in mitotic recombination. Support for this idea has also come from immunofluorescence analysis of Rad51 and Rad52 foci of cells undergoing *MAT* switching, which shows that Rad52 foci persist longer, consistent with not only a role in loading Rad51 but also with a later, Rad51-independent annealing step. Once annealing has occurred, the second 3′ end can promote a second round of primer extension.

In the case of *MAT* switching, the annealed second end has a long, nonhomologous 3′ ssDNA tail (the Y sequence); this must be clipped off to expose a 3′ end that is base paired to its template and can therefore be elongated by new DNA synthesis. The "clippase" appears to be the same as used in SSA, requiring Rad1–Rad10 and presumably Msh2–Msh3, Rad59, Slx4, and Saw1. Recent analysis by Eric Alani's lab suggests that the first strand can be ligated to the *MAT*-proximal sequences even if this cleavage doesn't take place, so that after DNA replication, two daughter cells are produced, one of which has completed *MAT* switching and one of which has a lethal chromosome break (**Figure 7.13**).

The completion of the repair event must involve filling-in ssDNA regions that are beyond the points where new synthesis was initiated. Genetic evidence suggests that these regions may be filled in by several different DNA polymerases, but a significant role appears to be played by the error-prone translesion polymerase Rev3 (Polζ). This conclusion is based on the finding by Jeffrey Strathern's lab that there is a 1000-fold, Rev3-dependent increase in mutations that arise in these ssDNA regions during DSB repair. However, deleting Rev3 does not impair completion of *MAT* switching. *MAT* switching is also not impaired in a double mutant lacking both DNA Polζ and Polη (*rev3Δ rad30Δ*) that also ablates another translesion DNA polymerase, Polη, which has been implicated in gene conversions in vertebrate cells.

There must also be ligation of the segments of new and old DNA. Only one experiment has been carried out to address the identity of the ligase involved in this step and it yielded an ambiguous answer. In a strain lacking DNA ligase 4 and temperature-sensitive for the essential DNA ligase 1, *MAT* switching was still completed, as judged from analysis of the DNA strands on denaturing gels. It is possible that the ts DNA ligase 1 retained a small but sufficient activity to complete the job at the restrictive temperature. Alternatively there could be an unknown third DNA ligase in budding yeast, or else the nick could have been "nick-translated" outside the regions embraced by the restriction endonucleases used to create the fragments that could be analyzed.

7.9 A SMALL FRACTION OF ECTOPIC GENE CONVERSION EVENTS ARE CROSSOVER ASSOCIATED

The SDSA mechanism should yield noncrossover outcomes. However, about 1% of *MAT* switching events are associated with crossovers. Such crossovers, between *MAT* and *HML* or between *MAT* and *HMR*, would produce inviable outcomes—a circular chromosome carrying *MAT/HML* or a large deletion fusing *MAT* with *HMR*. If one induces switching in a diploid where only one chromosome III can undergo switching it is indeed possible to recover a low percentage of these "Strathern circles" and "Hawthorne deletions" in cells that are hemizygous for essential regions of the chromosome (**Figure 7.14**).

A more convenient way to analyze crossovers associated with HO-induced gene conversion is to study interchromosomal ectopic recombination. The *MAT* switching system was modified by deleting *HML* and *HMR* and inserting, on a different chromosome, a copy of *MAT* sequences carrying an "inc" mutation that prevents HO cleavage (**Figure 7.15**). The great majority of HO-induced gene conversions occur without a crossing over, but about 4% show novel restriction fragments consistent with a reciprocal crossover (CO) occurring between the donor and recipient chromosomes.

How could such crossovers arise? As mentioned in Chapter 6, one possibility would be that gene conversion still proceeds by SDSA but the small moving D-loop would be trapped by annealing of the opposite end's ssDNA, producing a dHJ intermediate. Alternatively a small proportion of DSB ends could be processed directly into a dHJ pathway. In this case most of the newly copied DNA would remain base paired with the old template strands.

dHJs can be resolved in two distinct ways. First, a combination of the BLM helicase and topoisomerase IIIα can "dissolve" dHJs so that they are resolved as noncrossover (NCO) outcomes. Second, HJ resolvases could create crossovers. If dissolving dHJs is important in reducing crossover outcomes, then one would expect an increase in crossovers in the yeast ectopic system by deleting the BLM homolog, Sgs1, or its associated Top3 and Rmi1 partners. Indeed this is the case; there is about a fourfold increase in crossovers when any of these proteins is deleted (Figure 7.15). These results suggest that as much as 12% of the intermediates of ectopic *MAT* switching involve a dHJ. It is possible—even likely in view of meiotic experiments to be discussed in a later chapter—that HJ resolvases always cleave dHJs to give a crossover outcome; noncrossovers would only arise by SDSA or by "dissolving" dHJs by Sgs1–Top3–Rmi1.

But how would the DSB ends be guided into an SDSA pathway most of the time and a dHJ pathway in other instances? As noted above, this

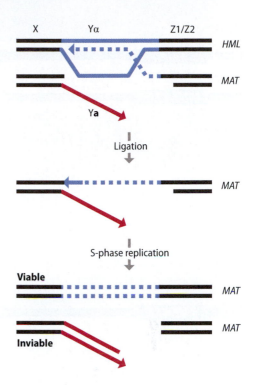

Figure 7.13 *MAT* switching may yield viable cells in the absence of Rad1–Rad10. Rad1–Rad10 is required to clip off the nonhomologous Y tail after second end capture. In the absence of such tail clipping in a *rad1*Δ mutant, ligation of the invading strand to the resected end of the opposite side of the DSB, followed by DNA replication, is apparently able to produce a viable, switched cell and an inviable sister cell. (Adapted from Lyndaker AM, Goldfarb T & Alani E [2008] *Genetics* 179:1807–1821. With permission from the Genetics Society of America.)

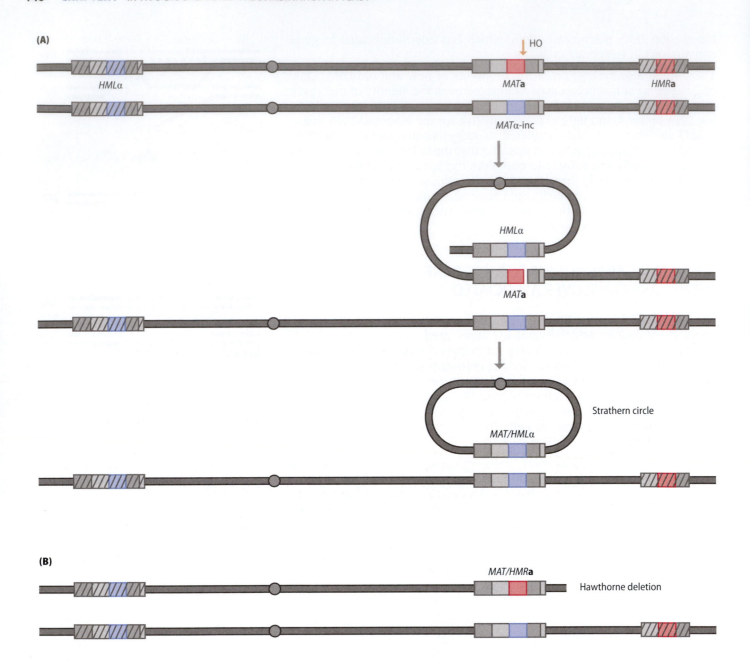

Figure 7.14 A rare crossover during *MAT* switching can produce a stable circular derivative. (A) A crossover during recombination between *MAT*a and *HML*α can produce what is termed a Strathern circle, with the loss of distal sequences that contain essential genes. This derivative is only viable in diploid cells with an intact copy of chromosome III. (B) An analogous *MAT/HMR* deletion, called a Hawthorne deletion, can also be recovered; this deletion also removes essential genes and is only seen in a diploid with an intact chromosome III. (A, adapted from Strathern JN, Newlon CS, Herskowitz I & Hicks JB [1979] *Cell* 18:309–319. With permission from Elsevier. B, adapted from Haber JE, Rogers DT & McCusker JH [1980] *Cell* 1:277–289. With permission from Elsevier.)

would not be hard to imagine if in fact the dHJs were the consequence of a late annealing event during SDSA. But in fact the roles of two other DNA helicases suggest that most dHJs arise from the authentic dHJ pathway.

7.10 TWO OTHER HELICASES REGULATE GENE CONVERSION OUTCOMES

In addition to the role of the Sgs1 helicase in dissolving dHJs, two other 3′ to 5′ helicases play important roles in *MAT* switching. If new DNA synthesis occurs in a small, moving replication D-loop, there is likely to be a helicase that follows the replication bubble, to dissociate the newly synthesized strand. A candidate for this helicase is the FANCM homolog, Mph1. In *mph1*Δ strains undergoing HO-induced ectopic recombination the proportion of crossover outcomes rises from about 4% to more than 10% (Figure 7.15). This increase is independent of the effect of *sgs1*Δ discussed above, and an *sgs1*Δ *mph1*Δ double mutant has more than 25%

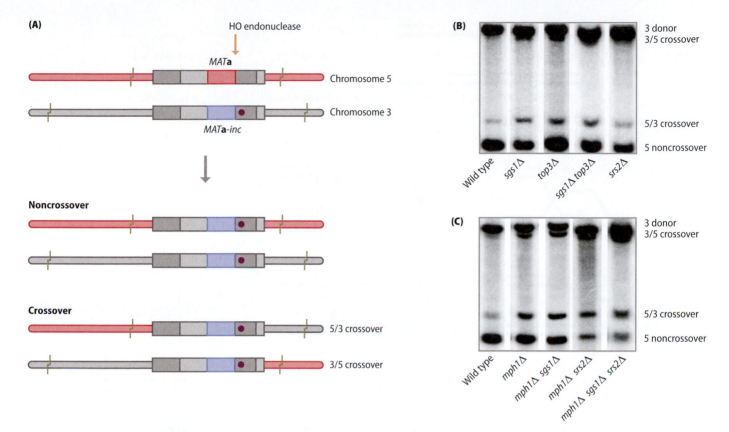

crossovers. A reasonable explanation is that Mph1 is important in displacing the newly made strand from its template; in *mph1Δ* the newly synthesized DNA would be more likely to remain semi-conservatively base paired to the template strand, thus promoting the formation of more dHJs (**Figure 7.16**).

A second helicase, Srs2, appears to function primarily in the SDSA pathway. *In vitro*, Srs2 has been shown to displace Rad51 from ssDNA. This activity has been invoked to explain how Rad51 might be prevented from binding to ssDNA at a stalled replication fork, thus allowing alternative solutions to resume replication. But in HO-induced SDSA, the consequence of deleting Srs2 is the loss of about two-thirds of all the gene conversion events. There is no significant change in the amount of crossover products relative to total DNA, but the percentage of crossovers rises threefold because of the loss of completed SDSA events (Figure 7.15). At what stage the block occurs is not yet clear, though one idea is that Srs2 is needed to remove Rad51 after the strand exchange step. Still, why doesn't the absence of Srs2, like the lack of Mph1, just shunt DSB repair toward dHJs? Is the intermediate shunted to some irreversible structure? This is a problem needing a solution.

7.11 DSB BREAK REPAIR IS SURPRISINGLY DIFFERENT FROM GAP REPAIR

Many of the assays defining the basic outcomes of DSB repair by homologous recombination—transformation and *MAT* switching—were in fact gap repair events. The original dHJ model also was based on the assumption that not only were 5′ ends shortened by 5′ to 3′ resection but the 3′ ends were also trimmed back, so that there was a gap requiring repair. Later experiments showed convincingly that the 3′ ends are incredibly stable and not trimmed back into gaps. For example, the end

Figure 7.15 Interchromosomal gene conversion, using *MAT* sequences can be associated with crossing over. (A) Ectopic recombination between a cloned *MAT*a sequence inserted on chromosome 5 (Chr 5) and a *MAT*a-*inc* mutation (*red dot*) that cannot be cut by HO endonuclease on Chr 3 can occur with or without an associated crossover. In this strain the normal donors, *HML* and *HMR*, are deleted. Restriction endonuclease sites are shown as vertical lines. A crossover produces a pair of reciprocal translocation chromosomes that can be assayed on Southern blots by changes in restriction fragment length. (B) The level of crossing over is seen most clearly for the lower of the two crossover bands (5/3 crossover) in a Southern blot. Crossovers account for only about 4% of the wild-type events. The percentage crossing over increases to greater than 10% in *sgs1Δ* or *top3Δ* mutants that eliminate "dissolving" a dHJ. In contrast, the level of exchange is not actually increased when the 3′ to 5′ *SRS2* helicase is deleted, but the recovery of the noncrossover outcome (presumably by SDSA) is markedly reduced. (C) The effects of deleting the *MPH1* helicase alone and in combination with deletions of the two other 3′ to 5′ helicases are shown. Defects in these three helicases are additive. (A, B, adapted from Ira G, Malkova A, Liberi G et al. [2003] *Cell* 115:401–411. With permission from Elsevier. C, adapted from Prakash R, Satory D, Dray E et al. [2009] *Genes Dev* 23:67–79. With permission from Cold Spring Harbor Laboratory Press.)

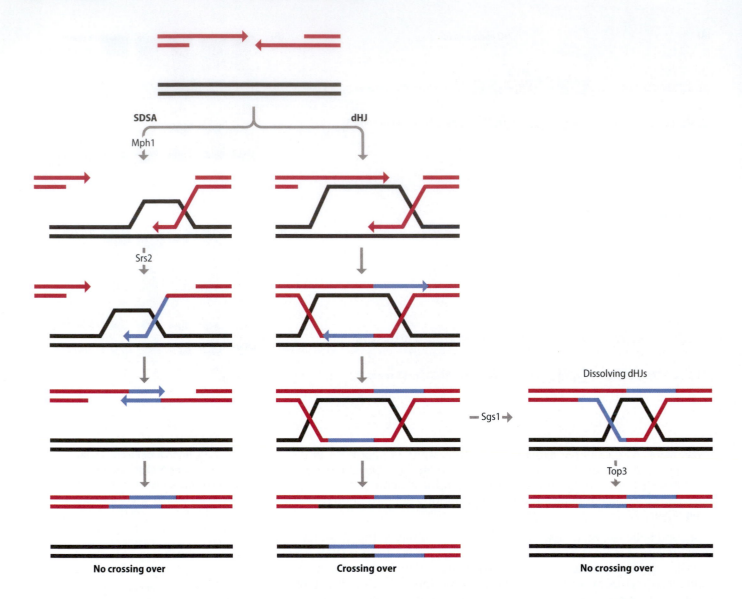

Figure 7.16 Possible roles for Sgs1, Mph1, and Srs2 helicases in regulating crossover and noncrossover outcomes. Sgs1 (along with Top3 and Rmi1) is required to dissolve dHJs. When this pathway is blocked by deleting Sgs1 or Top3, crossovers increase because there is an increase in the abundance of dHJs. In contrast, Mph1 appears to channel early recombination intermediates away from the dHJ pathway so that when Mph1 is absent more intermediates become dHJs. Srs2 acts in a way that is required for completion of SDSA but not the dHJ pathway so that in its absence the abundance of dHJs does not change but many SDSA events fail to be completed. (Adapted from Ira G, Malkova A, Liberi G et al. [2003] *Cell* 115:401–411, and from Prakash R, Satory D, Dray E et al. [2009] *Genes Dev* 23:67–79. With permission from Elsevier and Cold Spring Harbor Laboratory Press, respectively.)

of an HO-induced DSB can engage in SSA events in which one of the homologous sequences is very close to the DSB and the other sequence is far away, requiring hours of resection before it is exposed. Second, the long, 3'-ended resection products can be detected on Southern blots of denatured DNA many hours after the DSB was created, as described in Chapters 4 and 5. But now we realize that gap repair is often carried out in a very different way from true break repair. Gaps can be repaired either by SDSA as we have discussed above or by a modified form of BIR. Two observations drove home this conclusion.

One striking difference between gene conversion and BIR is that initiation of primer extension is greatly delayed—by hours—in BIR compared to gene conversion. This same delay was seen when repair requires gap repair, but only when the gap was 5 kb or larger (**Figure 7.17**). These results suggested that the two ends of a DSB must be in close proximity with respect to the donor template in order to initiate rapid repair of the DSB. The second finding was that repair of gaps ≥5 kb were almost completely Pol32 dependent. Pol32, a nonessential subunit of DNA polymerase δ, has been shown to be required for BIR but have little effect on gene conversions (even in *MAT* switching where there is a gap of about 700 bp). A gap of only 1.2 kb showed a >50% reduction without Pol32 (**Figure 7.18**). These results raise the likely possibility that gap repair frequently occurs

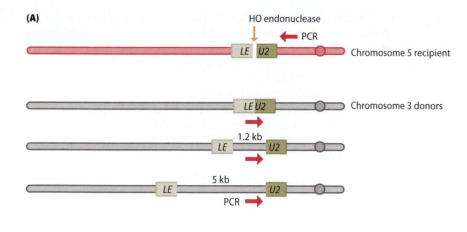

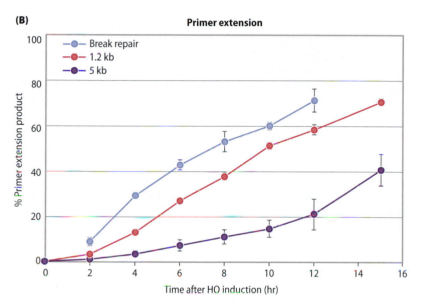

Figure 7.17 Differences between break repair and gap repair reveal a recombination execution checkpoint. (A) An HO-induced DSB in a *LEU2* gene containing a cloned HO cleavage site, located on Chr 5, can be repaired by a donor locus on Chr 3. Three donors that differ in the amount of space between the common *LE* and *U2* homologies were tested. The initiation of repair is assayed by using a pair of PCR primers that will only yield a product when the *U2* end of the DSB invades the donor and initiates new DNA synthesis. Thus the differences in kinetics do not reflect the length of time to complete repair but only to start new DNA synthesis. (B) Whereas the kinetics of repair with a gap of 1.2 kb is similar to that for break repair (the donor with no space separating *LE* and *U2*), the kinetics of repairing a 5-kb gap are markedly delayed, more resembling the kinetics of BIR (shown in Chapter 8). (Adapted from Jain S, Sugawara N, Lydeard J et al. [2009] *Genes Dev* 23:291–303. With permission from Cold Spring Harbor Laboratory Press.)

by the initiation of BIR. One possibility is that strand invasion of one end initiates both leading- and lagging-strand synthesis that meets the second DSB end after traversing the gap and the ends are then joined by a strand annealing event. Alternatively, both ends might initiate converging replication forks that would terminate when they collided.

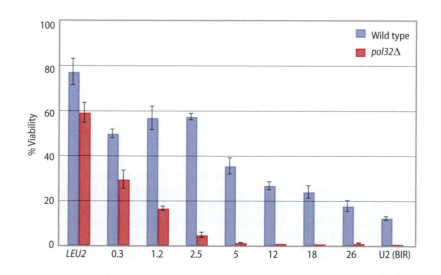

Figure 7.18 Pol32 is required for gap repair but has little effect on break repair. The experiment shown in Figure 7.17, including strains with still larger gaps, was performed in wild-type and *pol32Δ* cells, measuring cell viability after HO endonuclease induction. Pol32 has been shown to be essential for ectopic BIR events, but to have only a small effect on gene conversion (that is, break repair). For reference, a strain was included with only the *U2* homology present (hence only BIR is possible). Even with a gap of 1.2 kb, Pol32 appears to have an important role that becomes essential with a 5-kb or larger gap. (Adapted from Jain S, Sugawara N, Lydeard J et al. [2009] *Genes Dev* 23:291–303. With permission from Cold Spring Harbor Laboratory Press.)

These results led us to suggest that there was a *recombination execution checkpoint* (REC) that delays the initiation of new DNA synthesis if the ends of the DSB are not strand-invaded close together on the same template DNA. What might be the advantage of such a checkpoint? One conjecture is that if a DSB occurs in a repeated sequence, it would be dangerous if the two ends invaded different homologous sequences on different chromosomes and each initiated BIR. The REC stops rapid initiation of synthesis and provides time so that the ends might dissociate and repeat a search for homology. This would allow both ends to be aligned on the same template.

It should be noted that one can readily obtain budding yeast transformants in which a plasmid can only be repaired by gap-repairing 40 kb or more from a chromosomal template; however, the efficiency of gap repair also declines with the distance that polymerases must traverse. For example, when each of the ends of the DSB shares 2 kb homology with a site on another chromosome, the efficiency falls from >80% for a simple break repair to about 35% when the gap is 5 kb (Figure 7.17) and the events are nearly all Pol32 dependent (Figure 7.18). But efficiency drops further, to 18% when there is a 26-kb gap. When there is only one homologous end so that repair must be through BIR that must traverse 35 kb to the chromosome end, repair is only 12% (Figure 7.18). Despite the fact that BIR appears to use all of the well-studied replication factors that are needed in normal replication, it may still be inherently less processive. Alternatively, because of the delays imposed by the REC, cells are all arrested in G2/M before repair DNA synthesis is initiated. Under these circumstances, some normal DNA replication components may be inherently limiting; for example the MCM proteins are largely exported from the nucleus in G2/M cells.

These surprising results suggest that we should re-examine some of the events studied previously to support SDSA mechanisms. For example one experiment that lent strong support to the SDSA model was carried out both in *Drosophila* (where P-element excision produced the DSB) and in budding yeast, with an HO break. Repair required the copying of 8 tandem copies of a 375-bp *Drosophila* 5S DNA (see Figure 6.38). Half of the outcomes had 8 copies of the 5S sequence in the recipient, while the other half had a different number of repeats from 1 to 13 copies in the recipient; all of the donor sequences retained 8 copies. That all the changes were observed in the recipient is the expected result for SDSA; however, how copy number changes would occur was less certain. It is possible that the events that have 8 copies in the recipient represent instances where one end initiated copying and went across the entire region to anneal with the second DSB end. Those events with fewer or more than 8 copies could arise by replication slippage during SDSA, but they could also appear if each end independently initiated synthesis of one strand and then the two complementary, partially copied repeated sequences annealed with each other. However, those outcomes that do not have 8 copies in the recipient could instead be the result of two independent BIR events. In the case of two independent BIR events the joinings would occur by SSA of the partially copied segments. One wonders whether they would be eliminated in a *pol32Δ* strain.

The idea that events initiated by BIR could be completed by SSA was previously suggested by Lorraine Symington to explain the outcomes of DSB-initiated recombination between inverted repeat sequences located on a circular plasmid. Depending on which end of the DSB initiated new synthesis, an SSA event between the newly copied sequences and the second end could lead to a noncrossover or a crossover (**Figure 7.19**).

Figure 7.19 Repair of a plasmid with a DSB in inverted repeat sequences by a combination of BIR and SSA. A DSB in one repeat is repaired by BIR/SSA to yield both crossover and noncrossover outcomes. Strand invasion and BIR from one end replicates to nearly the end of the broken, now-linear molecule (shown in *dark blue*), stopping where resection has shortened the template. Extensive 5′ to 3′ resection from both ends leads to two alternative SSA events, one of which restores the original configuration (as seen by the position of the *gray* rectangle) while the other results in an apparent crossover. (Adapted from Kang LE & Symingston LS [2000] *Mol Cell Biol* 20:9162–9172. With permission from the American Society for Microbiology.)

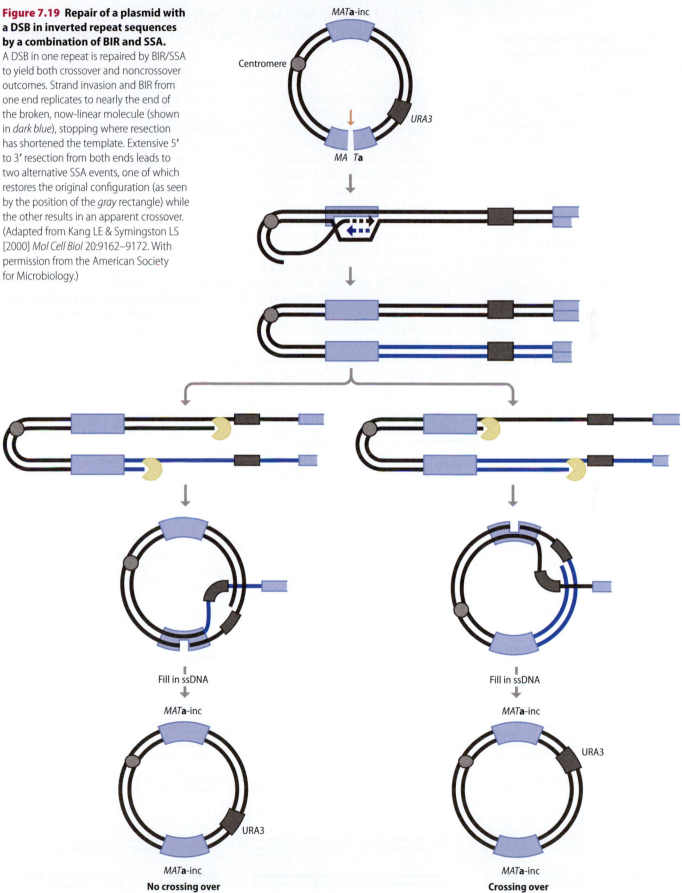

Figure 7.20 Rapid mismatch correction of a mutation during strand invasion. Gene conversion during *MAT* switching in a case where there is a base-pair difference (*red*) between the donor and recipient to the right of the HO cleavage site can occur in different ways. (i) SDSA in which there is no mismatch correction of the heteroduplex while *MAT* ssDNA is annealed with *HML* results in *MAT* switching in which the mutation in the donor sequence is not transferred to the recipient. (ii) SDSA in which there is rapid mismatch correction of the heteroduplex—and in which the invading strand is preferentially corrected to the genotype of the donor—results in *MAT* switching with the transfer of the donor sequence variant to *MAT*. (iii) If repair proceeds by forming a dHJ intermediate, then dHJ dissolution can produce a switched *MAT* locus with a heteroduplex that, after replication, will produce a sectored colony. Such sectored colonies arose when mismatch correction was eliminated by deleting the *PMS1* gene. (Adapted from Haber JE, Ray BL, Kolb JM & White CI [1993] *Proc Natl Acad Sci USA* 90:3363–3367. With permission from the National Academy of Sciences.)

7.12 STUDY OF MISMATCH CORRECTION DURING STRAND INVASION RAISES QUESTIONS ABOUT HOW *MAT* SWITCHING OCCURS

The involvement of BIR even in some modest gap repairs may also help explain another result that doesn't easily "fit" with simple SDSA: the fate of heteroduplex DNA during *MAT* switching. Strand invasion results in the formation of heteroduplex DNA in the Z region. Evidence for heteroduplex formation can also be seen in fate of a T → A base pair difference at position Z11, located 9 nt from the 3′ end of the DSB (**Figure 7.20**). In *MAT***a** cells the Z11 mutation creates a "stuck" (*stk*) phenotype in which HO cleavage is significantly reduced. The same base-pair change in *MAT*α results in a sterile (Matα1⁻) phenotype and an almost complete resistance to HO cutting. When *MAT***a**-stk is induced to switch, most cells switch to *MAT*α, where the *stk* mutation has been replaced [scenario (ii) below]; however, a minority are converted to *mat*α-*stk*. Moreover, 25% of the time the cells switched such that one of the daughter cells, produced after the switched DNA was replicated, was *MAT*α while the other was *mat*α-*stk* [scenario (iii) below]. These *sectored colonies* suggest that there was heteroduplex DNA remaining at *MAT* after gene conversion, such that one strand replicated to *MAT*α and the other to *mat*α-*stk*. The proportion of these colonies increased to 60% in the absence of the *PMS1* mismatch repair gene and another 25% resulted in two *mat*α-*stk* daughter cells. In wild-type cells the DNA sequence of the initial PCR product reflecting the earliest synthesis of new DNA from the *HML*α template showed that

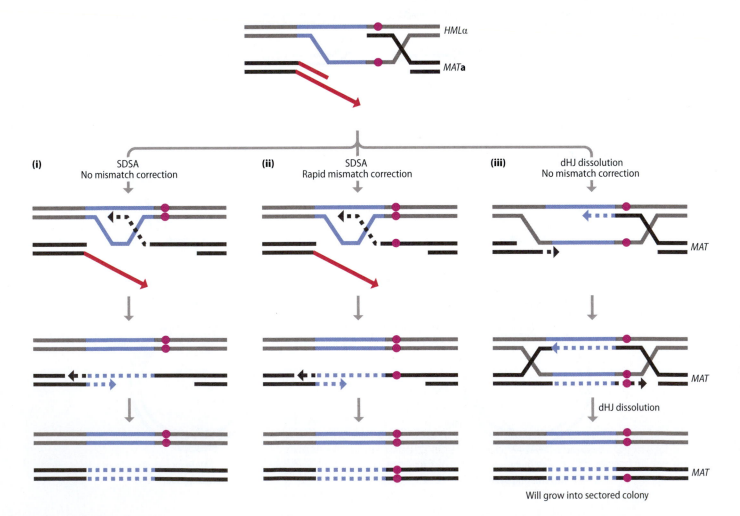

Will grow into sectored colony

the Z11 site had the wild-type T allele, whereas when the same analysis was done in the *pms1Δ* strain retention of the A (*stk*) allele was evident. These results suggest that heteroduplex DNA is rapidly corrected, before strand displacement of the newly synthesized strand (Figure 7.20, panel ii). However, further reflection raises an important question: if the second DNA strand during *MAT* switching is primer extended from the opposite DSB end after it has annealed with the newly made strand that was displaced from the donor template, how can there be sectored colonies? The SDSA model would predict that the two DNA strands at *MAT* should be both wild type or both mutant; there is no obvious way to have heteroduplex DNA at *MAT*.

One way that heteroduplex DNA could form in the *MAT-Z* region would be that the left end independently strand-invades and copies the same template, extending into the Y region and copying Z11T, as expected for a dHJ mechanism accompanied by dissolution of the dHJ (Figure 7.20, panel iii). Another possibility is that the two ends use BIR to copy at one or both ends. If this were the case, the unwinding of the template duplex from the newly made strands could produce heteroduplex DNA. However, a strong argument against two BIR events is that both kinetically and genetically, *MAT* switching with its small gap of about 700 bp is distinctly different from BIR. As noted above, BIR events are strongly dependent on the Pol32 subunit of DNA polymerase δ, but *pol32Δ* has little effect on *MAT* switching.

7.13 ANOTHER SPECIAL FEATURE OF *MAT* SWITCHING IS DONOR PREFERENCE

MAT switching is unusual because it demands recombination with highly heterochromatic donor sequences; essentially these are the only two such non-telomeric silenced regions in budding yeast. However, it turns out that—at least in wild-type cells—there is little change in the kinetics or outcomes of switching if the donor is unsilenced. Moreover, there is another aspect of *MAT* gene conversion that is remarkable: *MAT* switching exhibits strong *donor preference*, such that *MAT***a** recombines 90% of the time with *HML* whereas *MAT*α strongly chooses *HMR*. This preference is maintained even if *HML* carries Y**a** instead of the normal Yα or if *HMR* carries Yα.

Donor preference is controlled by the *cis*-acting ~275-bp recombination enhancer (RE) located about 17 kb centromere proximal to *HML* (**Figure 7.21A**). *MAT***a** cells lacking RE behave like *MAT*α and recombine almost always with *HMR*. In *MAT*α cells, the RE is inactivated by the binding of the Mcm1-Matα2 repressor complex (Figure 7.21B). Repressor binding sets up a small heterochromatic domain that prevents any other proteins binding and thus *MAT*α cells behave as if RE is absent. In *MAT***a** cells, RE binds three transcription factors—Mcm1, Fkh1, and Swi4/Swi6—even though there are no open reading frames nearby. One way donor preference might work would be to alter the silent chromatin structure of the nearby, preferred donor to make it more accessible; but in fact there is no change in donor preference when the "wrong" donor is unsilenced. Instead, it seems that RE causes a DSB-dependent change in the conformation of chromosome III, effectively reducing the distance between *HML* and *MAT***a** (Figure 7.21A). This large-scale chromosome restructuring can be seen even when recombination is prevented by deleting the actual homology between *MAT* and *HML*. This change in chromosome conformation depends on the FHA domains of the multiple copies of the Fkh1 protein that bind to RE. Presumably the FHA domains associate with phosphothreonine modifications of a protein bound near the DSB,

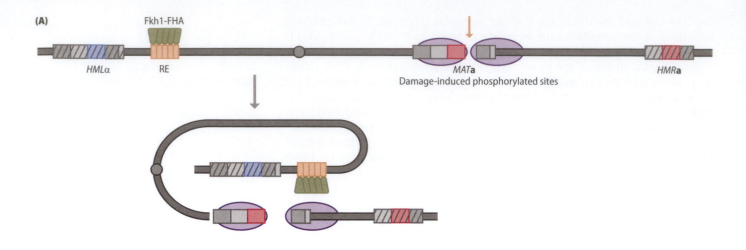

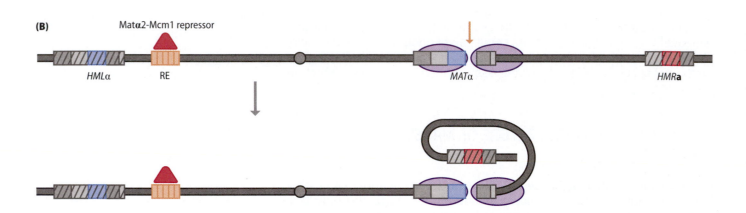

Figure 7.21 Donor preference in budding yeast. (A) The recombination enhancer (RE) contains multiple binding sites for the Fkh1 protein. The phosphothreonine-binding domain of Fkh1 (Fkh1-FHA) associates with DNA damage-induced modifications of a still unidentified protein that binds close to the ends of the DSB (shown as a *purple* area) and thus brings *HML* into proximity with the DSB at *MAT*. (B) RE is active in *MAT**a*** cells but repressed by the Matα2-Mcm1 repressor complex in *MAT*α cells, thus allowing *HMR* to act as the preferred donor, as it does when RE is deleted. (Adapted from Li J, Coï E, Lee K et al. [2012] *PLoS Genetics* 8:e1002630.)

but this casein kinase-II-dependent modification is unrelated to histone modifications such as γ-H2AX. Thus donor preference can be established in a strain in which the RE is replaced by four LexA binding domains, to which a LexA-FHA(Fkh1) fusion protein binds. ChIP has shown that the FHA domain will bind to the first several kilobases on either side of the DSB and will thus bring HML into proximity with the broken *MAT* locus. Whether this clever strategy is used to promote specific recombination events in metazoans is not yet evident, but there seems to be an analogous recombination enhancer that enforces donor preference in the very different mat switching system of *S. pombe*, which is described below.

7.14 MATING-TYPE SWITCHING IN *S. POMBE* IS SURPRISINGLY DIFFERENT FROM THAT IN *S. CEREVISIAE*

Fission yeast also switches mating-type genes by DSB-induced gene conversion. There are two alternative, silenced donors carrying opposite mating-type information. However, the *S. pombe* system has proven to be different in essentially almost every detail compared to *MAT* gene switching in budding yeast. To begin, there is no site-specific endonuclease that creates a DSB even though a DSB is created. The mating-type alleles in *S. pombe* are known as P (plus) and M (minus) and are expressed at the *mat1* locus, which contains the M or nonhomologous P DNA surrounded by the H1 and H2 sequences (**Figure 7.22A**). Approximately 15 kb away are two silent donors, usually *mat2-P* and *mat3-M*; each has H1 and H2 surrounding the M or P sequence as well as sharing a third homology block, H3.

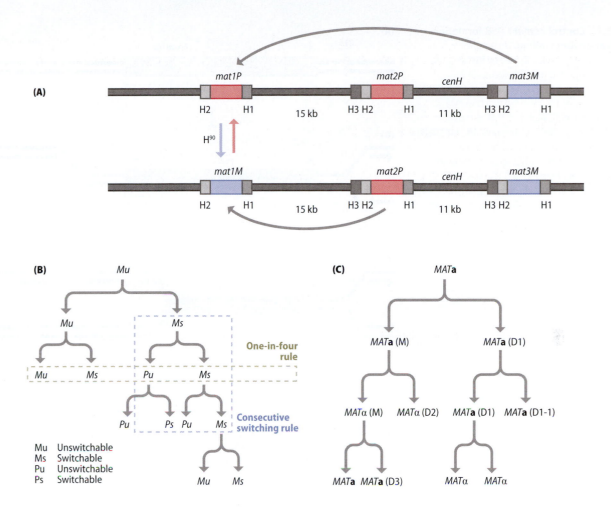

Both *mat2* and *mat3* lie within a transcriptionally silenced region surrounding the heterochromatic *cenH* region of repeated K sequences that exhibit homology to centromere sequences. Despite the proximity of the two donors, *mat* switching exhibits strong donor preference, with *mat2P* being the preferred donor for *mat1M* whereas *mat3M* is the partner for 90% of the replacements of *mat1P*.

In fission yeast, only one of the two sister cells switches while the other acquires the potential to switch in the following cell division (**Figure 7.22B**). Once a cell has switched, one of its daughters will be able to switch in every succeeding generation. This puzzling pattern became understandable when Amar Klar discovered that *mat1* acquires a strand-specific "imprint"—a single-strand nick that cannot be ligated—which is converted by DNA replication into a DSB on one chromatid and an intact chromatid, which again inherits the imprinted mark (**Figure 7.23A**). So, unlike budding yeast, there is no site-specific endonuclease that creates the DSB and the pedigree of switching is quite distinct from budding yeast (**Figure 7.22C**).

Precisely what this imprint is and how the single-strand nick is created prior to replication is still the subject of intense investigation and is discussed below, but it is clear that the DSB only appears in S-phase cells, whereas a nick can be detected in cells prior to replication, as seen on denaturing gel electrophoresis to separate the strands. The nick appears to lack a 5′ or 3′ OH so that it cannot be ligated *in vitro*. Some evidence suggests that the nick is derived from one or more RNA nucleotides inserted at that site, which could be the remnants of an incompletely processed Okazaki fragment. In any case, the starting point for *mat1* switching is a

Figure 7.22 Mating-type switching in fission yeast. (A) The *mat1* locus can contain either P or M information. The *mat2P* and *mat3M* loci share short regions of homology (H1 and H2 with *mat1* as well as H3 shared only by the donors). Homothallic switching, from one mating type to the other, can occur in strains designated h90. Unlike budding yeast the expressed *mat1* locus and its two unexpressed and heterochromatic donors are all close together. *mat2* and *mat3* are separated by a heterochromatic region with homology to centromere sequences (*cenH*). *S. pombe mat1* switching shows strong donor preference, with *mat1P* recombining with *mat3M* and *mat1M* using *mat2P*. (B) Pattern of *mat* switching in *S. pombe*. Only one in four cells from an unswitched mother will switch (the one-in-four rule). After switching has occurred one of the two daughter cells can switch (the consecutive switching rule). This pattern is distinctly different from that of budding yeast (C), where a mother (M) and its first daughter (D1) do not switch but the same mother and its second daughter (D2) switch in concert, while the first daughter becomes a mother and does not switch with its first daughter (D1-1). After the first generation, budding yeast cells that have already switched can switch again. (Adapted from Egel R [2005] *DNA Repair* 4:525–536. With permission from Elsevier.)

Figure 7.23 Control of *mat1* DSB formation by replication fork stalling. (A) A rightward-moving replication fork is blocked by Rtf1 and Rtf2 proteins bound to the *RTS1* locus. (B) A stable nick in the top strand at the edge of H1 region of homology is converted to a DSB by a leftward-moving replication fork that also pauses at a *Mat Pausing Site* (MPS1). Thus only one of the two daughter DNA strands carries a DSB. (C) At the same time, the other replicated strand is imprinted. Rather than repair this one-ended break by recombining with its sister chromatid, this DSB end is used to initiate recombination with either *mat2* or *mat3*. (Adapted from Arcangioli B & de Lahondes R [2000] *EMBO J* 19:1389–1396. With permission from Macmillan Publishers, Ltd.)

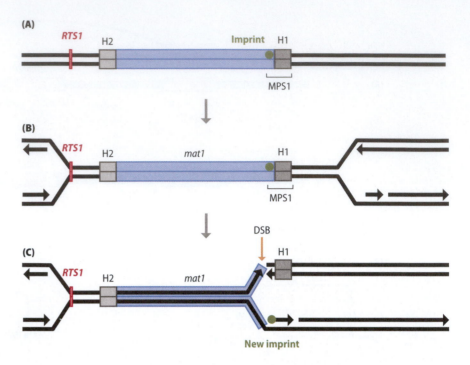

broken replication fork, the kind of intermediate we examined in Chapter 1 that is repaired by BIR. But the situation in *mat1* switching is much more complicated: most surprising is that the repair does not normally engage the intact (or maybe nicked) sister chromatid but instead recombines with one of the silent donors.

The creation of the nick and its subsequent conversion into a DSB depends on very precise controls of replication around the *mat1* locus. Replication proceeding rightward is strongly blocked by the directional fork-blocking *RTS1* site and its binding protein Rtf1 (**Figure 7.23B**). Replication proceeding leftward passes the *SAS1* site that is essential for creating the break and pauses at the *MPS1* site (**Figure 7.23C**). At some point the top strand of the second sister chromatid also becomes nicked. This process requires several "*swi*" (switching) genes. *SWI1* and *SWI3* are involved in replication fork stalling step. *SWI7* encodes DNA Polα, which could mean that creating the nick is tightly coupled to DNA replication. Richard Egel has pointed out that replication problems over the entire genome in the partially functional *swi7* strain could "distract" Swi1/Swi3 away from the programmed stalling events at *mat1*. This would mean that the lack of nicking could be attributed to a reduced presence of Swi1–Swi3. Benoit Arcangioli's lab found that the nick can be erased by inducing transcription across the site, thus allowing one to control the induction of *mat* switching (**Figure 7.24**). This trick makes it possible to follow the switching event more synchronously and to show that creation of the single-strand nick and replication fork stalling at *MPS1* appear to be coordinated (Figure 7.24).

When replication passes the nick made in the previous cell cycle, a break is made while the converging replication fork is blocked. As noted above, this break should be the perfect substrate for recombinational repair of a single-ended DSB by BIR, using the intact sister chromatid as a template; but this is not what happens under normal circumstances. Rather, the DSB end initiates recombination with either *mat2* or *mat3*, depending on which is the preferred donor. The details of the repair event are not known. The majority of data favor repair by SDSA, with one strand being synthesized before the other. An SDSA model is illustrated in **Figure 7.25**.

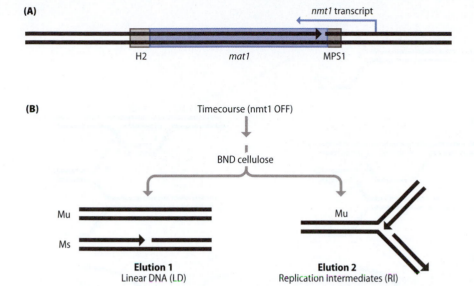

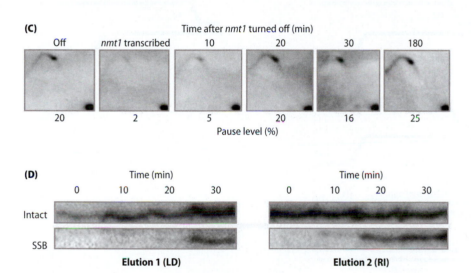

Figure 7.24 ***mat1* imprinting is repressed by transcription across the MPS1 site.** (A) When a transcript from a nmt1 promoter crosses the MPS1 region, imprinting is prevented. (B) Linear and branched DNA molecules can be fractionated using column chromatography with BND cellulose. The un-nicked Mu (unswitchable) and nicked Ms (switchable) forms can be analyzed. If the *nmt1* promoter is repressed so that a nick is produced, then one can see the rapid establishment of a stalled replication fork on two-dimensional gels (C) and the coincident formation of a single-strand break (SSB), which can be visualized by a Southern blot following denaturing gel electrophoresis (D). The SSB is first seen in the replicating fraction, demonstrating that its origin is tied to replication. (Adapted from Holmes AM, Kaykov A & Arcangioli B [2005] *Mol Cell Biol* 25:303–311. With permission from the American Society for Microbiology.)

The SDSA model is supported by a clever genetic experiment by Amar Klar that introduced a mutant *P* sequence at *mat3* in place of the normal M sequences. In this strain *mat1P* carries a different mutation (**Figure 7.26**). When the mutant *mat1P (P-BamHI)* recombines with *mat3P (P-BclI)* it was possible to recover wild-type *mat1P* sequences. The most likely source of such recombinants would be heteroduplex DNA containing an intermediate that can be mismatch corrected to a wild-type *mat1-P* allele.

It is possible to envision a similar outcome if the H1 end of *mat1* invades *mat3* and proceeds by BIR (Figure 7.25). Dissociation of the newly replicated DNA from the *mat2* or *mat3* template could be followed by re-joining with the other end of the *mat1* locus through an SSA event, similar to what was illustrated in Figure 7.19. In any case physical monitoring of switching by PCR makes it clear that the synthesis of an intermediate must at least sometimes extend into and even beyond the H3 region (sequences not at *mat1*), since a primer distal to *mat1* and a primer proximal to *mat3* will produce a product (**Figure 7.27**).

Clear evidence that both strands of the switched *mat1* locus were newly synthesized, either by SDSA or BIR-SSA, came from Benoit Arcangioli's important experiment using the "heavy–light" (HL) density transfer

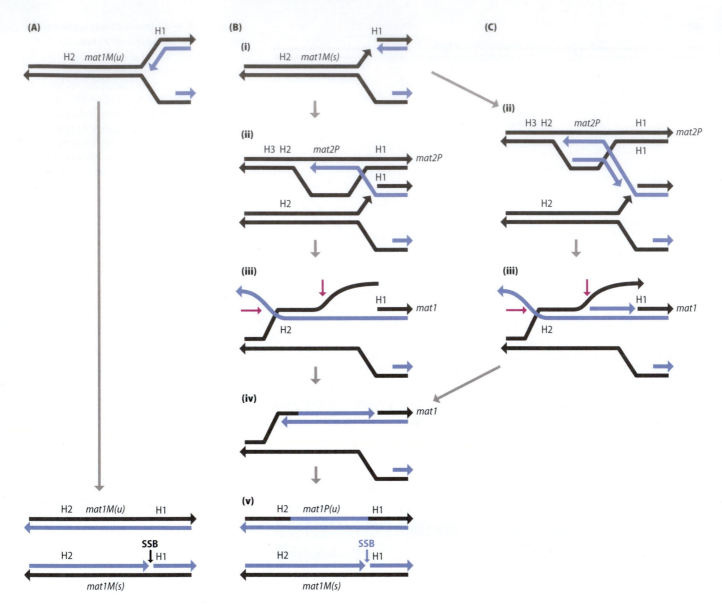

Figure 7.25 A possible recombination mechanism to produce *mat* switching in fission yeast. (A) DNA replication of the unswitchable Mu strain results in an intact Mu daughter cell and an Ms daughter that has an imprint, or nick, on one strand. Newly synthesized DNA is shown in *blue*. (B) An SDSA-like mechanism of *mat1* switching. (i) A DSB is created by replication arriving at the single-strand nick in *mat1M*. The one-ended chromosome break is then resected and—with Rad51—invades the *mat2P* donor to copy one strand that is displaced (ii), as in other SDSA events, and re-anneals with homology in the H2 region (iii). After the nonhomologous M sequences are clipped off (iii), the newly generated end can begin second strand synthesis (iv), creating a complete copy of *mat1P(u)* and an mat1M(s) sister cell (v). (C) An alternative possibility for *mat1* switching would be that the DSB end invades H1 sequences at *mat2-P* and initiates a BIR-like event (ii) that would be joined back to *mat1-H2* sequences by SSA after one strand was resected in a 5′ to 3′ direction (iii). In either case, how this annealing takes place without destroying the integrity of the bottom DNA strand, which remains unswitched, is not yet understood. The bottom strand eventually completes replication with the establishment of the single-strand nick imprint. (Adapted from Arcangioli [2000] *EMBO Rep* 1:145–150. With permission from Macmillan Publishers, Ltd.)

method. Cells were transferred from normal light medium to heavy isotope medium. After a single round of replication, all of the bulk DNA shifts to an HL density, but when the density-fractionated DNA was probed on a Southern blot it was clear that 25% of the *mat1* DNA was HH (**Figure 7.28**). Given that only one in four cells switch in any generation (Figure 7.21) this is the amount of DNA that one would expect to represent switching events. A control strain carrying a *mat1M-smt0* deletion

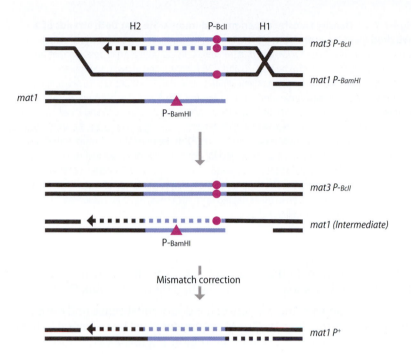

Figure 7.26 Evidence of a heteroduplex intermediate during *S. pombe mat* switching. In the special circumstances when a mutant *mat1P* (*P-BamHI*) recombines with a different mutation in the *mat3P* (*P-BcII*) donor, switching can produce a wild-type *mat1P* outcome. These recombinants suggest that switching must have generated a heteroduplex intermediate, subject to mismatch repair. (Adapted from Yamada-Inagawa T, Klar AJ & Dalgaard JZ [2007] *Genetics* 177:255–265. With permission from the Genetics Society of America.)

that prevented it from switching showed only the expected HL density. These data are different from budding yeast density transfer experiments (Figure 7.9) although in both cases the switched mating type locus had two newly synthesized strands. In the case of fission yeast, one strand of the *mat1* region must be heavy because switching takes place during

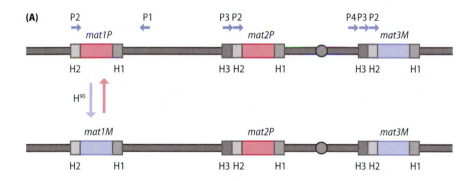

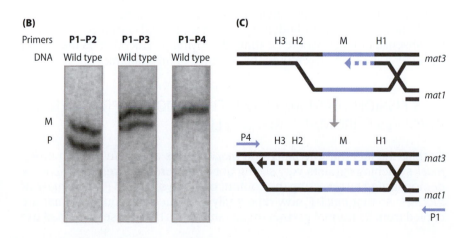

Figure 7.27 Physical monitoring of new DNA synthesis during *mat* switching in fission yeast. (A) In a continuously switching population (h[90]) PCR primers P1 and P2 detect both P and M sequences at *mat1*. PCR primer P3 is homologous to H3 sequences found at *mat2P* and *mat3M* but not at *mat1*, while P4 is *mat3M*-specific. (B) The appearance of P1–P3 PCR products suggests that DNA synthesis must initiate in the donor and extend to at least the H3 sequences that are not present at *mat1*. Similarly a P1–P4 product shows that new synthesis can extend beyond the boundaries of shared sequences. (C) Illustration of repair synthesis extending beyond H3. (A, B, adapted from Arcangioli B & de Lahondes R [2000] *EMBO J* 19:1389–1396. With permission from Macmillan Publishers, Ltd.)

(A)

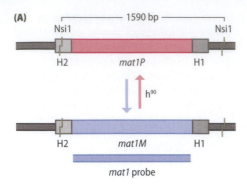

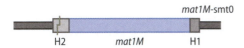

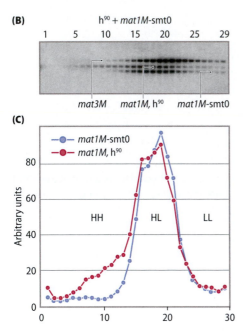

Figure 7.28 Density transfer experiments demonstrate that both strands of a switched *mat1* locus are new. (A) A mixture of homothallic (h^{90}) cells, which switch back and forth from *mat1P* to *mat1M*, and *mat1M*-smt0 cells carrying a small deletion that prevents switching were transferred to heavy (^{13}C and ^{15}N) medium and allowed to replicate once. DNA was fractionated by CsCl density gradient separation and the fractions (1–30) separated on a Southern blot after restriction endonuclease digestion. (B) A probe to M sequences hybridizes both to the *mat3M* donor and to the *mat1* locus, which is a mixture of M and P cells, as well as to nonswitching *mat1M*-smt0 DNA, which migrates as a smaller band because the smt0 mutation is a small deletion. (C) Graphical representation of (B). In the *mat1M*-smt0 mutant that cannot switch, all the *mat1* density is HL, as expected for normal replication, but in a strain in which switching can take place, 25% of the *mat1* sequences are HH, only expected if *mat* switching has resulted in the copying of both DNA strands in the newly recombined product. Note also that *mat3M* DNA also is HL as seen in (B). (Adapted from Arcangioli B [2000] *EMBO Rep* 1:145–150. With permission from Macmillan Publishers, Ltd.)

normal replication. In the budding yeast experiment, *MAT* switching produces two new strands at *MAT* even when there is no bulk DNA synthesis.

Whichever happens, the copying of the donor must cease and the extended DSB now carrying the new P or M sequences must be re-joined to the *mat1* region at the shared H2 homology region (Figure 7.27). Copying beyond H2 of the donor may be partially blocked by the presence of imperfect inverted repeat sequences in H3 that might promote the dissociation of the repair replication machinery, but as we saw in Figure 7.27, copying does go into and past H3 in some cases. In any case, there is apparently a need to clip off some nonhomologous sequences, possibly those extending into H3 or the donor region, because other "swi" genes encode the Msh2 and Msh3 proteins (*swi4* and *swi8*) that along with Rad1–Rad10 (*swi9* and *swi10*) are needed, apparently to remove a nonhomologous tail from recombination intermediates (**Figure 7.29A**). In budding yeast *MAT* switching, the role of this "clippase" including Rad1–Rad10 and Msh2–Msh3 is obvious: the original Y sequence has to be clipped off before the second strand of SDSA can proceed; but in fission yeast *mat1* switching, it is not yet clear just what is being removed. One result that suggests that the original *mat1-P* or M region is being clipped off (analogous to that in budding yeast) is that, in a *swi8* mutant lacking Msh3, *mat1P* is often replaced by a duplication that includes *mat2P*, the 11 kb spacer, and *mat3M*. This duplication can be explained if repair DNA synthesis started at *mat3M*, but in these mutant strains there is no way to clip off the nonhomologous *mat1P* sequences; consequently switching will not be completed. The only viable outcomes would be those instances in which copying that started at *mat3M* continued until *mat2P* was at least partly copied, so that the P sequences at *mat1P* and those just copied at *mat2P* could anneal with no requirement for the Rad1–Rad10, Msh2–Msh3-dependent clipping activity (**Figure 7.29B**). (Analogous duplications were shown to occur in budding yeast in a plasmid-based SSA assay in the absence of Rad1, where continued copying produced ends that could perfectly anneal and overcome the absence of tail clipping.)

7.15 FISSION YEAST *MAT1* SWITCHING DONOR PREFERENCE INVOLVES CHROMATIN REMODELING

Despite the fact that the *mat2* and *mat3* donors are close to each other, *mat1* switching exhibits very strong donor preference. Preference can be demonstrated by reversing the content of the two donors (that is, *mat2-M mat3-P*), so that *mat1-P* now repeatedly introduces P sequences that are copied from its normal partner, *mat3*. Mutations that cause increased use

(A)

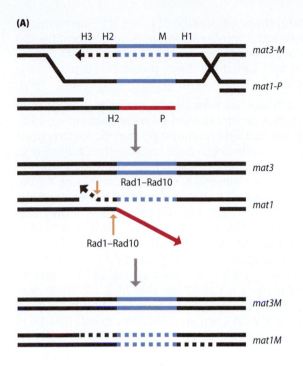

Figure 7.29 Removal of nonhomologous sequences during *mat* switching. (A) As in budding yeast *MAT* switching, the annealing of the newly copied sequences at *mat1* is followed by clipping off the nonhomologous tail by Rad1-Rad10 endonuclease with the participation of Mah2-Msh3. (B) In mutants lacking Msh2 or Msh3 (when the *swi4* or *swi8* genes are deleted) switching may occur in which both *mat3* and *mat2*, plus the intervening 11-kb region, are copied into the *mat1* locus. This event can occur presumably without the need for removal of nonhomologous DNA tails, so that the newly copied *mat2P* sequences can anneal with the original P sequences at *mat1*. (Adapted from Fleck O, Heim L & Gutz H [1990] *Curr Genet* 18:501–509. With permission from Springer Science and Business Media.)

(B)

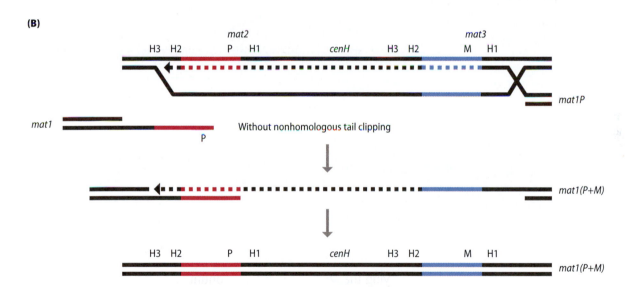

of the "wrong" donor were identified, including *swi2*, *swi5*, *swi6*, and a set of "*clr*" mutants. The Swi5 protein, discussed in Chapter 3, is a novel recombination protein that can act with Sfr1 in general DSB repair but in *mat1* switching it acts with another partner, Swi2. Swi2 does not affect general homologous recombination. Swi2 and Swi6 opened up another realm of study in *S. pombe*, the control of heterochromatin both in the *mat2/mat3* region and at centromeres.

Swi6 proved to be the *S. pombe* homolog of the chromodomain protein HP-1 in *Drosophila* that is responsible for establishing and maintaining heterochromatin. Clr4 is the *S. pombe* homolog of *Drosophila* Su(var)3-9, encoding a histone H3-K9 methyltransferase. These genes have no orthologs in budding yeast. Three other proteins, Clr3, Clr6, and Sir2 are histone deacetylases that are needed to silence the donors and to maintain donor preference. All of these mutations cause the unsilencing of the *mat2–cenH–mat3* region and cause expression of genes that are normally silenced when they are inserted into centromere heterochromatin. So donor preference can be overcome by unsilencing the *mat2–mat3* region. In contrast, budding yeast *MAT* switching donor preference is unaffected by unsilencing one of the donors.

More recently Shiv Grewal's lab has provided a clearer picture of how donor preference is affected. A recombination-promoting complex, RPC, consisting of Swi2 and Swi5, exhibits mating-type specific localizations in the donor region. In *mat1-P* cells, Swi2–Swi5 localize (by ChIP) only to a recombination enhancer, SRE, that is located adjacent to *mat3*, but in M cells, Swi2–Swi5 spreads over the entire silent region (**Figure 7.30A**). In *swi2Δ* cells, only *mat2* is used as a donor (**Figure 7.30B**). In contrast, *swi6Δ* results in the exclusion of *mat2*. The double mutant shows that both donors can be used. Swi2 physically interacts with Swi6, but Swi2 is not required to form heterochromatin. Conversely, in P cells, Swi2 localizes to a small region to the right of *mat3*, independently of Swi6; but in M cells, the spreading of Swi2 (and Swi5) across the region depends on Swi6. The binding site near *mat3* is called the *switching recombination enhancer 3* (SRE3). Swi2–Swi5 are part of a recombination-promoting complex (RPC), though how it works remains obscure. Recall that Swi5 also partners with Sfr1 to act as a mediator of *S. pombe*'s Rad51. Deletion

Figure 7.30 Donor preference in fission yeast. (A) Donor choice in fission yeast *mat1* switching involves a recombination-promoting complex (RPC) containing Swi2 and Swi5. In *mat1P* cells, the RPC localizes to a specific SR3E DNA element located adjacent to *mat3*, but in *mat1M* cells the RPC spreads across the silent mating-type region, including *mat2P* and—for as yet unknown reasons—promotes the preferential use of *mat2* as the donor. Preference for *mat2* usage depends on a newly identified SRE2 element. (B) In a *swi2Δ* mutant, *mat2P* becomes the preferred donor for both *mat1P* or *mat1M*, whereas in the absence of the heterochromatin associated Swi6 protein that silences the *mat2-cenH-mat3* region, RPC remains at the SRE3 element regardless of mating type and *mat3M* becomes the preferred donor. In the absence of both Swi2 and Swi6, either donor can be used. (Adapted from Jia S, Yamada T & Grewal SI [2004] *Cell* 119:469–480. With permission from Elsevier.)

(A)

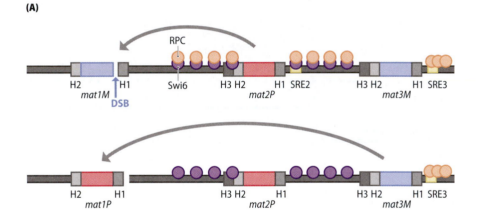

(B)   *swi2Δ* mutant

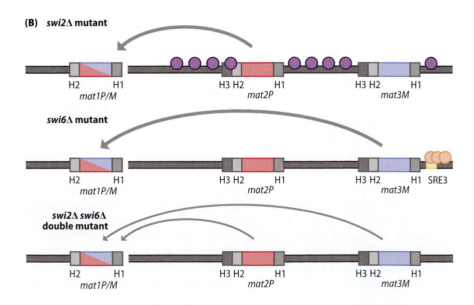

of the SRE3 has a phenotype similar to *swi2Δ* cells: only *mat2-P* is used as a donor. Thus SRE3 appears to function similarly to the budding yeast RE, although there is no evidence that any of the same proteins are involved. Recently Genevieve Thom's lab has discovered there is an equivalent enhancer adjacent to *mat2*P, named SRE2. Swapping the positions of SRE2 and SRE3 causes a reversal of donor preference for both mating types. How M or P expression influences the chromatin spreading is not well understood, but recently the *mat1M*-encoded factor Mc has been identified as a key factor that regulates the level of Swi2 expression. Loss of Mc reduces Swi2 and Swi5 to levels comparable to those in P cells and disrupts RPC spreading across the *mat2/3* region.

How these changes in Swi2–Swi5 binding across the donor region affect donor preference is still to be discovered. Grewal's lab has shown that the presence of the SRE or Swi2–Swi5 does not affect the nucleosome positioning in the two donors, so it is unlikely that the change in donor usage is simply caused by making one of the regions more accessible. This is also the conclusion that has been reached in *S. cerevisiae* donor preference.

7.16 A *MAT1* DSB IS NOT LETHAL IN THE ABSENCE OF THE DONORS

In budding yeast *MAT* switching, nearly all cells die when a DSB is created in strains lacking *HML* and *HMR*. But in fission yeast, cells lacking *mat2* and *mat3* are quite robust, even though there is no defect in the creation of the DSB. However, if one deletes the Rad52 homolog in *S. pombe* the cells die. Cells capable of switching (but not a derivative eliminating the nicking site) are also inviable in the absence of Rad51 or Rad54 or in the double mutant *swi5Δ rad57Δ*. These results suggests that the DSB created during replication can be repaired, not by NHEJ, but by homologous recombination, using the sister chromatid as the repair template. Presumably these HR proteins are required to allow the establishment of a recombination-dependent replication fork that can then replicate the region between the DSB end and the stalled replication fork at RTS1 (Figure 7.23).

Arcangioli's lab discovered that the Mus81–Eme1 proteins that can cleave nicked HJs and related structures are required for survival in the absence of donor. Deleting Mus81 has little effect on viability when donors are present. Through the use of two-dimensional gel electrophoresis that can resolve branched DNA structures, it became evident that the absence of Mus81–Eme1 causes the accumulation of a specific branched intermediate (**Figure 7.31**). This intermediate suggests there is some sort of sister chromatid recombination, producing either a Holliday junction or another strand-invasion structure. What kind of structure would be generated that is distinct from intermediates that arise during repair of the same DSB by an ectopic donor is not yet evident, but clearly the way repair happens must be significantly different.

A simple model to repair the DSB from the sister is shown in **Figure 7.31B**. The re-annealing and primer extension of the second broken end to seal up the last Okazaki fragment provides a dsDNA region where the 3' end of the DSB can invade and set up BIR. The presence of a nicked or sealed HJ could explain why Mus81–Eme1 would have a distinctive role, and could account for the branched structure that is seen in the absence of Mus81 on two-dimensional gels (Figure 7.31).

The fact that the allelic site on a sister chromatid can serve as an efficient donor to repair the DSB at *mat1* when the two normal donors are deleted

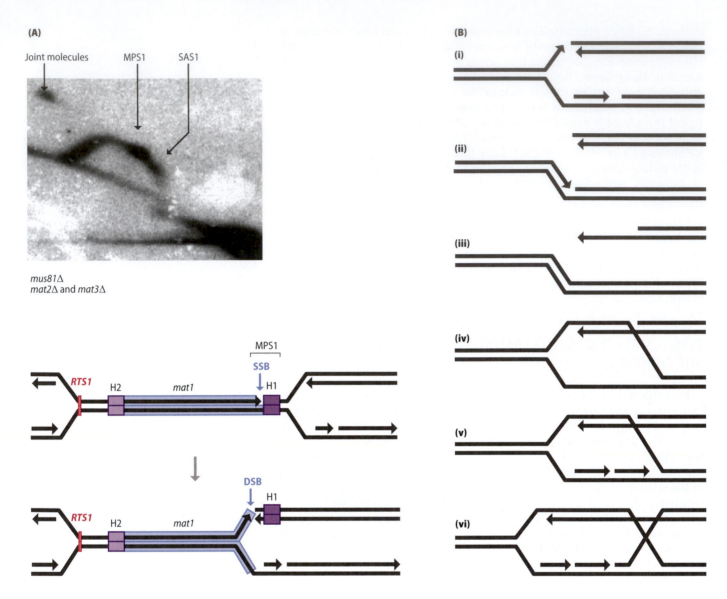

Figure 7.31 In the absence of *mat2* and *mat3* donors, the DSB at *mat1* is repaired by sister chromatid recombination. (A) A Southern blot of two-dimensional gel electrophoresis, probed to show the *mat1* region, shows the replication fork stalling imposed by the *MPS1* region as well as another stall at the *SAS1* site is the same as when donors are present. In the absence of Mus81, there is the appearance of branched joint molecules that suggest that the *mat1* locus—in the absence of *mat2* and *mat3* donors—recombines with its unbroken sister chromatid. (B) A possible molecular scheme for repair is shown. The stalled replication fork at MPS1 (i) apparently overcomes the nick/imprint and restores one intact chromatid (ii, iii). Strand invasion of the resected broken chromatid (iv) creates a D-loop that can be converted into replication fork by BIR (v). This process produces first a nicked Holliday junction (v) and then an intact HJ after branch migration (vi). Either of these branched intermediates may be the substrate for Mus81. (Adapted from Roseaulin L, Yamada Y, Tsutsui Y et al. [2008] *EMBO J* 27:1378–1387. With permission from Macmillan Publishers, Ltd.)

raises a fundamental question. Why isn't the sister chromatid normally the preferred partner? In budding yeast and mammals DSBs recruit additional cohesin complexes that should encircle the two sisters. We don't know if this happens at *mat1*. The elaborate alterations of chromatin structure of the donors suggests that they may have a distinctive architecture that could place one of them in a preferred position to outcompete the sister chromatid allele. Alternatively, the sister template strand may be incompletely replicated so that the 3′ end of *mat1* will have no intact partner with which to repair. Another possibility is that the sister chromatid's sequences homologous to the DSB may be inaccessible because of the binding of the proteins that are needed to create the imprinted nick in the sister chromatid. Perhaps the failure to repair from the normal donors triggers a DNA damage response that in turn allows the blocked left replication fork to fill in the gap, thus providing an intact template for sister chromatid repair. There are a number of experiments that recommend themselves: are the kinetics of normal *mat1* switching faster than sister chromatid repair? What is the state of the lagging-strand across from the DSB? Does the left replication fork bypass the RTS1 block? Does the replication-associated DSB trigger the formation of the phosphorylated histone H2A that would in turn lead to recruitment of cohesins?

7.17 ARE THERE RECOMBINATION ENHANCERS IN OTHER PROGRAMMED RECOMBINATION EVENTS?

To date no other genetically programmed chromosomal rearrangement appears to use a *cis*-acting recombination enhancer region that has apparently nothing to do with regulating transcription of nearby loci. We do know that enhancers controlling transcription of adjacent genes or gene segments play very significant roles in immunoglobulin gene rearrangements, which involve concerted cleavage events by the Rag1–Rag2 complex or during immunoglobulin class switching, where transcription appears to be a prerequisite for DSBs initiated by activation-induced cytidine deaminase (AID).

A more relevant example comes from studying gene conversions that are responsible for immunoglobulin gene diversity in the chicken. In chicken DT40 cells, a previously joined VJ region of the λ gene diversifies by repetitive recombination with 25 pseudogenes that lie upstream. The initiating lesion is a nick or DSB that is created by the AID cytosine deaminase and subsequent creation of a nicked, abasic site. Homologous recombination is required for introduction of mutations copied from the pseudogenes. When recombination is abolished, by ablating one of the several Rad51 paralogs or BRCA2, Polη, or other recombination proteins, there is still diversification, but this comes from an alternative somatic hypermutation mechanism and the mutations are not templated from the pseudogenes.

Recently, Nancy Maizels' lab tethered copies of HP-1 into the V region of an immunoglobulin gene and showed that this insertion created a pseudo-heterochromatic region with reduced histone acetylation. The presence of HP-1 impaired the use of these segments in VJ joining and thus greatly increased nontemplated somatic hypermutations. This result suggests that an "open" chromatin structure may be necessary for efficient use of a donor sequence in DT40 cells. The result contrasts with budding yeast donor preference, where unsilencing a "wrong" donor does not affect its use. It is also possible that the HP-1 mediated modification of the pseudogene region causes a relocalization of these sequences to a different nuclear compartment that precludes their use in recombination.

SUMMARY

The ability to induce a highly synchronous *MAT* switching process in budding yeast has made it possible to follow many of the early steps in DSB repair, which take place over the course of at least one hour. These steps include 5' to 3' resection, recruitment of RPA and then Rad51 to the ssDNA, homology searching and strand invasion, chromatin remodeling of the donor, initiation of new DNA synthesis, second end capture, and the removal of nonhomologous sequences. The effect of cell cycle stage and the important roles of a number of recombination proteins and DNA replication factors can be assessed even under conditions where cells are arrested. An analogous approach in fission yeast has revealed a quite different *mat* switching process, in which the DSB is intrinsically coupled to DNA replication and in which donor preference is enforced in a very different manner. Together these and related systems provide a way to study DSB repair in great detail. Analogous studies in metazoans should be possible as it becomes easier to induce efficient site-specific DSBs.

SUGGESTED READING

Arcangioli B (2000) Fate of mat1 DNA strands during mating-type switching in fission yeast. *EMBO Rep* 1:145–150.

Arcangioli B & de Lahondes R (2000) Fission yeast switches mating type by a replication-recombination coupled process. *EMBO J* 19:1389–1396.

Broach JR (2004) Making the right choice—long-range chromosomal interactions in development. *Cell* 119:583–586.

Dalgaard JZ & Klar AJ (1999) Orientation of DNA replication establishes mating-type switching pattern in *S. pombe. Nature* 400:181–184.

Egel R (2005) Fission yeast mating-type switching: programmed damage and repair. *DNA Repair* 4:525–536.

Haber JE (2012) Mating-type genes and *MAT* switching in *Saccharomyces cerevisiae. Genetics* 191:33–64.

Hicks WM, Yamaguchi M & Haber JE (2011) Inaugural Article: Real-time analysis of double-strand DNA break repair by homologous recombination. *Proc Natl Acad Sci USA* 108:3108–3115.

Jain S, Sugawara N, Lydeard J et al. (2009) A recombination execution checkpoint regulates the choice of homologous recombination pathway during DNA double-strand break repair. *Genes Dev* 23:291–303.

Jia S, Yamada T & Grewal SI (2004) Heterochromatin regulates cell type-specific long-range chromatin interactions essential for directed recombination. *Cell* 119:469–480.

Klar AJ (1987) Differentiated parental DNA strands confer developmental asymmetry on daughter cells in fission yeast. *Nature* 326:466–470.

Prakash R, Satory D, Dray E et al. (2009) Yeast Mph1 helicase dissociates Rad51-made D-loops: implications for crossover control in mitotic recombination. *Genes Dev* 23:67–79.

Thon G & Klar AJ (1993) Directionality of fission yeast mating-type interconversion is controlled by the location of the donor loci. *Genetics* 134:1045–1054.

Vengrova S & Dalgaard JZ (2006) The wild-type *Schizosaccharomyces pombe* mat1 imprint consists of two ribonucleotides. *EMBO Rep* 7:59–65.

Wang X & Haber JE (2004) Role of *Saccharomyces* single-stranded DNA-binding protein RPA in the strand invasion step of double-strand break repair. *PLoS Biol* 2:104–111.

Wang X, Ira G, Tercero JA et al. (2004) Role of DNA replication proteins in double-strand break-induced recombination in *Saccharomyces cerevisiae. Mol Cell Biol* 24:6891–6899.

White CI & Haber JE (1990) Intermediates of recombination during mating type switching in *Saccharomyces cerevisiae. EMBO J* 9:663–673.

Wolner B, van Komen S, Sung P & Peterson CL (2003) Recruitment of the recombinational repair machinery to a DNA double-strand break in yeast. *Mol Cell* 12:221–232.

CHAPTER 8
BREAK-INDUCED REPLICATION

Break-induced replication (BIR) was introduced in Chapter 1 as a way to understand the basic steps in homologous recombination. Here we examine it in more detail. The past decade has revealed that recombination-dependent DNA replication—BIR—is a major pathway of DSB repair. Its most obvious roles are in maintaining chromosome ends in the absence of telomerase and, it is presumed more than demonstrated, in restarting DNA replication at stalled and broken forks. The earliest indications of BIR came from studying bacteriophage recombination and replication, but it has more recently been invoked as a mechanism to generate chromosomal rearrangements and gene copy number variation in human disease.

8.1 BIR IS IMPORTANT IN THE MATURATION AND REPLICATION OF BACTERIOPHAGE

Recombination-dependent DNA synthesis as a mechanism to generate recombinant molecules was first invoked in 1961 by Matthew Meselson and Jean Weigle, who suggested that the broken end of one molecule could use a second, homologous molecule as a template to copy to its end. A more explicit connection between recombination and the initiation of DNA replication was postulated by Ann Skalka, whose ideas are still evident in the models we draw more than 35 years later. Chapter 16 reviews the historical development of these ideas in detail.

The replication and maturation of many bacteriophage requires the generation of molecules that are longer than a unit length. Two well-studied examples are recombination in bacteriophage λ and the late-phase replication events in phage T4. When phage λ infects cells, some of the circular molecules are opened by the terminase protein, a site-specific endonuclease that cleaves at the *cos* site (**Figure 8.1A**). Terminase remains bound to one end, protecting it from exonucleolytic digestion, so that effectively there is only one DSB end. The free end is resected by phage exonucleases, and the 3′ single-stranded end can then invade into an intact phage λ circle. This is a one-ended recombination event in which strand invasion creates a D-loop (**Figure 8.1B**). At least two outcomes are possible. First, the D-loop can be nicked, leaving a single Holliday junction (HJ) that can be resolved without initiating any new DNA synthesis [**Figure 8.1B** (**iii**)]. This "break–join" produces a recombinant molecule that can then be packaged into a phage head [**Figure 8.1B**(**iv**)]. A variation on this idea would be that the D-loop migrates so that a double HJ is formed [**Figure 8.1B**(**ii**)]; subsequently this structure can be resolved to give a break–join outcome. Alternatively, the D-loop serves as the site

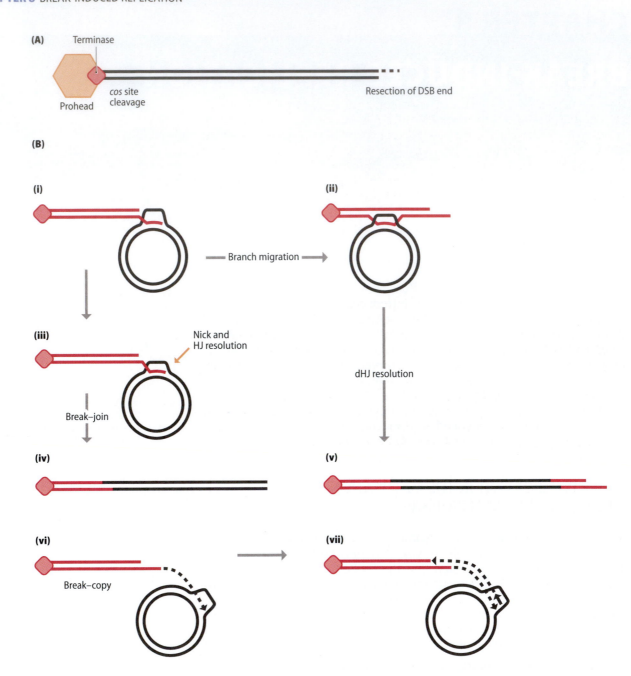

Figure 8.1 Creation of recombinant phage λ molecules.
(A) The *terminase* endonuclease sometimes cleaves phage λ, remaining covalently attached at one end, blocking end resection. Resection at the other end by RecBCD or λ exonuclease provides a ssDNA strand for recombination. (B)(i) RecA-mediated strand invasion creates a D-loop that can be cleaved by HJ resolvases (iii) to produce "spliced" DNA molecules (iv) with little or no new DNA synthesis (termed break–join). Strand invasion can be followed by branch migration to form a double Holliday junction (ii) that can be cleaved to form a different break–join outcome. Alternatively the strand invasion intermediate can mature into a replication fork that can copy one or both strands by rolling circle replication (vi and vii). This form of BIR, termed break–copy, can produce multiple copies of the circular template. (Adapted from Gumbiner-Russo LM & Rosenberg SM [2007] *PLoS One* 2:e1242.)

to initiate BIR. Again, one can imagine two variations on this theme. As first demonstrated in an *in vitro* experiment by Tim Formosa and Bruce Alberts, the end of the invading strand can be used as a primer to initiate leading-strand synthesis [**Figure 8.1B(vi)**]. If there is no lagging-strand synthesis, then a "bubble" can move down the DNA creating a long single-stranded recombinant product. Some later event would then fill in this ssDNA to make a conservatively replicated dsDNA recombinant molecule. However, the D-loop could also recruit both leading- and lagging-strand DNA polymerases and promote BIR [**Figure 8.1B(vii)**]. The longer-than-unit-length recombinant molecules can then be packaged into phage heads and can be recovered.

The products of phage λ recombination can be analyzed by density transfer methods analogous to those employed by Meselson and Frank Stahl in their classic study of *Escherichia coli* replication. This approach showed that about half of the outcomes are likely to have arisen by break–join events while the other half are by BIR. **Figure 8.2** presents the density gradient separation of recombinant progeny phage when two differently marked "heavy" phage are co-infected to select wild-type recombinants under conditions where repair DNA synthesis can occur but normal replication is prevented. About half of the recombinants are almost completely heavy (HH), the outcome expected for break–join recombination, while the others are consistent with progeny being half-heavy and half-light. But because these recombinants arose under conditions where normal replication is blocked, the HL do not have the one-strand H and one-strand L outcome expected from semi-conservative replication of a HH phage. Instead these recombinants arose by BIR, where one half of the molecule is HH and has been extended by BIR to create a LL half, as expected for break–copy outcomes (Figure 8.2C). If the infection is carried out in a temperature-sensitive mutant host so that even repair DNA replication is completely blocked, then only HH (break–join) progeny are obtained. Conversely if the host lacks two key HJ resolvases, RuvC and RecG, then the break–join HH outcomes are absent and only break–copy outcomes are seen. This result supports the idea that break–join requires the cleavage of HJs or related structures as shown in Figure 8.1B.

A similar experiment was carried out by Tokio Kogoma using cells carrying multiple copies of a plasmid containing a phage λ *cos* site. If only a low level of terminase is expressed, so that only a fraction of molecules are cut, then a terminase-cut DSB end is free to recombine with uncut, circular plasmids. The result is a tenfold amplification of plasmid DNA, most likely promoted by rolling-circle replication initiated by BIR.

BIR also plays a key role in the life cycle of phage T4. Gisela Mosig first demonstrated that the initial stages of phage replication occur using an origin of replication, but late T4 DNA replication is recombination dependent, by a BIR mechanism (**Figure 8.3A**). Kenneth Keuzer's lab has studied plasmid recombination in cells in which phage T4 infection has taken over the replication machinery, which results in the incorporation of hydroxymethylcytosine in place of cytosine, among other changes.

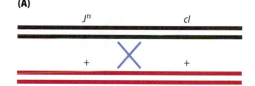

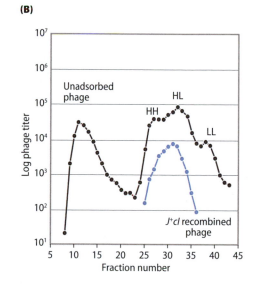

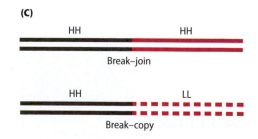

Figure 8.2 Observation of DNA replication during recombination. (A) Two genetically marked ^{15}N-labeled (heavy) phage were co-infected and grown in [14]N (light medium) under conditions that limited DNA synthesis. The phage particles were isolated and the particles were separated on a CsCl density gradient. (B) An aliquot of each sample was taken and plated to select for J^+ *cl* recombinants as well as to measure total phage titer. Unadsorbed phage have heavy DNA and a heavy protein coat and migrate as much heavier particles. Phage released by infection had light coats but can be separated by their DNA of different density. In wild-type cells, there are nearly equal amounts of HH and HL phage. Among the J^+ *cl* recombinants, the HH recombinants proved to result from break–join mechanisms whereas the HL phage arose by break–copy (C). When the same experiment was carried out in the absence of the host's two Holliday junction resolvases, RuvC and RecG, no break–join outcomes occurred; instead all of the phage apparently arose from a break–copy mechanism. If this experiment is carried out when all DNA replication is blocked, then only a HH peak of recombinants is found. (Adapted from Motamehdi MR, Szigety SK & Rosenberg SM [1999] *Genes Dev* 13:2889–2903. With permission from Cold Spring Harbor Laboratory Press.)

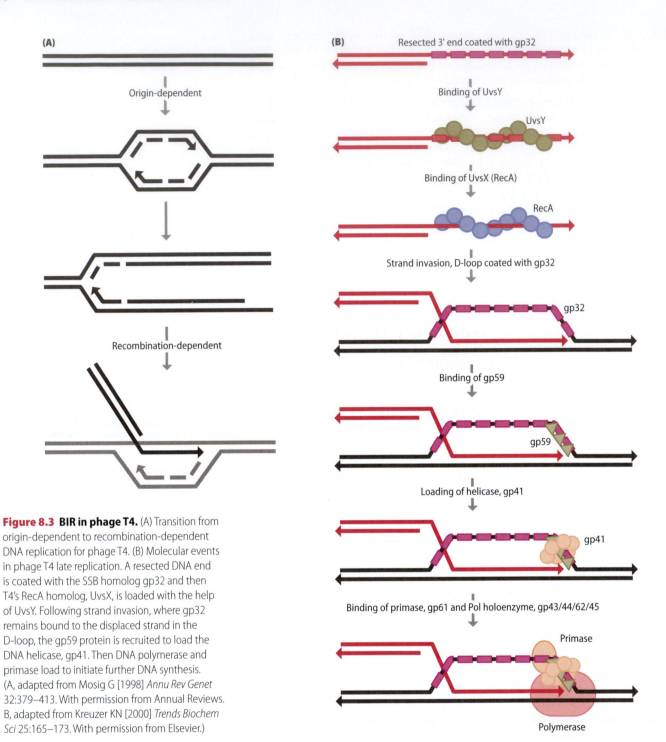

(A)

Origin-dependent

Recombination-dependent

(B)

Resected 3' end coated with gp32

Binding of UvsY

UvsY

Binding of UvsX (RecA)

RecA

Strand invasion, D-loop coated with gp32

gp32

Binding of gp59

gp59

Loading of helicase, gp41

gp41

Binding of primase, gp61 and Pol holoenzyme, gp43/44/62/45

Primase

Polymerase

Figure 8.3 BIR in phage T4. (A) Transition from origin-dependent to recombination-dependent DNA replication for phage T4. (B) Molecular events in phage T4 late replication. A resected DNA end is coated with the SSB homolog gp32 and then T4's RecA homolog, UvsX, is loaded with the help of UvsY. Following strand invasion, where gp32 remains bound to the displaced strand in the D-loop, the gp59 protein is recruited to load the DNA helicase, gp41. Then DNA polymerase and primase load to initiate further DNA synthesis. (A, adapted from Mosig G [1998] *Annu Rev Genet* 32:379–413. With permission from Annual Reviews. B, adapted from Kreuzer KN [2000] *Trends Biochem Sci* 25:165–173. With permission from Elsevier.)

A very informative experiment is shown in **Figure 8.4**. Two plasmids, sharing limited homology were introduced. One of them could be cleaved by the I-TevI restriction enzyme within the homologous region. This appears to be a gap-repair scenario that would yield the recombinant *Xba*I^S product illustrated in Figure 8.4A, but in fact this product was barely visible. Instead, there were much larger DNA molecules, some of which were too large to enter the gel (shown as an asterisk in Figure 8.4B). Further analysis demonstrated that these were the products of a rolling-circle replication process, as illustrated in Figure 8.4C. Analysis also showed that BIR in phage T4 requires the homologs of SSB (gp32) and RecA (UvsX) as well as a RecA loading factor (UvsY) and the gp59 protein

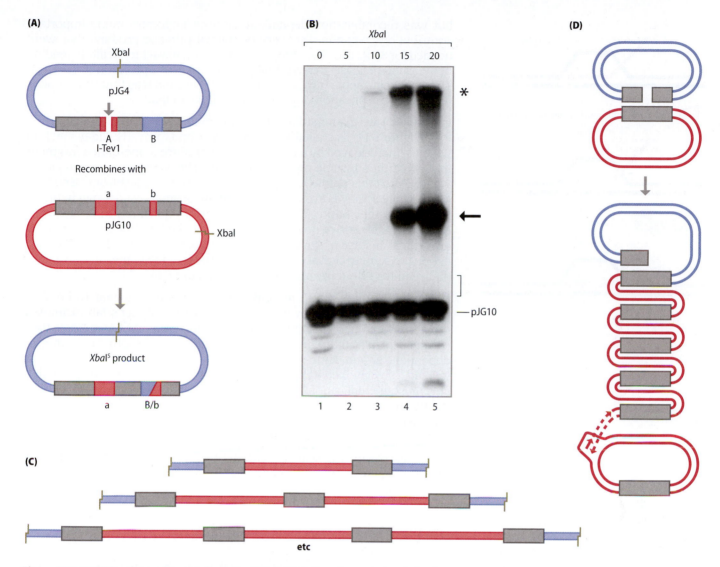

Figure 8.4 A DSB leads to rolling-circle replication in T4-infected *E. coli.* (A) Plasmid pJG4 shares a region of homology (*gray* boxes) with pJG10 in a region where pJG4 can be cleaved *in vivo* with the I-Tev1 endonuclease within sequence A. An XbaI site (*green*) in pJG4 can be cleaved in the plasmid DNA but, because phage T4-infected cells substitute hydroxymethyl cytosine for C in replicated DNA, the site cannot be cut after replication. The expected DSB repair products in XbaI-digested DNA are XbaI-sensitive plasmids carrying the "a" region of homology (*red bar*) and either B or b; but this is barely seen on a Southern blot probed with the "a" sequence (B), as indicated by the region marked by a bracket. Instead, much higher molecular weight DNA homologous to the probe is seen. Some of the products can migrate into the gel (*arrow*) while others are retained in the well (*). (C) Structure of higher molecular weight products in which the XbaI sites in the red sequences cannot be cleaved because they are modified during the replication associated with this rolling circle repair process. (D) An illustration of the rolling circle replication that produces the high molecular weight products. (Adapted from George JW, Stohr BA, Tomso DJ & Kreuzer KN [2001] *Proc Natl Acad Sci USA* 98:8290–8297. With permission from the National Academy of Sciences.)

needed to recruit a replicative helicase (gp41) to the incipient replication fork (**Figure 8.3B**). Once these factors have loaded, DNA polymerase and primase are recruited to initiate new DNA synthesis.

8.2 BIR IS ALSO IMPORTANT IN *E. COLI*

BIR also appears to play an important role in maintenance of genome stability in *E. coli*. Cells lacking the normal origin of replication of *E. coli* could engage in what Tokio Kogoma called "stable DNA replication" (SDR) that was independent of the oriC origin and the origin-initiator protein DnaA,

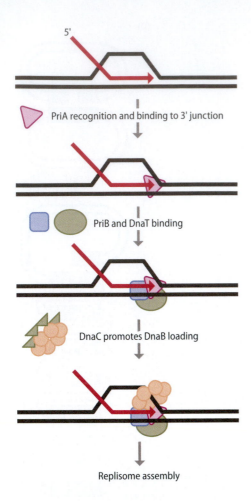

Figure 8.5 A model of PriA-mediated replication restart. PriA binds to the 3′ junction of an invading strand in a D-loop. PriA recruits PriB and DnaT, which in turn recruit DnaC, the loading factor for the replicative helicase, DnaB, which is needed for extensive DNA replication. (Adapted from Gabbai CB & Marians KJ [2010] *DNA Repair* 9:202–209. With permission from Elsevier.)

but was recombination dependent. Among Kogoma's many important contributions was the discovery of special primase proteins that were needed for SDR. The PriA, PriB, and DnaT proteins act to facilitate loading of the DnaB helicase proteins that are required for replication (**Figure 8.5**). These proteins apparently substitute for the normal way that DnaA opens up DNA at a replication origin so that DnaB can load.

In Chapter 1 we examined one mechanism to restart a stalled replication fork (Figure 1.11). A stalled fork can regress and form a Holliday junction which is then cleaved by a HJ resolvase to produce a one-ended fragment that could then resume replication by BIR. This is shown again in **Figure 8.6**. In *E. coli* this pathway requires PriA. Further studies from Bénédicte Michel's group have revealed that there are several distinct mechanisms to allow recombination-dependent restart of replication in *E. coli*; they differ in the details of how replication is blocked and in which proteins are critical. Instead of RuvC cleaving the HJ of the regressed fork, RecBCD exonuclease may chew back the end until the HJ becomes a Y-structure that requires PriA to resume replication (Figure 8.6).

BIR may also play a role in DSB repair events that appear to be simple break or gap-repair events. When Kenneth Kreuzer's lab examined repair between two plasmids, sharing a limited region of homology, they obtained results that could not be squared with the dHJ or synthesis-dependent strand annealing (SDSA) mechanisms of gene conversion. Instead, they invoked a BIR-like mechanism that they termed "extensive chromosome replication" (ECR). Here, the two ends participate independently and sequentially, in two consecutive BIR events (**Figure 8.7**). In the end there would be three complete molecules one of which would appear to be a simple break-repair outcome.

As we will see in discussing gene targeting (Chapter 11), there is evidence that a small linear DNA fragment can be used to completely replicate the entire *E. coli* chromosome, with each end initiating a BIR event.

8.3 BIR HAS BEEN WELL DOCUMENTED IN BUDDING YEAST

The most extensive studies of BIR in eukaryotes have come from studying it in budding yeast. Two general strategies have been used to understand how BIR takes place. First, linearized DNA containing a centromere and one telomere end can be transformed into yeast, such that DNA sequences near the uncapped end are homologous to a region of a yeast chromosome (**Figure 8.8**). Strand invasion and new DNA synthesis allow the copying of the template to its own telomere. Experiments by Philip Hieter and later by Lorraine Symington have demonstrated that hundreds of kilobases can be copied, creating a nonreciprocal translocation on the repaired, now-stable chromosome. BIR can progress through the entire array of more than one hundred 9-kb rDNA repeats. It seems, however, that BIR events are impaired if they must cross the very small centromere in budding yeast, which consists essentially of a single, modified nucleosome and some adjacent positioned nucleosomes. It is possible that the problem is not that the centromere is a barrier to BIR (after all it is not a barrier to normal replication); rather it could be that re-replicating a centromere region causes problems in reestablishing centromere function.

A second approach to study BIR in budding yeast employs HO endonuclease to create a DSB in situations where there is little or no homology shared by the centromere-distal end of the DSB but the centromere-proximal end shares homology to a region on a different chromosome arm or another chromosome (**Figure 8.9**). The sequences must be oriented such that repair will progress toward the template

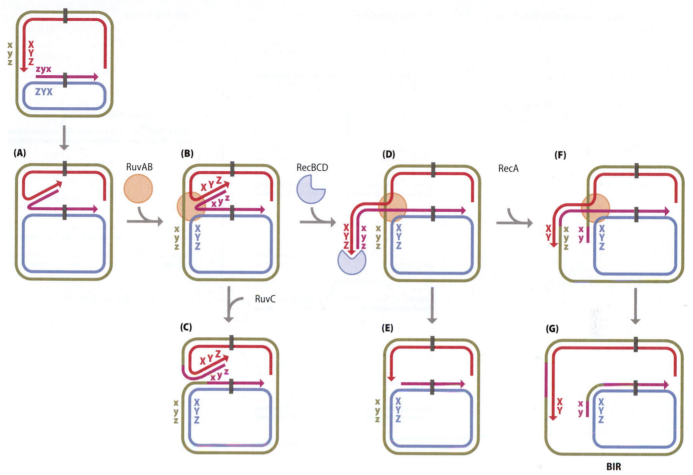

Figure 8.6 Models for replication restart. *Green* and *blue* lines are the template strands of the chromosome, *red* and *magenta* lines are the newly synthesized strands, the arrows indicate the 3′ end of the growing strands. A stalled fork blocked at XYZ can branch migrate (A) to form a "chicken foot" reversed replication fork (a Holliday junction) that can be cleaved by RuvABC (B, C). Either the regressed fork or the HJ-cleaved form can be degraded by exonucleases such as RecBCD (D, E). RecA and the recruitment of PriA can lead to replication restart (F, G). (Adapted from Michel B, Boubakri H, Baharoglu Z et al. [2007] *DNA Repair* 6:967–980. With permission from Elsevier.)

chromosome's telomere, to produce a nonreciprocal translocation. This ectopic BIR assay works with as little as 70-bp homology in a haploid where the break could invade a homologous sequence at the opposite end of the same chromosome. The efficiency of BIR in haploids is greater when the events are intrachromosomal than interchromosomal, and the efficiency increases substantially as the size of homology increases to several kilobases.

A third situation that appears to involve BIR is the maintenance of yeast telomeres in the absence of telomerase, the enzyme that adds RNA-templated short, repetitive sequences to chromosome ends. In many organisms the repeats are quite uniform, such as the TTAGGG sequences in mammals, but in budding yeast the sequence is degenerate, TG_{1-3}. If one deletes the yeast *TLC1* gene encoding telomerase RNA or any of three "ever shortening telomere" *EST* genes that encode telomerase's protein components, telomeres erode progressively until the vast majority of cells undergo senescence and cell death. However, a very small proportion of cells become "survivors" and proliferate. There are two distinct pathways for telomere maintenance (**Figure 8.10**). One pathway (Type 1) involves recombination with and amplification of subtelomeric sequences, most notably 5.5-kb and 6.7-kb variants of Y′ sequences; a second mechanism (Type 2) leads to long extensions of the telomere repeats themselves.

Figure 8.7 The extensive chromosome replication model of DNA repair. SDSA (A) is compared with ECR (B). In ECR, one end of a DSB engages the template and initiates BIR. Following cleavage of a HJ-containing intermediate, the second end carries out a similar process, resulting in the creation of three rather than two intact DNA molecules. (Adapted from Stohr BA & Kreuzer KN [2002] *Genetics* 162:1019–1030. With permission from the Genetics Society of America.)

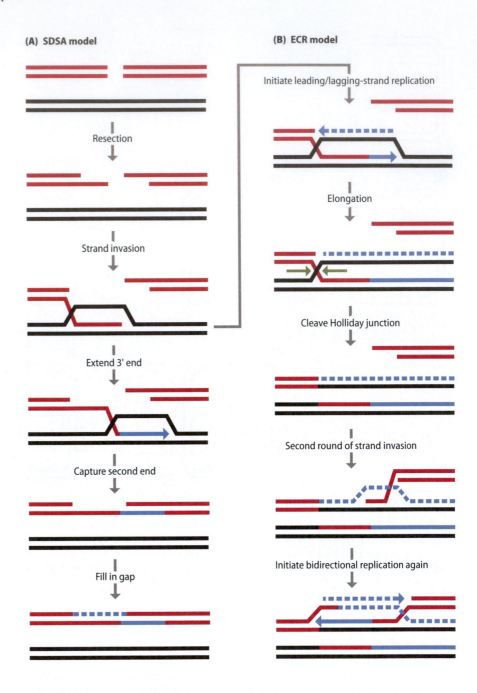

These two mechanisms have different genetic requirements, though both appear to involve BIR, as both processes are both Rad52 and Pol32 dependent. However, Type 1 survivors are Rad51 dependent whereas Type 2 survivors are Rad51 independent but require the MRX complex, Rad59, and Sgs1.

We presume that BIR is important as a mechanism to restart replication at collapsed or broken replication forks in eukaryotes, but to date there is not a tractable system to study these kinds of defects and the role of BIR in detail. The development of site-specific enzymes that cleave only one strand of DNA should make it possible to explore these repair events. Such experiments will be of great interest, because so far all the model BIR systems exhibit *very* long delays whereas BIR in the context of fixing a broken replication fork must be able to act within a much shorter time. Perhaps the presence of cohesins holding sister chromatids together facilitates rapid repair.

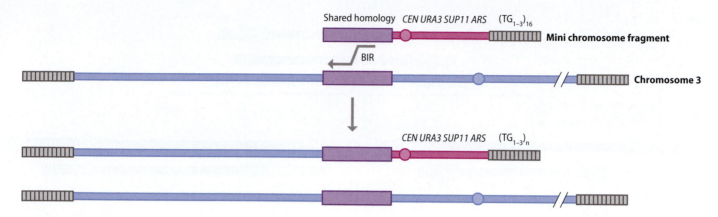

Figure 8.8 Assay of BIR in budding yeast by plasmid transformation. Budding yeast is transformed with a linearized plasmid D8B having a centromere, a selectable marker (*URA3*), and an origin of replication (ARS). The fragment has telomere sequences at one end and a DSB end sharing a region of homology with a chromosomal site on Chr 3.

BIR produces a supernumerary chromosome with partial duplication of a chromosome arm. (Adapted from Davis AP & Symington LS [2004] *Mol Cell Biol* 24:2344–2351. With permission from the American Society for Microbiology.)

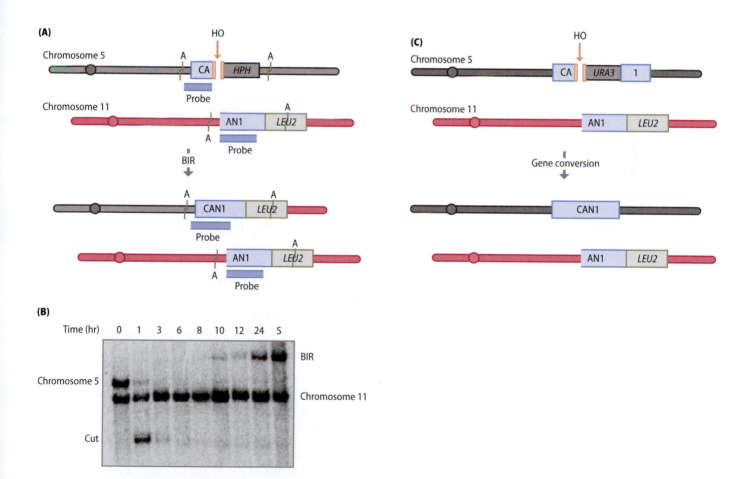

Figure 8.9 An interchromosomal BIR assay in haploid yeast.
(A) An HO cleavage site was inserted in the terminal, dispensable region of Chr 5, removing the 3' part of the *CAN1* gene, and leaving "CA." A 1-kb "AN1" segment of *CAN1* was inserted in Chr 11, such that BIR will produce an intact *CAN1* gene as part of a nonreciprocal translocation. (B) Southern blot of a chromosome-separating gel, probed with *CAN1* sequences shows

HO-mediated cleavage of Chr 5 and the appearance of a larger Chr 5;11 translocation product. (C) Configuration of a comparable gene conversion assay. (A, B, adapted from Lydeard JR, Jain S, Yamaguchi M & Haber JE [2007] *Nature* 448:820–823. With permission from Macmillan Publishers, Ltd. C, from Lydeard JR, Lipkin-Moore Z, Sheu YJ et al. [2010] *Genes Dev* 24:1133–1144. With permission from Cold Spring Harbor Laboratory Press.)

(A)

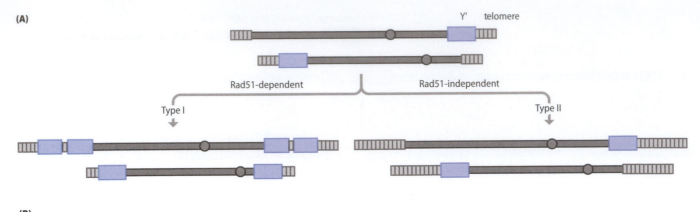

(B)

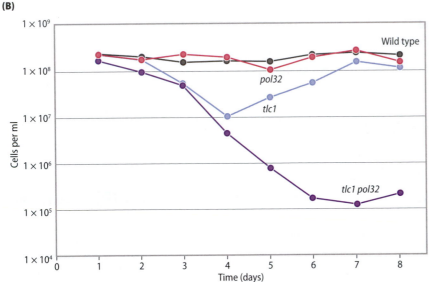

Figure 8.10 Budding yeast telomere maintenance in the absence of telomerase requires Pol32. (A) Two pathways of survivors arising in the absence of telomerase. Type I Rad51-dependent events frequently involve recombination between subtelomeric Y' elements, whereas Type II Rad51-independent events result from recombination between the divergent telomere sequences themselves. (B) Haploid yeast lacking the telomerase RNA gene, *TLC1*, were cultured overnight and a small aliquot used to inoculate the next overnight culture. By day 4, cells with ever-shortening telomeres undergo senescence and cell death, but a small proportion of survivors emerge and take over the culture at later times. Both pathways are blocked by the absence of Pol32. (From data in Lydeard JR, Jain S, Yamaguchi M & Haber JE [2007] *Nature* 448:820–823. With permission from Macmillan Publishers, Ltd.)

All BIR events in yeast require the Rad52 protein, but there are both Rad51-dependent and Rad51-independent BIR pathways. Most events require Rad51.

8.4 BIR IS USUALLY Rad51 DEPENDENT

HO-mediated Rad51-dependent BIR occurs efficiently and can be analyzed by Southern blots and PCR, and chromatin immunoprecipitation (ChIP), much like gene conversions. In one well-studied haploid system (Figure 8.9), a DSB is generated in the nonessential terminal region of Chr5, so that loss of the more distal sequences is inconsequential. The DSB is created adjacent to the 5' half of *CAN1* (called "CA"). The DSB can be repaired by BIR using a 5' truncated "AN1" segment, placed 30 kb from the end of Chr11 or some other site. The CA and AN1 segments share 1.15 kb of homology. HO induction of BIR leads to the restoration of the *CAN1* gene and loss of a distal HPH marker in about 20% of cells. In comparison, a gene conversion assay at the same location (by restoring the "1" sequences distal to the DSB) (Figure 8.9C) is more than 70% efficient.

A diploid BIR system was created by modifying a diploid (**Figure 8.11A**), by using transformation to create a truncation of the sequences distal to the HO cleavage site on one chromosome (**Figure 8.11B**). This diploid, which retains only 46 bp of homology distal to the DSB on the intact homologous chromosome, proved to be highly informative. Repair can occur either by

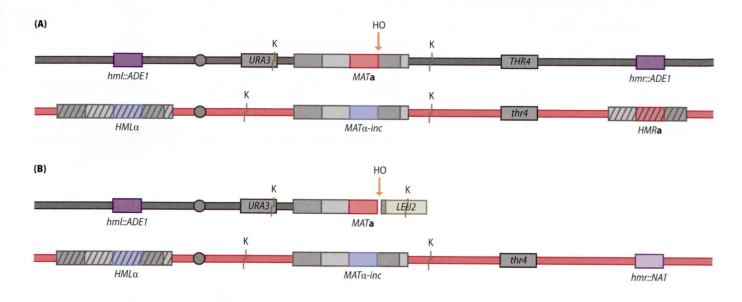

(A)

hml::ADE1   URA3   HO   MATa   THR4   hmr::ADE1

HMLα   MATα-inc   thr4   HMRa

(B)

hml::ADE1   URA3   HO   MATa   LEU2

HMLα   MATα-inc   thr4   hmr::NAT

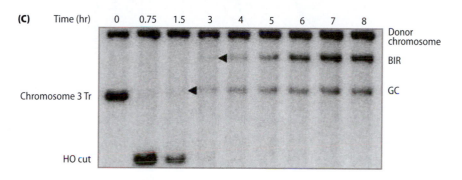

(C)   Time (hr)   0   0.75   1.5   3   4   5   6   7   8

Donor chromosome

BIR

Chromosome 3 Tr   GC

HO cut

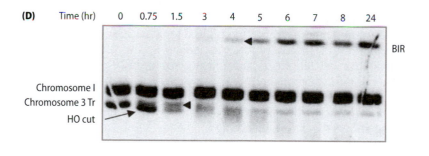

(D)   Time (hr)   0   0.75   1.5   3   4   5   6   7   8   24

BIR

Chromosome I
Chromosome 3 Tr
HO cut

Figure 8.11 A diploid BIR assay in budding yeast. (A) An interchromosomal gene conversion assay in a diploid in which a DSB at *MAT*a is repaired by a homologous *MAT*α-*inc* locus that cannot be cleaved. *Kpn*I (K) restriction sites are shown. (B) The diploid in (A) was truncated by the addition of telomere sequences so that only 46 bp of homology remain on the distal end of the HO cleavage site at *MAT*a. (C) A Southern blot of *Kpn*I-digested DNA, probed with *MAT* sequences, shows cleavage of the truncated chromosome (Chr 3 Tr) and repair of the DSB by both GC and BIR. The first appearance of products is marked by arrowheads. Note that the kinetics of GC precede BIR by several hours. (D) BIR and GC assayed by a Southern blot of a chromosome-separating gel, probed with *ADE1* which hybridizes to Chr 1 (unchanged) and to the broken Chr 3. Again gene conversion, restoring the size of Chr 3 Tr, is seen earlier than the BIR product. (From data in Malkova A, Naylor M, Yamaguchi M et al. [2005] *Mol Cell Biol* 25:933–944. With permission from the American Society for Microbiology.)

gene conversion (GC) or by BIR. Despite the very limited distal homology, GC accounted for about 10% of the events, whereas BIR occurred in about 80% (**Figure 8.11C, D**). There was also some loss of the broken chromosome. In contrast, in a diploid where the template chromosome was intact (**Figure 8.11A**), gene conversion was the overwhelming outcome and there were almost no BIR outcomes or chromosome losses. So, BIR is reasonably efficient but is relegated to a back-up role when both ends of the DSB are present and GC dominates. One explanation for the relative inefficiency of BIR is that there is a long delay before leading/lagging DNA synthesis is triggered. In a second version of this experiment, all the homology distal to the DSB was removed (**Figure 8.12**). Although Rad51-mediated strand invasion, monitored by ChIP (Figure 8.12B), occurs with similar kinetics to that seen for gene conversion events, the next step—the initiation of new DNA synthesis, monitored by PCR—is greatly

Figure 8.12 Early events in BIR. (A) The diploid assay system in Figure 8.11 was modified to remove any homology distal to the DSB by the insertion of a *KAN* gene. (B) Induction of HO-mediated cleavage is followed by resection (not shown) and recruitment of Rad51. The Rad51 filament promotes strand invasion into the donor chromosome, which can be assayed by ChIP (using *blue* PCR primers), as graphed after normalization in (B). Following strand invasion, the appearance of new DNA synthesis can be assayed by PCR (*purple* PCR primers), also graphed in (B). This step is delayed relative to strand invasion and is coincident with completion of repair, as assayed by Southern blot of separated chromosomes (B). (From data in Jain S, Sugawara N, Lydeard J et al. [2009] *Genes Dev* 23:291–303. With permission from Cold Spring Harbor Laboratory Press.)

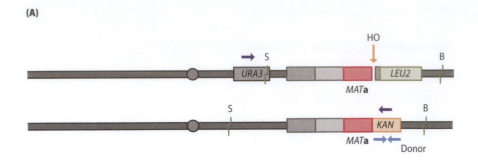

(A)

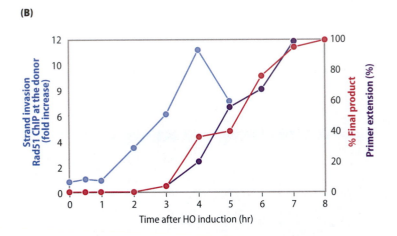

(B)

delayed (Figure 8.12B). The process can also be visualized by a Southern blot of a chromosome-separating gel as shown in Figure 8.11D. After HO induction, the truncated chromosome is cut, after which it is repaired either to a chromosome of nearly the original length (GC) or to full length (BIR). It is evident that the GC product appears with the same kinetics as in a strain with a full-length broken chromosome, but the BIR product does not appear for several hours. This delay turns out not to be the time it takes to traverse 100 kb to the end of the chromosome, because a Southern blot of diagnostic restriction fragments surrounding the site of the DSB also shows this delay, reflecting only a few kilobases of new DNA synthesis (Figure 8.11C). In fact, the kinetics of repair, once replication begins, is quite rapid, traversing the 100 kb template in less than 30 min. This observation leads to the important conclusion that the delay in replication does not occur after the elongation of the first strand (analogous to the first steps in SDSA) that might then pause to ensure that lagging-strand synthesis can be organized; rather, the initiation of leading- and lagging-strand synthesis is coordinated and neither occurs as rapidly as in gene conversion. This conclusion is bolstered by results examining the role of different DNA polymerases, described in Section 8.5.

The long delay in initiating BIR is reminiscent of the situation during repair of gaps larger than about 1–2 kb, as discussed in Chapter 7. This delay, the result of a recombination execution checkpoint (REC), appears to depend on whether the cell can determine if both DSB ends can strand invade the same template. How this constraint is imposed remains unsolved. In some yet-undefined way the cell can sense the distance between the sites of strand invasion of the two ends (or if there are two strand invasions), possibly by monitoring if both ends have encountered the same, enlarged D-loop. Recent studies have suggested that this signaling is controlled by the 3′ to 5′ helicases, Sgs1 and Mph1. The double deletion of Mph1 and Sgs1 dramatically accelerates the rate of both large gap-repair and BIR. How these helicases regulate this step is not yet known.

8.5 Rad51-DEPENDENT BIR REQUIRES ALL THREE MAJOR DNA POLYMERASES

The mechanism by which the strand-invasion D-loop is converted into a replication fork with leading- and lagging-strand synthesis is not yet clear; there is no homolog of PriA in eukaryotes. A striking finding is that nearly all ectopic BIR events require Pol32, a nonessential subunit of the DNA polymerase δ complex. It is possible that Pol32 in some way acts in conjunction with other proteins to recruit a replicative helicase, as PriA does in prokaryotes.

To examine the role of various DNA replication factors in BIR, without impairing normal replication, cells are arrested after replication, by nocodazole treatment, after which a thermosensitive allele of a DNA replication component can be inactivated. Experiments of this type have shown that BIR uses the same helicase complex as does normal replication. In addition to the Mcm2–7 heterohexamer, helicase activity requires Cdc45, Cdt1, and the GINS complex; all of these components are needed for initiating new DNA synthesis in BIR as they are at origins. In fact, it seems that all of the components of DNA replication are required for BIR except for those such as Cdc6 and the ORC complex that are specifically needed to recruit proteins to the pre-replicative complex at origins. BIR involves both leading- and lagging-strand synthesis (**Figure 8.13**); consequently BIR requires Polα-primase and two factors that help Polα act processively: Mcm10 and Ctf4. BIR also needs components that were previously identified as pre-replication origin-loading factors, such as Dpb11–Sld2–Sld3. Interestingly, without these factors even the initial, presumably leading-strand, extension from the 3′ end of the invading strand does not occur. That no *leading*-strand synthesis occurs without Polα or its co-factors again suggests that leading- and lagging-strand synthesis are tightly coordinated and that there is not an initial step of primer extension (as in gene conversion) that is later converted into a full replication fork. However, one cannot rule out the possibility that the initiation of even the first strand synthesis in BIR needs Polα-primase.

The initiation of BIR is also blocked by using temperature-sensitive mutations of DNA Polδ, but, surprisingly, Polε is not required for the initial DNA synthesis (Figure 8.13). However, Polε is required for replication of the template to continue after the first several kilobases. This separation of functions between Polδ and Polε is strikingly different from what has been seen in replication, where Polδ appears to direct most lagging-strand synthesis and Polε is concerned with leading-strand copying. However, it is known that S phase replication can be completed in the absence of catalytic activity of Polε so long as the C-terminal portion of the Pol2 (Polε) protein is present. The nonessential subunits of Polε (Dpb3 and Dpb4) are not required for BIR.

The idea that there is a transition between Polε-independent and -dependent replication complexes after the first several kilobases of new DNA synthesis resonates with the finding from Symington that there are frequent template switches in a BIR assay in which a linearized fragment can initiate BIR with either of two polymorphic homologous chromosomes of a diploid (**Figure 8.14**). These template switches are confined to the first several kilobases, after which copying is apparently processive. BIR can copy more than a megabase of DNA including rDNA.

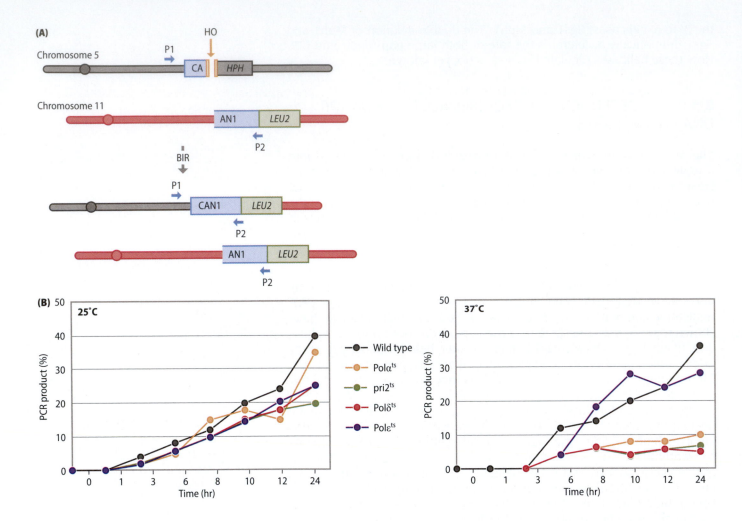

Figure 8.13 Roles of DNA polymerases in BIR. (A) In the assay system shown in Figure 8.9, cells were modified to carry a temperature-sensitive mutation of DNA Polα primase (Pri2), Polδ or Polε. The initiation of new BIR DNA synthesis was assayed by PCR between primers P1 and P2. (B) Initiation of BIR is equally efficient for all the mutants and the wild type at the permissive temperature of 25°C (*left*), but at 37°C (*right*), neither ts-Polα nor ts-Polδ supports the initiation of leading-strand synthesis. Surprisingly Polε is not required at the beginning of BIR that is measured by this PCR assay, but is required to complete the >30 kb replication to the chromosome end (not shown). (From data in Lydeard JR, Jain S, Yamaguchi M & Haber JE [2007] *Nature* 448:820–823. With permission from Macmillan Publishers, Ltd.)

8.6 MANY OTHER GENES AFFECT THE EFFICIENCY OF BIR

Several other recombination-related proteins influence the efficiency of BIR. As noted above, deleting either of two 3′ to 5′ helicases, Sgs1 or Mph1, causes significant increases in BIR efficiency in the haploid HO-induced system where there is only about 1 kb of homology with the template (**Figure 8.15A**). One possible role of Sgs1 would be in discouraging strand invasion and synapsis with short regions of homology; thus in *sgs1Δ* cells, strand invasion intermediates would be more stable and able to persist long enough to assemble a replication fork. Another possibility could be related to the finding that *sgs1Δ* mutants show a reduced rate of 5′ to 3′ resection of DSB ends once resection has proceeded for a few kilobases. Reducing resection may increase the amount of Rad51 protein bound to sequences near the DSB as there will be less ssDNA. Consistent with this idea is that a deletion of Exo1, which encodes another exonuclease involved in 5′ to 3′ resection, also increases BIR levels. Conversely, overexpression of either Sgs1 or Exo1 has the opposite effect, severely reducing survival by BIR.

Another key factor in BIR is a 5′ to 3′ helicase, Pif1. Pif1 has the opposite polarity from Sgs1 or Mph1. Grzegorz Ira has found that deleting *PIF1* has little effect on gene conversion but profoundly impairs BIR after the step of strand invasion. It thus appears that Pif1 is needed for extensive DNA synthesis in BIR. How Pif1, with 5′–3′ polarity on ssDNA, interacts with the replicative 3′–5′ CMG (Cdc45–MCM–GINS) helicase complex is not yet known. As CMG progresses down the leading-strand template, possibly

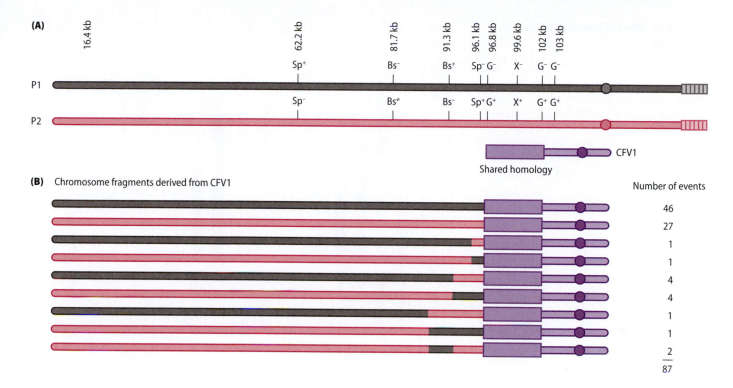

Figure 8.14 Template switching during the initial stages of BIR. (A) A linearized chromosome fragmentation vector (CFV) with telomere sequences on the right end was transformed into a diploid strain carrying many restriction site polymorphisms on the two template chromosomes. (B) BIR could start with roughly equal frequency on either of the two homologs but in 14 of 87 cases there were one or more template switches as BIR proceeded, seen as changes in the color of the sequences on the repaired CFV1 chromosome. These jumps are mostly confined to the first several kilobases after BIR is initiated. (Adapted from Smith CE, Llorente B & Symington LS [2007] *Nature* 447:102–105. With permission from Macmillan Publishers, Ltd.)

Pif1 is operating on the opposite strand. Alternatively, Pif1 might track behind the migrating D-loop, facilitating the unwinding of the newly copied strand.

BIR is also less efficient in the absence of proteins involved in ubiquitylation and SUMOylation of PCNA at amino acid K164 (**Figure 8.15B**). These modifications are associated with enabling DNA replication past photodimer lesions and in facilitating post-replication repair of DNA. Monoubiquitylation of PCNA (Pol30-K164) by Rad6/Rad18 facilitates recruitment of error-prone translesion DNA polymerases, while further polyubiquitylation by Rad5/Mms2/Ubc13 channels DNA repair into an error-free pathway that appears to involve a template switch during repair synthesis between sister chromatids. SUMOylation of the same K164 residue by the Siz1 SUMO ligase facilitates recruitment of Srs2 to a stalled replication fork. Srs2 is then able to strip off Rad51 from an ssDNA end and thus to enable an alternative bypass of a DNA lesion by one of the translesion DNA polymerases. In BIR, the K164A mutant reduces BIR by about 50% (Figure 8.15B); similar reductions are accomplished by deleting either Rad18 or Siz1 or in the *rad18Δ siz1Δ* double mutant.

Recently, BIR-specific mutations of PCNA (*POL30*) were uncovered by an analysis of *pol30* mutations that suppress the cold-sensitivity of *pol32Δ*, which is unable to proliferate at 16°C. Surprisingly, two *pol30* mutations that suppress *pol32Δ*'s cold sensitivity do not suppress *pol32Δ*'s BIR defect; in fact the *pol30* mutations, by themselves, are defective in BIR! The defect of *pol30-FF248,249AA* is much more severe than mutations that prevent its ubiquitylation and SUMOylation (**Figure 8.15C**). These mutations could be defective in the steps of establishing a full replication fork. Moreover, these mutations are dominant, suggesting that even one such mutant subunit of the PCNA trimer may be sufficient to disrupt BIR.

As noted in Section 8.5, the initial stages of BIR appear to be the same as for strand invasion in gene conversion: 5′ to 3′ resection of DSB ends, the loading of RPA and then its displacement by Rad51, aided by Rad52, Rad55, and Rad57. However, there seems to be another set of

Figure 8.15 Effect of several mutations on BIR efficiency. (A) Both Sgs1 and Exo1 inhibit BIR in the interchromosomal haploid assay shown in Figure 8.9. Overexpression of RAD51 from a strong promoter (PGK::Rad51) also significantly affects the efficiency of BIR. Statistical significance of p < 0.01 is indicated by *. (B) A K164R mutation that blocks SUMOylation of PCNA inhibits BIR but not as severely as a *pol30-FF248,249AA* mutation in the replicative clamp PCNA (C). (From data in Lydeard JR, Lipkin-Moore Z, Jain S et al. [2010a] *PLoS Genet* 6:e1000973.)

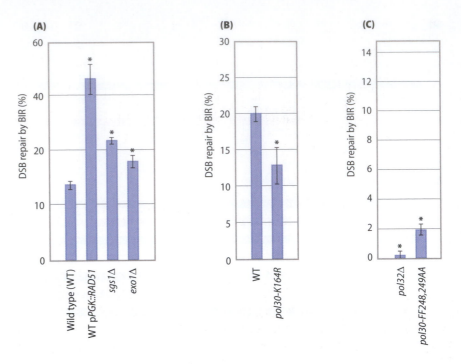

Rad51-aiding proteins, the Psy3–Csm2–Shu1–Shu2 (PCSS) that were found to be important in meiotic recombination, and in mitotic cells, as discussed in Chapter 4. Deleting one of the PCSS proteins reduces HO-induced BIR about twofold, while *rad55Δ* or *rad57Δ* reduce repair five- to tenfold. But *rad55Δ shu2Δ* double mutants are nearly as defective in BIR as *rad51Δ* mutants. The idea that the Shu proteins might therefore be an alternative Rad51-facilitating pathway is supported by finding that overexpressing Rad51 suppresses this defect both in *rad55Δ shu2Δ* and in *shu2Δ*. It is possible that the SHU complex only functions in the context of replication-related events.

There are two other curious aspects to these findings. First, deleting the PCSS proteins has little impact on gene conversion, where *rad55Δ* has a much greater impact than on BIR. Second, overexpressing Rad51 has a remarkable facilitating effect on BIR, whereas it has no effect on (the inherently more efficient) gene conversions studied in the same context. Why having more Rad51 would specifically facilitate BIR is not clear, but possibly the long delay in initiating new DNA synthesis means that more DNA is resected and made single-stranded and some of the Rad51 gets soaked up in regions that cannot perform BIR.

8.7 REPLICATION DURING BIR IS FAR MORE MUTAGENIC THAN NORMAL REPLICATION

Despite the fact that BIR uses the complete replication machinery that is employed in normal S phase, the rate of mutation accompanying BIR is as much as 2800 times higher than spontaneous events. By placing a *lys2* gene containing an out-of-frame run of adenines (for example, AAAAAAA) at several locations along the template chromosome arm, Anna Malkova's lab demonstrated that frameshift mutations to restore Lys+ activity depended on the number of A residues and to some extent on the position relative to the site where BIR initiated (**Figure 8.16**). Another set of reversion events involved a template jump between quasi-palindromic sequences. Mutations were independent of the error-prone DNA

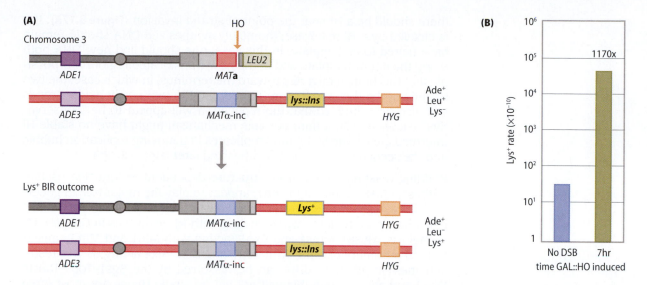

Figure 8.16 BIR is highly mutagenic in budding yeast. (A) The diploid assay illustrated in Figure 8.11 was modified to insert a *lys2::Ins* mutant on the template arm. The insertion is an AAAA sequence that creates a frameshift mutation. Lys⁺ recombinants arise when replication slippage occurs during BIR, deleting one A and restoring the open reading frame. (B) A dramatic increase in repair-associated mutation is seen in the *LYS2* reversion rate when cells were plated 7 hr after inducing HO endonuclease. (Adapted from Deem A, Keszthelyi A, Blackgrove T et al. [2011] *PLoS Biol* 9:e1000594.)

polymerase Polζ or on another translesion polymerase, Polη. However, ≥90% of possible frameshift mutations were corrected by the mismatch repair machinery, since the rate was tenfold higher in a *msh2Δ* mutant. Similarly, in a Polδ proofreading mutant (*pol3-5DV*) there is a significantly higher level of BIR-associated frameshifts, suggesting that Polδ is a major player in creating, but then correcting, errors. Finally, it appears that one contribution to the high mutation rate is that cells arrested by a DSB and undergoing BIR experience a dramatic rise in dNTP pools, which causes DNA polymerases to be error-prone. Mutant cells in which DNA damage does not trigger a DNA damage checkpoint-dependent increase in such pools have a fivefold lower level of mutations.

It is noteworthy that these results are strikingly different in many respects from the high level of mutations seen in gene conversion. The high rate of frameshift mutagenesis and base-pair substitutions measured in GC were independent of mismatch repair (see Chapter 6). Moreover, a different proofreading-defective Polδ mutation, *pol3-01*, had a dramatically opposite effect on mutations from that seen for *pol3-5DV* in BIR, namely the elimination of essentially all the template jumps, be they −1 frameshifts or quasi-palindrome events or interchromosomal template switches.

Very recently Anna Malkova has shown that virtually all of the newly generated *lys2* frameshift mutations are found on the repaired chromosome. This result suggests that all of the newly-copied DNA is found on the repaired chromosome while the template chromosome remains unaffected. This evidence of asymmetric inheritance of newly-copied strands is supported by BrdU incorporation experiments described in the following section.

8.8 HOW BIR IS FINALLY RESOLVED IS NOT YET KNOWN

A major unsolved question concerns the resolution of Holliday junctions and the final disposition of newly synthesized strands during BIR. If BIR proceeds by formation of a unidirectional replication fork inside a D-loop, there should be a HJ near the point of strand invasion (**Figure 8.17A**). If it is cleaved by a HJ resolvase, the newly synthesized DNA should remain base paired to its template. In this case, one should find new DNA both along the donor template and in the new translocated region. It is possible that this branch migrates toward the terminus, in which case, the two strands of the donor will be re-paired and the two newly made strands will also be paired, so that the replication will appear to be "conservative" (**Figure 8.17B**). A third possible mechanism might have no stable HJ intermediates, if the first strand replicates in a moving replication bubble and the second strand is only synthesized later (**Figure 8.17C**).

Budding yeast has at least four structure-dependent endonucleases: two of these, Mus81–Mms4 and Yen1, appear to play the major role in generating crossovers in interchromosomal mitotic recombination. This will be discussed in more detail when dealing with HJ resolvases in Chapter 14. In meiosis, Mus81–Mms4 and Exo1 (acting with Mlh1 and Mlh3) appear to be the primary players, with Yen1 and Slx1–Slx4 acting as last-resort activities. In addition, dHJs can be dissolved by the Sgs1–Top3–Rmi1–Dna2 complex, though this will not act on single HJs as are most often imagined in BIR intermediates. To date, only the roles of Mus81 and Yen1 have been examined in BIR; neither the single mutant nor the *mus81Δ yen1Δ* double mutant had any effect on the success of BIR. It will be necessary to ablate the remaining possible resolvases to establish firmly that there is an HJ that must be resolved for BIR to be completed.

The question of whether BIR leads to semi-conservative or conservative DNA synthesis can be directly answered by DNA combing or by other methods in which base analogs (such as BrdU) can be incorporated into DNA during DNA repair and then detected using immunofluorescence. A combination of these approaches has led Kirill Lobachev and Anna Malkova to favor the mechanism shown in Figure 8.17C. In fact, these investigators were able to identify intermediates of BIR by

Figure 8.17 Alternative modes of resolving intermediates in BIR. (A) Resolution of a non-migrating HJ leaves both copies of the replicated region with one old and one newly copied strand. (B) Migration of the replication bubble without HJ resolution results in one conservatively replicated chromosome and one unaltered template chromosome. (C) Discoordinated synthesis of leading and lagging strands will also produce the outcome where all newly copied DNA is on the recipient.

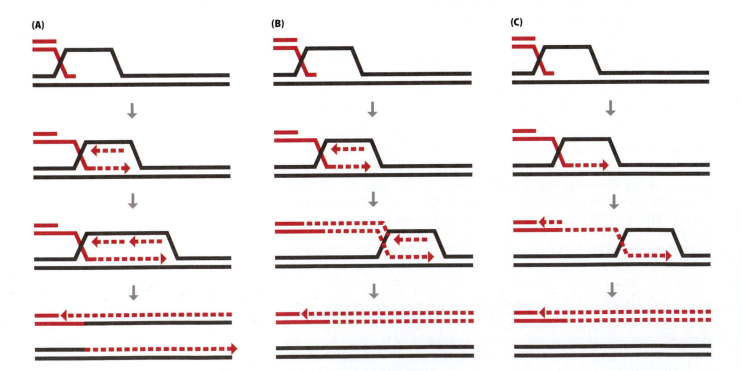

two-dimensional gel electrophoresis that are best explained by the formation of a migrating D-loop with a long single-stranded region that has not yet been filled in by second-strand synthesis. Such a mechanism can account for the increased rate of mutagenesis associated with BIR and the template switchings (Figures 8.14 and 8.16), because leading- and lagging-strand synthesis are not coupled and ssDNA regions are susceptible to mutation that cannot be mismatch-corrected. This mechanism does not yet account for the roles of all three DNA polymerases that were shown to be essential in the ectopic BIR system (Figure 8.13), unless Polα-primase would be needed for initiating one or the other sequential steps in DNA synthesis.

8.9 THERE IS ALSO A Rad51-INDEPENDENT BIR PATHWAY

Although most homologous recombination depends on Rad51, there are several examples of Rad52-dependent but Rad51-independent recombination events. One is the maintenance of telomeres in the absence of telomerase. Without telomerase there are two types of survivors, in which Type I is Rad51 dependent while Type II is Rad51 independent but dependent on Rad59, the MRX complex and Sgs1–Top3. Here, the telomere sequences themselves become greatly elongated (see Figure 8.10).

The requirements for Rad59 and MRX have also been shown for Rad51-independent repair events in the *rad51Δ/rad51Δ* diploid illustrated in Figure 8.11A. Tid1 (Rdh54) is also required. Unexpectedly, for telomere maintenance without telomerase, *tid1Δ* depresses both Type I and Type II survivors. Also difficult to understand, when BIR is studied by transforming in a linear plasmid, there seems to be little requirement for the MRX proteins or Rad59 either for Rad51-dependent or -independent BIR. The different genetic requirements may reflect some special features of each system, such as the length of homology available, the rate of end-degradation, the degree of mismatch, or whether there are non-homologous sequences at the DSB end that need to be excised before new synthesis can occur. When transformation is used, there may be induction of stress or other DNA damage responses that are different from creating a single break within a chromosome.

Another system that has provided a number of insights employs a plasmid containing an inverted repeat in which one copy contains an HO cleavage site. This plasmid was presented in Figure 3.20. When the amount of homology is about 1 kb on either side, most events are Rad51 dependent and are gene conversions without crossing over, but as the amount of homology is reduced more and more, the nature of the repair events changes. Repair still occurs at a surprisingly high rate even when there is only 33-bp homology on either side of the DSB; but now the events are entirely Rad51 independent. Most likely these Rad51-independent events arose by BIR followed by SSA, as illustrated in Figure 7.19.

How strand invasion is initiated without Rad51 when homology is very short remains a great mystery. Rad59 shares some homology with Rad52 and can carry out strand annealing *in vitro*, but *in vivo* Rad59 plays a secondary role to Rad52, as described in Chapter 5. Rad50, part of the MRX complex, appears to have some strand-opening activity that may be able to facilitate D-loop formation. Tid1 may be able to help remodel chromatin for strand invasion, though it does not do so in Rad51-dependent events, where only Rad54 appears to act. But it remains to be seen if there is an undiscovered component of Rad51-independent BIR that will truly take the place of Rad51. Malkova has suggested that the 3′-ended ssDNA created by resection could—without Rad51—establish a strand invasion structure by strand annealing (as in SSA) with homologous regions that transiently become single-stranded as a result of transcription, or replication. There has not been a careful study of the effect of transcription of a donor sequence on the efficiency of Rad51-independent BIR, but since most Rad51-independent BIR occurs between Ty elements it should be possible to inactivate their transcription by *ste12Δ*.

The role of Sgs1 in Type 2 telomere survival without telomerase is also confusing. Sgs1 is important in limiting SSA between homeologous sequences and in preventing efficient initiation of Rad51-dependent BIR. Consequently, one might have assumed that Sgs1 would also discourage Rad51-independent recombination between significantly mismatched substrates such as budding yeast telomeres, which are imperfect stretches of $TG_{1-3}(TG)_{2-3}$. Strand invasion by one telomere into another would be expected to form mismatched strand invasion intermediates. But in fact *sgs1Δ* markedly reduces Type 2 outcomes. Possibly this reduction merely reflects the fact that *sgs1Δ* markedly increases Type 1 events. An improvement in the survival of yeast lacking telomerase is seen in cells lacking the Msh2 mismatch repair gene, but it has not been shown that the increased survivors were particularly Rad51 independent.

It should be noted that Rad51 is surprisingly dispensable for gene targeting by an "ends-out" linear DNA, whereas even short gap-repair, when the DSB ends are oriented inward, is Rad51 dependent (**Inline Figure 8.1**). This finding emphasizes the differences in the way strand invasions occur in what would seem rather similar ways to start recombination.

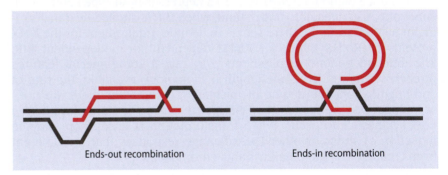

Ends-out recombination   Ends-in recombination

One striking feature of Rad51-independent BIR is that the DSB made on a truncated chromosome III (Figure 8.11) does not readily recombine near the DSB with homologous sequences on the other chromosome. Instead the DSB appears to be resected back 30 kb or more, until particularly recombinogenic sequences are exposed. Thus whereas the great majority of Rad51⁺ repair events occur within a few kilobases of the DSB (thus retaining an inserted *URA3* marker), nearly all the repair events in *rad51Δ* diploids take place 30 kb from the DSB, where there are a pair of dispersed, repeated Ty elements (**Figure 8.18**). Indeed most of the Rad51-independent repair events occur between one of these Ty sequences and one of 30 Ty elements and even more solo LTR sequences dispersed on other chromosomes. These repeated sequences provide a large number of targets for possible BIR events. Whether there are other features of Ty that make them exceptionally recombinogenic is not known; however, in the particular arrangement studied in Figure 8.18, the two Tys downstream from the DSB are in inverted orientation, so that extensive 5′ to 3′ resection can lead to the formation of an inverted hairpin end of a truncated chromosome that subsequently can go through rounds of breakage and recombination. It is interesting that the few chromosomal rearrangements that have occurred in evolutionary time among various *Saccharomyces* species all have arisen between Ty elements.

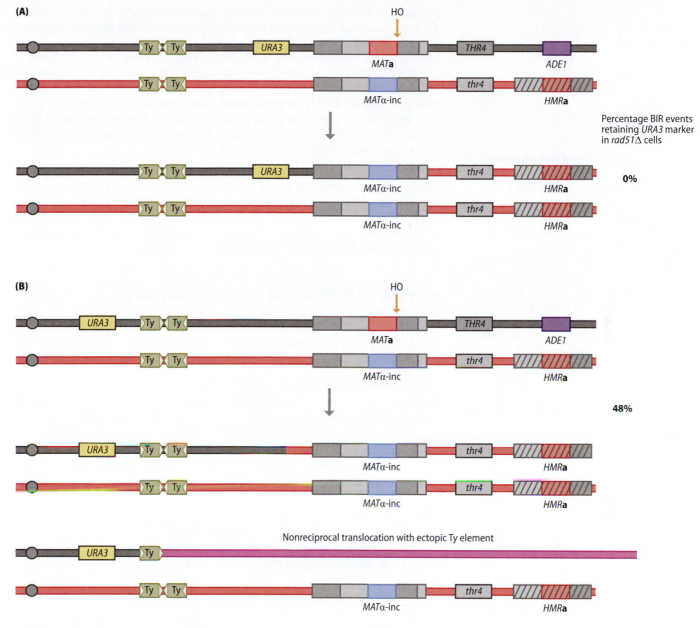

Figure 8.18 Rad51-independent BIR preferentially uses Ty repeated elements distant from the DSB site [C1]. (A) In the absence of Rad51, none of the DSBs induced by HO are repaired by gene conversion, but BIR events, resulting in LOH of the right arm of Chr 3 are seen. However, few repair events occur to retain a *URA3* marker several kilobases to the left of the DSB. (B) A large fraction of the repair events can retain a *URA3* marker [C2] inserted more than 30 kb from the DSB, centromere proximal to a pair of 6-kb inverted Ty retrotransposon elements. Some repair events have copied the intact donor chromosome but many proved to involve recombination between one of the Ty sequences and a copy of Ty located on a different chromosome. Many of the remaining repair events use another pair of Ty sequences lying closer to the centromere. (Adapted from Malkova A, Signon L, Schaefer CB et al. [2001] *Genes Dev* 15:1055–1160, and from VanHulle K, Lemoine FJ, Narayanan V et al. [2007b] *Mol Cell Biol* 27:2601–2614. With permission from Cold Spring Harbor Laboratory Press and the American Society for Microbiology, respectively.)

8.10 HALF-CROSSOVERS ARE AN ALTERNATIVE PATHWAY PRODUCING NONRECIPROCAL TRANSLOCATIONS

When only one end of a DSB successfully strand-invades a template, the branched structure can be used to initiate BIR. But if BIR is prevented, this intermediate can be altered by branch migration to create a single complete Holliday junction and a half-exchange, or a complete dHJ

Figure 8.19 Generation of half-crossovers.
(A) If a D-loop formed by strand invasion is impaired in initiating new DNA synthesis, it can branch migrate "backwards" to form a dHJ that can be resolved as a half-crossover, with one recombined intact chromosome and a chromosome fragment. (B) In the BIR transformation assay shown in Figure 8.8 and 8.14, if POL32 is deleted, the expected BIR outcomes are greatly reduced and instead one finds instances where there are the expected BIR products (CF1 or CF2) but one of the template chromosomes has been lost. These are half-crossover outcomes. The normal outcomes—using one or the other or a mixture of the template chromosomes—may represent normal BIR events but they may be examples where the broken chromosome fragment was partitioned to the sister cell, allowing the selected cell with a repaired CF to have both homologs. (Adapted from Smith CE, Lam AF & Symington LS [2009] *Mol Cell Biol* 29:1432–1441. With permission from the American Society for Microbiology.)

(**Figure 8.19A**). This structure can be resolved to yield a half-crossover (half-CO)—where there is one intact chromosome and a broken fragment that would be lost. Indeed, in the absence of Pol32, events that normally would be recovered as BIR in budding yeast appear as nonreciprocal crossovers; that is, where there is the selected crossover chromosome there is concomitant loss of what ought to have been the template chromosome (**Figure 8.19B**). Recent work from Lorraine Symington implies that both the Mus81–Mms4 and Yen1 endonucleases are needed to obtain crossovers in a *pol32Δ* strain. Additional situations resulting in half-crossovers are discussed in Section 8.13.

One interesting result from Symington is that half-COs also appear when BIR is impaired by using a defective DNA Polδ-ct mutation that is defective in processive DNA synthesis during repair but is competent for normal replication. Apparently this mutant is able to establish normal strand invasion but becomes slowed during replication, possibly allowing cleavage of the D-loop to be resolved as a half-CO.

8.11 BIR IS OBSERVED IN OTHER ORGANISMS

Much less is known about BIR in organisms other than *E. coli* and budding yeast, but it is evident that all higher eukaryotes have similar pathways that play important roles.

Telomere elongation in Kluyveromyces lactis

Another budding yeast, but evolutionarily quite distant, *Kluyveromyces lactis* has 25-bp telomeric repeats that are largely identical in sequence. Mutation of telomerase RNA or protein components leads to senescence. From among the population of dying cells emerge survivors. These viable cells have experienced a lot of BIR-like events that involve recombination among homologous subtelomeric sequences that are located at the

(A)

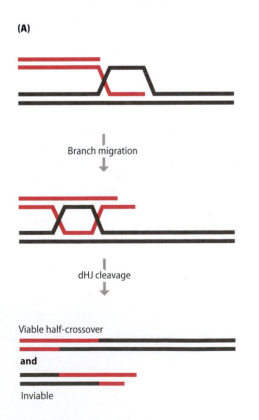

(B)

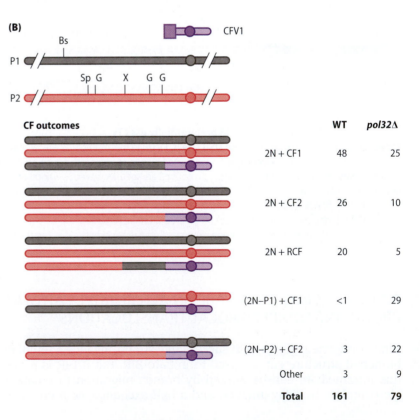

CF outcomes		WT	pol32Δ
	2N + CF1	48	25
	2N + CF2	26	10
	2N + RCF	20	5
	(2N–P1) + CF1	<1	29
	(2N–P2) + CF2	3	22
	Other	3	9
	Total	**161**	**79**

ends of different chromosome arms. Most survivors have long telomere sequences similar to Type II events in *Saccharomyces cerevisiae* cells lacking telomerase, but Michael McEachern's lab found that about 50% of telomeres in a group of survivors had replaced all their telomeric repeats and acquired a copy of a *URA3* gene placed ~120 bp from one telomere. Moreover, losses of a subtelomeric copy of *URA3* (presumably also by BIR) are largely Rad51 dependent and Rad50 independent, but survival is essentially knocked out in a *rad51Δ rad59Δ* double mutant although not in *rad50Δ rad51Δ* double mutants. So there are some differences between the survival mechanisms in *S. cerevisiae* (where *rad51Δ rad50Δ* and *rad51Δ rad59Δ* give similar lack of survivors) and *K. lactis*.

McEachern's studies in *K. lactis* have provided strong evidence that telomere elongation can occur by BIR, in a rolling-circle variant in which the telomere end invades a circle of telomeric DNA. Such "t-loops" have been well documented in mammalian cells and autonomously replicating circles that include telomeric sequences and adjacent subtelomeric sequences have been demonstrated in budding yeasts. By introducing a marked circular plasmid containing telomeric sequences into a *ter1Δ* strain that lacks telomerase activity, McEarchern was able to demonstrate a "roll and spread" mode of telomere elongation that incorporated the marked circles in tandem arrays (**Figure 8.20**).

BIR in **Schizosaccharomyces pombe**

BIR has been documented in *S. pombe* by Tim Humphrey's lab, using an HO-induced DSB on a minichromosome that was derived from Chr 3 (**Figure 8.21**). In addition to gene conversion events that we examined in Figure 6.37, it was possible to recover repair events in which the *distal* portion of the minichromosome was lost, resulting in loss of the *his3*+ marker, an example of loss of heterozygosity (LOH). Some of the LOH

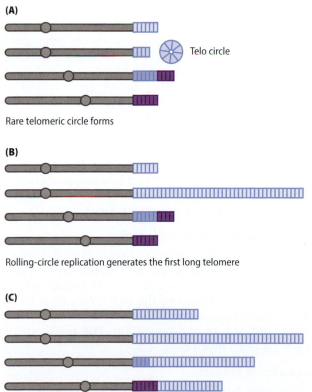

(A)

Telo circle

Rare telomeric circle forms

(B)

Rolling-circle replication generates the first long telomere

(C)

Sequence from long telomere spreads to all other telomeres by BIR events

Figure 8.20 BIR can mediate the amplification of telomeric sequences in the absence of telomerase. (A) Sequences at different telomeres are shown in different colors. A rare telomeric circle can be used as a rolling-circle template after recombination with the telomere sequences on a chromosome to generate a long telomere (B). Subsequently this long telomere can be used as a BIR template to elongate other short, recombinogenic telomeres (C). (Adapted from McEachern MJ & Haber JE [2006] *Annu Rev Biochem* 75:111–135. With permission from Annual Reviews.)

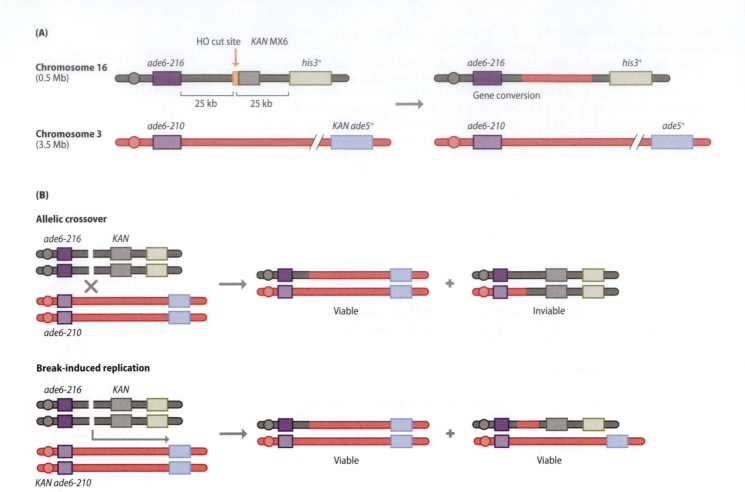

Figure 8.21 **BIR repair in *S. pombe* between a minichromosome 3 (Ch[16]) and Chr 3.**
(A) A minichromosome derived from chromosome 3 can be maintained on medium in the absence of adenine by the presence of intragenically complementing *ade6* markers. After HO induction, gene conversion without crossing-over can produce a viable outcome. (B) In contrast, gene conversion accompanied by a crossover yields a Ade+ His− G418S strain, whose sister cell is dead because it lacks essential segments of Chr III distal to the end of the minichromosome. (C) Similarly, BIR yields a sectored colony derived from an Ade+ His− G418S BIR derivative and a viable cell that could either have lost the sister broken chromosome (not shown) or repaired the DSB by gene conversion (*right*). (Adapted from Cullen JK, Hussey SP, Walker C et al. [2007] *Mol Cell Biol* 27:7745–7757. With permission from the American Society for Microbiology.)

events proved to arise by reciprocal crossovers between the minichromosome and Chr 3; in this case the sister cell is inviable, being homozygous for the truncated arm of the minichromosome. In other cases, however, both sister cells were viable and could be shown to have duplicated the template chromosome arm. These events required the homologous recombination machinery including fission yeast homologs of Rad51, Rad52, Rad55–Rad57, Rad54, and the MRX proteins. Unlike budding yeast, BIR in fission yeast also needs the branched structure resolvase Mus81. It is interesting that essentially all crossovers in meiosis in fission yeast also require Mus81, whereas deleting Mus81 in budding yeast has little consequence unless another crossover pathway is eliminated. This result concerning BIR suggests that budding yeast have a number of resolvases that act redundantly whereas fission yeast may have but one. Indeed, fission yeast lack the HJ resolvase Gen1.

Drosophila melanogaster *telomere elongation*

Drosophila differs from most other well-studied model organisms in that it lacks telomerase and maintains its chromosome ends by frequent transpositions of a retrotransposon, HetT-A. However, when a P-element excision creates a DSB and lops off the HeT-A repeats at the chromosome end, both new transpositions and BIR using a homologous chromosome as a template can restore end-protection. The requirements for BIR in flies have not yet been much explored.

BIR in Xenopus laevis *extracts*

Vincenzo Costanzo has exploited the ability to establish replication and some aspects of recombination in cell-free extracts of *Xenopus*. By nicking

DNA enzymatically, his lab created a situation where an oncoming replication fork was stalled and broken. Replication is blocked when the GINS subunits of the leading-strand CMG (Cdc45-MCM-GINS) helicase dissociate (**Figure 8.22**). Their reloading and resumption of DNA synthesis depends on both Rad51 and the MRN proteins. A double mutation in PCNA that is analogous to the *pol30-FF248,249AA* double mutant in budding yeast that prevents BIR but not normal replication also prevents replication restart in the *Xenopus* system. This is an excellent model system to study replication restart and BIR.

8.12 BIR-LIKE EVENTS ARE IMPORTANT FOR HUMANS

Nonreciprocal translocations are a hallmark of human cancer cells; whether any of these arise by BIR is hard to judge. First of all, the enormous lengths of mammalian chromosome arms seem impossibly long to be copied by a single recombination-induced replication fork. Second, DNA sequencing of translocation junctions are much more consistent with the rearrangements arising by nonhomologous end-joining (NHEJ.) Nevertheless there is at least one example that seems a good candidate for BIR. Facioscapulohumeral muscular dystrophy is frequently associated with nonreciprocal translocations involving diverged 3.3-kb repeated sequences near the telomeres of chromosomes 4 and 10. These could arise by BIR, as the distance that a repair replication fork would have to travel is not out of the question. Several clinical studies have shown that these rearrangements arise at high levels in the normal population, but no cell line has been shown to have this instability. This rearrangement could be the basis of a way to study BIR in human cells, other than events involving telomeres.

As noted in the Introduction, Thanos Halazonetis, Vassilis Gurgolis, and Jiri Bartek have argued that precancerous and cancer cells with activated oncogenes often exhibit a high level of replication fork stalling and breakage. Recently Halazonetis' lab has shown that depletion of PolD3, the human cell homolog of yeast Pol32, impairs the growth of cells that are under oncogene-induced replicative stress (here caused by the overexpression of cyclin E), but there is no effect when PolD3 is depleted in normally growing cells. These data imply that BIR is relevant in higher eukaryotes and is a key pathway for repair of the collapsed DNA replication forks that are present in human cancers.

The main focus of attention in thinking about BIR in humans involves the maintenance of telomeres by alternative lengthening of telomeres (ALT). Whereas most cancer cells avoid senescence by reactivating telomerase, many transformed cell lines and some cancers avoid cell

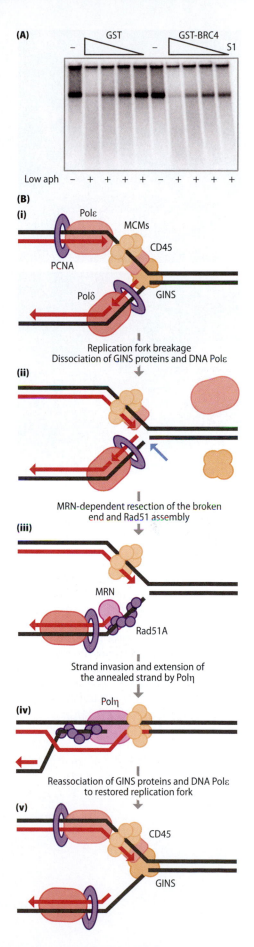

Figure 8.22 BIR in *Xenopus* extracts. (A) Replication in *Xenopus* extracts was slowed by adding aphidicolin (aph) and ssDNA breaks were created by treating the chromatin with S1 nuclease. The relative amount of S1 used is indicated by the triangle. The extent of replication is measured by ^{32}P incorporation into sperm DNA. At high concentrations, S1 severely impairs replication. Rad51 depletion was accomplished by binding Rad51 to the BRC4 protein segment of BRCA2 fused to the GST protein. An equivalent concentration of GST protein was used as a control. The depletion of Rad51 resulted in a significant decrease in ^{32}P incorporation into DNA as replication proceeds in the presence of S1 nuclease. These results argue that Rad51 is required to permit DNA replication when S1 nuclease has collapsed replication forks. (B) A summary of the observations in (A). Replication stalling and strand breakage (*blue* arrow) is accompanied by dissociation of the GINS subunits of the CMG helicase (ii). Reestablishment of replication depends on both the Rad51 and MRN proteins (iii, iv) and reloading of GINS (v). (Adapted from Hashimoto Y, Puddu F & Costanzo V [2012] *Nat Struct Mol Biol* 19:17–24. With permission from Macmillan Publishers, Ltd.)

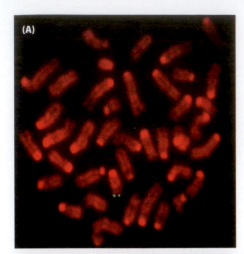

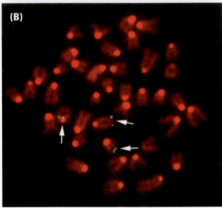

Figure 8.23 Proliferation of a sequence inserted at one telomere into other telomeres. (A) A novel DNA sequence marked by a *neo* gene (*green*) was inserted within telomere sequences on chromosome 15 of a mouse embryonic stem cell, as seen by FISH in metaphase-arrested cells. The *red*-stained mouse chromosomes are all acrocentric with a densely staining centromeric region at one end. (B) An example of the proliferation of the *neo*-marked sequence to two additional chromosomes, including a telomere close to a centromere. (From Neumann AA, Watson CM, Noble JR et al. [2013] *Genes Dev* 27:18–23. With permission from Cold Spring Harbor Laboratory Press.)

death by telomere shortening by what appears to be a recombination-dependent mechanism. The evidence supporting recombination in ALT is of three sorts. First, a marker integrated into a single telomeric region can be found at multiple chromosomal sites in ALT cells after they have grown for many generations. This observation argues that there is frequent recombination among telomeres that is not simply a reciprocal exchange of ends but a replicative process that duplicates the marker to new telomeres. Recently, Roger Reddel's group has shown that one can see the same kind of telomere × telomere recombination in normal mouse embryonic stem cells as well (**Figure 8.23**).

The second line of evidence concerning recombination in ALT cells is that there is a tenfold elevated rate of sister chromatid exchange (SCE) relative to SCE rates elsewhere in the genome; this high rate is not seen in telomerase-positive cells. These telomere-SCE (T-SCE) events are measured by a technique called chromosome orientation fluorescence *in situ* hybridization (CO-FISH), in which BrdU-labeled strands are preferentially excised from fixed samples after UV-irradiation by enzymatic removal (**Figure 8.24**). If the chromosomes are then hybridized with a strand-specific probe, normally there would only be one strand that can hybridize with the strand-specific probe at each end of a replicated chromosome; but in T-SCE the probe is found to hybridize to telomere regions on both sister chromatids. Thus there must be some form of SCE that accompanies the maintenance of telomeres in ALT cells. An example of T-SCE is shown in **Figure 8.25**. However, there are no data that yet demonstrate that such T-SCE events arise during the elongation of a given chromosome end by ALT, which may only occur rarely at a given end as it becomes too short to maintain its integrity. Interestingly, Mus81 HJ resolvase appears to be required for ALT, analogous to the BIR dependence on Mus81 in *S. pombe*. Depletion of Mus81 in ALT cells reduced telomere recombination and caused their growth arrest. Mus81 apparently interacts with the key telomere-protecting protein Trf2.

Third, telomeres in ALT cells are associated with "ALT-associated promyelocytic leukemia (PML) bodies" (APBs) that contain—in addition to telomere DNA sequences—Rad51, the MRN proteins, and the BLM and Werner's (WRN) helicases, all of which are identified with homologous recombination. As shown in Figure 8.25, a BLM-related helicase, WRN, appears to suppress ALT and T-SCE. Reddel's lab has shown that overexpressing Sp100, a constituent of PML bodies but not ALT-associated APBs, caused the relocalization of MRN proteins away from APBs and resulted in the inactivation of ALT and the progressive shortening of telomeres. This result certainly argues that the recombination machinery of APBs plays a key role in ALT. Given that budding yeast MRX proteins are essential for Type II telomere maintenance, it is of course tempting to imagine that MRN proteins are carrying out a similar role in mammalian ALT. One other finding that could implicate BIR is that depletion of the replication-associated flap endonuclease FEN1 causes the senescence of ALT cells but not those expressing telomerase. Still, at the moment, it is not clear if BIR plays a role in ALT.

8.13 THERE IS A Rad52-INDEPENDENT HOMOLOGOUS RECOMBINATION PATHWAY

Although budding yeast Rad52 is required for almost all homologous recombination, *rad52Δ* cells can, at a low frequency, give rise to half-crossover outcomes. How DSBs can be repaired without Rad52 in mitotic cells remains mysterious because Rad52 seems to be absolutely required for the loading of Rad51, and also for Rad51-independent BIR. However,

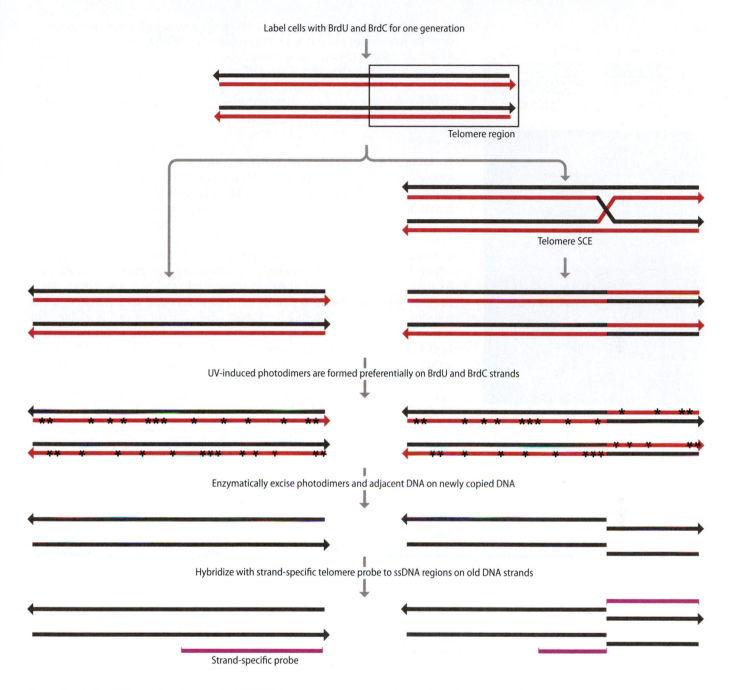

Figure 8.24 CO-FISH reveals telomere-SCE (T-SCE). Chromosome orientation FISH (CO-FISH) is carried out by growing cells for one generation in the presence of BrdU and BrdC. UV irradiation preferentially creates photodimers on the Br-containing pyrimidines; these newly replicated strands can then be digested away by UV repair enzymes derived from *E. coli*. This treatment leaves single-stranded regions of "old" DNA that can hybridize with a strand-specific fluorescent probe under nondenaturing conditions where dsDNA will not hybridize. Under normal circumstances, without a crossover in the telomere region, a strand-specific telomere probe should hybridize to only one of the two sister chromatids at a given end. T-SCE places some old DNA on each sister chromatid and both sisters will now hybridize to the strand-specific probe. (Adapted from Nabetani A & Ishikawa F [2011] *J Biochem* 149:5–14. With permission from Oxford University Press.)

as noted in Chapter 5, some SSA events—especially when the length of homology is long—are Rad52 independent. In a diploid heteroallelic for *his4* mutations, and homozygous for a *rad52* mutation, His⁺ recombinants arose at a frequency of about 1% relative to wild type (**Figure 8.26**). Genetic analysis showed that the His⁺ cells mostly had become 2n-1 aneuploids

(A)

Nascent C-strand labelling by CO-FISH

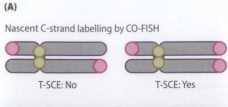

T-SCE: No   T-SCE: Yes

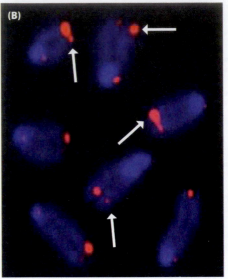

Figure 8.25 T-SCE is seen in ALT cells. (A) T-SCE is revealed when the probe lights up both sister chromatids at the same end, as described in Figure 8.24. (B) T-SCE is evident in ALT cells with a defect in the Werner's (WRN) helicase. Arrows point to examples where a strand-specific probe hybridizes to each of the two sisters at a given chromosome end. (A, adapted from Nabetani A & Ishikawa F [2011] *J Biochem* 149:5–14. With permission from Oxford University Press. B, from Laud PR, Multani AS, Bailey SM et al. [2005] *Genes Dev* 19:2560–2570. With permission from Cold Spring Harbor Laboratory Press.)

carrying a crossover His+ chromosome but missing its reciprocal product. These *RAD52*-independent recombinants required the MRX complex, but were surprisingly largely independent of both Rad51 and Rad59. These could arise by half-crossovers (Figure 8.26A). Alternatively, long regions of ssDNA can anneal without Rad52, so it is possible that half-CO events could arise by SSA between breaks at different locations on two different molecules (Figure 8.26B). It is possible to estimate the frequency with which DSBs arise spontaneously by following the fate of *rad52Δ* cells as they progress from G1 through replication and mitosis. About 8% of *rad52Δ* cells give rise to a pair of cells of which one dies because it has a lesion that requires Rad52 for repair while the other cell is viable. The lesions most likely are DSBs of a sister chromatid arising during replication. Given that the yeast genome is about 10,000 kb, the rate of breakage would be about 0.08% per 100 kb. This rate seems too low to account for 1% of cells with an HO-induced DSB on Chr 3, which is 100 kb from its centromere, to anneal with a spontaneous DSB on a homologous chromosome, unless the rate of breakage were tenfold greater in that interval.

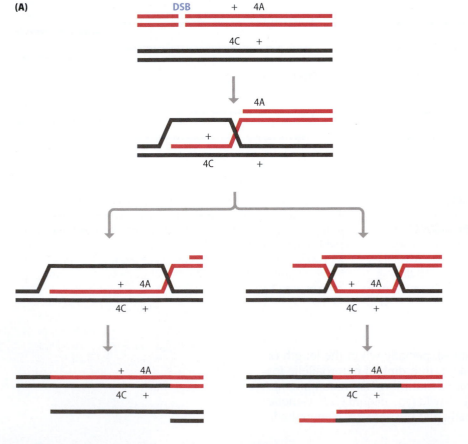

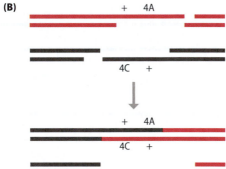

Figure 8.26 Rad52-independent heteroallelic recombination leading to a half-crossover. (A) Strand invasion without Rad52 is imagined to create a short stretch of hDNA that can be extended by branch migration. The hDNA can subsequently be mismatch repaired to give a His+ recombinant (*not shown*). Subsequent HJ resolution will produce a half-crossover. (B) Two distant DSBs can be resected to allow Rad52-independent SSA that can be mismatch repaired to His+. The opposite ends do not share homology and cannot re-join. (Adapted from Coïc E, Feldman T, Landman AS & Haber JE [2008] *Genetics* 179:199–211. With permission from the Genetics Society of America.)

8.14 BIR CAN PRODUCE COPY NUMBER VARIATION

A major factor in cancer ontology is gene *copy number variation* (CNV). Recent advances in DNA sequencing of cancer genomes have revealed many examples of duplications, triplications, and quite often very complex amplifications of chromosome regions. BIR can produce duplications as nonreciprocal translocations, and may also be implicated in the formation of *segmental duplications* (SD).

In humans, many recurrent chromosomal duplications or deletions are found in which the endpoints are >10 kb (sometimes much longer) regions of homology. Rearrangements are believed to occur by "nonallelic homologous recombination" (NHAR), although the precise mechanisms are difficult to deduce.

Segmental duplications arise in budding yeast

One way to study the origin of such duplications in yeast is to select for increases in gene dosage to compensate for the presence of a weak, hypomorphic allele of a gene that is essential for growth or is required under certain growth conditions. One strategy is to select for improved growth of a cell in which there is a deletion of one of two dispersed copies of an essential gene, such as a ribosomal protein gene or the histone H2A gene. Cells can compensate for the absence of one member of such gene pairs by the formation of whole chromosome disomy, where the chromosome carrying the remaining gene is duplicated; this duplication most likely is caused by mitotic chromosome nondisjunction. One survey found that 8% of the genes deleted in yeast gave rise to such disomy, apparently for dosage compensation. However, duplications of a segment of a chromosome that carries the remaining gene copy are also found.

When one of the two unlinked copies of the histone H2A–H2B gene pair is deleted (*hta1Δ hta2Δ*), yeast cells grow slowly, but they regain better growth by duplicating the remaining pair, *HTA2 HTB2*. One novel outcome was the result of a fortuitous arrangement of Ty elements flanking these genes, on opposite sides of the centromere, such that a crossover between the two Tys produced a circular minichromosome carrying the HTA2–HTB2 genes (**Figure 8.27**).

One well-studied case is a variety of spontaneous gene duplications arising when the *RPL20A* gene is deleted in budding yeast, giving rise to duplications of the *RPL20B* gene. Such duplications can be as tandem

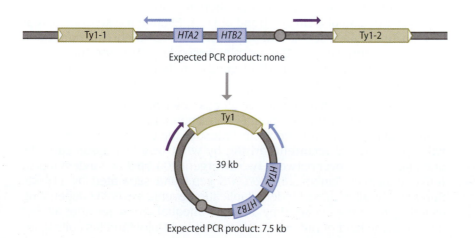

Figure 8.27 Amplification of H2A–H2B genes by Ty-mediated recombination. Cells grow slowly in the absence of one pair of histone H2A and H2B genes. Faster-growing colonies carry a circular, centromeric, autonomously replicating chromosomal segment carrying an extra copy of the H2A and H2B genes (*HTA2 HTB2*), through recombination between flanking Ty elements. Arrows indicate location of PCR primers that would yield a product in a circular rearrangement. (Adapted from Libuda DE & Winston F [2006] *Nature* 443:1003–1007. With permission from Macmillan Publishers, Ltd.)

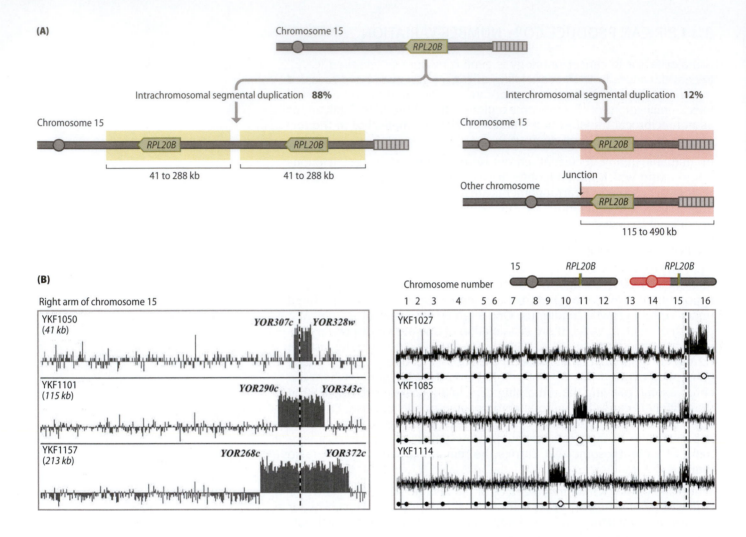

Figure 8.28 Duplication of RPL20B in a strain lacking the homologous gene *RPL20A.* (A) Various sizes of both intra- and interchromosomal segmental duplications of *RPL20B* gene were found as faster-growing colonies in cells lacking the *RPL20A* gene that encodes the same ribosomal protein. Intrachromosomal duplications were usually tandem duplications, whereas interchromosomal events were usually nonreciprocal translocations, most likely arising by BIR. (B) Evidence for duplications shown by comparative genome hybridization (CGH) in which sequences that are present in greater than one copy are shown as dark regions above the baseline. In the three examples of segmental duplications shown on the left, the sizes of the duplicated regions range from 41–213 kb. The names of yeast genes on chromosome 15 at the boundaries of the duplications (YOR307c, etc.) are noted. Nonreciprocal translocations are shown on the right involving the sequences on chromosome 15 and those on chromosomes 11 and 10. (A, from Payen C, Koszul R, Dujon B & Fischer G [2008] *PLoS Genet* 4:e1000175. B, from Koszul R, Caburet S, Dujon B & Fischer G [2004] *EMBO J* 23:234–243. With permission from Macmillan Publishers, Ltd.)

duplications of a region containing this gene or as segmental duplications in which a segment of the chromosome containing *RPL20B* is inserted into another chromosome (**Figure 8.28**). These segmental duplications (SD) arise spontaneously as often as 1×10^{-7} cells; when there are defects in DNA replication, for example, in a *clb5Δ* strain lacking a nonessential cyclin B, the incidence of SD rose more than 700 times. SDs also arise frequently after treatment of cells with camptothecin, which creates stalled and broken replication forks where topoisomerase I is covalently bound to one end of a single-strand of DNA. Even in the absence of replication inhibitors, it seems that sequences that naturally stall replication fork progression, such as the complex secondary structure that can be adopted by strands encoding tRNA genes, are hot spots of chromosome breakage leading to SDs.

The kinds of rearrangements that appear can be classified as "recurrent" (meaning that the end-points of independently isolated events are found frequently) or "nonrecurrent." Most of the tandem duplications in wild-type cells are recurrent, arising by what was first assumed to be unequal crossing over between two divergent 300-bp long tandem repeat (LTR) sequences flanking the *RPL20B* gene and separated by 115 kb. Surprisingly, all of these tandem duplication events are Pol32 dependent, as if they have not occurred by a simple unequal crossover after all, but involve some kind of BIR event. When Bernard Dujon and his colleagues artificially created a pair of flanking homologies lacking mismatches, a class of Pol32-independent SDs were recovered; these presumably arose by unequal crossing over, which must have been suppressed by the divergence of the flanking repeats. The interchromosomal SDs also proved to be mostly recurrent, frequently using an LTR centromere proximal to the *RPL20B* gene as the junction point of a nonreciprocal translocation with another LTR on another chromosome. Note that for these events to occur, for example by BIR, the junction on the second chromosome must not have deleted any essential genes. In some cases the second copy of *RPL20B* was found on an extra chromosome that might have been generated by BIR but was accompanied by a nondisjunction event that added the translocation chromosome to a full haploid set of chromosomes.

There were other SDs, both intrachromosomal and interchromosomal, that were nonrecurrent. Most of these did not involve LTR sequences and had junctions made up of short regions of microhomology less than 10 bp. Although there is a pathway of microhomology-mediated NHEJ (MMEJ), these rearrangements were not eliminated by deleting components required for MMEJ or NHEJ. The fact that these events were Pol32 dependent suggests that they, too, arise by a replication-dependent mechanism.

Analyses of SDs in various mutant backgrounds have revealed other distinct mechanisms of SD formation. Deletion of Rad51 led to an increase in interchromosomal BIR events that appeared to occur between the LTR centromere proximal to *RPL20B* and divergent LTRs elsewhere in the genome. This finding fits a pattern of observations that Rad51 discourages ectopic recombination between diverged LTR sequences, but in *rad51Δ* strains the great majority of events use these diverged sequences. As noted before, in a model system involving inverted repeats on a plasmid, Rad51-independent recombination was able to use much smaller regions of homology (about 30 bp) versus roughly 100 bp needed for Rad51-dependent gene conversions.

An even more striking discovery is that there is a remarkably efficient Rad52-*independent* pathway of SD formation. Deleting Rad52 completely eliminates intrachromosomal and interchromosomal rearrangements involving the LTR repeats, but in their place are found very short regions of microhomology located at both intrachromosomal tandem duplications and interchromosomal insertions. Also surprising is that all of these events are Pol32 dependent. These outcomes seem most compatible with some sort of perturbation of DNA replication.

Nonrecurrent SDs in human disease may involve BIR

Microhomology-mediated duplications were also documented by James Lupski's lab and others in a number of human genetic disorders. Two of the most thoroughly documented examples are the case of the demyelinating Pelizaeus-Merzbacher disease (PMD), in which the *PLP1* gene copy number is increased, and amplifications of the terminal part of the X chromosome containing the *MECP2* gene, which is associated with mental retardation in males. Here the complexity of the rearrangements often involves the assembly of multiple segments of originally noncontiguous DNA located megabase pairs apart (**Figure 8.29**). Regions with extreme versions of these types of rearrangements have been seen in whole-genome sequencing of human cancer cells. A variety of breakpoint junctions are found, many with microhomology that might be consistent with end joining, but increasingly there is the sentiment that most of these junctions involve a form of BIR—microhomology-mediated BIR—that can use very short regions of homology.

Figure 8.29 Example of long-distance joinings of chromosomal segments with microhomology at the junctions in humans.
(A) A duplication around the *PLP1* gene is associated with multiple joinings of initially distant sequences, presumably by a series of three jumps (shown as *dotted lines*) of the DNA replication machinery during microhomology-mediated BIR (MM-BIR). These sites can be >than 1 Mb apart. (B) Microhomologies at the junctions of these translocations. (Adapted from Lee JA, Carvalho CM & Lupski JR [2007] *Cell* 131:1235–1247. With permission from Elsevier.)

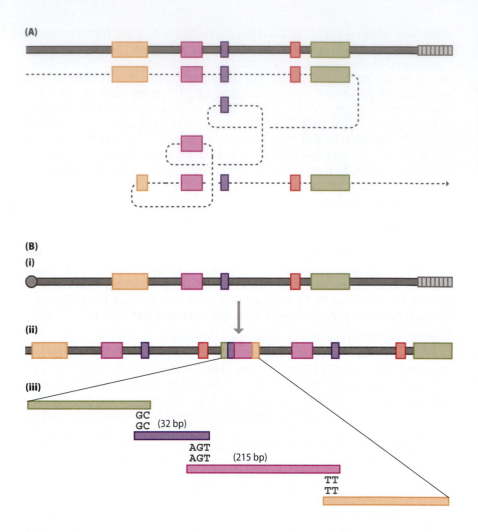

8.15 CNV MAY ARISE FROM MICROHOMOLOGY-MEDIATED BIR

Several different versions of microhomology-mediated BIR (MM-BIR) have been proposed. Gilles Fischer interpreted the yeast rearrangements as MM-IR (microhomology-induced replication) and noted that mutations that stalled replication greatly increased their frequency. James Lupski proposed a mechanism called "fork stalling and template switching" (FoSTeS) to account for the mammalian events: later Lupski and colleagues broadened the notion of FoSTeS to MM-BIR. In all of these models, the key feature is that recombination induced replication is not processive and may dissociate a partially replicated strand and promote its annealing at another location, possibly involving another replication bubble. These exits and entries may happen many times. Whether these dissociations and invasions involve still-attached DNA polymerases or would involve a homologous recombination step (but recognizing unusually tiny lengths of homology) is not at all evident.

As noted in Chapter 6, yeast *MAT* switching can be exploited to study interchromosomal microhomology-mediated repair, in which there is a template jump during DSB repair to copy sequences that are only 73% identical (Figure 6.34). Two jumps are required and the sequences at the junctions share only 5–11 bp perfect match. However, these events are Pol32 independent and may represent MM-SDSA rather than MM-BIR.

8.16 CHROMOTHRIPSIS IS AN UNEXPECTED TYPE OF GENOME INSTABILITY THAT MAY REQUIRE BIR

In 2011, Peter Campbell's lab reported an analysis of the genomes of some cancer cells in which there had been massive rearrangements on a single chromosome—pulverizing, as one report characterized the event. Rearrangements involved tens or hundreds of breaks and re-joinings with, at best, microhomologous junctions (**Figure 8.30**). The authors suggested that these changes likely occurred in a single catastrophic event. An initial characterization suggested that the results could be explained by the breakage and reassembly of a single replicated chromatid and the reassembly into a scrambled arrangement with many regions having two

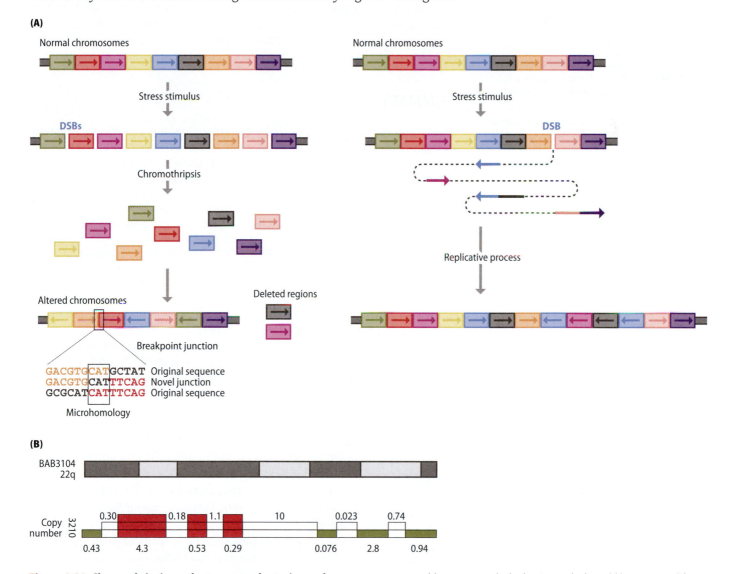

Figure 8.30 Chromothripsis, or chromosome shattering and rejoining. Many rearrangements occur within a single chromosome. (A) Two possible mechanisms causing multiple intrachromosomal rearrangements after exposure to an unspecified stress stimulus are shown. (*Left*) Chromosome fragmentation and re-joining by NHEJ can scramble sequence order. If both sister chromatids were broken it is possible to generate a chromosome with duplications of some regions, while other segments could be lost. (*Right*) Rearrangements could be triggered by DSB repair leading to microhomology-mediated break-induced replication (MM-BIR) that involves rampant template switching. In some instances this process could create not only duplications, which could be accounted for by joining pieces of a pair of sister chromatids into a single chromosome (*left*), but triplications, which must have involved some sort of replication. (B) An example of one region in chromosome 22 in which there are some segments with only one copy (*green*) and others with three copies (*red*). *Gray* and *dark gray* bars represent regions stained by Geimsa, a standard cytological reference. (A, adapted from Maher CA & Wilson RK [2012] *Cell* 148:29–32. With permission from Elsevier. B, adapted from Liu P, Erez A, Nagamani SC et al. [2011] *Cell* 146:889–903. With permission from Elsevier.)

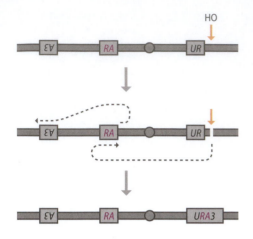

Figure 8.31 A yeast model to test aspects of MM-BIR. A DSB in the nonessential terminal end of one chromosome can initiate BIR with a homeologous "RA" sequence located several hundred kilobases distant. In turn, the replication can proceed by another microhomology-mediated jump to another diverged "A3" target, yielding a complete *URA3* gene as part of a complex nonreciprocal translocation. (Based on work by Ranjith Anand, Brandeis University.)

copies of some sequences. Leaving aside why such fragmentation would occur on one chromosome, this outcome could be explained by efficient nonhomologous end-joining of the many fragments. A subsequent investigation by Lupski and colleagues found many other examples of chromothripsis some of which contained not only duplications but also triplications of some regions (Figure 8.30B), a result that would be hard to explain by simply rearranging the segments of a single replicated chromatid. Instead it was proposed that microhomology-mediated BIR played an important role. This is a new and exciting area of research. Much more will probably be understood by the time this book is published.

This process, too, can be modeled in yeast, by requiring the assembly of a gene (*URA3*) from overlapping, homologous—or homeologous—segments (**Figure 8.31**). Our preliminary results have found that the initial Rad51-mediated BIR step (joining UR and RA) is very sensitive to the degree of divergence between the sequences, but that a second jump, from RA to A3, is remarkably insensitive to a high degree of sequence mismatch.

SUMMARY

Recombination can occur in many instances when only one end of a DSB is able to find a homologous template or when the two ends become uncoordinated, either in time or space. In these situations BIR becomes a powerful mechanism to repair a break by initiating extensive copying of a template until it reaches a telomere or until it meets a converging replication fork. In phage recombination and late replication, BIR can spin off copies through rolling circle replication. In budding yeast, there are two distinct Rad52-dependent BIR pathways, one requiring Rad51 and its associated mediators and a second that is Rad51 independent but requiring Rad59 and the MRX proteins. BIR appears to be an important mechanism in maintaining telomere ends in the absence of telomerase and in creating complex copy number variation in human disease. Based on work in budding yeast, BIR requires essentially all of the proteins needed for normal replication, except a small number specifically devoted to firing origins; but BIR needs additional functions that are not needed for normal replication. To date, the nonessential Pol32 subunit of Polδ and a region of PCNA have both been implicated in initiating BIR.

SUGGESTED READING

Deem A, Keszthelyi A, Blackgrove T et al. (2011) Break-induced replication is highly inaccurate. *PLoS Biol* 9:e1000594.

Haber JE (1999) DNA recombination: the replication connection. *Trends Biochem Sci* 24:271–275.

Hastings PJ, Lupski JR, Rosenberg SM & Ira G (2009) Mechanisms of change in gene copy number. *Nat Rev Genet* 10:551–564.

Llorente B, Smith CE & Symington LS (2008) Break-induced replication: what is it and what is it for? *Cell Cycle* 7:859–864.

Lydeard J, Jain S, Yamaguchi M et al. (2007) Break-induced replication and telomerase-independent telomere maintenance require Pol32. *Nature* 448:820–823.

Lydeard JR, Lipkin-Moore Z, Sheu YJ et al. (2010) Break-induced replication requires all essential DNA replication factors except those specific for pre-RC assembly. *Genes Dev* 24:1133–1144.

McEachern MJ & Haber JE (2006) Break-induced replication and recombinational telomere elongation in yeast. *Annu Rev Biochem* 75:111–135.

Michel B, Boubakri H, Baharoglu Z et al. (2007) Recombination proteins and rescue of arrested replication forks. *DNA repair* 6:967–980.

Michel B, Grompone G, Flores MJ & Bidnenko V (2004) Multiple pathways process stalled replication forks. *Proc Natl Acad Sci USA* 101:12783–12788.

Mosig G (1998) Recombination and recombination-dependent DNA replication in bacteriophage T4. *Annu Rev Genet* 32:379–413.

Neumann AA, Watson CM, Noble JR et al. (2013) Alternative lengthening of telomeres in normal mammalian somatic cells. *Genes Dev* 27:18–23.

Zhang F, Carvalho CM & Lupski JR (2009) Complex human chromosomal and genomic rearrangements. *Trends Genet* 25:298–307.

CHAPTER 9

SISTER CHROMATID REPAIR

The most frequent source of chromosome breaks is the process of DNA replication, and the most common template with which a DSB can be repaired is the sister chromatid. Most likely, recombination mechanisms first arose in this repair context; however, we know less about sister chromatid repair than some other forms of recombination because there have not been good model systems in which there is a DSB on only one of two sisters. Nevertheless, from what we have learned we can see that there are several competing pathways of repair.

9.1 HOMOLOGOUS RECOMBINATION IS REQUIRED AFTER IONIZING RADIATION

X-rays create DSBs. In a haploid yeast cell such breaks are lethal if not repaired, and in diploids the failure to repair a break will at best lead to chromosome loss and the formation of an aneuploid cell. Nearly all repair depends on the *RAD52* gene, which is needed for almost all homologous recombination (**Figure 9.1A**). Very similar results are seen if cells are treated with radiomimetic chemicals such as methylmethane sulfonate (MMS; **Figure 9.1B**) or bleomycin (**Figure 9.1C**). There is little impact in deleting genes such as *YKU70* that are required for nonhomologous end-joining (NHEJ), although *rad52Δ yku70Δ* double mutants are somewhat more sensitive to DSBs (Figure 9.1C). Human cells also show similar survival curves, and repair is largely dependent on the BRCA2 gene (**Figure 9.1D, E**). Ionizing radiation dosage is usually presented in units of kilorads or gray (Gy) (**Box 9.1**).

If one looks at the viability of a haploid strain with increasing doses of radiation, there is an evident "shoulder" on the curve indicating that some cells are more resistant to irradiation. These are post-replication, G2 cells that have a sister chromatid with which to repair the break by homologous recombination, whereas cells prior to replication can only use NHEJ (**Figure 9.2**). But even if one compares a G1 diploid (where any broken chromosome should have a homologous chromosome to act as a template) with a G2 haploid (which only has a sister chromatid) the G1 diploid is less radio-resistant than the G2 haploid. Thus, there is a distinct advantage when the template for homologous repair is the sister chromatid.

9.2 SISTER CHROMATID REPAIR IS PREFERRED OVER RECOMBINATION WITH A HOMOLOG

In diploid budding yeast cells, DNA damage created during S phase is most often repaired by recombination between sister chromatids rather than by using identical sequences on the other homologous chromosome. An important experiment that demonstrated this point was carried out by Lisa Kadyk and Lee Hartwell using a budding yeast diploid. One striking finding was that the rate of X-ray-induced interhomolog gene

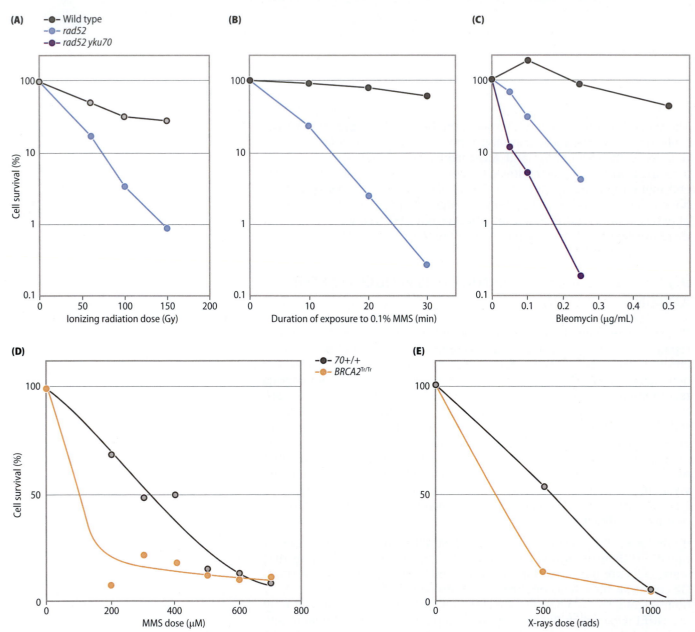

Figure 9.1 Lethality caused by ionizing radiation and radiomimetic drugs. (A) Increasing doses of ionizing radiation kill budding yeast haploid cells. Most resistance is Rad52 dependent. Similar effects are seen for yeast cells treated with MMS (B) or bleomycin (C). Although deleting genes required for NHEJ, such as *YKU70*, has little effect by itself, a strain lacking both *RAD52* and *YKU70* is much more sensitive than one lacking *RAD52* alone. Mammalian diploid cells are killed by MMS (D) or X-rays (E). Cells carrying a truncation of the *BRCA2* gene (*BRCA2*^Tr) are very sensitive to these agents. (A, adapted from Matuo Y, Nishijima S, Hase Y et al. [2006] *Mutat Res* 602:7–13. With permission from Elsevier. B, from Odagiri N, Seki M, Onoda F et al. [2003] *DNA Repair* 2:347–358. With permission from Elsevier. C, adapted from Tam AT, Pike BL, Hammet A & Heierhorst J [2007] *Biochem Biophys Res Commun* 357:800–803. With permission from Elsevier. D, E, adapted from Patel KJ, Yu VP, Lee H et al. [1998] *Mol Cell* 1:347–357. With permission from Elsevier.)

> **Box 9.1 Radiation dosage and radiation resistance**
>
> Ionizing radiation (IR) damage is measured in terms of gray (Gy) defined as 1 joule of energy absorbed by 1 kg of matter. Often radiobiologists convert Gy to rads, where 100 rad = 1 Gy. The effect of X-rays and other IR damage on DNA is quite constant across organisms of different complexity. Measurements of chromosome breakage in yeast and mammals have come up with a very similar value: a dose of 1 Gy causes 1 DSB in approximately 150 Mb (150,000,000 bp). For every DSB there are many single-strand nicks and other sorts of damage including the creation of "dirty ends" that do not break DNA at a phosphate bond but leave a fragmented base at the end that may require special end-processing.
>
> It is surprising to note that the dose required to kill most diploid mammalian cells—on the order of 1 kilorad (10 Gy)—barely touches even a haploid yeast cell. Different organisms display a huge range of radiation resistance even though the number of DSBs per megabase is quite similar. Some of the difference may reflect variation in the efficiency of DSB repair mechanisms, but recent work by Michael Daly has suggested that much of the intrinsic radio-sensitivity of different organisms comes from non-DNA damage inflicted by ionizing radiation, caused by reactive oxygen species that damage proteins as much or more as they damage DNA. These reactive oxygen species are quenched with vastly different efficiency in different organisms using a number of small-molecule antioxidants such as glutathione or superoxide dismutase; but at least one factor that seems to be very important is the amount of manganese relative to magnesium within the cell.

conversion markedly *decreased* in G2 cells, even though there were twice as many copies of the sequences that could recombine (**Figure 9.3**). This decrease can be explained if the DSB is repaired by a sister chromatid (which cannot produce a Leu+ recombinant). Kadyk and Hartwell concluded that a DSB in a G2 cell will be repaired 90% of the time by a sister chromatid instead of a homolog.

Most likely this strong preference comes from the fact that sister chromatids are held together after replication by cohesins and cohesin-like Smc5–Smc6 proteins, so that the ends of a DSB are constrained to interact with the sister (**Figure 9.4A**). In budding yeast, and most likely in higher organisms, this preference is enhanced by the DNA damage-dependent recruitment of additional cohesin and Smc5–Smc6 proteins around the DSB and its adjacent sister. The assembly of these additional encirclements around the damaged chromatid and its sister is triggered by the extensive phosphorylation of histone H2AX (γ-H2AX) by the ATM and ATR DNA damage checkpoint kinases. In mammals, γ-H2AX can extend over a megabase along the damaged chromosome. In budding yeast, γ-H2AX phosphorylation is confined to about 100 kb around the DSB, and both cohesin and Smc5/6 are recruited over this entire domain. This recruitment requires the formation of γ-H2AX and depends on the Scc2 cohesin loading factor and also on the recruitment of the Mre11–Rad50–Xrs2 (MRX) complex.

Sister chromatid repair (SCR) is probably the most critical task of the recombination machinery. But studying SCR is difficult because the recombination between two genetically identical molecules usually does not produce a genetic outcome that can be scored. Nevertheless, SCR can be assessed by physical analysis of chromosomes and—with some tricks—even by genetic assays.

9.3 SCR CAN BE VISUALIZED ON CHROMOSOME-SEPARATING GELS

In haploid yeast, SCR can be monitored by observing the reconstitution of full-length chromosomes after ionizing radiation (IR). For example, intact *Schizosaccharomyces pombe* chromosomes, ranging from 3.5 to 5.6 kb,

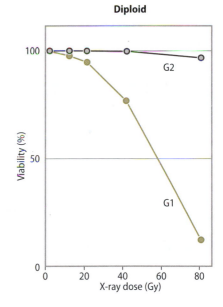

Diploid

Figure 9.2 Effect of cell cycle stage on radiation sensitivity. Diploid cells were arrested at the G1 or G2 phases of the cell cycle by treatment with mating pheromone α-factor or by nocodazole treatment, respectively. Cells were irradiated and plated for viability as a function of dose. Even though G1 diploids have pairs of homologous chromosomes, repair is much more efficient when a radiation-induced DSB can be repaired from a sister chromatid, in G2, after DNA replication. (Adapted from Kadyk LC & Hartwell LH [1992] *Genetics* 132:387–402. With permission from the Genetics Society of America.)

Figure 9.3 Cell cycle dependence of interchromosomal recombination. Recombination between two *leu2* alleles in a budding yeast diploid (A) was measured after cells were arrested in the G1 and G2 phases of the cell cycle as described in Figure 9.2. (B) The rate of radiation-induced gene conversion was much lower in G2 cells, because DNA damage is preferentially repaired from sister chromatids, thus lowering the frequency with which heteroallelic recombination occurs. (Adapted from Kadyk LC & Hartwell LH [1992] *Genetics* 132:387–402. With permission from the Genetics Society of America.)

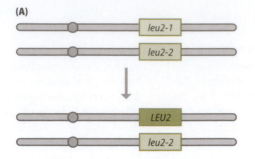

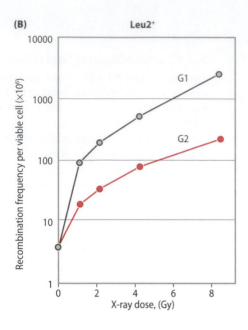

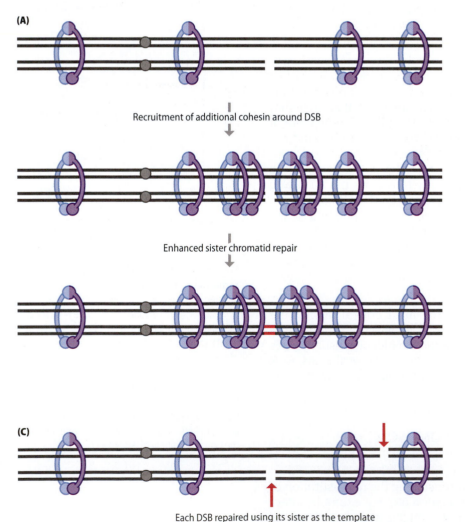

Recruitment of additional cohesin around DSB

Enhanced sister chromatid repair

Each DSB repaired using its sister as the template

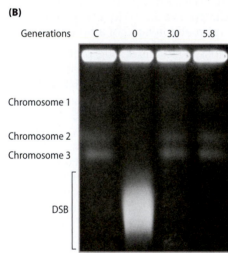

Figure 9.4 Cohesins facilitate sister chromatid repair. (A) Cohesins hold sister chromatids together and a DSB recruits additional cohesins to the region surrounding the break. Sister chromatid cohesion facilitates sister chromatid repair (SCR). (B) Intense IR creates multiple breaks on every chromatid in the three *S. pombe* chromosomes, as seen in an ethidium-bromide-stained chromosome-separating gel in lane 2 (0 generations). (C) Since random IR-induced breaks are in different locations, sister chromatid repair is possible to restore the shattered chromosomes to full length. (B, from Takeda J, Uematsu N, Shiraishi S et al. [2008] *DNA Repair* 7:1250–1261. With permission from Elsevier.)

can be separated by pulsed-field gel electrophoresis (**Figure 9.4B**). After sufficiently intense IR to break each chromatid in a few places, the intact chromosome bands disappear, replaced by a smear of broken chromosomes spread out over a much smaller size range. But since sister chromatids are broken by IR in different locations, each sister can serve as the template to repair the other, using homologous recombination—gene conversion or possibly BIR—to patch up the breaks and restore the full-length molecules (Figure 9.4). In yeasts, in the absence of Rad52 there is no repair.

In budding yeast, where the chromosomes range in size from 230 kb to >1 Mb, the absence of sister chromatid cohesion (here by inactivating the Scc1/Mdc1 cohesin subunit with a temperature-sensitive mutation) dramatically impairs repair (**Figure 9.5**). The defect is seen even if Scc1 is inactivated after S phase is complete (but before IR). Note that even without cohesins, sister chromatids may still be held by other proteins and simply by the intertwining of the sister chromatids along their length, so it is not surprising that there is still some repair without cohesin.

If one simply blocks the *damage-induced* cohesin assembly (by inactivating a temperature-sensitive Scc2 cohesin loading factor in cells that have already assembled—and retain—their normal cohesin) there is a fourfold reduction in repair efficiency. A similar defect is seen when one of the MRX proteins is ablated, because MRX is required for damage-dependent, γ-H2AX-dependent cohesin loading. The finding that MRX is required for loading damage-associated cohesin can account for the findings that spontaneous recombination between heteroalleles in a diploid increases about seven- to tenfold in *mre11Δ*, *rad50Δ*, or *xrs2Δ* mutants. If

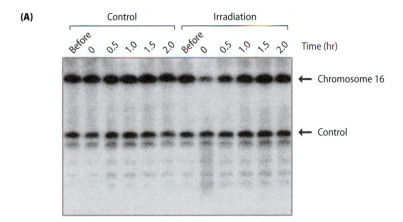

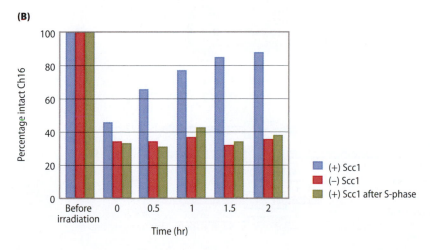

Figure 9.5 Sister chromatid repair in budding yeast. (A) Chromosome-separating gel electrophoresis followed by Southern blot probing for Chr 16 shows that most copies are broken into smaller fragments after X-irradiation. Restitution of full length Chr 16 is seen within 30 min. (B) Efficient SCR is dependent on the loading of cohesins during DNA replication. Inactivation of the cohesin subunit Scc1 impairs SCR, even if it is re-expressed after cells have completed S phase. (Adapted from Sjögren C & Nasmyth K [2001] *Curr Biol* 11:991–995. With permission from Elsevier.)

damage arises in S phase and would normally be repaired by a sister, the loss of damage-induced additional cohesin should decrease the chance the lesion will be repaired by a sister and increase the chance that recombination will occur between homologous chromosomes (**Figure 9.6**); so preventing cohesin recruitment in MRX mutations should increase the frequency that the damaged chromatid will recombine with the homolog and increase heteroallelic gene conversion.

We should note that some of the increase in heteroallelic recombination could alternatively be explained by the fact that loss of MRX proteins will also slow down 5′ to 3′ resection, thus making shorter regions of ssDNA. Shorter ssDNA ends will reduce the chance that heteroduplex DNA formed during strand invasion will cover both alleles (**Figure 9.7**). When nearby mismatched sites are included in heteroduplex, their co-correction by mismatch repair enzymes will tend to result in co-conversion and thus will not lead to a wild-type recombinant.

Analogous monitoring of broken DNA can be carried out in mammalian cells using a "comet assay," more formally known as single-cell gel electrophoresis. IR-damaged cells can be encased in low-melting agarose and then lysed. The agarose is then subjected to an electric field that allows smaller fragments of DNA to spread out (like a comet's tail) from the cell, as visualized by a fluorescent DNA dye (**Figure 9.8**). This assay is more qualitative rather than quantitative in demonstrating the extent of DNA fragmentation and the ability of cells to repair the broken chromosomes.

9.4 SOME SISTER CHROMATID REPAIR IS SEEN AS SISTER CHROMATID EXCHANGE

In addition to observing the restitution of broken chromatids to their full size, the most convenient assays to examine repair of sister chromatids all look at crossovers associated with repair. It is not known what proportion of sister chromatid repair (SCR) events are accompanied by crossing over, but as we will see below, mutations in proteins such as the RecQ

Figure 9.6 Interhomolog heteroallelic gene conversion is in competition with SCR. Heteroallelic gene conversion is increased when cohesins are not recruited to sister chromatids after damage. Increased interhomolog recombination is seen in MRX mutants or in the absence of γ-H2AX. A similar effect is seen when yeast diploids are assayed for loss of heterozygosity by mitotic recombination. (Adapted from Bressan DA, Baxter BK & Petrini JH [1999] *Mol Cell Biol* 19:7681–7687. With permission from the American Society for Microbiology.)

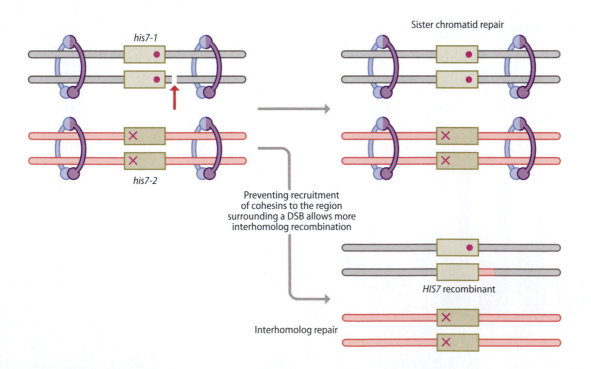

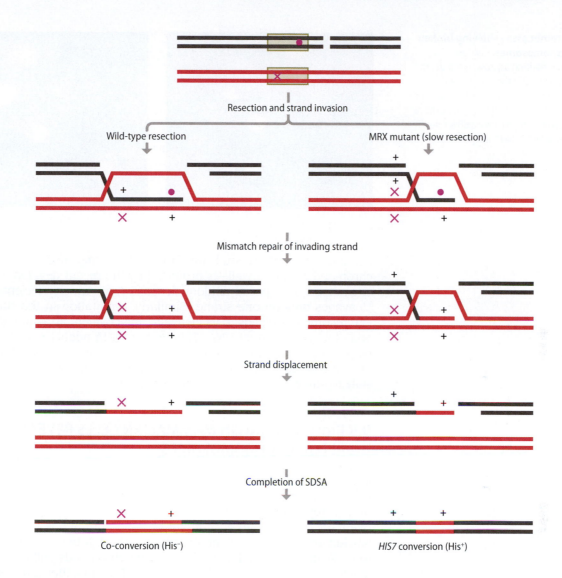

Co-conversion (His⁻)

HIS7 conversion (His⁺)

homolog BLM (Bloom Syndrome) helicase cause a marked increase in sister chromatid exchange (SCE). SCE assays have provided many of our insights about how sister chromatid breaks are repaired.

In budding yeast, where the chromosomes are too small to use cytological methods, it is possible to monitor SCE by analyzing the formation of duplications of a circular chromosome or centromeric plasmid that arise by crossing over. For this approach to be useful, one needs to create a DSB on only one sister. Aguilera's lab has found that a suboptimal, 21-bp HO endonuclease recognition site is poorly cut and a large fraction of the rare cleavage events are single-strand nicks. If such nicks are induced in G1 in a centromeric plasmid, DNA replication will produce one molecule with a DSB and an intact sister, which can serve as a template for repair. The product of gene conversion with an associated crossover is a dicentric dimer circle (**Figure 9.9**). One limitation on studying the fate of such nicks generated in G1 cells is that they may re-ligate, thereby reducing the proportion of events that can be monitored; still a few percent of the cells exhibit an SCE event. These SCE events are dependent on Rad52 and Rad51, although there appears to be a Rad51-independent pathway that requires Rad59. The MRX complex is also important. Confirming the results observed with γ-irradiation, Aguilera's group showed that these SCE events depend on sister chromatid cohesion (Figure 9.9B).

Figure 9.7 Reduced resection in MRX mutants could also explain increased heteroallelic gene conversion. Long heteroduplex regions will frequently result in co-conversions that do not yield prototrophic recombinants. Reducing 5′ to 3′ resection in MRX mutants creates shorter heteroduplex and thus increases the probability that mismatch repair of a heteroduplex will yield a prototroph. In this SDSA model, the invading strand is corrected by mismatch repair in favor of the intact strand sequence. If the heteroduplex is longer, there is a greater chance of co-converting the markers, thus reducing the formation of prototrophs.

Figure 9.8 A comet assay showing broken mammalian chromosomes. Cells are embedded in low-melting agarose, lysed gently in alkaline conditions, and then subjected to electrophoresis. Broken DNA (stained *yellow*) from damage induced cells (*right*) will migrate to form a tail. (From Tuteja N, Ahmad P, Panda BB & Tuteja R [2009] *Mutat Res* 681:134–149. With permission from Elsevier.)

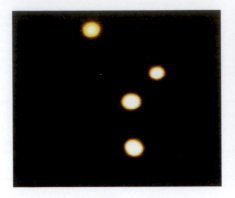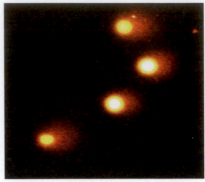

An improvement in studying SCE events—and noncrossover sister chromatid repair as well—is promised by the recent development of an engineered site-specific endonuclease, I-AniI, which efficiently makes cleavages only on one strand. Similarly, a mutation in the site-specific FLP recombinase leads to its cleavage and covalent attachment to the 3' end of a specific target DNA. This protein–DNA adduct mimics the trapping of topoisomerase I on DNA by camptothecin, but here all the events can be created at one specific site. These enzymes should make it possible to study SCR and SCE in greater detail.

9.5 BrdU LABELING IN MAMMALIAN CELLS REVEALS "HARLEQUIN" CHROMOSOMES

In mammalian cells, the most widely used approach to examine SCE is by following the fate of BrdU incorporated into DNA during one round of DNA replication, as discussed in Chapter 1. Normally, semi-conservative replication will produce a pair of sister chromatids with one strand labeled and the other, older strand, unlabeled. If the cells are permitted to replicate again, each pair of new sister chromatids will have only one labeled strand (**Figure 9.10**). However, if SCE occurs, then a region of BrdU extending to the telomere will be found on the otherwise unlabeled sister and the equivalent region on its sister will be unlabeled. In a wild-type strain, such exchanges are not rare; there are several events per cell division. This observation suggests that replication-associated breaks occur and must be repaired every cell cycle.

Exposure of cells to DNA damaging agents such as mitomycin C dramatically increases the incidence of SCE, producing "harlequin" chromosomes, so named because of the black-and-white checked eighteenth century costume of the *commedia dell'arte* clown Harlequin. Presumably there are many more SCEs because there are many more DSBs or other lesions that promote SCE.

The incidence of SCE is also significantly increased in the absence of BLM helicase. In this case, it is quite possible that the increase in SCE does not come from an increase in the number of damaged sites but from an increase in the proportion of recombination intermediates that are resolved as crossovers. As we discussed in Chapter 6, BLM acts to "dissolve" dHJs and thus remove them from the pool of events that could be resolved as crossovers. There are similar SCE increases in cells with mutations in the Rmi2 protein that interact with BLM; presumably both TopIIIα and Rmi1 defects would also produce this phenotype. Interestingly, deletions of other RecQ homologs in vertebrates do not show similar increases in SCE; although deleting Werner's syndrome (WRN) helicase increases SCE specifically in telomere regions. Curiously, the absence of WRN reduces SCE in combination with the BLM deficiency. Double mutants lacking both BLM and RecQL1 or BLM and RecQL5 enhance SCE even more than the BLM single mutant even though, mutations in these two homologs have no SCE phenotype by themselves.

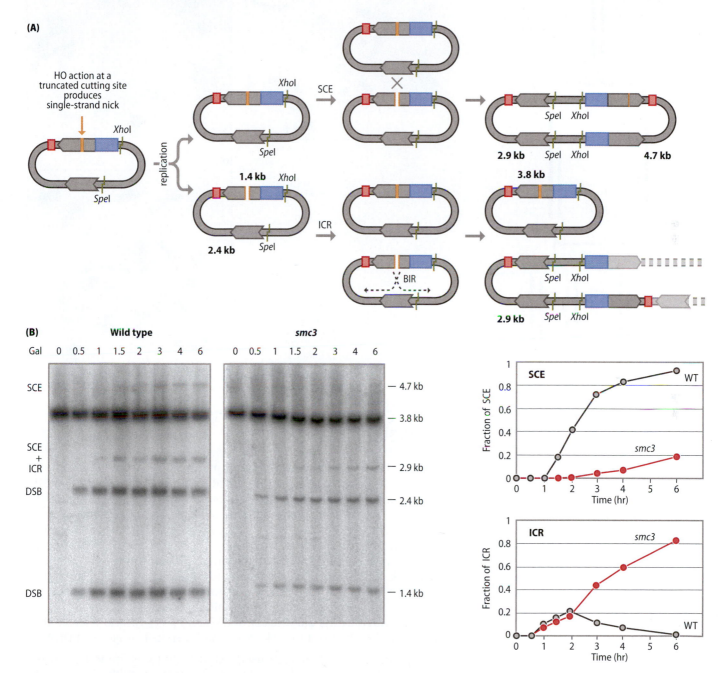

Figure 9.9 *In vivo* **analysis of sister chromatid repair on a plasmid containing a single-strand nick.** (A) A plasmid in which there is a smaller-than-normal recognition site for HO endonuclease suffers infrequent single-strand nicks in G1 cells that are converted to DSBs only after replication. Two modes of repair are possible: intersister (SCE) repair and intrachromosomal repair (ICR), which occurs by BIR. These pathways yield different outcomes that can be distinguished after cleavage of the DNA with *Xho*I and *Spe*I restriction endonucleases. In particular SCE yields a novel 4.7-kb fragment homologous to the probed region (*blue*). BIR can continue indefinitely (indicated by *dashed lines*), by rolling-circle replication, to produce outcomes of variable size, but all of which will only contain the 3.8 kb indistinguishable from the parent plasmid. (B) The kinetics of DSB formation and repair are visualized on a Southern blot. HO expression in G1 cells (time = 0) does not produce DSBs but release of cells into S phase produces DSBs. Repair can be seen by the appearance of 2.9 and 4.7 kb bands. This analysis, graphed on the right, reveals that SCE is reduced while ICR is enhanced in a temperature-sensitive mutation of the Smc3 subunit of cohesin, at its restrictive temperature. (Adapted from Cortés-Ledesma F & Aguilera A [2006] *EMBO Rep* 7:916–926. With permission from Macmillan Publishers, Ltd.)

(A)   BrdU-labeled chromosome transferred
to normal medium for two generations

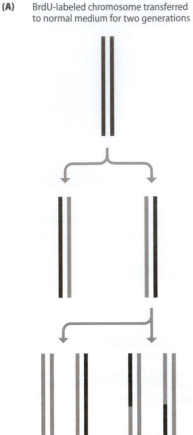

(B) **(C)** **(D)**

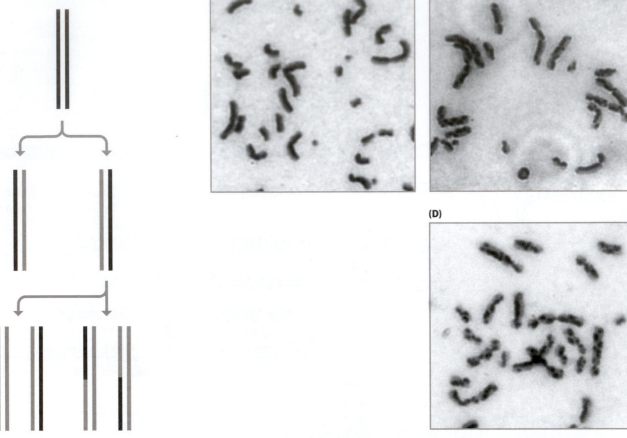

Figure 9.10 BrdU labeling reveals SCE.
(A) Cells grown in BrdU are transferred to medium lacking BrdU for two generations so that only one DNA strand per pair of sister chromatids is labeled. (B) In wild-type cells, SCEs are rare, so that the labeled strand, detected by antibody labeling of fixed and partially denatured chromosomes, will appear unbroken, but if there has been an SCE then the staining will jump from one chromatid to the other. (C) Addition of a DNA damaging agent, mitomycin C, creates multiple SCEs per chromosome pair (creating a "harlequin" pattern). (D) Greatly elevated SCE is also seen without adding exogenous DNA damage in human cells lacking the BLM helicase. (B–D, courtesy of Shriparna Sarbajna and Stephen West, Cancer Research UK.)

9.6 SCE CAN ALSO BEEN SEEN BY USING THE CO-FISH TECHNIQUE

The BrdU labeling methods have been modified by a procedure known as chromosome orientation fluorescence *in situ* hybridization (CO-FISH), as discussed in Chapter 8. BrdU is incorporated for a single replication cycle, but then the metaphase spreads of sister chromatids are exposed to UV irradiation, which creates nicks in BrdU-substituted DNA much more than in normal strands. Following UV exposure, the BrdU-labeled strand is preferentially excised by treating the chromosomes on a microscope slide with the *E. coli* enzyme ExoIII, which chews away the nicked BrdU strand. Thus, at the end of this process, each sister chromatid has only its original template strand, one with its original Watson strand and the other with only the old Crick strand (**Figure 9.11**). Strand-specific DNA probes can then be used to look at the fate of specific regions of DNA.

This type of analysis has been used extensively to examine SCE in telomere regions, by hybridizing the chromosomes with fluorescently labeled strand-specific probes. In the absence of any mitotic exchanges, each strand-specific probe lights up one terminus of each sister chromatid, at opposite ends where the strands of the same polarity are located (Figure 9.11), but if there has been an exchange somewhere along the chromosome arm, one chromatid will have both signals. If there has been exchange within the telomere region itself, some of the hybridization will be found on both sister chromatids at the same end (Figure 9.11). Tumor cells that maintain their chromosome ends by the recombination-dependent ALT mechanism frequently display telomeric SCE.

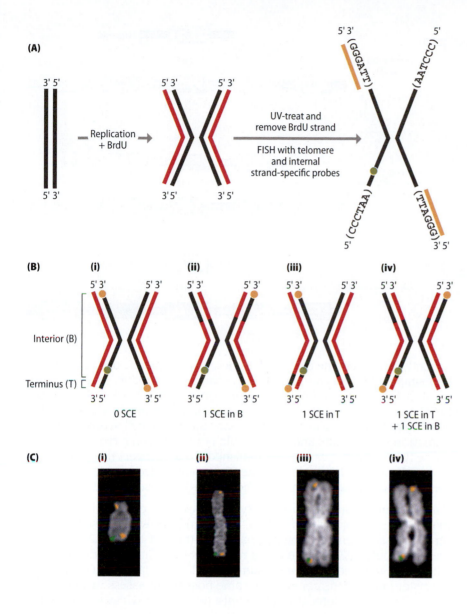

Figure 9.11 CO-FISH detection of SCE.
(A) As described in Figure 8.24, SCE can be visualized by hybridization of a strand-specific (Watson- or Crick-specific) probe to chromosomes where newly copied BrdU-labeled DNA has been excised. (B) An SCE in the interior of the chromosome (i) can move a strand-specific *green* fluorescent probe on the opposite chromatid from both telomere markers (ii), whereas SCE close to the terminus (T) will place both *orange* fluorescent telomere probes on one sister (iii). A more complex event is shown in (iv). (C) Examples of each category are shown. (Adapted from Rudd MK, Friedman C, Parghi SS et al. [2007] *PLoS Genet* 3:e32.)

9.7 SCE CAN BE ASSAYED GENETICALLY BY STUDYING UNEQUAL SCE AND LONG-TRACT GENE CONVERSION

Sister chromatid recombination is normally genetically silent, as the template and the recipient have exactly the same sequence. To score these events genetically, one can measure unequal sister chromatid exchanges (uSCEs). These are studied in cells that contain artificially constructed nearby repeated sequences that are arranged so that crossovers (and some gene conversion events) will yield a genetically identifiable outcome. Michael Fasullo developed an often-used system in budding yeast by placing two overlapping segments of the *his3* gene in tail-to-head arrangement that cannot be fused by SSA to give a *HIS3* prototroph (**Figure 9.12A**). There are two different mechanisms that can yield identical His+ recombinants. The most-cited mechanism is termed a uSCE event, in which a simple crossing over between the tail part on one chromatid with the head part of its sister will yield a triplication of *his3* segments, the middle one being *HIS3*. The reciprocal crossover product has a single *his3* segment lacking both the 5′ and 3′ ends, but generally this product would not be recovered.

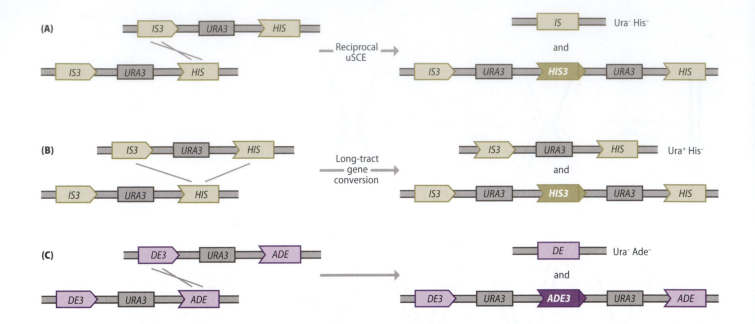

Figure 9.12 Two sister recombination outcomes yielding an apparent uSCE. (A) Reciprocal unequal SCE (uSCE) between overlapping 3' and 5' segments of the *HIS3* gene yields a His⁺ product, while the reciprocal product lacks the intervening *URA3* marker. (B) Long-tract gene conversion (LTGC) produces the same His⁺ product but the sister chromatid retains the original construct. (C) A similar assay used by Kadyk and Hartwell to study sister chromatid repair. (A, B, adapted from Fasullo, MT & Davis RW [1987] *Proc Natl Acad Sci USA* 84:6215–6219. With permission from the National Academy of Sciences. C, adapted from Kadyk LC & Hartwell LH [1992] *Genetics* 132:387–402. With permission from the Genetics Society of America.)

Recently it has become evident that there is a frequently used alternative to a simple uSCE that will produce the same His⁺ product in a nonreciprocal fashion, termed long-tract gene conversion (LTGC) (**Figure 9.12B**). LTGC can occur by SDSA if the two ends of a DSB invade the two adjacent homologous regions. In this case, the sister cell would have an unrearranged template and there would not have been a crossover at all. These two outcomes can be distinguished if one recovers these events as sectored colonies. In simple SCE, the His⁻ side will also have lost the intervening *URA3* marker, but in LTGC, both sides of the sector are still Ura⁺.

Using an analogous tail-to-head system based on overlapping parts of the *ADE3* gene, separated by *URA3*, Kadyk and Hartwell devised a visible screen for apparent SCE events, based on the fact that *ade2 ade3* cells are white but *ade2 ADE3* cells are red (**Figure 9.12C**). By examining red/white sectored colonies that presumably reflect the growth of each original sister into half the colony, it became possible to assess what fraction of events were reciprocal uSCE events (where the white sector would be Ura⁻ and the red sector Ura⁺) versus those that arose by LTGC (where both sectors would be Ura⁺). The frequency of such sectors was too low to be assayed among spontaneous events but among those induced by X-rays, where recombination was stimulated 100-fold, roughly two-thirds were long-tract gene conversions and one-third were reciprocal exchanges. Thus, the majority of apparent uSCEs induced by DSBs are not the expected simple crossing over that one detects by BrdU labeling.

uSCE is also induced by UV irradiation, which will block replication. Kadyk and Hartwell also showed, in a series of clever experiments, that the stimulation of SCR by UV required cells to pass through DNA replication, where stalling of the replication fork at UV damage, especially in a *rad1* strain, provoked a big increase in LTGC. Whether these events were Rad51 independent was not established and whether *rad1Δ* did more than prevent nucleotide excision repair (such as affect the clipping of nonhomologous tails or the cleavage of a nascent Holliday junction) is not known.

A variation on the uSCE/LTGC assay was devised by Sue Jinks-Robertson in which a *HIS3* gene containing an artificial intron was broken up such that the 5' and 3' coding regions were in inverted orientation and separated by a 350-bp sequence that itself shares homology with a second sequence located nearby and in inverted orientation (**Figure 9.13**). Crossing over between the inverted repeats results in the inversion of the 3′ exon, allowing mRNA splicing and the expression of *HIS3*. This could happen either by intrachromosomal crossing over or by LTGC. (In this instance, if crossing over happened by uSCE, the resulting His⁺ recombinant would be either an acentric fragment or a dicentric chromosome and would not be recovered.) Most of the events appeared to result from LTGC, again suggesting that crossovers are rare.

9.8 THE GENETIC REQUIREMENTS FOR SPONTANEOUS SCR DIFFER FROM THOSE SEEN IN INTERHOMOLOG REPAIR

Spontaneous uSCEs in budding yeast arise reasonably frequently, at a rate of about 1×10^{-6} per cell division. Although these events require Rad52, they are surprisingly independent of Rad51, Rad55, or Rad54; in fact, *rad51Δ* gives a modest increase in the number of recombinants. The lack of importance of Rad51 is clearly different from its role in spontaneous heteroallelic recombination, where recombination is reduced tenfold. Even *rad51Δ rad59Δ* has only a two- to threefold reduction in SCR, whereas this combination is nearly as defective as *rad52Δ* for spontaneous intrachromosomal heteroallelic recombination, for example between a pair of *ade2* heteroalleles placed in a chromosome in an inverted repeat arrangement, the same *ade2* gene used in the SCR experiments. *Spontaneous SCR is reduced 25-fold in a rad51Δ rad1Δ rad50Δ strain*, similar to the reduction in heteroallelic recombination. It is quite likely that many spontaneous uSCE events arise during replication by a route that does not require homologous recombination (but still needs Rad52); possibly these events arise by template switching of DNA polymerases during replication. Someone needs to figure out how this works.

The efficiency of sister chromatid recombination (lumping together both uSCE and LTGC) declines in cells lacking the components of the "clippase" that can remove 3′-ended nonhomologous flaps from recombination intermediates (that is, Rad1, Msh2, and Rad59, as discussed in Chapter 5). In the absence of knowing what lesions initiate spontaneous recombination events it is hard to specify the mechanisms by which recombination is proceeding. Masayuki Seki has suggested that the role of Rad1, Msh2, and Rad59 can be accounted for by a Rad51-independent gap-repair mechanism that would require SSA-mediated annealing of homologous segments from different chromatids (which would still need Rad52) (**Figure 9.14**). If the annealed regions have nonhomologous tails, then their clipping will require Rad59, Msh2, and Rad1.

Similar to the findings in mammalian cells that SCR is elevated by the absence of BLM, deletion of yeast's single BLM homolog, Sgs1, also stimulates spontaneous uSCE almost tenfold. Because one role of the many-faceted Sgs1 is to act with Top3 and Rmi1 to "dissolve" dHJs, it would be easy to explain the increase in SCR by the accumulation of more dHJs, which would lead to more crossovers. However, this conclusion does not sit entirely well with several additional observations. First, it seems that the majority of SCR events are not crossovers but rather LTGC, but unfortunately the analysis of sectored colonies has not been done with *sgs1Δ* or *sgs1Δ rad51Δ* to see which pathway is most affected. Second, the increase in SCR is largely Rad51 independent and we would have thought that forming dHJs would require Rad51. A clear Rad51 dependence for the formation of replication-associated HJs that should form between sister chromatids was seen by looking at the accumulation of a "spike" of X-shaped molecules on two-dimensional gels of DNA from yeast cells whose replication is retarded by the addition of a

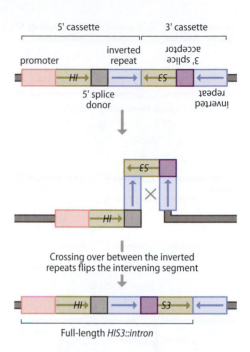

Figure 9.13 An intrachromosomal assay mimicking uSCE. An inversion caused by an intrachromosomal reciprocal exchange produces a functional *HIS3* gene. The *HIS3* gene has been modified so that an intron is placed within the ORF and then the distal exon and part of the intron have been inverted. A crossover between two inverted regions of homology, one within the intron and one to the right of the gene, results in the establishment of a functional *HIS3* gene containing a functional intron. (Adapted from Datta A, Adjiri A, New L et al. [1996] *Mol Cell Biol* 16:1085–1093. With permission from the American Society for Microbiology.)

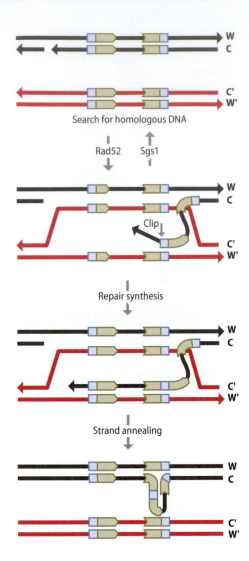

Search for homologous DNA

Rad52 ← → Sgs1

Clip

Repair synthesis

Strand annealing

Figure 9.14 A model for Rad51-independent SCE by gap repair. A gap during replication can be repaired by template switching. Mis-annealing of one of the flanking regions of homology (*blue*) by Rad52-mediated SSA leads to a structure in which the protruding 3'-ended nonhomologous tail can be clipped off by a Rad1- and Msh2-dependent process. The clipped end can then be used as a primer for fill-in synthesis. Re-annealing of the strands back to their original templates results in a heteroduplex DNA on the top DNA molecule that contains a product that is indistinguishable from a long-tract gene conversion. (Adapted from Onoda F, Seki M, Wang W & Enomoto T [2004] *DNA Repair* 3:1355–1362. With permission from Elsevier.)

low concentration of MMS (Figure 4.16). In the absence of Sgs1 or other components of the dHJ dissolving activity, there is a marked increase in the formation of HJ intermediates. So, why Rad51 is often dispensable for SCE remains a mystery.

9.9 SCE CAN BE ANALYZED GENETICALLY IN MAMMALIAN CELLS

Similar constructs have been developed in mammalian cells. Maria Jasin's lab developed an assay based on G418-resistance by repairing an I-SceI cut in the *SCneo* gene or (as shown) by creating a functional *GFP* gene (**Figure 9.15**). Expression of I-SceI leads to the creation of two types of outcomes that can express *GFP*. The first is a simple gene conversion, with no chromosomal rearrangement; these gene conversions could be either intra- or interchromosomal (Figure 9.15). The second outcome involves either uSCE or LTGC to create a triplication of GFP segments the middle one expressing functional *GFP* (Figure 9.15). Ralph Scully created a more versatile version in which an mRNA splice donor and splice receptor needed to yield a functional blasticidin-resistance BsdR (blasticidin S deaminase) gene that can only be made after uSCE or LTGC (**Figure 9.16A**). The displaced BsdR gene segments are themselves nested between a promoterless GFP gene segment and a *GFP* gene disrupted by an I-SceI cleavage site. In this arrangement (**Figure 9.16B**), the triplication also places one copy of the 5' end of the blasticidin S deaminase gene ahead of the 3' end, with its functional intron now containing the functional *GFP* gene; consequently this triplication will allow the cells to express both GFP and blasticidin resistance and uniquely identifies sister chromatid recombinants, either uSCE or LTGC.

Most of the sister chromatid repair events are not via uSCE, but by LTGC. Among GFP+ colonies following I-SceI induction 75% were from intra- or interchromosomal short patch gene conversion; that is, there was no change in the overall arrangement of sequences (**Figure 9.17A**). The remaining events had the triplication structure. One ambiguity in using this construct is that the I-SceI site in one of the sisters must not have been cut or else it had to be re-ligated; otherwise the GFP- and blasticidin-expressing chromatid would still have a DSB needing repair (**Figure 9.17C**). Indeed, there is one class of outcomes from inducing I-SceI, in which there are four, five, or more copies of the GFP sequences. These gene amplifications may be the result of continued presence of I-SceI and continued cutting and recombination, since the triplication itself still harbors a functional I-SceI site; alternatively these amplifications may reflect some sort of iterative event during a single DSB repair event involving sisters. Such amplifications could partly account for telomere elongations by recombination in the absence of telomerase.

Another, and possibly related, anomaly among the repair products is that a significant proportion of the GFP+ events fail to incorporate the entire template into the recipient (**Figure 9.17B**). These events most likely initiated DNA repair synthesis at one end in what ought to be a LTGC event and succeeded in copying enough of the region to become GFP+ but terminated before the end. These events have been generally interpreted as aborted repair replication events that join to the other DSB end by NHEJ, but it is also possible that they represent large deletions created by the second I-SceI site.

The recombination proteins required for SCR are only beginning to be studied. Consistent with a role for γ-H2AX and cohesin recruitment in SCR inferred in yeast, the absence of γ-H2AX reduces the incidence of

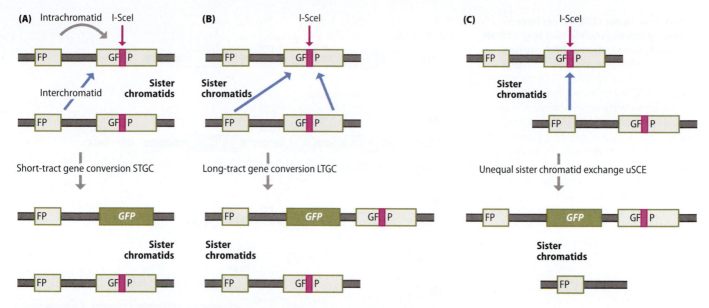

Figure 9.15 Possible outcomes from an unequal sister chromatid exchange assay. An I-SceI cleavage of a site in one of two sister chromatids results in the creation of a GFP-expressing recombinant by several possible pathways. (A) Short tract gene conversion can occur both by intrachromatid and interchromatid gene conversion. (B) Long tract gene conversion occurs when the "GF" and "P" ends of the DSB-strand invade different homologous sequences. (C) True unequal crossing-over results in a reciprocal exchange in which there is a functional *GFP* gene in the middle of a triplication, along with a reciprocal product that lacks most of the sequences. Note that in each instance a second I-SceI site either was not cleaved or was repaired by nonhomologous end-joining. (Adapted from Johnson RD & Jasin M [2000] *EMBO J* 19:3398–3407. With permission from Macmillan Publishers, Ltd.)

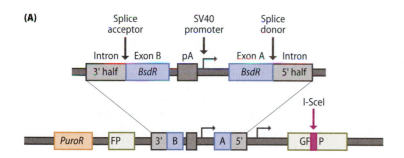

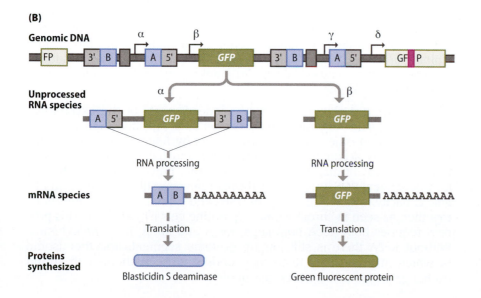

Figure 9.16 I-SceI-induced sister chromatid repair assays in mammalian cells.

(A) Recombination between an I-SceI cut *GFP* gene and a 5'-deleted GFP fragment produces a GFP+ cell either by intrachromosomal or unequal SCE or by LTGC. Note that if both sister chromatids are cleaved by I-SceI, the site not involved in the event that produces GFP must be re-sealed by NHEJ. (B) A modification of this assay places the 3' and 5' blasticidin resistance gene segments in between GFP regions. uSCE and LTGC create a duplication of the *BsdR* gene such that a long mRNA can now be spliced to give a functional BsdR product as well as GFP. (Adapted from Puget N, Knowlton M & Scully R [2005] *DNA Repair* 4:149–161. With permission from Elsevier.)

Figure 9.17 Some LTGC events have a nonhomologous junction that might result from early termination of DNA copying. In addition to the expected short-tract gene conversions (A) and long-tract gene conversions (C), some LTGC events have terminated early lacking the complete triplication structure (B.) Alternatively, these outcomes could also reflect the fact that I-SceI can cut both sisters and that the second site must be repaired in some way to allow recovery of the product. For example, the sequences around the right-hand I-SceI site might be resected and end-joined, creating a large deletion. (Adapted from Puget N, Knowlton M & Scully R [2005] *DNA Repair* 4:149–161. With permission from Elsevier.)

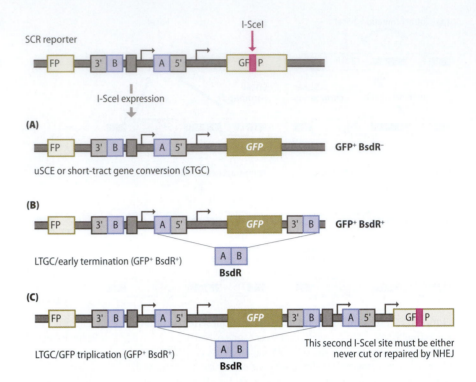

uSCE and LTGC recombinants. An interesting, but unexplained finding is that a deficiency in the Rad51C paralog resulted in a bias toward recovering LTGC over the short-patch events.

9.10 RADIO-RESISTANCE IN *DEINOCOCCUS RADIODURANS* INVOLVES BOTH EFFICIENT DNA REPAIR AND RESISTANCE TO OTHER OXIDATIVE DAMAGE

Some organisms, including the bacterium *Deinococcus radiodurans* and the eukaryotic bdelloid rotifers, exhibit extraordinary resistance to X-ray, and to desiccation. For example, whereas humans are killed by ionizing radiation at doses of 10 Gy, and *E. coli* by about 200 Gy, *D. radiodurans* can survive >10,000 Gy. For a long time it was assumed that the extreme resistance of these organisms reflected some way to avoid suffering chromosomal breaks or to some remarkably different mechanism of DNA repair; however, it now seems clear that the amount of DNA damage per Mb of DNA is roughly constant—about 0.005 DSBs/Gy per Mb. In fact diploid budding yeast can fully restore its genome even with IR-induced 250 DSBs. But budding yeast is far less radio-tolerant than another yeast, *Ustilago maydis*, or the rotifers.

Recent work has shown that much—indeed most—of the exceptional resistance of some organisms reflects their ability to squelch the damage to proteins and other cellular components by reactive oxygen species (ROS). This is accomplished through the unusual millimolar concentration of divalent manganese molecules that prevent the formation of iron-dependent ROS. But it is true these organisms still do have a remarkable capacity to suffer hundreds of DSBs and yet put their genomes back together, as seen in chromosome-separating gels (**Figure 9.18**). This process requires both RecA homologs, RecA and RadA, in *D. radiodurans*. Without RecA there is still joining of some segments together (some of which may occur by RecA-independent SSA). Without both strand exchange proteins, there is no assembly.

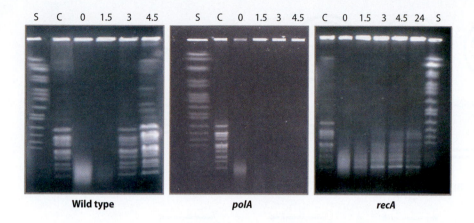

Figure 9.18 **Chromosome breakage after 7000 Gy IR exposure in *D. radiodurans*.** Wild-type cells can restore chromosome integrity after 7000 Gy exposure in about 4.5 hr. *D. radiodurans* DNA was cleaved with the rare-cutting *Not*I restriction enzyme (lane C) and compared with intact *S. cerevisiae* chromosomes (S) on a pulse-field gel. DNA is visualized by ethidium bromide staining. This process is dependent on both DNA polymerase I (PolA) and RecA, to different extents. (From Zahradka K, Slade D, Bailone A et al. [2006] *Nature* 443:569–573. With permission from Macmillan Publishers, Ltd.)

Part of the explanation for their survival is that these organisms carry multiple colinear genomes that can serve as mutual templates to repair damage even in G1 cells. Miroslav Radman's lab has proposed that the actual mechanism of repair is a hybrid between SDSA and SSA, called "extended SDSA," or ESDSA (**Figure 9.19**). ESDSA also requires both the main replicative DNA polymerase PolIII as well as PolI (PolA), which normally functions to fill in gaps for complete ligation (Figure 9.18). One end of a broken segment strand invades into a different segment sharing homology, and the invading end is primer-extended to the end of the template. This long ssDNA tail can then invade a different segment having overlapping homology to make even a longer ssDNA region. Eventually ssDNA regions can anneal by SSA and gaps in the annealed structure can be filled in by a DNA polymerase. Finally, long segments are joined together by additional SSA events or by homologous recombination leading to a crossover to create the intact molecule.

An alternative to ESDSA would be BIR, followed by SSA, but one important observation argues that ESDSA is distinct from BIR. If cells are pulse-labeled with BrdU, they take up nearly all the label initially into ssDNA rather than into dsDNA (**Figure 9.20**). This conclusion is based on the use of an anti-BrdU antibody that only recognizes the label when the DNA is single-stranded, or in denatured DNA. BrdU is first taken up in ssDNA and then, with a half-life of about 25 min, converted into dsDNA. These experiments represent a powerful example of using real-time analysis of repair in a bacterial system to work out the replication and repair proteins involved in DSB repair.

9.11 SISTER CHROMATID REPAIR LEAVES SEVERAL IMPORTANT QUESTIONS UNANSWERED

Genetic assays to look at uSCE turned out mostly to produce LTGC, which appear to result from mechanisms that do not result in a crossover. These studies argue that reciprocal uSCE, especially in mammals, is relatively rare. In contrast, BrdU labeling suggests that reciprocal SCE occurs multiple times every cell cycle in mammalian cells. There are clearly big differences between the short-range, ectopic recombination events, with 1–2 kb of shared homology, and "real" SCE events where the sister chromatids are aligned and share essentially infinite homology to either side. It is, of course, quite possible that only a small fraction of all SCR events are crossover associated. This hypothesis is supported by the phenotype of the BLM mutant that gives many more exchanges without (apparently) increasing the number of lesions. It is also possible that

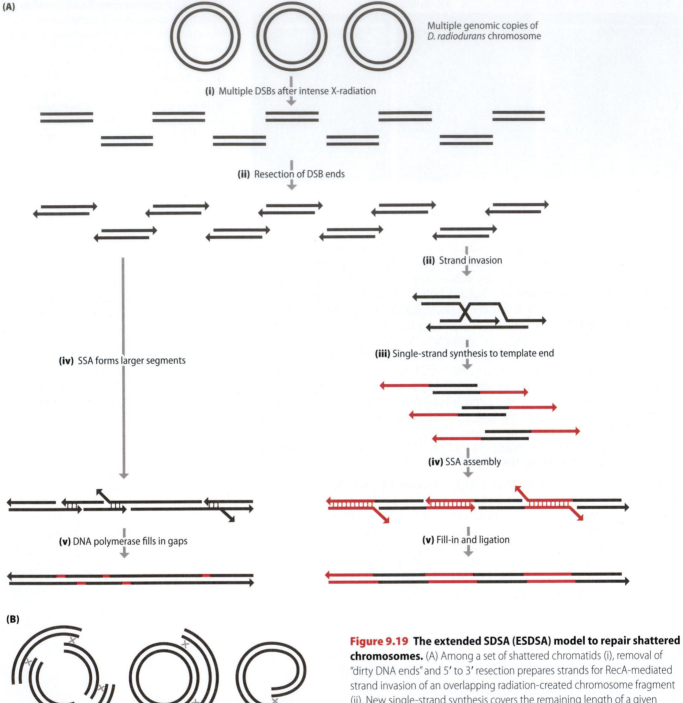

(A)

Multiple genomic copies of
D. radiodurans chromosome

(i) Multiple DSBs after intense X-radiation

(ii) Resection of DSB ends

(ii) Strand invasion

(iv) SSA forms larger segments

(iii) Single-strand synthesis to template end

(iv) SSA assembly

(v) DNA polymerase fills in gaps

(v) Fill-in and ligation

(B)

Figure 9.19 The extended SDSA (ESDSA) model to repair shattered chromosomes. (A) Among a set of shattered chromatids (i), removal of "dirty DNA ends" and 5' to 3' resection prepares strands for RecA-mediated strand invasion of an overlapping radiation-created chromosome fragment (ii). New single-strand synthesis covers the remaining length of a given segment (iii). This extended ssDNA strand can in turn invade and extend further from a different, overlapping segment. Eventually, complementary single-strand annealing allows the formation of larger and larger pieces (iv). PolI then fills in the gaps (v). (B) Large re-joined segments can then be joined into a complete molecule by single-strand annealing or possibly other homologous recombination processes. (Adapted from Zahradka K, Slade D, Bailone A et al. [2006] *Nature* 443:569–573. With permission from Macmillan Publishers, Ltd.)

(A)

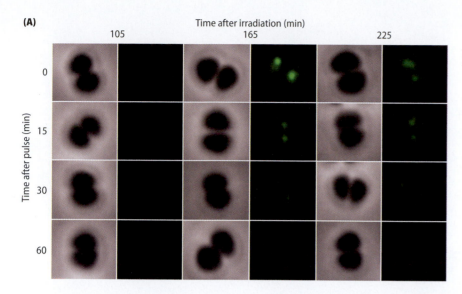

(B)

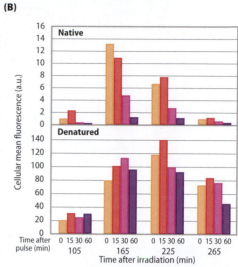

because BLM-deficient cells have reduced rates of resection of DSB ends, that the proportion of events resolved as crossovers would be altered, for example by changing the proportion of events that are repaired by SDSA versus dHJ.

Of course it is also possible that events initiated by site-specific endonucleases are quite different from those arising during replication. A single-strand gap can be repaired by RecA/Rad51-mediated repair that is associated with the formation of a dHJ (Figure 2.2), so that these events could be intrinsically more likely to be resolved with crossovers than the SDSA or LTGC mechanisms induced by a DSB. We must await further investigations to understand this subject more thoroughly.

SUMMARY

Sister chromatid repair plays a key role in repairing the large proportion of DNA damage arising during replication. "Equal" SCE preserves genetic integrity by preventing interhomolog or ectopic recombination that could result in LOH or translocations. There are large gaps in our understanding of these events, including why so many of them arise apparently in the absence of Rad51 and, even more profoundly, what are the lesions that promote these repair events.

Figure 9.20 BrdU incorporation into single- and double-stranded DNA during ESDSA.
(A) BrdU was administered for 15 min at the times indicated and then "chased" by removal of BrdU from the medium. Incorporation into ssDNA is shown by indirect immunofluorescence using an antibody that only recognizes BrdU in single-stranded DNA. This signal disappears with a half-life of approximately 25 min. (B) Fixed cells were imaged either without DNA denaturation (*top*) or after DNA denaturation (*bottom*) for the time points shown in (A). The amount of fluorescence is given in arbitrary units (a.u.). These data show that as BrdU is lost from ssDNA it is incorporated into dsDNA structures. (Adapted from Slade D, Lindner AB, Paul G & Radman M [2009] *Cell* 136:1044–1055. With permission from Elsevier.)

SUGGESTED READING

Daly MJ (2009) A new perspective on radiation resistance based on *Deinococcus radiodurans*. *Nat Rev Microbiol* 7:237–245.

Daly MJ (2012) Death by protein damage in irradiated cells. *DNA Repair* 11:12–21.

Dong Z & Fasullo M (2003) Multiple recombination pathways for sister chromatid exchange in *Saccharomyces cerevisiae*: role of RAD1 and the RAD52 epistasis group genes. *Nucleic Acids Res* 31:2576–2585.

Hartlerode AJ & Scully R (2009) Mechanisms of double-strand break repair in somatic mammalian cells. *Biochem J.* 423:157–168.

Helleday T (2003) Pathways for mitotic homologous recombination in mammalian cell. *Mutat Res.* 532:103–115.

Slade D & Radman M (2011) Oxidative stress resistance in *Deinococcus radiodurans*. *Microbiol Mol Biol Rev* 75:133–191.

Wilson DM 3rd & Thompson LH (2007) Molecular mechanisms of sister-chromatid exchange. *Mutat Res* 616:11–23.

CHAPTER 10
GENE TARGETING

One of the most important applications of homologous recombination is in gene targeting. The ability to knock out or modify genes revolutionized classical genetics, where obtaining well-defined mutations was usually a matter of chance and only achieved after a great deal of effort to isolate and characterize the mutations. Gene targeting depends on one old technology—transformation of DNA fragments that can recombine with homologous genomic sites—and the much newer recombinant DNA technology to create precise mutations. Transformation of DNA fragments in bacteria to create a new genotype was accomplished by Frederick Griffiths in the 1930s, before it was clear that DNA was the hereditary material. But it has only been in the past 30 years that it became possible to exploit genome sequence information to create precisely defined mutations that could be introduced to replace genomic sequences. Ironically, efficient gene targeting in bacteria took much longer to accomplish than in eukaryotes. But even in eukaryotes efficient gene targeting has proven to be quite difficult in many organisms, though incredibly easy in others. So, we start with the easy one: budding yeast.

10.1 BACTERIAL TRANSFORMATION PROVIDED THE FIRST EVIDENCE OF GENE TARGETING

The key to demonstrating that DNA was the genetic material was Griffith's transformation experiment showing that a heat-killed smooth-colony virulent strain of *Pneumococcus* could transform a live rough-colony avirulent strain into a virulent recombinant. A decade later, Oswald Avery, Colin MacLeod, and Maclyn McCarty used this approach to show that DNA was the hereditary material. In essence a DNA segment is taken up by the bacterium and a mutation or variant in the host chromosome is replaced by homologous recombination with the transforming trait. DNA can be introduced as "naked" DNA or efficiently injected into bacterial cells by phage-mediated transduction in which part of a host bacterium's DNA has been packaged in a phage head. We will examine more recent gene targeting strategies in bacteria in Section 10.7.

10.2 GENE CORRECTION AND MODIFICATION IN A EUKARYOTE WAS FIRST ACCOMPLISHED IN BUDDING YEAST

Correction of a yeast mutation by transformation was accomplished by Gerald Fink's lab in 1978, using a cloned copy of the yeast *LEU2* gene inserted in a pBR322 vector that propagates in *E. coli*. John Carbon's lab had inserted fragments of yeast genes into pBR322 and identified one carrying *LEU2* because it complemented an *E. coli leuB* mutation. The pBR322 plasmid cannot replicate in yeast, but when plasmid DNA was introduced into a *leu2* mutant yeast strain, there were a few Leu+ recombinants. In these rare transformants, the plasmid containing *LEU2* had integrated by a crossover between the homologous *leu2* sequences, creating a *leu2*–plasmid–*LEU2* integration (**Inline Figure 10.1**). Some integration events occurred at random locations but the great majority were at *leu2*.

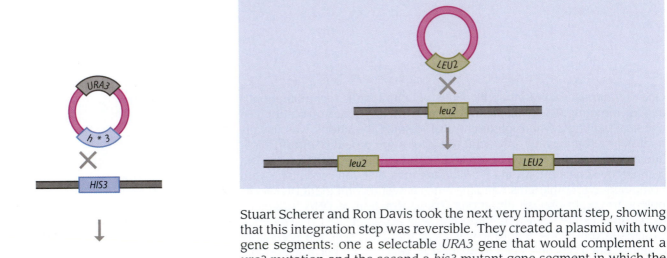

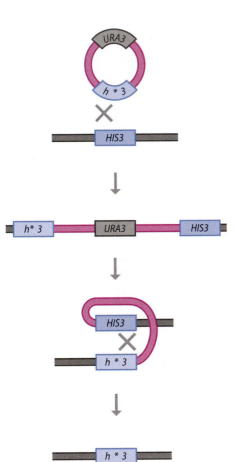

Figure 10.1 Integration and excision of a circular plasmid to introduce an *in-vitro*-generated mutation. Reciprocal recombination to the left of a deletion in the center of the *HIS3* gene carried on a plasmid (*h* * 3) results in a tandem duplication of the mutant and wild-type genes. Excision of the plasmid occurs by the reverse process. If the second point of crossing over is on the opposite side of the deletion from the site of integration, then the chromosome will now harbor the mutation. The excised copy, lacking an origin of replication, is lost. (Adapted from Sherer S & Davis RW [1979] *Proc Natl Acad Sci USA* 76:4951–4955. With permission from the authors.)

Stuart Scherer and Ron Davis took the next very important step, showing that this integration step was reversible. They created a plasmid with two gene segments: one a selectable *URA3* gene that would complement a *ura3* mutation and the second a *his3* mutant gene segment in which the central section was deleted (*h*3) (**Figure 10.1**). Some of the Ura+ transformants proved to be integrations at the *HIS3* locus, with the structure *h*3–plasmid–*URA3*–*HIS3*. (There were also integrants at the *ura3* locus, with the structure *ura3*–plasmid–*h*3–*URA3*.) If the transformants with *his3* duplications were grown without selection for many generations, about 1 in 100 colonies became Ura−, about half of these restored the strain to its original *HIS3* genotype, but the others had become *his3Δ*. In essence, the "popping-out" of the duplication occurred by a crossover in the homologous region on the opposite side from that used in the integration (Figure 10.1). The inefficient step of screening for such pop-outs was later replaced by selecting for the loss of the *URA3* marker, by growing cells in the presence of the inhibitor 5-fluoroorotic acid (5-FOA). Thus it was possible to ablate gene function in a highly specific fashion. With knowledge of the gene's precise sequence it became imaginable to introduce deletions and even single amino acid substitutions.

A way to introduce many gene modifications without having to select for pop-outs was developed by David Shortle, James Haber, and David Botstein in 1982. If one created a *URA3*-containing pBR322 plasmid with homologous gene segment lacking both the N- and C-terminal parts of a gene, integration selecting for *URA3* can produce a duplication in which each copy of the targeted gene was truncated: gene-3′Δ–*URA3*–5′Δ-gene (**Figure 10.2A**). Assuming both N- and C-terminal segments were required for function, this one-step integration event creates a gene knockout. This trick proved to have other uses. One could start with a plasmid in which the gene is 3′-truncated, so that integration produces only one expressed copy of the gene: gene-3′Δ–*URA3*–complete gene. By this strategy it was possible not only to mutate the amino acid sequence (**Figure 10.2B**) but also mutate or even replace the promoter region, since the original promoter drives expression of a nonfunctional, truncated gene (**Figure 10.2C**).

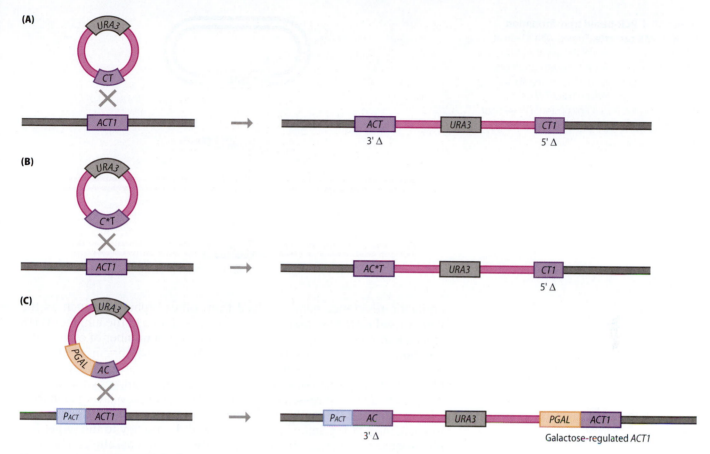

Figure 10.2 Strategies to knock out or modify a gene by integrative transformation of a circular plasmid. (A) Integration by homologous recombination of a gene fragment missing the 5′ and 3′ ends of the actin *ACT1* gene creates a mutation by generating two incomplete copies of the gene, one lacking the 5′ end and one the 3′ end. This type of disruption is easily reversible by screening for pop-out events. (B) In a similar fashion, mutations within the gene can be introduced. (C) By attaching a different promoter to the 5′ end of a gene, one can create a single functional copy under the control of a different promoter. (Adapted from Shortle D, Haber JE & Botstein D [1982] *Science* 217:371–373. With permission from the American Association for the Advancement of Science.)

10.3 DOUBLE-STRAND BREAKS GREATLY IMPROVE GENE TARGETING

We saw in Chapter 6 that a double-strand break within a cloned gene strongly directs integration of a plasmid to that locus. If one has a plasmid carrying both *URA3* and *HIS3*, a restriction endonuclease cleavage within the *HIS3* gene would ensure that nearly all the integrations would be at the *HIS3* locus and not at *URA3*. A major advance was made by Rodney Rothstein, who showed that a linearized restriction fragment, derived from a plasmid, will efficiently promote replacement of a gene region with a selectable marker (**Figure 10.3**). *"Ends-out"* transformation yielded high levels of transformation (about 1 in 10,000 cells) almost all of which had accurately replaced an open reading frame (ORF) with a selectable marker.

Initially, the ends-out fragments used in gene replacement shared several kilobases of homology with the chromosomal target; however, accurate gene replacements in budding yeast can be made with an astonishing small 35 bp of homology on either side of a selectable marker, if the selectable marker itself shared no homology with the genome. Thus a strain completely deleted for the chromosomal *URA3* gene could be used with *URA3* as a selectable marker. Alternatively one can use

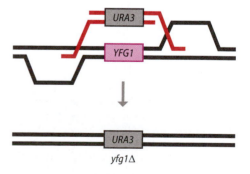

Figure 10.3 Ends-out recombination can be used to replace a gene with a selectable marker. Gene targeting directed by the DSB ends promotes homologous recombination with sequences flanking your favorite gene (*YFG1*). (Adapted from Rothstein RJ [1983] *Methods Enzymol* 101:202–211. With permission from Elsevier.)

Figure 10.4 PCR-based transformation of a KAN-MX cassette. Primers with 35–50 nt of homology to one side of the target and 18 bp overlapping the KAN-MX cassette produces a short ends-out fragment that can replace a chromosomal gene. (Adapted from Wach A, Brachat A, Pohlmann R & Philippsen P [1994] *Yeast* 10:1793–1808. With permission from John Wiley & Sons, Inc.)

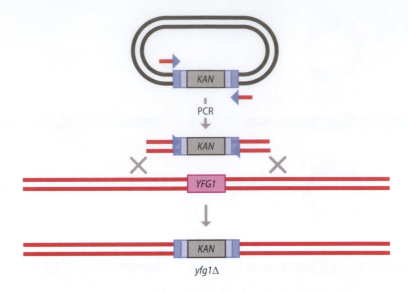

antibiotic-resistance genes derived from other organisms, such as the widely used KAN-MX cassette conferring resistance to the inhibitor G418 (called kanamycin) (**Figure 10.4**). There are now a number of other drug-resistance cassettes [hygromycin B (HPH), nourseothricin (NAT), and bialaphos (PAT)], along with homeologous versions of common selectable markers (e.g. *URA3* or *LYS5* derived from *Kluyveromyces lactis* or other fungi that can be expressed in budding yeast but are too diverged in their DNA sequence to recombine efficiently with Sc-*ura3* or *lys5* sequences). In each case, PCR primers carrying 35–50 nt homology to the target site plus sequences overlapping the ends of the marker cassette can be used to amplify a fragment that will integrate with high reliability at the desired locus. Some variants of the MX cassettes have direct repeats, such as the *loxP* sites for Cre recombinase (see Chapter 11), so that the selectable marker can be excised, leaving only a 33-bp *loxP* site. Recently another marker, *amdSYM*, has been introduced that can be selected both positively (growth on acetamide as a nitrogen source) and negatively (loss of the marker makes cells resistant to fluoroacetamide), thus joining *URA3* as a marker one can select both for its introduction and then for its loss.

Rothstein's ends-out gene targeting strategy has been adapted and modified to create a large number of different types of gene modifications. Many of these have been done systematically, so that there are sets of >5000 yeast strains in which each ORF has been modified, including the following:

• Precise deletion of the ORF of each gene in the genome, as well as other chromosome features such as origins of replication (**Inline Figure 10.2**). More than 80% of yeast knockouts are viable, although in some cases—for example when one deletes one of a pair of duplicated genes—the knockout of one often results in chromosome duplication genes (disomy) of the chromosome carrying the second copy.

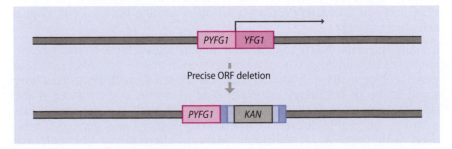

Deletions made in this way can be much larger than the removal of a single gene. It is possible to extirpate tens of kilobases by introducing a fragment with a selectable marker flanked by homologies from distant regions. In fact, it is possible in this way to create translocations.

- Precise replacement of the ORF with a reporter gene such as *URA3* or *GFP*, thus enabling one to compare the level of gene expression of different genes by monitoring expression of the same protein controlled by many different promoters (**Inline Figure 10.3**).

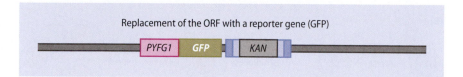

Replacement of the ORF with a reporter gene (GFP)

- Insertion of N- or C-terminal epitope tags such as HA, Myc, or the tandem-affinity purification (TAP) tag (**Inline Figure 10.4**). These modifications have made it possible to use common antibodies to monitor protein levels by Western blots. TAP tags have also enabled labs to carry out systematic proteomic analysis of protein complexes associated with the tagged protein.

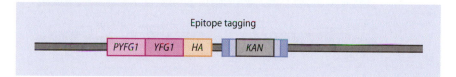

Epitope tagging

A clever variation on this theme affixes two different epitopes that are separated by a stop codon. The construct also has two in-frame *loxP* sites so that induction of the site-specific Cre recombinase deletes both the first epitope and the stop codon, allowing expression of the second tagged version of the same protein, from the same promoter (**Figure 10.5**). This approach has been used to follow the fate of old and newly synthesized histone proteins during replication and transcription.

- Replacement of a gene promoter by a different promoter (**Inline Figure 10.5**). This approach makes it possible to place a gene under different control.

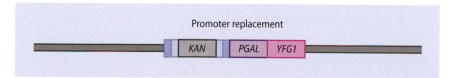

Promoter replacement

For example, a gene can be expressed under the control of a galactose-inducible promoter and turned on or off by changing carbon sources. Similarly, a tetracycline-inducible promoter can be regulated by addition of the antibiotic to the medium. Alternatively, the time of expression of a gene during the cell cycle can be modulated by replacing its promoter. One such approach changed the time of expression of a mismatch repair gene so that it was only expressed after DNA replication was largely complete (**Figure 10.6A**). A powerful way to study genes in meiosis has been to place them under the control of the *CLB2* promoter, as this gene is expressed in growing cells but is not expressed in cells entering sporulation and meiosis (**Figure 10.6B**).

Figure 10.5 Conditional replacement of an epitope tag. (A) A budding yeast histone H3 gene was modified at the protein C-terminus by the insertion of a hygromycin-resistance (Hyg) cassette such that translation will express H3 tagged with the HA epitope as well as a protein linker that is encoded by the *loxP* site. Induction of the Cre recombinase causes site-specific recombination between the two *loxP* sites, removing HA, a stop codon, and Hyg, resulting in the creation of H3 fused to the T7 epitope. (B) Cells expressing H3-HA were arrested in the G1 phase of the cell cycle and Cre was expressed to switch epitopes. The Cre enzyme is fused to an estrogen receptor domain (EBD) that only translocates to the nucleus in the presence of β-estradiol. Then cells were released and grown for several generations. (C) The ratio of HA to T7 tag at many sites along the chromosome was determined several generations after inducing the switch from one epitope to the other, to measure how stable nucleosomes were carrying H3-HA in different gene and non-gene regions. (Adapted from Radman-Livaja M, Verzijlbergen KF, Weiner A et al. [2011] *PLoS Biol* 9:e1001075.)

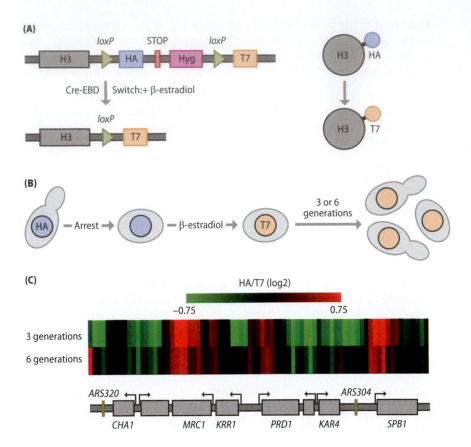

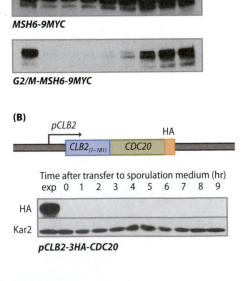

Thus it is possible to remove expression of an essential gene such as *CDC20* only in meiotic cells when it will not be expressed in the *pCLB2* construct.

- Creation of reduced-expression (hypomorphic) alleles of essential genes by inserting a selectable marker within the 3′ untranslated region (**Inline Figure 10.6**). These DAmP (diminished allele mRNA perturbation) alleles have allowed analysis of genetic interactions of essential gene mutations with other nonessential mutations of other genes.

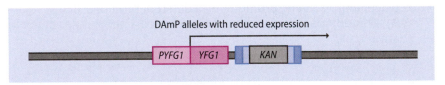

Figure 10.6 Replacing promoters changes the time of expression of a gene. (A) Replacing the normal promoter of the *MSH6* gene with the *CLB2* promoter restricts expression to the later part of the cell cycle as measured by the time that α-factor-arrested G1 cells are released to begin S phase. The abundance of the Msh6 protein tagged with the MYC epitope was measured by a Western blot, using anti-MYC antibody. Exponentially growing (exp) cells show the average level of expression of the tagged protein. (B) The same *CLB2* promoter is expressed in exponentially growing cells (Exp) but is not expressed when cells are shifted to the nitrogen-starvation conditions that promote sporulation and meiosis. Here the Cdc20 protein is tagged with the HA epitope. An unregulated protein (Kar2) is monitored by Western blot as a control of even loading of the lanes. (A, adapted from Hombauer H, Srivatsan A, Putnam CD & Kolodner RD [2011] *Science* 334:1713–1716. With permission from the American Association for the Advancement of Science. B, adapted from Lee BH & Amon A [2003] *Science* 300:482–486. With permission from the American Association for the Advancement of Science.)

- Creation of temperature-sensitive mutations by attaching a thermo-labile degron sequence to the N-terminus (**Inline Figure 10.7**). At high temperature the amino acids of the degron segment are ubiquitylated, which leads to proteasome-mediated degradation of the fusion protein. Here, the insertion of the DNA encoding the N-terminal segment of the mammalian DHFR protein degron sequence at the N-terminus of the ORF is accompanied by insertion of the KAN-MX (G418-reistance) cassette as well as the *CUP1* promoter to drive expression, because the insertion may interfere with normal transcriptional controls. Often, efficient destruction of the fusion protein at the restrictive temperature of 37°C also requires the overexpression of the E3 ubiquitin ligase protein Ubr1.

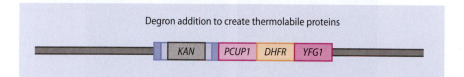

Degron addition to create thermolabile proteins

- More recently a new degron system has been developed, which uses the binding of a module to the plant hormone auxin to trigger degradation without having to shift temperatures. Here the 229-amino-acid IAA17 protein from *Arabidopsis* is fused to the C-terminus of the target protein (by creating a gene fusion). In addition, expression of an auxin-regulated Skp1-AtTIR1 E3 ubiquitin ligase results in the poly-ubiquitylation and degradation of the fusion protein after adding auxin to the medium.

- Insertion of GFP or other fluorescent protein sequences as a C-terminal fusion protein, with a selectable marker further downstream (**Inline Figure 10.8**). N-terminal fusions are also used.

Fluorescent fusion protein to study stability and localization

This strategy has been used to demonstrate the localization and abundance of many yeast proteins. An example is shown in **Figure 10.7**, where the separase Cdc6-GFP protein is delocalized from the nucleus when the nuclear localization sequence (NLS) is mutated. Fluorescently tagged proteins can also be exploited to examine their mobility and stability by techniques such as fluorescence recovery after photobleaching (FRAP).

- Finally, ends-out transformation can be used to introduce single base pair alterations into an ORF (**Inline Figure 10.9**).

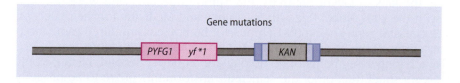

Gene mutations

One can introduce a modified version of the gene with sufficient additional homology to include a positively selectable marker beyond the 3' end of the gene so that some fraction of the transformants will not

Figure 10.7 Monitoring cellular localization of a GFP-tagged protein. (A) Cdc6 normally localizes to the nucleus as seen in the coincidence of GFP and DNA (DAPI-stained) signals compared with the light microscope (DIC) image. (B) When the nuclear localization signal (NLS) is deleted from the Cdc6 protein, it fails to accumulate in the nucleus. (From Luo KQ, Elsasser S, Chang DC & Campbell JL [2003] *Biochem Biophys Res Commun* 306:851–859. With permission from Elsevier.)

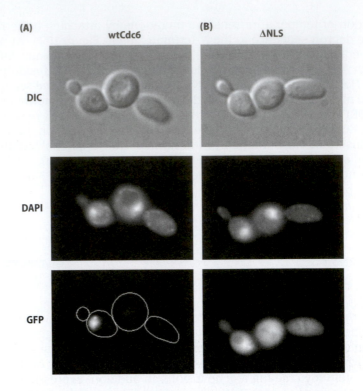

only have introduced the selectable marker but will also have replaced the gene deletion with the modified sequence (**Figure 10.8A**). In this case, one needs longer segments of DNA homologous to the region including the ORF as well as sequences outside the transcription unit. This approach will even work if one transforms the construct into cells with a wild-type ORF. Selection of transformants by the selectable marker confers a 10–30% chance of co-introducing the point mutation into the ORF, which can be verified by PCR. An alternative approach is to first delete the ORF with a selectable marker and then introduce the entire gene with its *in vitro*-generated mutations and select or screen for loss of the original deletion. One example is shown for a diploid situation in **Figure 10.8B**. Alternatively, if one first replaces the entire gene of interest with a counter-selectable marker such as *URA3*, then one can transform with the modified gene sequence, selecting for 5-FOA-resistant (Ura⁻) cells, which should ensure the replacement of the gene with a modified version.

10.4 GENE TARGETING IS MORE DIFFICULT IN MAMMALS THAN IN YEAST

Gene targeting in mammalian cells is more difficult than in yeast. The ratio of accurate gene knockouts to random integrations is often 1 in 1000. The first knockouts using ends-out fragments were accomplished by the labs of Oliver Smithies and Mario Capecchi, who were honored with the 2007 Nobel Prize in Physiology and Medicine for their success. Successful gene knockouts in mouse or human cells required some modifications of the basic idea promoted by Rothstein in budding yeast. First, much longer segments of homology are required. Second, because only a very small fraction of the selected transformants prove to be properly targeted, a simultaneous positive and negative selection screen is generally applied (**Figure 10.9**). The targeting fragment has a selectable *neo* gene (G418-resitance) in the middle but in addition has a gene at one end that can be used to poison cells that incorporate it. One such gene is the

(A)

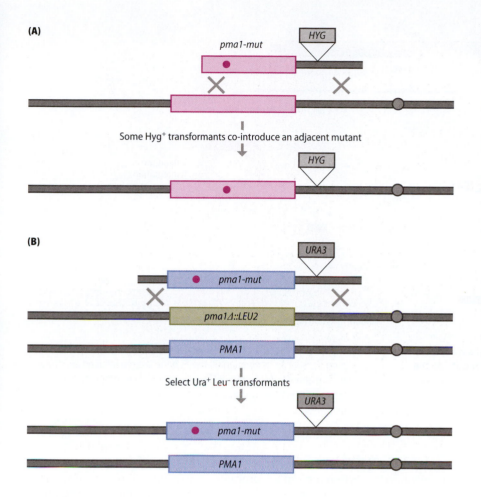

(B)

Figure 10.8 Allele replacement adjacent to a selectable marker located outside the gene. (A) A haploid strain carrying a wild-type yeast *PMA1* gene was transformed with an *in vitro*-mutagenized segment of the *PMA1* gene. Selection of Hyg$^+$ transformants could then be screened for the presence of mutations in the adjacent gene. About 30% of transformants also introduce an adjacent mutation. (B) Replacement of a *pma1Δ::LEU2* gene deletion in a diploid with a mutant gene. Selection of Ura$^+$ Leu2$^-$ transformants ensures that gene replacement occurred at the proper site. A diploid carrying the introduced mutation could be sporulated to obtain haploid segregants, to test if the mutation is viable or displays a recessive phenotype. (A, adapted from Na S, Perlin DS, Seto-Young D et al. [1993] *J Biol Chem* 268:11792–11797. With permission from the American Society for Biochemistry and Molecular Biology. B, adapted from Harris SL, Na S, Zhu X et al. [1994] *Proc Natl Acad Sci USA* 91:10531–10535. With permission from the National Academy of Sciences.)

herpes simplex thymidine kinase (HSV-TK) gene that kills cells that are otherwise TK$^-$ in the presence of the antiviral agent, gancyclovir. Cells that carry out the desired gene replacement will lop off the HSV-TK segment because it is not flanked on one side by sequences homologous to the target; but nonspecific integrations, at least most of them, should insert the entire fragment, including the HSV-TK region, and hence will be killed when gancyclovir is present. A number of other negative selection genes have been used, including HGPRT and cytosine deaminase; toxins such as diphtheria or ricin can be used as well. The use of a negative selection removes most, but not all, of the mis-targeted fragments (some escape because end-degradation will have destroyed the negative selection gene). But suffice it to say that it has now become possible to knock out genes in cultured cells and to take mutated embryonic stem cells and obtain mice carrying gene knockouts. Breeding then allows the construction of offspring with multiple knockouts. Currently the record for the largest number of gene modifications introduced into one mouse stands at six homozygous gene knockouts or modifications.

10.5 GENE TARGETING IS IMPROVED BY CREATING A CHROMOSOMAL DSB AT THE TARGET LOCUS

An alternative to simply using ends-out transforming fragments is to create a DSB at a designated site on a chromosome. In this case, the locus to be modified suffers the break and uses either an intact template that is transformed into cells or an ends-out fragment with the modified sequences (which could be base-pair substitutions, deletions, or other

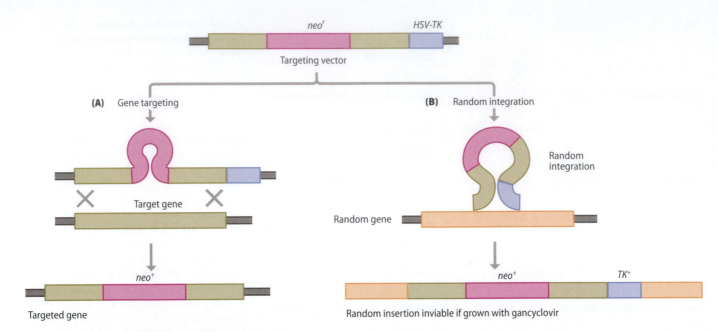

Targeting vector

(A) Gene targeting

Target gene

neo⁺

Targeted gene

(B) Random integration

Random integration

Random gene

neo⁺   TK⁺

Random insertion inviable if grown with gancyclovir

Figure 10.9 Gene targeting in mammalian cells can be improved by the presence of a counter-selectable marker outside the region of homology. The herpes simplex virus thymidine kinase gene (HSV-TK) will kill TK⁺ mouse cells in the presence of a drug such as gancyclovir. (A) When integration occurs at the desired site the cells are G418-resistant because of the *neo* gene, but the TK gene is lost; however, during nonspecific integration (B) the TK gene is often retained and these transformants can be eliminated by their gancyclovir sensitivity. (Based on experiments by Mario Cappecci and Oliver Smithies and reviewed in Capecchi MR [2005] *Nat Rev Genet* 6:507–512.)

modifications). When a circular template is introduced, repair of the DSB should be by gene conversion, which sometimes will integrate the entire plasmid; but when the template is a linear fragment it is likely that the replacement occurs by SSA as both the ends of the DSB and of the fragment are resected and can anneal (**Figure 10.10**). This approach is highly efficient in budding yeast and provides a great improvement in mammalian cells. Recently, this approach has been extended by Francesca Storici, who showed that single-stranded templates of DNA—or even RNA!—can be used to accurately repair and replace the sequences at the chromosomal DSB or to create deletions in yeast.

The limiting step in this approach is how to create the DSB in any given gene of interest. To carry out "genetic surgery" on chromosomes, the key is to be able make a DSB at a specific chromosomal site. One strategy is to take advantage of a small number of site-specific endonucleases (sometimes called *meganucleases*) that have very large recognition sites. The most commonly used enzymes are HO endonuclease and the I-SceI nuclease, which has an 18-bp recognition site, but there are a number of other similar enzymes. In yeast (or in mammals) there is no natural I-SceI site in the nuclear genome, but the 18-bp sequence can be introduced

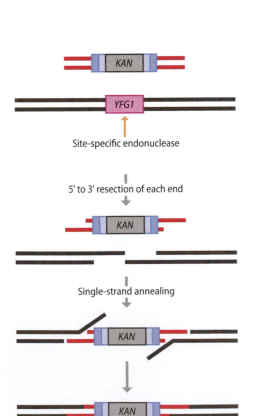

KAN

YFG1

Site-specific endonuclease

5' to 3' resection of each end

KAN

Single-strand annealing

KAN

KAN

Figure 10.10 Improved gene targeting by creating a DSB within the targeted locus. If a DSB can be created in the target locus, it can be repaired with a co-transformed linear fragment sharing homology at or near the DSB ends. The DSB can be replaced by gene conversion with an intact template on a circular plasmid, but is more efficiently replaced by introducing a linearized fragment sharing homology at or near the chromosomal DSB ends. Resection of the DSB ends on the chromosomal site and at the ends of the transforming fragment lead to integration by SSA. (Adapted From Choulika A, Perrin A, Dujon B & Nicolas JF [1995] *Mol Cell Biol* 15:1968–1973. With permission from the American Society for Microbiology.)

by standard gene targeting methods at any location. Once the cleavage site is integrated, expression of I-SceI allows conditional induction of a DSB. Transformation of a template fragment at the same time provides the means to repair and modify the locus. This approach is particularly useful if you wish to insert a series of different modifications, all at the normal genomic locus. However, the downside of this approach is that one must first create a precisely targeted introduction of the cleavage site. An alternative approach to create a DSB at a specific site is to create a library of meganucleases, each able to cleave a different genomic target so that modifications could be directly introduced to a specific locus. The idea of "designer meganucleases" has become a reality. Four different approaches have been successfully undertaken.

Modifications of an existing site-specific endonuclease

A combination of rational design and high throughput screening of several natural meganucleases, such as I-CreI, has been used to obtain enzymes that no longer cleave the original recognition site but instead cut a very different locus. Although I-CreI acts as a dimer and cleaves a palindromic site, it is possible to design new enzymes that form heterodimers of two differently modified I-CreI proteins, allowing the cleavage of non-palindromic sites. Successful gene targeting and ablation of the *RAG1* gene in mammalian cells has been accomplished by this approach.

Zinc-finger nucleases

A different approach is to use the specificity of zinc fingers—protein domains that each recognize a specific 3-bp sequence—and to combine zinc fingers to bind a specific 9- or 12-bp sequence. This construct is fused to the endonuclease domain of the FokI enzyme (**Figure 10.11**). Combining two such zinc-finger nucleases (ZFNs) again will provide the specificity needed to cleave a single target in the mammalian genome. This approach has proven highly successful in several tests in a range of organisms, including *Drosophila*, *Caenorhabditis*, zebrafish, several mammalian cells, viral DNA, and tobacco. In many of these examples, ZFN-mediated cleavage was used to create small deletions in an ORF, to create efficiently knockouts of interesting genes. It is now possible to order a commercially produced custom meganuclease, although the cost is still high.

TALE nucleases (TALENs)

Very recently a new and much less expensive strategy for creating site-specific endonucleases has burst upon the scene and has taken over the market for this approach. These TALE nucleases (TALENs) are based on a remarkable family of plant pathogen proteins called transcription activator-like effectors (TALEs) (**Figure 10.12A**). Remarkably, a TALE is composed of up to 30 tandem repeats of a 33–35 amino acid modules. Each module binds to a single base pair. Only two amino acids in the middle of each module need to be changed to recognize a different base pair (namely, NI recognizes A, NN finds G, HD binds C, and NG binds T). Thus, by fusing tandem arrays of TALE repeats to a FokI nuclease "head" one can obtain a site-specific nuclease. The modular nature of these proteins makes it possible to rapidly and cheaply assemble them into large proteins that recognize and cleave a particular DNA sequence. These enzymes appear to have very good specificity and already they have been shown to be comparable to ZFNs in creating ablations of genes by cleavage and NHEJ and in directing targeted gene modification (**Figure 10.12B**). ZFNs and TALENs cut leaving 4-bp 5′ overhangs, as opposed to HO or I-SceI that leave 4-bp 3′ extensions. This difference may account for the fact that ZFNs and TALENs appear to give rise to deletions around the cleavage site more readily than the natural meganucleases.

Figure 10.11 A zinc-finger nuclease (ZFN).
(A) Each zinc finger can recognize 3 consecutive specific base pairs; strung together multiple zinc fingers can recognize 9 or 12 bp. The set of recognition motifs are hooked to a nonspecific endonuclease domain from the FokI endonuclease. Two such ZFNs can cleave unique sites in the mammalian genome, leaving 5′ overhanging ends. (B) Efficient gene targeting within the *IL2Rγ* gene in which a ZFN cleaves within exon 5 and is repaired by a linearized DNA template carrying a novel *Bsr*B1 restriction site. Exponentially growing or G2-arrested cells were transfected with the indicated plasmids at two different concentrations ("low" and "high"). Unselected cells were assayed by PCR-amplifying the exon 5 region and cleaving with *Bsr*B1. As many as 18% of all the exons have been modified by incorporating the restriction site. (Adapted from Urnov FD, Miller JC, Lee YL et al. [2005] *Nature* 435:646–651. With permission from Macmillan Publishers, Ltd.)

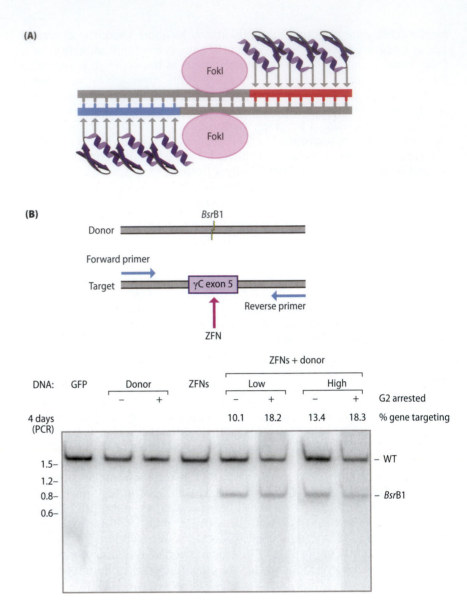

Figure 10.12 TALE nucleases. (A) TALE nucleases (TALENs) are composed of multimers of 33 amino acid units each recognizing a single base pair, joined to the FokI endonuclease "head." (B) When the TALENs were introduced into human embryonic stem cells along with a targeting sequence carrying the puromycin-resistance selectable marker, transformants could be obtained in which both chromosomal targets were replaced (those with only the T SphI restriction fragment) or only one site was replaced (those with both the WT and T bands on the Southern blot). (Adapted from Hockemeyer D, Wang H, Kiani S et al. [2011] *Nat Biotechnol* 29:731–734. With permission from Macmillan Publishers, Ltd.)

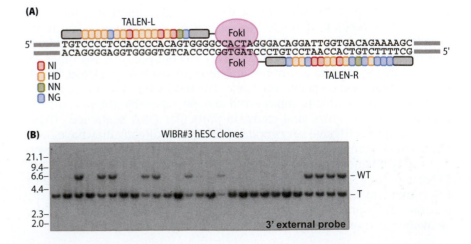

CRISPR nucleases

In 2013 several labs have exploited a nuclease system found in some bacteria and archaea that is based on an RNA guide sequence that both recognizes a DNA target and binds to an endonuclease. The "clustered regularly interspaced short palindromic repeats" (CRISPR) system uses genomically encoded RNA arrays to guide the cleavage of foreign DNA by a CRISPR-associated (Cas) nuclease (**Figure 10.13**). The Cas9 enzyme from *Streptococcus pyogenes* was engineered to be a nuclear enzyme in both mammals and budding yeast. A guide RNA (gRNA) can be conditionally expressed, to target cleavage of a specific sequence. The Cas9 enzyme makes a blunt-end cut and has been shown to promote both deletion formation by NHEJ and accurate gene targeting at very high efficiency. The guide RNA is targeted to a 23-bp sequence that begins with G and ends in GG. Figures 10.13B and C illustrate correction of a "stop" mutation within a *GFP* gene in human cells by gRNA compared with the use of a pair of TALENs that cleave the same region. Moreover, a mutation in Cas9 has made it possible to produce single-stranded nicks that will become double-stranded only after DNA replication. Creating nicks rather than DSBs appears to increase the frequency of gene targeting by diminishing NHEJ outcomes. CRISPRs are *very* inexpensive (one merely has to insert a different 23-nt segment into the guide RNA to obtain a new specificity).

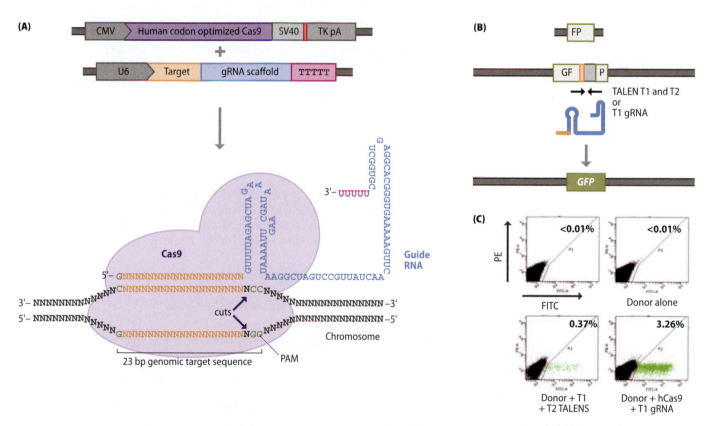

Figure 10.13 CRISPR-mediated gene targeting. (A) The *Streptococcus* Cas9 protein ORF was modified for expression in human cells, including the addition of an SV40 nuclear localization signal (SV40). Expression of both Cas9 and a guide RNA (gRNA) are sufficient to create a blunt-ended DSB adjacent to a protospacer-adjacent motif (PAM). The guide RNA specifies a 23-bp sequence beginning with G and ending with GG. (B) Design of an assay to evaluate CRISPR and TALEN-mediated gene correction. A gene expressing GFP is interrupted by a stop codon and some adjacent sequence. A pair of TALENs (T1 and T2) cutting near the stop codon can be used to induce a DSB that will be corrected by recombination with a linear dsDNA homologous fragment to produce GFP-expressing cells (C). The efficiency of gene replacement is higher in this instance using a gRNA recognizing the same sequences as T1. (Adapted from Mali P, Yang L, Esvelt KM et al. [2013] *Science* 339:823–826. With permission from the American Association for the Advancement of Science.)

10.6 DESIGNER MEGANUCLEASES MAKE IT POSSIBLE TO THINK ABOUT GENETIC SURGERY

The use of designed meganucleases raises the possibility that one could perform precise "genetic surgery" at least on cells that are accessible in patients as well as in cell culture, but there are still a number of issues to resolve before this approach is likely to be approved in humans. Of greatest concern is the finding that the custom nucleases studied to date are not truly specific and will cleave a number of degenerate sequences elsewhere in the genome. Still, it is hard to imagine that genetic surgery will not be common in the next generation.

The use of site-specific nucleases has the potential to improve dramatically our ability to knock out and knock in gene modifications and open up the real possibility of correcting defects in some human diseases, notably those arising in circulating cells derived from blood stem cells. There are still many hurdles to overcome, however. First, can these enzymes cope with mammalian chromatin structure? Several experiments have suggested that the accessibility of these nucleases varies in different cell lines. In some instances, treating the cells with a histone deacetylase inhibitor, such as TSA, results in a "loosening" of chromatin structure and increased ability to cleave a target site. Still, chromatin organization poses a formidable barrier to the use of the nucleases in many circumstances. Second, are these new nucleases sufficiently precise in the sites they can cleave, or will many other degenerate sites be cleaved? Two different approaches have been used to explore the "landscape" of nuclease cleavage. Fred Alt's lab has used deep sequencing to look for translocations in which the I-SceI enzyme was used to generate a specific cleavage at a known target site. The I-SceI cut could be joined—by NHEJ—to some other DSB. In pre-B cells many of these translocations were to the immunoglobulin heavy chain genes that were being cleaved by RAG proteins: some were to the c-Myc locus, which is known to form IgH-Myc translocations in Burkitt's lymphoma, and some were to degenerate I-SceI sites. These sites had between one and four mismatches compared to the canonical I-SceI site. Thus, this endonuclease is not as site-specific as one had hoped.

Christof von Kalle's lab used a different approach to study the fidelity of two ZFNs. Here they provided a linear, modified integrase-defective lentivirus containing a selectable marker flanked by homologies to the ends of the DSB created by the ZFN (**Figure 10.14**). Accurate gene targeting would

Figure 10.14 A DSB at the target locus ensures most events are accurate gene replacements. A ZFN creates a chromosomal DSB within a region of homology shared by a linearized lentiviral sequence. About 90% of the selected transformants integrate precisely by homologous recombination. Another 4% integrate at the designated site but by NHEJ. Only about 6% of the integrations are at ectopic sites that prove to have partial homology to the ZFN target. (Adapted from Gabriel R, Lombardo A, Arens A et al. [2011] *Nat Biotechnol* 29:816–823. With permission from Macmillan Publishers, Ltd.)

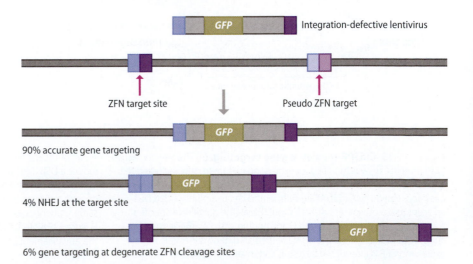

recombine with the homologies and integrate the marker; this happened nearly 90% of the time. Another 4% of the events proved to be NHEJ captures of the linearized lentivirus, so that the homologous regions were duplicated along with the marker. Most of the remaining 6% were NHEJ-mediated insertions elsewhere in the genome, and most of these were at sites that shared 88–96% identity with the designed ZFN cleavage site. So here again, there is a substantial level of "off-target" cleavage, though it is confined to sites that are similar to the cleavage site. Only a subset of *in-silico*-identified possible degenerate sites were cleaved, however, perhaps illustrating the issue of accessibility.

10.7 CONDITIONAL GENE KNOCKOUTS MAKE IT POSSIBLE TO ANALYZE ESSENTIAL GENES

A complicated but powerful variation on gene targeting, especially in mammalian cells, is used to create conditional knockout alleles. For essential genes, conditional alleles are required. The targeted modifications of the gene leave all exons intact, but at least one essential exon is flanked by sites within introns that can be used to delete a region by site-specific recombination. Most commonly, deletions are created using the Cre recombinase, which will delete sequences between two *lox* sites. If the cells (or animal) carry an inducible (or tissue-specific) Cre recombinase gene, recombination between the *lox* sites will produce a deletion in the chromosome, creating a knockout. This strategy is discussed in more detail in Chapter 11. The creation of conditional alleles has been of enormous power when the gene of interest is essential in embryonic development. In practice, cells or animals are created that have one allele of a gene knocked out by conventional gene targeting and that carry a "floxed" second allele that can be converted to a null mutation by induction of Cre recombinase (**Figure 10.15**).

Figure 10.15 Gene targeting to produce a "floxed" allele. Gene replacement places *lox* sites on either side of an important exon in a gene. Expression of the Cre recombinase results in deletion of the exon and loss of function of the gene. Because the selectable marker within an intron can interfere with transcription and splicing, it is usually removed. Two strategies are used. (A) The targeting vector actually has three *lox* sites. Brief expression of the Cre recombinase will sometimes excise the selectable marker, generating a single *lox* site, which later can be used with the more distant *lox* site to delete the exon when Cre is expressed in a tissue- or temporally specific manner. (B) The selectable marker is surrounded by FRT sites that will create a deletion when the site-specific FLP recombinase is expressed. Then the exon can be deleted by expression of Cre.

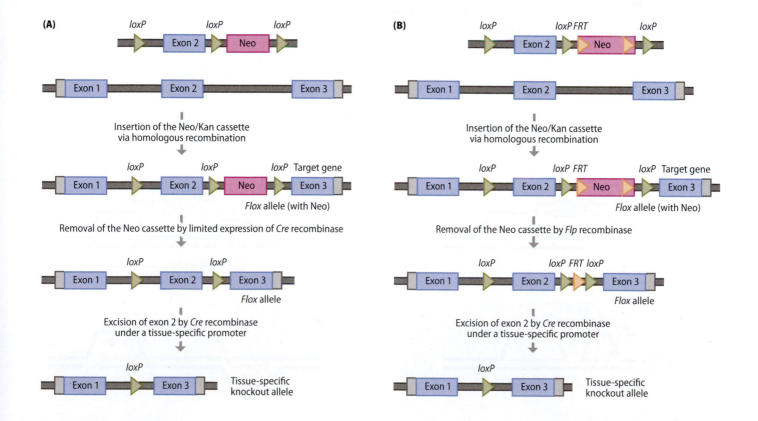

The immense size of most mammalian genes means that most knockouts limit themselves to ablations of only a few of many exons. Whether one could create complete gene knockouts if two homologous segments of linearized DNA were derived from the 5' and 3' ends of a 250-kb gene is not known. Large deletions have been created in yeasts and DT40 cells.

10.8 ENDS-OUT TRANSFORMATION LIKELY HAPPENS THROUGH SEVERAL DIFFERENT PATHWAYS

Most cartoons of gene replacement show a homologous fragment engaging in a double crossover with the chromosome, resulting in a replacement of both strands of the intervening chromosomal sequence with the incoming fragment. But this is unlikely to be an accurate picture of what occurs. Surprisingly, there is as yet no consensus on what does happen, and it is likely that there are several different pathways.

At least some events may be simply described as two independent strand invasions that both lead to two crossovers (**Figure 10.16**). This simple picture would explain cases where each of the two regions of homology flanking a selectable marker can promote huge deletions of 50 kb or more. In fact, in one case, a large chromosomal inversion was created by transforming in two fragments with ends homologous to two distant chromosome locations. In these instances there may be a D-loop created at each end that is (we assume) then cleaved by nucleases that either recognize branched intermediates or Holliday junctions (Figure 10.16 panel (i)). One nuclease that may be involved in this resolution step is the Rad1–Rad10 nuclease that removes 3'-ended nonhomologous tails and cleaves some insertion–deletion heteroduplexes in SSA or in gene conversions. Deletion of Rad1 in budding yeast markedly reduces targeted deletions of an ORF and, as discussed more below, appears to alter the mechanism by which transformants are produced. In mouse cells with deletion of Ercc1, the homolog of Rad10, gene targeting is completely eliminated.

Figure 10.16 Fate of ends-out transforming DNA. A linear fragment carrying a selectable marker flanked by targeting sequences is transformed into the cell and one or both ends can engage in Rad51-mediated strand invasion. (i) Both ends can engage in strand invasion and the formation of Holliday junctions that can be resolved to replace the gene with the selectable marker. (ii) One or both ends can invade and set up divergent BIR replication forks that could terminate by encountering other replication forks. (iii) One strand of the fragment may be assimilated, creating a heteroduplex region that can be subject to mismatch correction.

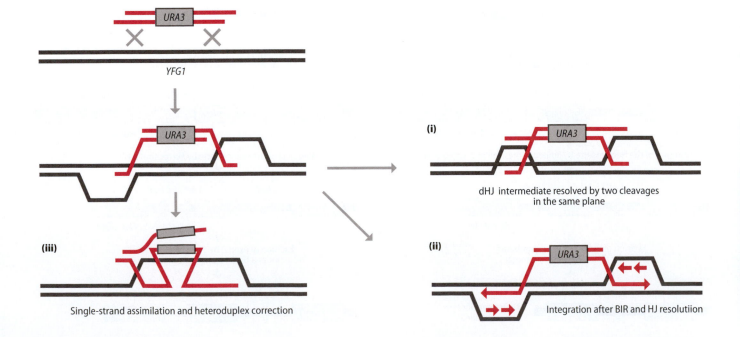

Hit-and-run transformation

One indication that not all transformation events involve simple cross-overs is the recovery of "hit-and-run" events. Here, a fragment end encounters homology, apparently initiates strand invasion but then rather than resulting in a crossover, seems to initiate BIR. After a time, the fragment, now extended by copying homology on at least one side, dissociates and integrates somewhere else in the genome (**Figure 10.17**). Gerald Adair first noted these events in studying correction of an APRT mutation in mammalian cells. A fragment (designated APR) was used to correct an *APRT* gene mutation (A*RT). Some of the APRT+ cells were unusual because they still carried the original mutation but also had a complete APRT gene and additional adjacent sequences, but now inserted at a foreign location.

Such events have also been found in budding yeast; in some instances they can be the majority of the attempts to replace a gene segment. For example, when a *URA3*-marked fragment is introduced to replace *ADE2*, nearly half the Ura+ transformants ended up with an autonomously replicating plasmid that contained at least part of the *ADE2* gene and some of the surrounding sequences, but the cells were still Ade2+ because the original locus was still present. The replicating circles appear to have arisen after one homologous end of the transforming fragment had strand-invaded but then apparently set up a BIR event that replicated along the template chromosome until it copied a nearby origin of replication. Then, somehow, the BIR event stopped and the ends of the now-extended linear fragment ligated to form a replicating circle (**Figure 10.18**). These hit-and-run events can be easily misunderstood if one merely tests for the presence of a proper knockout by using a pair of PCR

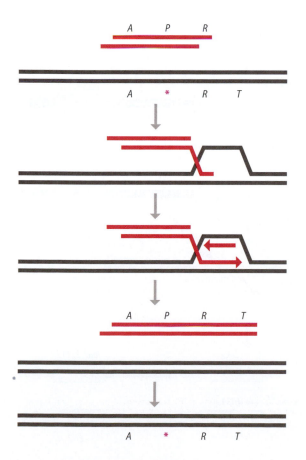

Hit-and-Run transformant

Figure 10.17 Hit-and-run transformation in mammals. An attempt to correct a mutation in the *APRT* gene (*A*RT*) with a fragment covering the mutation sometimes results in BIR followed by integration of an elongated, intact *APRT* gene that integrates illegitimately at another location. (Adapted from Scheerer JB & Adair GM [1994] *Mol Cell Biol* 14:6663–6673. With permission from the American Society for Microbiology.)

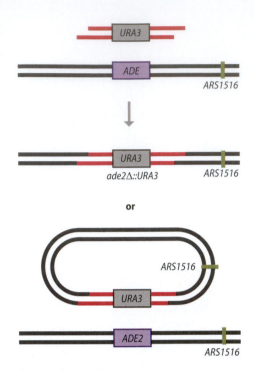

Figure 10.18 Hit-and-run transformation in budding yeast. Instead of integrating by ends-out targeting at the desired locus, some transformants engage in BIR from one or both ends, which can result in the copying of a nearby ARS sequence. The extended sequence is then joined by NHEJ to the opposite end of the linear transforming fragment, thus creating a circular extrachromosomal element that carries the selectable marker and is able to replicate autonomously. (Adapted from Kraus E, Leung WY & Haber JE [2001] *Proc Natl Acad Sci USA* 98:8255–8262. With permission from the National Academy of Sciences.)

primers, one in *URA3* and one outside the region of homology shared by the targeting fragment and the chromosome. PCR yields the expected size fragment! Disruptor beware: test not only for the presence of the knock-out you want to make, but also for the absence of the original sequences.

The existence of hit-and-run transformants strongly suggests that at least some of the time there is not simply a crossover at each homologous boundary; instead the ends are used to set up replication forks (namely, BIR) that might move a considerable distance, after which they may still undergo a crossover at each end to replace the original sequences (Figure 10.16 panel (ii)). How this intermediate would become integrated as a gene replacement is not obvious—or maybe these events are simply abortive and never progress to replacing the original sequences.

Assimilation of a single strand of transforming DNA

Another mechanism that may be frequent but nonproductive appears to involve the assimilation of only one strand of the transforming DNA into a long heteroduplex region, subject to mismatch correction (Figure 10.16 panel (iii)). Normally one cannot follow the fate of all the strands involved in gene targeting because the number of molecules that enter a given cell is unknown and one usually cannot recover any chromosome strands that do not encode the expression of the selectable marker. The Haber lab used HO endonuclease to liberate within the nucleus a single copy of a chromosomal region containing a *LEU2* gene that could then be used to recombine and correct a pair of *leu2* mutations, 400 bp apart (**Figure 10.19**). The *LEU2* segment was released from a chromosome by a pair of HO cuts that were positioned so that the flanking regions could repair the chromosomal break by SSA, which was nearly 100% efficient. Thus one could ask about the fate of the liberated *LEU2* fragment in essentially all the cells of a population. About 2 in 10,000 cells remained Leu2+, the result of correcting the two mutations in the resident *leu2* gene; this number is not very different from the frequency of transformants obtained by standard transformation procedures.

Figure 10.19 Gene correction by liberating a DNA fragment from a different site on a chromosome. HO endonuclease cleavage at both ends of a wild-type *LEU2* gene liberates a single copy of the sequence that can recombine with a *leu2-3,112* double mutant to produce a stable *LEU2* derivative. The cells are viable because the outside ends of the two DSBs can rejoin by SSA to produce a functional *HIS4* gene. In the absence of mismatch repair, a His+ colony can be sectored *LEU2/leu2* when heteroduplex DNA at *leu2* is not corrected. (Adapted from Leung W, Malkova A, Haber JE [1997] *Proc Natl Acad Sci USA* 94:6851–6856. With permission from the National Academy of Sciences.)

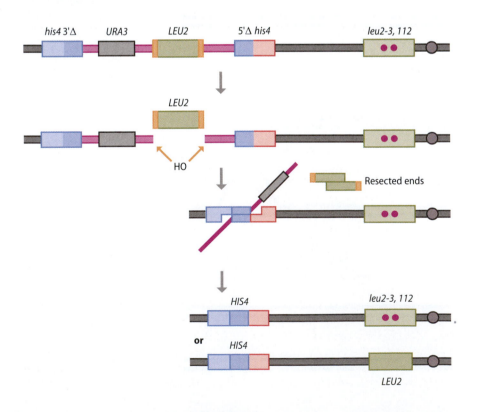

The mechanism by which these recombination events occurred appears to involve mismatch repair, because the frequency of Leu2$^+$ events was increased as much as 25-fold by knocking out the *PMS1* mismatch repair gene. Without Pms1, 70% of the colonies were sectored: *LEU2/leu2*. These results suggest that many of the gene correction events arose by the assimilation of only one of the two strands of the *LEU2* fragment, creating a long heteroduplex DNA that would be preferentially repaired in favor of the resident, unbroken strands (Figure 10.16 panel (iii)). Without *PMS1*, the heteroduplex would remain uncorrected and, after the first cell division, would give rise to one *LEU2* and one *leu2* cell that grew up into the sectored colony. A similar result was obtained when the target *leu2* gene contained not only the point mutations but also a 1.7 kb insertion of *ADE1* near one end. These experiments argue that one way of carrying out gene replacements involves single-strand assimilation, long hetero-duplex DNA formation and mismatch repair, which, however, usually operates to defeat the replacement of the chromosomal sequences.

A similar model for strand assimilation during transformation has been invoked by George Church's lab in studying gene targeting by "recom-bineering" in *E. coli*, as discussed in more detail in Section 10.10.

Independent strand invasions

Lorraine Symington's lab confirmed that such single-strand assimilations occurred, but they argued that most gene targeting events actually involved two independent strand invasions at the two ends. They placed 26-bp palindromes in the arms of the targeting DNA which replaced the *TRP1* gene with another selectable marker; these palindromes escape mismatch repair surveillance, so that heteroduplex DNA should be retained. If there was no repair of the central marker then only one strand of the heteroduplex would be able to grow into a colony. More than two-thirds of Trp$^+$ transformants grew up into colonies in which either one or both ends were heteroduplex for the palindromic and chromosomal markers: evidence of heteroduplex DNA. The key finding was that when heteroduplex was detected on both arms, an analysis of individual cells showed that the arrangement of the palindromes was nearly always in *trans*, the arrangement expected for the invasion of two 3'-ended regions into the target chromosome (**Figure 10.20**). In the absence of the mismatch repair gene *PMS1*, very few transformants showed sectoring for either palindromic site, as expected if only one strand, carrying *TRP1*, could replicate and grow into a Trp$^+$ colony. Moreover, most transformants now had the flanking palindromic insertions arranged in *cis*, as expected if only one strand had been assimilated (Figure 10.20 (iii)). Similar roles were found for deletions of Msh2, Msh3, Rad1, or Rad10, suggesting that they could be involved together, as they are in clipping of nonhomologous 3' tails in both strand invasion and SSA, and in correcting large insertion/deletions in heteroduplex DNA. In their absence, assimilation of one strand predominates (Figure 10.20iii). But the actual sequence of events may be more complex: neither nonhomologous tail clipping nor correction of large insertion/deletions in heteroduplex DNA depends on Pms1, and yet this mismatch repair protein seems also to be required to prevent assimilation of a single strand. Moreover the outcomes were different when the central transformation event involved only replacing a small mutation of the *MET17* gene with wild-type sequences. Here, the absence of Rad1 nearly eliminated sectoring of either palindromic marker site, but now there was a dramatic bias in recovering only the palindromic marker (91%) as if repair of the *met17-sna* mutation to wild type affected the repair of heteroduplex on both sides. But in the rare cases where there was heteroduplex on both sides, the palindromes were found to be in *trans* configuration. In contrast, *msh2* and *msh3*

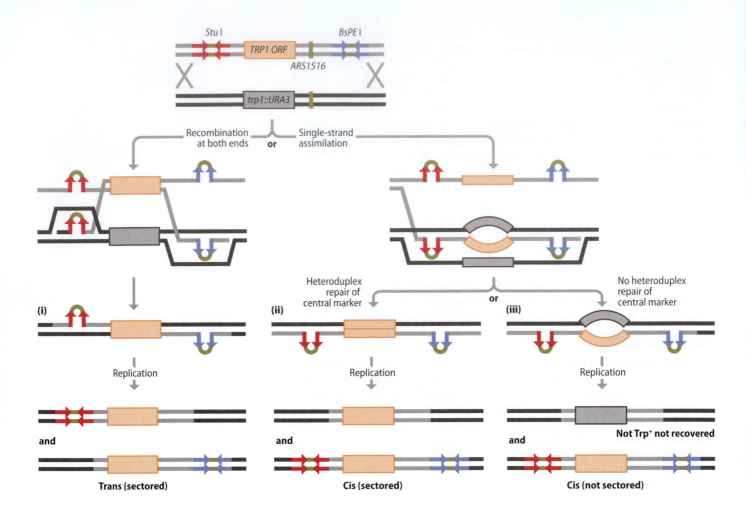

Figure 10.20 Analysis of DNA strand inheritance during ends-out transformation.
A transforming fragment has two short palindromic sequences that resist mismatch correction when they are in heteroduplex with an unaltered sequence. Replacement of the *URA3* marker inserted at the *TRP1* locus by *TRP1* involves removal of a large heterology. Three possible outcomes are shown. (i) Both ends of the fragment establish a crossover structure, replacing both strands of *trp1::URA3* with *TRP1*. After replication, a Trp⁺ colony should be sectored with the palindromic markers on the opposite halves of the colony, that is, in *trans*. (ii) Only one strand of DNA is assimilated and then the heteroduplex is corrected by mismatch repair enzymes, leaving the two palindromic markers in *cis*, but in a sectored colony where the opposite half has no palindromic markers. (iii) One strand is assimilated but there is no mismatch correction of the *URA3* vs *TRP1* heteroduplex. Consequently after replication only one of the two molecules will contain the *TRP1* marker and the colony will be uniformly *cis*. (Adapted from Langston LD & Symington LS [2004] *Proc Natl Acad Sci USA* 101:15392–15397. With permission from the National Academy of Sciences.)

mutations (which should affect both mismatch repair and branched intermediate clipping) both reduced the formation of hDNA on both sides and showed a *cis* configuration when there was hDNA on both sides. So there must be several interrelated correction processes going on in these transformation events.

10.9 ENDS-IN TARGETING IS MUCH MORE EFFICIENT THAN ENDS-OUT TARGETING

No matter what the exact mechanism of gene targeting is, it seems to be different from ends-in events that integrate a linear molecule at a region of homology, even if the initial strand invasion structures are the same (**Figure 10.21**). When these two modes of transformation are compared using similar homologous sequences, there are about three times as many integrations by ends-in recombination than ends-out. This difference may be an underestimation, because not all—in fact only a minority—of ends-in gene conversions are accompanied by crossing over that is required to be scored as an integration event. Experiments using a linearized plasmid with an ARS sequence show that only about 10–15% of transformants integrate the ends-in fragment; the rest have repaired the DSB on the autonomously replicating plasmid. So ends-in recombination seems to be on the order of 10–30 times more efficient than ends-out events. When the ends point inward, the repair event is limited to copying sequences in between; when they face outward, replication could go on without resulting in a double crossover that would yield a stable outcome. One indication that undaunted BIR would continue in an ends-out situation comes from the studies of transformation of *E. coli* by Patrick Dabert and Gerry Smith, who argue that ends-out fragments may replicate the entire circular 5 Mb *E. coli* chromosome!

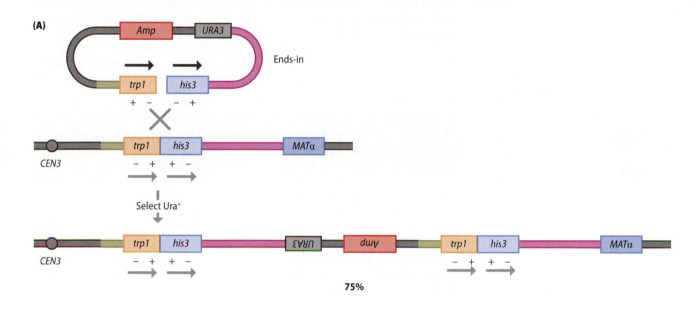

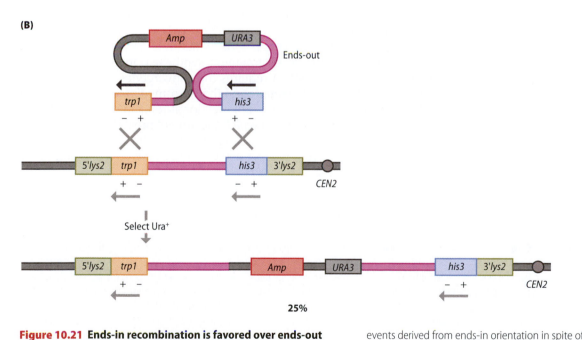

Figure 10.21 Ends-in recombination is favored over ends-out integration. A budding yeast strain was constructed with two targets for plasmid pGM54 which was linearized between the *trp1* and *his3* alleles. One target can be transformed in the ends-in configuration and will result in the integration of the *URA3* and other sequences (A). In a different chromosome location the same *his3* and *trp1* alleles are arranged in a different orientation that will allow ends-out recombination (B). Arrows indicate the orientation of transcription and the relative position of *his3* and *trp1* alleles are given by + and − symbols. As shown, 75% of the recovered events derived from ends-in orientation in spite of the fact that many such events could have repaired the DSB without an associated crossover and therefore would be lost, as the plasmid lacks an origin of replication. About 15% of the integration events also were accompanied by *HIS3* or *TRP1* prototroph formation by recombination between the heteroallelic homologous flanking sequences. This value did not change, whether the substrates were ends-in or ends-out. (Adapted from Hastings PJ, McGill C, Shafer B & Strathern JN [1993] *Genetics* 135:973–980. With permission from the Genetics Society of America.)

10.10 GENE KNOCKOUTS AND GENE MODIFICATION ARE EFFICIENT IN MODIFIED BACTERIA

Although transformation of bacterial fragments provided the basis for showing that it was DNA that encoded genes, it turned out to be less simple than expected to get designed gene knockouts. Bacteria such as *E. coli* are armed with strong defenses against incoming fragments of DNA, including the very aggressive RecBCD nuclease that rapidly degrades linear DNA to small fragments. Dabert and Smith overcame this limitation by placing properly oriented Chi (GCTGGTGG) sequences near the ends of the transforming DNA fragment. As discussed in Chapter 4, Chi sequences inhibit the rampant nuclease activity of RecBCD and convert it into an enzyme that creates 3′ ssDNA ends that can engage in strand invasion; consequently the entire fragment can be integrated with greatly improved efficiency.

A powerful approach to modify sequences in *E. coli*, termed "*recombineering*", has been developed by Don Court, who took advantage of three recombination genes encoded by phage λ that can carry out recombination independently of the recombination machinery of *E. coli*. Redα is a 5′-to-3′ exonuclease; Redβ is an annealing protein that also protects ssDNA; Redγ is an inhibitor of the nuclease activity of RecBCD of *E. coli*. The expression of these three proteins makes it possible to transform *E. coli* with DNA with homologies as short as 30–40 bp at each end and to design knockout constructs for all the nonessential genes. Site-directed mutants can also be introduced by recombineering, using ssDNA oligonucleotides of 20–70 nt. Recombineering has proven to be a powerful technique to modify mammalian genes that can then be recovered and introduced into the mammalian system.

Court and Anthony Poteete have each proposed mechanisms in which a dsDNA fragment with resected 3′ ssDNA ends anneals to the exposed lagging strand at the replication fork (which has more single-stranded character because of Okazaki fragment synthesis). In Court's proposal, the assimilation of the ssDNA requires the convergence of a second replication fork. However, Poteete found that Red-mediated transformation could occur in a plasmid with a unidirectional replication and thus proposed a modified mechanism in which the annealing of the ssDNA region led to its use as the primer for leading-strand synthesis. A second invasion and replication event would need to occur at the opposite end of the fragment (**Figure 10.22**).

Figure 10.22 Recombineering in *E. coli*. (A) The introduction of DNA sequences by bacterial recombineering can be carried out with linear DNA fragments bound to the Exo and Beta proteins of bacteriophage phage λ, comprising the Red system of recombination. Two models are shown, by Donald Court (B) and Anthony Poteete (C), for the efficient assimilation of a single strand at a replicating region in bacteria, catalyzed by the phage λ Red system. Phage λ Beta protein binds to ssDNA ends and facilitates base pairing with a complementary sequence which is exposed during DNA replication. In both mechanisms the recombination intermediate is proposed to be a dsDNA core flanked on either side by 39 nt ssDNA overhangs. The Court mechanism (B) posits that the λ Beta protein facilitates annealing of one 39 nt overhang to the lagging strand of the replication fork (i). This replication fork then stalls and backtracks so that the leading strand can template switch onto the synthetic dsDNA (ii). The heterologous dsDNA blocks further replication from this fork. Once the second replication fork reaches the stalled fork, the other 39 nt end of the integration cassette is annealed to the lagging strand in the same manner as before (iii). Finally, the crossover junctions must be resolved by unspecified *E. coli* enzymes. The Poteete mechanism (C) suggests that λ Beta facilitates ssDNA overhang annealing to the lagging strand of the replication fork (iv) and positions the invading strand to serve as the new template for leading-strand synthesis (v). This structure is resolved by an unspecified host endonuclease (*red triangle*), after which the synthetic dsDNA becomes template for both lagging and leading-strand synthesis (vi). A second template switch must then occur at the other end of the synthetic dsDNA. (Adapted from Mosberg JA, Lajoie MJ & Church GM [2010] *Genetics* 186:791–799, from Court DL, Sawitzke JA & Thomason LC [2002] *Annu Rev Genet* 36:361–388, and from Poteete AR [2008] *Mol Microbiol* 68:66–74. With permission from the Genetics Society of America, from Annual Reviews and from John Wiley & Sons, Inc., respectively.)

An alternative view of the mechanism of Red-mediated recombineering has been suggested by Church. Here, the strand complementary to the ssDNA exposed on the lagging strand is fully assimilated to form a duplex region containing mismatches or heterologies. Supporting this idea were data showing that if one used transforming fragments with mismatches in homologous regions, there was a strong bias to incorporating the mismatches on the strand complementary to the lagging strand (**Figure 10.23**). Moreover, exploiting the observation that phosphorothioate bonds impair

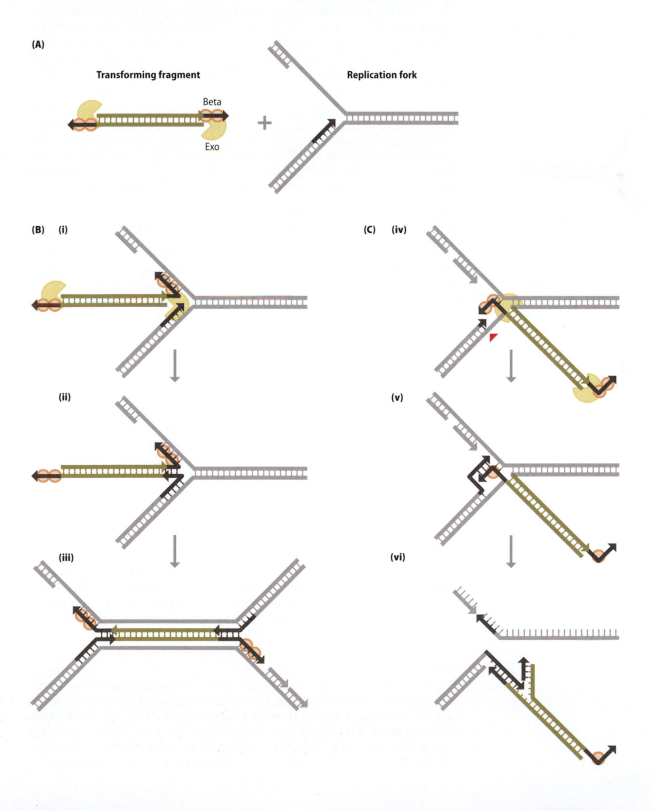

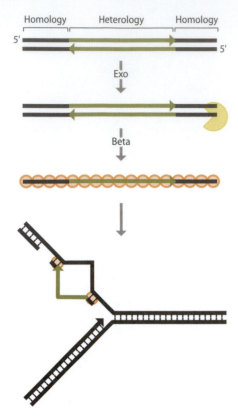

Figure 10.23 A single-strand assimilation model to account for bacterial recombineering. Phage λ Exo removes one strand, which is then annealed by the Beta protein at a ssDNA region exposed at a replication fork. (Adapted from model proposed by Mosberg JA, Lajoie MJ & Church GM [2010] *Genetics* 186:791–799. With permission from the Genetics Society of America.)

the ability to many exonucleases from degrading DNA ends, Church's lab showed that blocking degradation of the strand complementary to the lagging-strand template increased targeting efficiency tenfold, while blocking the opposite strand had no effect. This single-strand assimilation mechanism is reminiscent of that observed in budding yeast. Also similar to what was seen in yeast, blocking mismatch repair (MMR) can increase the efficiency of recombineering dramatically, up to hundredfold, consistent with the assumption that mismatch repair will preferentially convert a mismatch on the invading/assimilated strand in favor of the unbroken template. One way to accomplish this has been to use oligos containing chemically modified bases (2′-fluoro-uridine, 5-methyl-deoxy-cytidine, 2,6-diaminopurine or isodeoxyguanosine) in place of T, C, A, or G to avoid mismatch detection and repair. Without MMR, the efficiency of transformation can be as high as 25% of the cells, so that there is no need for the use of a selective marker to find colonies with modified genomes.

As in budding yeast, Red-mediated recombination in *E. coli* is versatile. It is possible to introduce very large deletions into the genome by providing homology at the two ends to distant parts of the genome. The efficiency of making deletions is roughly constant for a few base pairs up to several hundred base pairs but then falls off with size; but deletions of several kilobases still occur at about 1% of the rate for small deletions.

Church's lab has also developed high-throughput multiplexed systems to globally alter the *E. coli* genome. One project—simply unimaginable a decade or less ago—is to replace all the 314 TAG translation termination triplets in the entire genome with TAA stop signals so that there would be an available codon, which could be used to introduce novel amino acids into proteins. It is this kind of remarkable engineering that makes the idea of changing all the 400,000 codons that distinguish an elephant from a mammoth almost within the realm of science and out of science-fiction.

10.11 GENE TARGETING STRATEGIES CAN BE ADAPTED TO OTHER ORGANISMS

Ends-out transformation has proven to be successful, to varying degrees, in only some of the well-studied model systems. For example, in *Schizosaccharomyces pombe*, a larger fraction of transformants integrate the selectable marker nonspecifically, but accurate targeting occurs sufficiently often that it has been possible to make a library of fission yeast with the ablation of each nonessential gene. Some technical roadblocks have made it difficult to get rapid, efficient induction of I-SceI or HO nucleases, so that these approaches have not yet been well exploited to improve targeting.

In some other fungi, the ratio of accurate to inaccurate gene targeting can be improved by knocking out components of nonhomologous end-joining, which seems to compete with desired targeting both by creating end-to-end concatomers of the linearized fragment and by facilitating the insertion of the linear fragment at DSBs that have arisen during replication (or by the shocks of transformation).

Among vertebrate cells, the DT40 chicken cell line, derived from an immature immune cell, is almost as facile as budding yeast in allowing accurate gene targeting. The laboratory of Shunichi Takeda has been able to create cell lines homozygous for deletions of two (or more) genes, altogether requiring a minimum of four separate knockout events. As with budding yeast, it is possible to create deletions of at least 50 kb by using a linear fragment with two distant regions of homology flanking a selectable marker. The relative ease with which gene alterations can be made in DT40 has even led to a strategy of introducing a human chromosome into DT40, modifying it, and transfecting it back into a mammalian cell.

Gene modification in Drosophila

To get ends-out transformation to work in *Drosophila* required a series of complicated tricks, because simple injection of linearized DNA into the germ line was not successful (although this technique allows the introduction of modified P-element transposons that integrate at random). Yikang Rong and Kent Golic accomplished both ends-in and ends-out gene targeting by taking advantage of several different tricks. The strategy involved the use of both the site-specific FLP recombinase and the meganuclease I-SceI (**Figure 10.24**). For ends-in integration of a gene at a specific locus, a P element is constructed carrying a pair of directly oriented FRT sites, within which is the target gene (*pug*) containing an I-SceI cleavage site in an intron. The P element is inserted at random in the chromosome and then crossed with a parent carrying two other P-element constructs, expressing FLP and the I-SceI endonuclease. Expression of FLP results in the popping-out of the gene-targeting region from the P element as a circular fragment with a single FRT site. Then, I-SceI-mediated cleavage of the circular DNA in *pug* creates a linear fragment that can recombine in an ends-in fashion to disrupt the *pug* gene. So, although directly trying to inject linearized DNA to get targeted integration does not work, it seems that homologous recombination is relatively efficient in flies once a chromatin fragment is liberated in the nucleus.

In a similar fashion it is possible to disrupt a gene by ends-out recombination, by changing the arrangement of the homologous segments in the P element. The 5' and 3' segments are each inverted relative to their normal gene orientation. FLP again produces a circle with a single FRT site and I-SceI linearizes this ring to produce an ends-out recombining fragment.

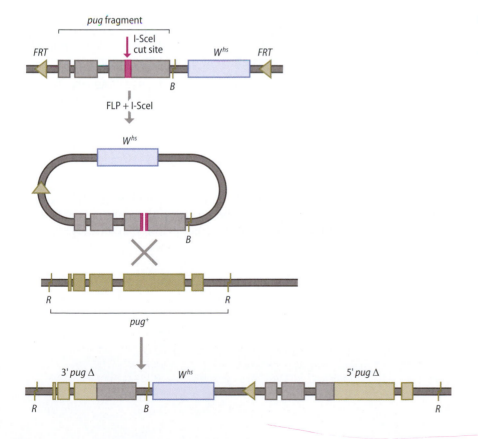

Figure 10.24 Gene targeting in *Drosophila*. A modified P element is created, carrying a segment homologous to a chosen target and harboring an I-SceI site as well as a pair of flanking FRT sites. After the P element integrates at random and becomes part of a *Drosophila* chromosome, expression of FLP excises the P element from the genome as a circular molecule. I-SceI expression then creates (in this case) an ends-in transformation event, disrupting the target gene by creating 3' and 5' deletions of the target gene. Variations on this approach can produce ends-out fragments for gene replacement. (Adapted From Rong YS & Golic KG [2000] *Science* 288:2013–2018. With permission from the American Association for the Advancement of Science.)

Variations on this theme allow the introduction of a mutant copy of the targeted gene, as has been previously shown for budding yeast (see Figure 10.2B). Ends-in recombination between an *in-vitro*-mutated targeting sequence and the chromosomal site can introduce a mutation in one of the duplicated copies. When the mutation was within 250 bp of the I-SceI-induced DSB, the mutation was apparently lost during recombination. Rong and Golic suggest this might be because the DSB is enlarged by exonuclease digestion to a gap, so that the mutation will be chewed away and gap repair will put wild-type sequences in the supplicated copy. However, it is also possible that the region is simply single-stranded and that mismatch repair, acting on heteroduplex DNA, preferentially restores the incoming mutation to the genotype of the unbroken template strand, as appears to be the case in yeast. However, mutations 1 kb or more from the DSB appear to be frequently retained and one ends up with a gene duplication with one copy carrying the mutation. Rong and Golic then took another page from the yeast playbook and induced a different DSB made by another meganuclease, I-CreI, at a site between the duplicated segments. This DSB promotes deletion by SSA, leaving the chromosome with a single copy of the gene, often containing the mutation. For small enough genes, the yeast trick of inserting a deletion of promoter and/ or the 5′ end would leave the cell with only a single copy that can be expressed and which could carry the *in-vitro*-generated mutation.

Very recently Dana Carroll's lab has returned to the question of why simply directing an endonuclease cleavage and introducing a DNA fragment in *Drosophila* embryos is so refractory to promoting gene targeting. They confirmed that adding a ZFN and a homologous, linear dsDNA fragment was indeed unsuccessful, but much to their delight, gene targeting was successful when they introduced single-stranded DNA oligonucleotides. When DNA is injected, it is also possible to get gene conversion at the site of nuclease-induced DSB using circular templates carrying homology.

Plants have lagged behind in gene modification

Simple transformation approaches have generally failed in plants. Neither tobacco nor *Arabidopsis* has been modified by simple ends-out transformation. The one quite remarkable exception is the moss *Physcomitrella patens*, where gene targeting is as efficient as budding yeast.

At present the way most plants are modified is by the use of *Agrobacterium* to genetically transform plants by transferring a region of plasmid DNA, T-DNA, into host plant cells. A combination of host proteins and bacterial virulence effector proteins enable the DNA to traverse the plant cytoplasm, enter the nucleus, and eventually allow integration of T-DNA into the plant genome, permitting stable expression of T-DNA-encoded transgenes.

Tobacco and presumably other plants can be modified by creating NHEJ-mediated gene knockouts by using the strategy of creating a DSB at a specific site on the chromosome using a customized meganuclease.

SUMMARY

The ability to accurately target gene sequences has revolutionized the study of gene regulation and cell biology. There is still much to learn about how gene targeting actually occurs and how to achieve efficient gene modifications in many model organisms. The development of site-specific endonucleases has made it possible to make modifications in mammalian cells at almost yeast-like levels.

SUGGESTED READING

Capecchi MR (2005) Gene targeting in mice: functional analysis of the mammalian genome for the twenty-first century. *Nat Rev Genet* 6:507–512.

Carroll D (2011) Genome engineering with zinc-finger nucleases. *Genetics* 188:773–782.

Erdeniz N, Mortensen UH & Rothstein R (1997) Cloning-free PCR-based allele replacement methods. *Genome Res* 7:1174–1183.

Hastings PJ, McGill C, Shafer B & Strathern JN (1993) Ends-in vs. ends-out recombination in yeast. *Genetics* 135:973–980.

Maggert KA, Gong WJ & Golic KG (2008) Methods for homologous recombination in *Drosophila*. *Meth Mol Biol* 420:155–174.

Mali P, Yang L, Esvelt KM et al. (2013) RNA-guided human genome engineering via Cas9. *Science* 339:823–826.

Mortensen R (2006) Overview of gene targeting by homologous recombination. *Curr Protoc Mol Biol* 23.1.1–23.2.12.

Munoz IG, Prieto J, Subramanian S et al. (2011) Molecular basis of engineered meganuclease targeting of the endogenous human RAG1 locus. *Nucleic Acids Res* 39:729–743.

Rothstein R (1991) Targeting, disruption, replacement, and allele rescue: integrative DNA transformation in yeast. *Meth Enzymol* 194:281–301.

Shortle D, Haber JE & Botstein D (1982) Lethal disruption of the yeast actin gene by integrative DNA transformation. *Science* 217:371–373.

Swingle B, Markel E, Costantino N et al. (2010) Oligonucleotide recombination in Gram-negative bacteria. *Mol Microbiol* 75:138–148.

Wang HH & Church GM (2011) Multiplexed genome engineering and genotyping methods applications for synthetic biology and metabolic engineering. *Meth Enzymol* 498:409–426.

CHAPTER 11
SITE-SPECIFIC RECOMBINATION

A powerful way to change the genome uses enzymes that recognize pairs of specific DNA sequences and catalyze a reciprocal crossover between them. These site-specific exchanges have become a well-used part of the arsenal to manipulate eukaryotic genomes. In this brief chapter, we will look at three examples that have had particular utility: the Cre and FLP recombinases and the pC31 integrase. But we will begin with the first well-studied example that illustrates a remarkable bit of biological regulation.

11.1 PHASE VARIATION IN *SALMONELLA* DEPENDS ON A SITE-SPECIFIC RECOMBINASE

One strategy for a parasite to proliferate in a host is to avoid or outwit immune surveillance. The most stunning example of this phenomenon is the antigenic variation displayed by the trypanosomes, which cause sleeping sickness, and by plasmodia, which cause malaria. In this section, we will look at a simpler, but elegant, system of antigenic variation, involving two alternative subunits of the cell's flagella (called flagellin), the major surface protein of the bacterium *Salmonella typhimurium*.

At one site on the *Salmonella* chromosome are the co-transcribed genes for the H2 flagellin protein and for the RH1 repressor of the alternative H1 flagellin gene, which is located elsewhere on the chromosome (**Figure 11.1**). Adjacent to the H2 and RH1 gene open reading frames (ORFs) is their promoter, which is situated between two short, inverted, repeated sequences, named *hix* sites. In addition to the H2 and RH1 promoter, this region carries the gene for another protein, the H-region invertase (Hin). Hin is expressed at very low levels so that it only acts in a small fraction of cells. The Hin enzyme can pair up the two inverted *hix* regions and catalyze a reciprocal crossover between them, such that the region in between becomes inverted. Consequently, the promoter that drives expression of the H2 and RH1 genes is pointed in the wrong direction and these genes cannot be transcribed. Lacking the RH1 repressor protein, the distant H1 gene is expressed and cells that were being attacked by anti-H2 antibodies made by the infected host escape until the organism can mount a response to this new antigen. The Hin gene continues to be expressed at a low level and thus the control region can again be inverted, restoring the expression of H2 and causing the repression of H1. Thus *Salmonella* can occasionally alternate back and forth between expressing H2 and expressing H1. This strategy enables the bacterium to escape the host's immune response to the presence of the original

Figure 11.1 Phase variation in *Salmonella*.
(A) Hin recombinase causes the inversion of the promoter adjacent to the H2 flagellin and Repressor of H1 (RH1) genes at a pair of *hix* sites. Inversion prevents H2 and RH1 expression. The absence of RH1 allows expression of the alternative flagellin H1. Hin expression is quite low and thus most descendants of cells expressing H2 continue to express H2, but occasionally a cell switches to express H1, which continues to be expressed by most of that cell's offspring, but once in a while a cell will switch back to express H2. (B) The *hix* sites are imperfect 13 bp inverted repeats, whose consensus sequence is shown below the two sites. (C) A consensus sequence derived from comparing the *hix*$_L$ and *hix*$_R$ sites with those of three other bacterial site-specific inversion systems known as Gin (G region inversion), Pin (P region inversion), and Cin (C region inversion). (Adapted from Johnson RC, Bruist MB, Glaccum MB & Simon MI [1984] *Cold Spring Harb Symp Quant Biol* 49:751–760. With permission from Cold Spring Harbor Laboratory Press.)

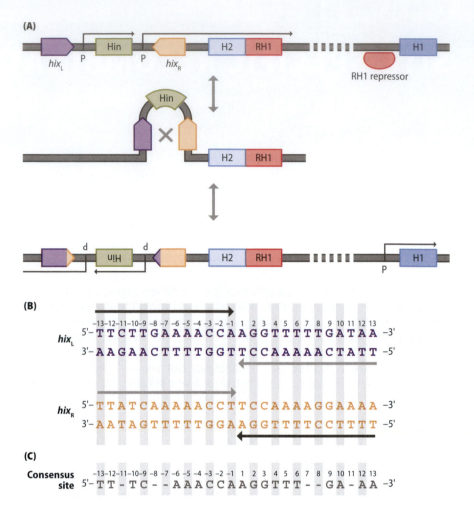

(B)

hix$_L$
5'- T T C T T G A A A A C C A A G G T T T T T G A T A A -3'
3'- A A G A A C T T T T G G T T C C A A A A A C T A T T -5'

hix$_R$
5'- T T A T C A A A A A C C T T C C A A A A G G A A A A -3'
3'- A A T A G T T T T T G G A A G G T T T T C C T T T T -5'

(C)

Consensus site
5'- T T - T C - - A A A C C A A G G T T T - - G A - A A -3'

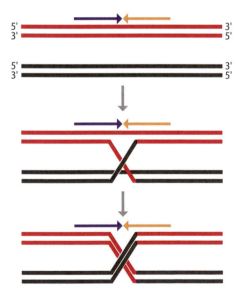

Figure 11.2 Mechanism of crossing over in site-specific recombination. Sequential strand exchange via a Holliday junction (HJ) intermediate leads to a crossover in the *hix* sites. Two strands of the same polarity are cleaved and the 5′ cut end of one strand is joined to the 3′ end of the other, producing a HJ. A similar exchange of the remaining two strands results in a crossover.

flagellar antigen and prolong its infection. There are a number of related site-specific inversion systems among bacteriophage; some of these "phase variations" affect the range of organisms that can be infected and some are designed to escape the host's defenses.

The *hix* sites are 26 bp, made up of a pair of inverted 13-bp repeats. The Hin enzyme works as a tetramer that binds to the 13-bp repeats and carries out two sequential break-and-join operations on two single-strands of the same polarity (**Figure 11.2**). The first exchange creates a Holliday junction (HJ), but instead of allowing the HJ to be cleaved by an HJ resolvase, the Hin enzyme goes on to break and join the other two strands, thus completing a crossover by a different mechanism. Each strand crossover involves the covalent formation of a bond between the 5′ end of the break-point to a tyrosine of the Hin enzyme, followed by a rotation and re-ligation to the opposite cut end. In the end, the two *hix* sites have undergone a reciprocal exchange and the region in between is inverted.

One interesting feature of the Hin enzyme is that it does not efficiently catalyze exchanges between *hix* sites that are located on different chromosomes, or even if the two sites are in direct orientation on the same DNA. Hin requires a highly orchestrated wrapping of DNA strands and the energy derived from DNA supercoiling to carry out the inversion (**Figure 11.3**). The reaction is facilitated by a "recombination enhancer" in the inverted segment to which Fis proteins can bind. However, there are other, evolutionarily related site-specific recombinases that do not have this constraint and can therefore catalyze crossovers not only to

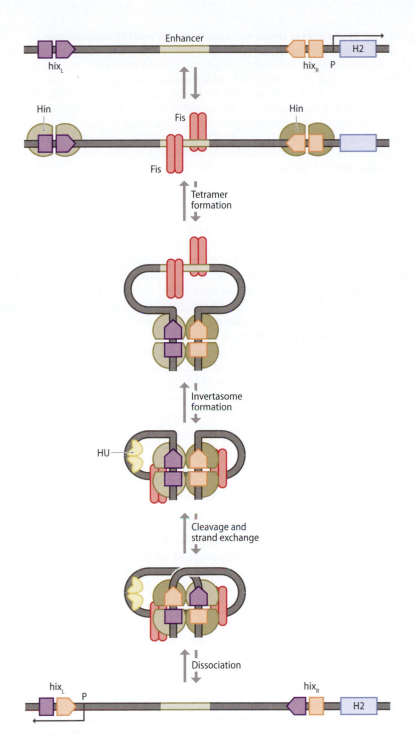

Figure 11.3 Folding of the chromosome facilitates Hin-mediated recombination. Efficient action of Hin recombinase depends on an enhancer DNA sequence that binds copies of the Fis protein that assists in creating a tetramer of two Hin dimers and creating the proper topology for Hin to catalyze strand exchange. Strand exchange involves a rearrangement of the subunits of the Hin enzyme. Another protein, HU, assists in this folding. (Adapted from Nanassy OZ & Hughes KT [1998] *Genetics* 149:1649–1663. With permission from the Genetics Society of America.)

create inversions but also to produce intrachromosomal deletions as well as reciprocal translocations between chromosomes. These enzymes include the functionally similar Cre and FLP recombinases.

11.2 Cre RECOMBINASE RECOMBINES AT A PAIR OF *LOX* SITES

Cre recombinase is a versatile enzyme that promotes both intra- and interchromosomal recombination. Its target sites are called *loxP* sites, which—like *hix* sites—have 13-bp inverted repeats but here they are

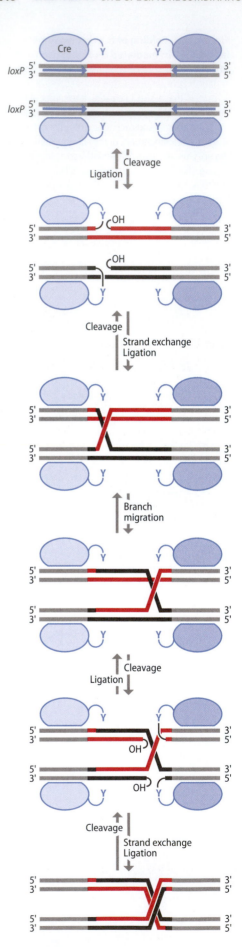

Figure 11.4 Mechanism of strand exchange in the Cre and FLP recombinases.
The Cre enzyme, acting on *loxP* sites, is shown. Each site contains an inverted pair of 13-bp repeats separated by an 8-bp spacer. The recombinase subunits bind to each of the 13-bp repeats. Initially two subunits (*light blue*) cleave strands of the same polarity and form covalent linkages with a catalytic tyrosine (Y) residue. A rotation of the complex occurs and ligation follows after strand exchange. Branch migration through the 8-bp region then facilitates the second, similar strand exchange involving the second pair of strands, leading to a crossover. An analogous mechanism occurs in FLP-mediated site-specific recombination between *FRT* sites. (Adapted from Jayaram M, Crain KL, Parsons RL & Harshey RM [1988] *Proc Natl Acad Sci USA* 85:7902–7906. With permission from the National Academy of Sciences.)

separated by 8 bp of DNA. Cre acts as a dimer, with each subunit bound to a *loxP* site. As with Hin-*hix*, Cre-*lox* recombination involves a sequential break-and-join mechanism between two strands of the same polarity (**Figure 11.4**). The first exchange occurs at one end of the 8-bp spacer. The resulting HJ then migrates across these 8 bp, after which the second exchange takes place between the remaining pair of strands. Each sequential reaction involves cleavage that results in a transient covalent bond between a phosphate at the "crossing point" of the recognition sequence and a tyrosine of the Cre enzyme. This bond is then reversed by ligating the DNA at the second site. Detailed X-ray crystallography and biochemical studies have revealed that the two recombining sites are complexed in opposite orientation and that the exchange reaction involves not only an exchange of Cre subunits, but an isomerization of the DNA, as shown in **Figure 11.5**.

11.3 Cre-MEDIATED RECOMBINATION CAN BE USED TO CREATE GENOME MODIFICATIONS

The most potent use of Cre-*lox* recombination has been to create conditional knockout alleles of genes in higher eukaryotes. A gene can tolerate the insertion of *lox* sites if they are located within introns. To accomplish the conditional inactivation of a gene, *lox* sites must be introduced on either side of one or more exons by gene targeting (**Figure 11.6**). This creates what is often called a "floxed" allele.

Cre-mediated gene ablations can be created in the germ line, so that all cells in the resulting offspring lack gene activity; but expression of Cre can be temporally regulated in a variety of ways to enable researchers to knock out gene function at a particular developmental stage or in specific tissues. Cre can be introduced by viral vectors, by placing the Cre gene ORF under the control of an inducible promoter, such as one regulated by tetracycline or one that is expressed only in certain tissues. Cre can also be regulated indirectly, by placing it downstream of an upstream-activating sequence (UAS$_{GAL}$) and expressing the Gal4 regulatory protein that is itself expressed under the control of a tissue-specific or developmentally specific promoter. In this way a gene that may be essential in early embryonic development can be inactivated in specific tissues of an adult.

Although the insertion of the small *lox* site into an intron is not problematic, it turned out that the presence of the selectable gene marker used to introduce the *lox* sites often interferes with normal gene expression. Consequently the selectable marker itself is usually flanked by its own site-specific recombination sites—usually the *FRT* sites of the *FLP recombinase* that has very similar properties to Cre, but with different sequence specificity. After the *loxP* sites have been introduced, expression of the FLP recombinase removes the selectable gene, as shown in **Figure 11.7**.

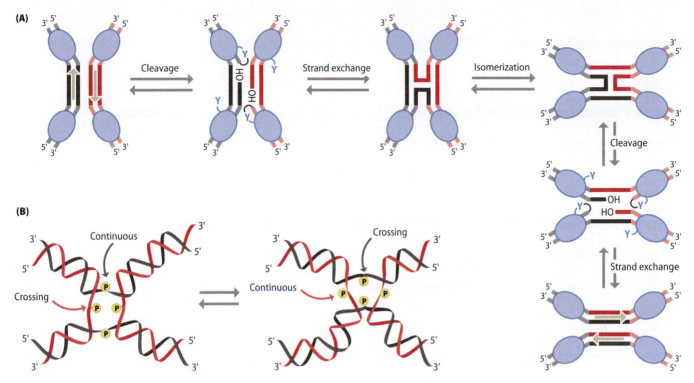

Figure 11.5 A more detailed view of Cre-mediated site-specific recombination. (A) Each *lox* site is bound by a dimer of Cre. Prior to exchange, the two *lox* sites are arranged in opposite orientation as shown by the orientation of the unique 8-bp sequence between the 13-bp inverted repeats. An exchange of two strands via a tyrosine (Y)-bound protein intermediate results in a Holliday junction (HJ) that undergoes branch migration and isomerization to facilitate the crossing over of the second pair of strands. Strand exchange is accompanied by an exchange of subunits. (B) X-ray crystallography of intermediates of Cre recombination shows that the phosphodiester bonds of the dark strands undergoing the first exchange are equivalent to the positions of the phosphates in the second pair of strands after isomerization. (Adapted from Gopaul DN, Guo F & Van Duyne GD [1998] *EMBO J* 17:4175–4187. With permission from Macmillan Publishers, Ltd.)

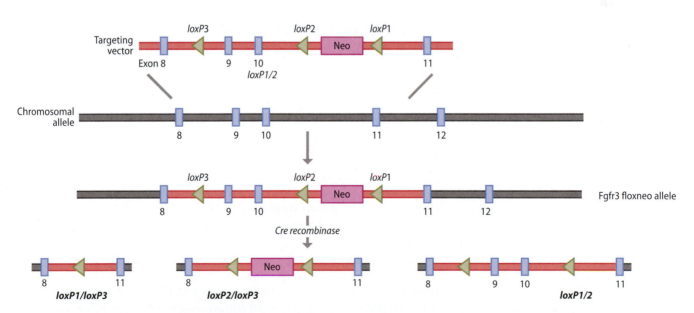

Figure 11.6 Generation of conditional knockout floxed allele.
A cloned segment of the Fgfr3 gene was modified by the insertion of three *lox* sites and a selectable Neo marker into two introns between exons 8 and 11 and introduced into a mouse chromosome. When a mouse carrying this floxed allele was crossed to a mouse expressing Cre from the EIIa promoter, which is expressed in the germ line, offspring were recovered that had deletions between some or all of the *lox* sites. Deletions between *lox* sites 1 and 3 or between 2 and 3 remove exons 9 and 10 whereas a deletion between sites 1 and 2 only removes the Neo marker. (Adapted from Su N, Xu X, Li C et al. [2010] *Int J Biol Sci* 6:327–332. With permission from Ivyspring International Publisher.)

(A)

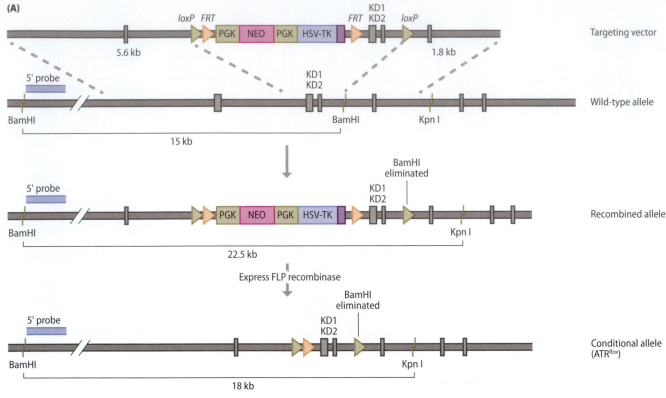

(B)

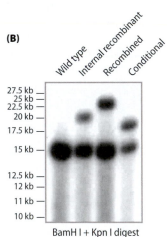

BamH I + Kpn I digest

Figure 11.7 Example of creating a floxed allele and removing a selectable marker by combining Cre- and FLP-mediated recombination. (A) The essential ATR gene is modified in a mouse chromosome by inserting *lox* sites on either side of two exons encoding the essential kinase domain (KD1 and KD2). After creation of ES cells, the selectable *NEO* marker of the modified allele is removed by expression of FLP recombinase. The still-functional ATR^{flox} allele is then made inactive by expression of Cre to remove the KD1 and KD2 exons. (B) Southern blots of restriction endonuclease-digested DNA confirm the structures shown in A. Another transformant, not useful in the experiment, can be recovered in which recombination on the right side of the targeting sequence occurs within the KD1/KD2 region, failing to integrate the distal *loxP* site (internal recombinant.) (Adapted from Brown EJ & Baltimore D [2003] *Genes Dev* 17:615–628. With permission from Cold Spring Harbor Laboratory Press.)

Cre-*lox* recombination has also been adapted to allow the conditional expression of genes, by removing a block to transcription or translation. One example is shown in (**Figure 11.8**). In a similar way, Cre-mediated recombination can also be used to turn on or turn off RNA interference (**Figure 11.9**).

Cre-*lox* recombination can also generate chromosomal inversions, as well as translocations. One clever system selects reciprocal translocations in mice by using Cre expression to join two segments of the HPRT gene, with a single *lox* site in an intron (**Figure 11.10**). A remarkable use of site-specific recombination is in the generation of a "brainbow." Here multiple copies of different fluorescent proteins are variably turned on or off by expression of Cre recombinase, generating a rainbow of colors that can mark the different "wiring" of adjacent neurons in the mammalian brain (**Figure 11.11**).

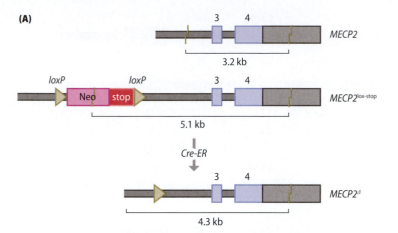

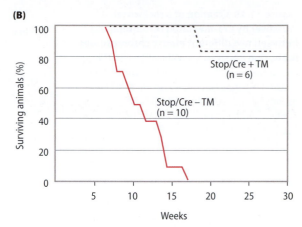

Figure 11.8 Turning on a gene in mouse by Cre-mediated excision of a block to protein expression. (A) An artificial "stop" cassette that impairs both transcription and translation was inserted between two *lox* sites into intron 2 of the MECP2 gene on the mouse X chromosome, creating a null mutation. A Cre protein fused to the estrogen receptor (ER) is constitutively expressed but Cre-ER cannot enter the nucleus until tamoxifen (TM) is injected into the mice. Once Cre is active in the nucleus, the stop cassette is deleted and MECP2 is expressed. (B) The remarkable result shown here is that the mortality in male mice lacking MECP2 expression is reversed by expressing MECP2 weeks after the animals were born. (Adapted from Guy J, Gan J, Selfridge J et al. [2007] *Science* 315:1143–1147. With permission from the American Association for the Advancement of Science.)

11.4 FLP RECOMBINASE EXCHANGES BETWEEN FRT SITES

Another widely used site-specific recombination system is based on the FLP recombinase of budding yeast that uses a pair of FRT sites (again, with 13-8-13 interrupted inverted symmetry) to cause site-specific exchanges. In budding yeast, FLP mediates an exchange between two inverted repeats on a naturally occurring nonessential plasmid, but it has been widely exploited both in *Drosophila* and in mammalian cells. FLP–FRT has been used extensively in *Drosophila* to cause deletions and other chromosomal rearrangements (**Figure 11.12**), but it has been especially useful

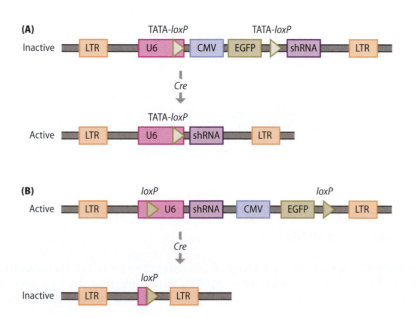

Figure 11.9 Site-specific recombination can turn on or off RNA interference. A retrovirus carrying a small hairpin RNA (shRNA) can be either turned on (A) or turned off (B) by Cre-mediated deletion of sequences. (Adapted from Ventura A, Meissner A, Dillon CP et al. [2004] *Proc Natl Acad Sci USA* 101:10380–10385. With permission from the National Academy of Sciences.)

Figure 11.10 Creation of a reciprocal crossover by Cre-mediated recombination between two different mouse chromosomes harboring *lox* sites. Recombination results in the formation of a functional HPRT gene at the junction of a translocation. (Adapted from Liu P, Jenkins NA & Copeland NG [2002] *Nat Genet* 30:66–72. With permission from Macmillan Publishers, Ltd.)

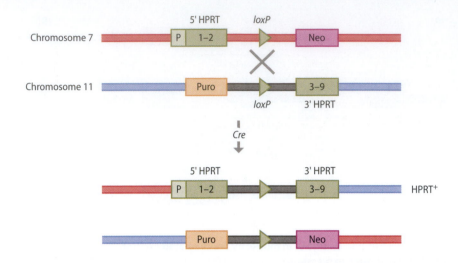

to promote a high level of mitotic crossing over and loss of heterozygosity (LOH). As we have discussed in Chapter 6 (for example, Figure 6.28), a crossover between homologous chromosomes in a diploid can produce a "twin-spot," in which an originally heterozygous genetic marker distal to the crossover point will become homozygous, one cell and its descendants being homozygous wild type and the other cell and its descendants being homozygous recessive. Historically, this was done by irradiating *Drosophila* larvae to generate such twin spots on the wing or other adult tissue, thus allowing researchers to examine the defect of an embryonically lethal mutant in adult tissue. This approach can be made far more efficient by placing FRT sites at the same position on homologous chromosomes and expressing FLP under control of a heat shock-inducible promoter or a tissue-specific promoter. An example is shown in **Figure 11.13**, in which one sector becomes a homozygous mutant (here used to test the neurophysiology of the eye) while the other half of the recombination event is eliminated by a homozygous lethal mutation. The generation of twin spots is also a powerful way to examine interactions between mutations, for example to create tissue homozygous for two mutations beginning with cells homozygous for one mutation and—on a different chromosome arm—heterozygous for a second mutation. FLP recombinase has also been exploited in *Drosophila* as part of a scheme to promote efficient gene targeting, as we discussed in Chapter 10.

Recently a modified FLP enzyme has been exploited to create site-specific DNA nicks in budding yeast. The FLP-H305L protein is blocked in the catalytic cycle so that the FLP enzyme binds to an FRT site and creates the initial cut, leaving the 3' end of the DSB attached to tyrosine 343, but it cannot complete the re-ligation step. Hence, the DNA has a nick with a protein-bound 3' end and a free 5' end (**Figure 11.14**). The arrival of a replication fork then results in the conversion of a nick into a DSB. This system should allow the study of recombinational repair of specific breaks by sister chromatid recombination.

11.5 ΦC31 INTEGRATION CAN TRANSFER LARGE CHROMOSOMAL SEGMENTS

Another site-specific recombinase, ΦC31 integrase (PhiC31), has become invaluable in another strategy to modify genomes, both in flies and in mammals. ΦC31 acts similarly to the well-studied integration of phage

(A)

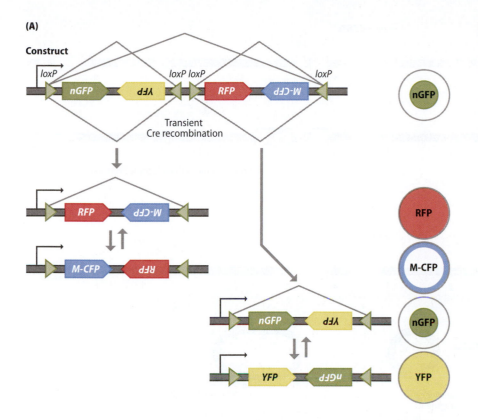

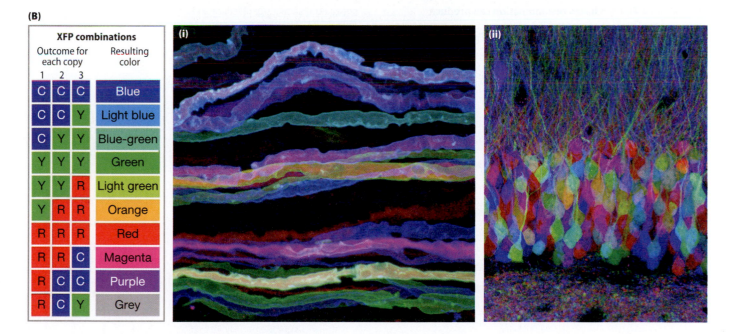

Figure 11.11 Cre-mediated recombination involving a series of fluorescent proteins can generate a "brainbow." (A) Brief Cre expression catalyzing deletions or inversions will create a series of different colors in adjacent cells by allowing expression of different combinations of four fluorescent proteins. Different fluorescent proteins are expressed either over the whole cell (RFP and YFP), in the nucleus (nGFP), or at the membrane (mCFP.) (B) The combinations for three different colors are shown. (i) Oculomotor axons. (ii) Dentate gyrus showing many different neurons. (Adapted from Livet J, Weissman TA, Kang H et al. [2007] *Nature* 450:56–62. With permission from Macmillan Publishers, Ltd.)

λ sequences by λ integrase, in which a circular phage λ molecule is integrated into the *E. coli* chromosome by a reciprocal crossing-over between an *attP* site on phage DNA and an *attB* site on the bacterial chromosome. The ΦC31 recombinase, derived from a different bacteriophage, can also promote integration of circular DNA sequences containing *attB* sequences irreversibly into an *attP* site located somewhere in the

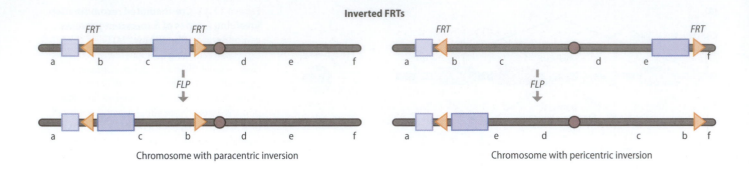

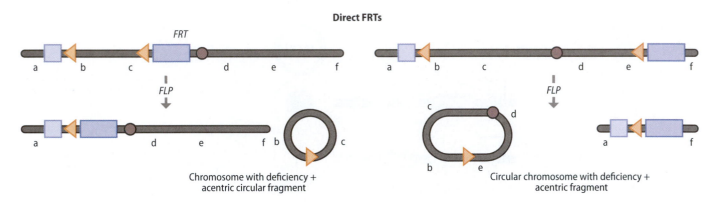

Figure 11.12 FLP-mediated recombination can produce chromosomal rearrangements. FRT sites in opposite orientation can create pericentric and paracentric inversions, while FRT sites in direct orientation can produce deletions and circular derivatives. (Adapted from Golic KG & Golic MM [1996] *Genetics* 144:1693–1711. With permission from the Genetics Society of America.)

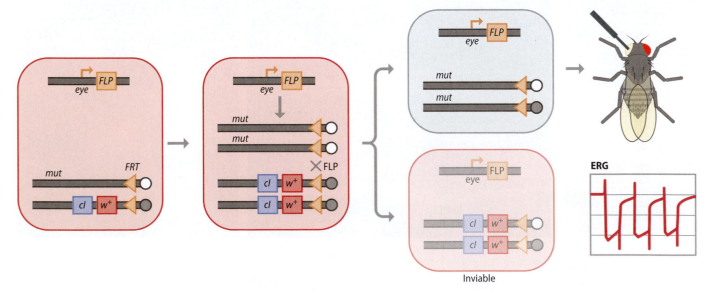

Figure 11.13 An example of FLP-mediated somatic crossing over in *Drosophila* generating reciprocal twin-spots, each homozygous for markers on one chromosome arm. In this figure, most cells in the fly are red-eyed because of a *w⁺* transgene inserted on the chromosome arm (they are homozygous *w* on the X chromosome). The fly is also heterozygous for a mutant of interest (*mut*) and for a recessive cell-lethal (*cl*) mutation. A FLP site-specific recombinase gene is present, expressed under the control of an eye-specific promoter (*eye*). Mitotic recombination driven by FLP-mediated recombination between FRT sites on the two homologous chromosomes produces a patch of white-eyed, homozygous mutant (*mut*) eye cells, while the reciprocal sector is homozygous for (*w⁺*), but these cells are eliminated because they are also homozygous for a recessive cell-lethal (*cl*) mutation. The patch of white-eye cells can then be tested for the mutant's effect on an electroretinogram (ERG). (Adapted from Venken KJ, Simpson JH & Bellen HJ [2011] *Neuron* 72:202–230. With permission from Elsevier.)

genome (**Figure 11.15**). Bacterial artificial chromosomes (BACs) engineered in *E. coli* to carry *attB* can be injected along with purified ΦC31 integrase into *Drosophila* or mouse cells so that the BAC will integrate at a "safe harbor" *attP* site that has been previously introduced into the target genome. Alternatively the BAC can be integrated at a small number of pseudo-*attP* sites that already exist in the target genome. Placing the inserted sequences in a defined location avoids the problems of variable gene expression that are encountered when different constructs integrate at quite different chromosomal locations. Thus, in flies, a P element or other transposable element carrying the *attP* sequences can be inserted into the genome and mapped so that subsequently any number of BACs can be integrated at the same locus, allowing one to test the effects of several alterations of sequences within the BAC. Hugo Bellen's lab and others have developed a large number of fly stocks in which the *attP* site is located at different sites. DNA segments as large as 100 kb can be incorporated in this fashion, after having been modified in *E. coli*. A second phage system, Bxb1, has also been developed, so that one can create successive alterations of a region within a chromosome.

In a variation on this theme, Bellen's lab has also constructed a small transposon based on the Minos transposable element, carrying two inverted *attP* sites as well as a selectable marker. This "Minos mediated integration cassette" (MiMIC) element can be integrated at many locations by using its transposon properties (**Figure 11.16**). The transposon also carries an inverted pair of *attP* sites flanking two markers (EGFP and yellow+). Into these cells can be introduced plasmid DNA carrying a pair of inverted *attB* sites flanking other DNA sequences. When ΦC31 is expressed, a pair of reciprocal exchanges between one *attB* and one *attP* site will integrate the new DNA sequences and lose the original markers. Thus a variety of modifications of a sequence of interest (a promoter or the amino acids within an exon of a protein, for example) can be introduced.

Michelle Calos has made similar modifications in mammalian cells, including embryonic stem (ES) cells that can be used to create transgenic mice. In mammals, there are a number of pseudo-*attP* sites in the

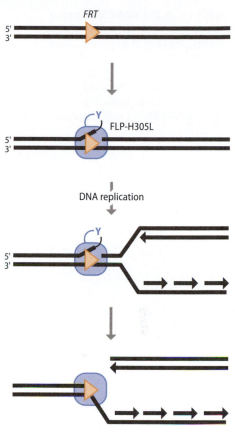

Figure 11.14 A mutant FLP enzyme can create a site-specific nick. The FLP-H305L mutant protein can carry out the first (cleavage) step within a FRT site, leaving the enzyme covalently linked via Tyr343 to the 3′ end of the cut DNA. This nick will be converted to a broken replication fork by the arrival of a replication fork. (Adapted from Nielsen I, Bentsen IB, Lisby M et al. [2009] *Nat Methods* 6:753–757. With permission from Macmillan Publishers, Ltd.)

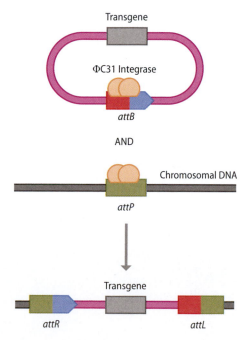

Figure 11.15 Site-specific ΦC31 recombination. ΦC31 recombination can be used to generate large inserts in *Drosophila* or mammals by recombination between an *attB* site on a circular plasmid and an *attP* site inserted in a host chromosome. (Adapted from Keravala A, Chavez CL, Hu G et al. [2011] *Gene Ther* 18:842–848. With permission from Macmillan Publishers, Ltd.)

Figure 11.16 The MiMIC transposon system. A Minos element, carrying inverted repeats (L and R) is the target of ΦC31 *attP* integration. The original insertion can be replaced by a second ΦC31 integration of a DNA segment with two inverted *attB* sites. (Adapted from Venken KJ, Schulze KL, Haelterman NA et al. [2011] *Nat Methods* 8:737–743. With permission from Macmillan Publishers, Ltd.)

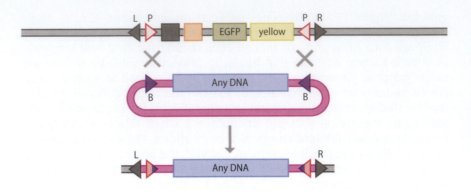

genome, so that only about 50% of the integrations occur at a specific, engineered *attP* site. Nevertheless, ΦC31-mediated transgenic insertions offer dramatic possibilities in gene therapy.

SUMMARY

Site-specific recombination has been exploited to create novel ways to rearrange chromosomes and to develop conditional mutations and conditional gene expression. These systems offer important ways to modify mammalian genomes.

SUGGESTED READING

Birling MC, Gofflot F & Warot X (2009) Site-specific recombinases for manipulation of the mouse genome. *Meth Mol Biol* 561:245–263.

Groth AC & Calos MP (2004) Phage integrases: biology and applications. *J Mol Biol* 335:667–678.

Perrimon N (1998) Creating mosaics in *Drosophila*. *Int J Dev Biol* 42:243–247.

Venken KJ & Bellen HJ (2007) Transgenesis upgrades for *Drosophila melanogaster*. *Development* 134:3571–3584.

CHAPTER 12

CYTOLOGY AND GENETICS OF MEIOSIS

The study of meiosis established the foundation for our understanding of mechanisms of homologous recombination. But meiotic recombination, involving dozens or hundreds of enzymatically induced double-strand breaks (DSBs), poses a much more complex problem than the repair of a single break in a somatic cell. Consequently meiotic recombination is highly choreographed and requires the participation of a large number of meiosis-specific proteins to ensure that repair is primarily effected between homologous chromosomes rather than sister chromatids and to regulate the number and distribution of crossovers.

12.1 RECOMBINATION IS REQUIRED TO GENERATE DIVERSITY BUT ALSO TO ENSURE CHROMOSOME SEGREGATION

Recombination was first studied in the context of meiosis, long before anyone understood its connection with chromosome damage repair. The basic ideas of meiotic recombination came from studying *Drosophila* in the 1910s, where the frequencies of exchange between heterozygous mutations at different loci gave rise to the idea of linkage and the creation of a genetic map. A comparison of the expected and recovered frequencies of double crossovers along a chromosome produced the idea of crossover interference. Many more insights came in the early mid-century with the use of fungi, where one could characterize all four products of meiosis. Observations in budding and fission yeasts and in the filamentous fungi *Neurospora* and *Ascobolus* demonstrated the non-reciprocal nature of gene conversion and the connection between gene conversion and crossing over.

Meiosis, of course, is responsible for generating genetic diversity, so that gametes have arrangements of alleles not found in either parent. But recombination during meiosis serves another critical function: crossovers are necessary for the proper segregation of homologous chromosomes at the first meiotic division. The proper alignment of the pairs of homologous chromosomes requires that the pairs of sister centromeres be under tension from microtubules pulling each pair of sisters toward the opposite ends of the spindle, but this state cannot be achieved if there are no connections between the pairs of chromosomes. Crossover structures, called chiasmata, held in place by sister chromatid cohesion, provide the necessary links that allow the alignment of homologs to occur (**Figure 12.1**). Chromosomes that fail to recombine and form chiasmata exhibit a high risk of *chromosome nondisjunction*, as is seen in the nonhereditary form of Down syndrome in humans.

Figure 12.1 Evidence of exchange between chromatids at chiasmata.
This photomicrograph shows a pair of homologous chromosomes, each with two chromatids, during prophase I of meiosis in a salamander. Two chiasmata are visible. (Adapted from Sadava D, Hillis DM, Heller NC et al. [2008] *Life: The Science of Biology*, 8th ed. With permission from Sinauer Associates, Inc.)

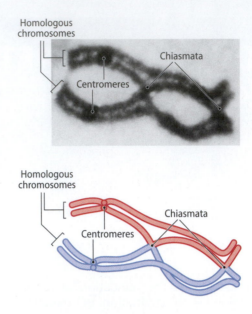

In mice or worms or flies, meiosis in the female produces a single egg that has inherited one of the four sets of chromosomes segregating after meiosis. The remaining sets of chromosomes are sequestered in a diploid and a haploid polar body (**Figure 12.2**). Curiously, in male *Drosophila*, there is no meiotic recombination, but males have a specialized way of segregating non-exchange chromosomes to avoid nondisjunction. In mammals and worms, the rate of crossing over in male meiosis is reduced relative to females.

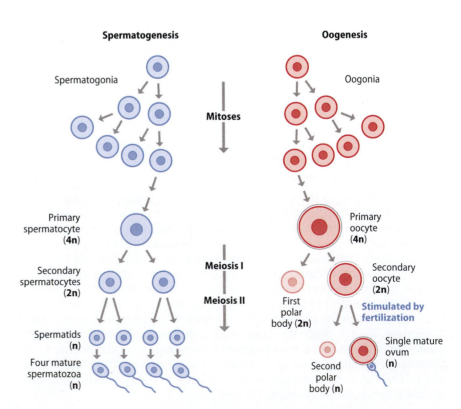

Figure 12.2 Oogenesis and spermatogenesis in mammals. Meiosis occurs after the diploid replicates its genome, yielding a 4n DNA content. In females only one of the four meiotic products is recovered in an egg, the other segregants being lost as a diploid (2n) polar body in meiosis I and a haploid (n) polar body at meiosis II. All four meiotic segregants can form sperm in males.

12.2 EARLY MEIOSIS IN *DROSOPHILA* STUDIES SHOWED THAT RECOMBINATION OCCURS AFTER CHROMOSOME REPLICATION

In 1913, Alfred Sturtevant, an undergraduate student in Thomas Hunt Morgan's lab, was the first to construct a genetic map based on the segregation of three markers along one chromosome. Morgan's lab had already shown that pairs of these markers were genetically linked on the X chromosome because there were more parental offspring than recombinant offspring. If two markers were not linked there should be equal numbers of all four types of offspring (**Figure 12.3**). For a *Drosophila* female heterozygous for the X-linked markers *y w min* crossed to a hemizygous *y w min* male, Sturtevant's results showed that *y* was closely linked to *w* and furthest from *min*. Recombination frequency gives an indication of the physical distance between genes, assuming that crossovers occur more or less randomly in all intervals. The percentage of meiotic recombination is expressed in cM (centimorgans), honoring Morgan, the founder of the *Drosophila* lab at Columbia University, where many of the fundamental findings about genetic linkage and mapping were made. Thus 1% recombination is defined as 1 cM. Note that in the example in Figure 12.3, *w* must be the central marker because the very infrequent double crossovers have the configuration *+ w +* and *y + min*.

The recovery of only one set of chromatids from meiosis allows one to deduce the location of most crossovers and their distribution along a chromosome, but by itself it is insufficient even to know if crossing over occurred before or after DNA replication (that is, at the two-strand or four-strand stage). However, Ernest Anderson found a way to answer this question in 1925 with fruit flies by taking advantage of a chromosomal translocation in which both X chromosomes of a female were joined to a single centromere. Two recessive X-linked markers, *f* and *cut*, were heterozygous on this *attached-X* chromosome. Anderson argued that if crossing over occurred between the two homologous chromosome arms prior to chromosome replication (remember, he didn't know it was DNA) then the linkage of distal markers (further from the centromere) would change relative to ones close to the centromere, but all the offspring

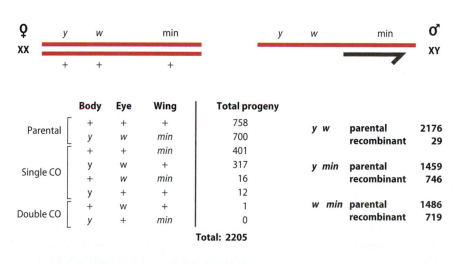

	Body	Eye	Wing	Total progeny
Parental	+	+	+	758
	y	w	min	700
Single CO	+	+	min	401
	y	w	+	317
	+	w	min	16
	y	+	+	12
Double CO	+	w	+	1
	y	+	min	0
			Total: 2205	

y w	parental	2176
	recombinant	29
y min	parental	1459
	recombinant	746
w min	parental	1486
	recombinant	719

y – *min* **linkage** [401 + 317 + 16 + 12]/2205 = 33.8%

y – *w* **linkage** [16 + 12 + 1]/2205 = 1.3%

w – *min* **linkage** [401 + 317 + 1]/2205 = 32.6%

Figure 12.3 An example of determining a genetic map from a three-factor cross in *Drosophila*. Alfred Sturtevant's analysis of T. H. Morgan's data, demonstrating linkage between three markers on the X chromosome in a heterozygous female crossed to a hemizygous triple recessive male. The total parental and recombinant outcomes for each gene pair are noted and the percent recombination calculated in each case. Note that the two least frequent outcomes among the eight possible outcomes represent double crossovers and uniquely define which marker is in the middle (in this case w). (Adapted from Johnson GB [1995] How Scientists Think. With permission from McGraw-Hill.)

Figure 12.4 Crossing over in flies must occur at the four-strand stage, after chromosome replication. E. G. Anderson used an attached-X female fly heterozygous for the *cut* and *forked* genes. (A) If crossing over occurred before replication, then all attached-X chromosomes would still be heterozygous for both markers although the linkage arrangement has changed from the parent. (B) If exchange happens after replication, then half of the time an exchange will create one attached-X chromatid that is homozygous for the distal marker (*cut*) or homozygous wild type. The other half of the exchanges will leave each attached-X heterozygous for both markers. Note that all offspring are still heterozygous for *forked* because this marker is very close to the centromere and shows only rare crossovers. (Adapted from Anderson EG [1925] *Genetics* 10:403–417. With permission from the Genetics Society of America.)

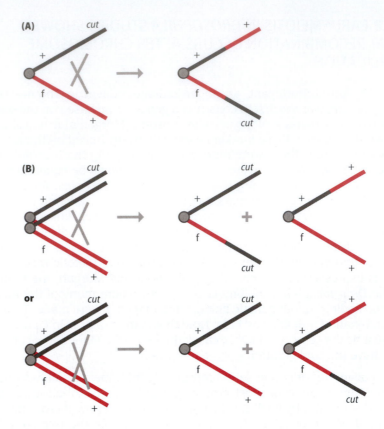

inheriting the attached X would still be phenotypically wild type for both traits (**Figure 12.4**). If, however, the chromosomes first replicated and only then did crossing over between the two markers occur, between one of the two chromatids of each arm, some of the eggs would be heterozygous *f/f⁺* but homozygous *cut/cut*. An equivalent percentage should be homozygous *cut⁺/cut⁺* but this genotype could not be distinguished from the phenotypically identical *cut⁺/cut*. The majority of offspring would have experienced no crossing over in the interval and would have the parental genotypes. This analysis also demonstrated that the *f* marker was closer to the centromere, as there were almost no cases where this marker had crossed over to give *f/f* attached-X offspring. Thus crossing over occurs after chromosome replication and there is more exchange for markers farther from the centromere.

12.3 IN FUNGI, ALL FOUR PRODUCTS OF A SINGLE MEIOSIS CAN BE RECOVERED

Analysis of the fate of the four chromatids during meiosis became much easier in fungi, where all four products of meiosis—spores—could be recovered and germinated to grow as haploid segregants. By replica plating, each spore colony can be tested for the segregation of different nutritional genetic markers, for example a defect in tryptophan biosynthesis or resistance to the amino acid analog, canavanine (**Figure 12.5**). The tetrads of *Saccharomyces* are usually found as a tetrahedron, so there is no positional order of the spores relative to the way chromatids lined up at the first meiotic division (Meiosis I); nevertheless one can trace the positions of crossovers along a chromosome arm carrying several genetic markers (as described in Section 12.15).

Figure 12.5 Budding yeast tetrad analysis. From a sporulated diploid heterozygous for canavanine resistance (*can1*) and requirement for tryptophan in the medium (*trp1*), the four spores contained within an ascus are separated on an agar plate containing rich medium. Nine tetrads are shown. Cells germinate and grow into colonies. (Note that spore 5C, marked X, failed to grow.) The spore colonies were then replica plated to a plate containing canavanine and to a "drop-out" plate lacking tryptophan. Most tetrads show 2:2 (4:4) segregation for each marker (an exception is tetrad 4, marked *, which displays a gene conversion for *trp1* segregation). The two genetic markers appear to be unlinked, given that tetrads 1, 2, 7, 8, and 9 are tetratypes (TT) while tetrad 2 is a parental ditype (PD) and tetrad 6 is a nonparental ditype (NPD). (Courtesy of Qiuqin Wu, Brandeis University.)

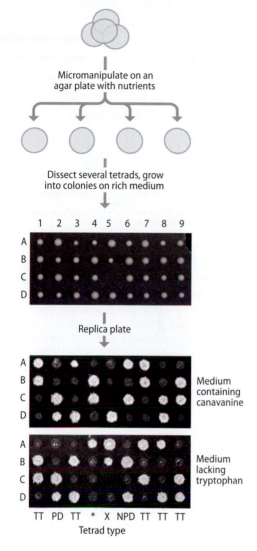

Even more information could be obtained from filamentous fungi that produce an ascus with spores in an *ordered tetrad*. First, the four meiotic products are arrayed in a linear reflection of the way the chromatids lined up prior to Meiosis I. Second, prior to spore maturation, each chromatid undergoes a single post-meiotic chromosome replication and mitotic segregation. This process produces a total of eight spores (**Figure 12.6**). In these ordered tetrads one can score the genetic map distance between the marker and its centromere. If there is no crossover, the spores are arranged in their parental configuration, ++++ − − − − or − − − −++++. This arrangement is termed *first division segregation* (FDS) because all the (+) spores are separated from all the (−) spores after the first meiotic division. If there is a crossover between a pair of non-sister chromatids, then there could be arrangements such as ++ − − − − ++ or ++ − − ++ − − or − − ++++ − − (**Figure 12.7**); these are called *second division segregation* (SDS). The proportion of SDS among all octads is a measure of the frequency with which the marker crosses over relative to the centromere, which is the anchor of the chromatids and itself has no crossing over. If a marker is very close to its centromere, the proportion of SDS will be close to zero.

The arrangement of markers among meiotic segregants, relative to the centromere or relative to each other, can be classified in one of three ways. Parental ditypes (PD) exhibit no visible exchange event and the segregants have the same marker arrangement as the two parental chromosomes (**Figure 12.8**). Tetratype (TT) asci have two spores with a parental arrangement of alleles and two spores displaying the consequences of a crossover between the two genetic markers. Non-parental ditypes (NPD) again display only two types of segregants, but all of them are recombinant (that is, not parental). In ordered, eight-spore asci, the pair of spores that came from the same chromatid before the post-meiotic division are thought of as one unit, so one can still discuss the segregation of markers in terms of a tetrad. Relative to the centromere, a SDS is equivalent to a tetratype outcome. Compared to the parental arrangement of the markers, a TT represents a single crossover and directly demonstrates that crossing over happened after chromosome replication. A NPD tetrad must have involved two exchange events in the same interval, each crossover involving a different pair of chromatids.

Figure 12.6 Meiosis in the filamentous fungus *Sordaria* produces ordered octads.
Octads are the result of a single post-mitotic replication of the four meiotic products. Segregation is ordered, so that markers very close to the centromere almost always exhibit first-division segregation (FDS) with all four spores of the same genotype together. A crossover along a chromosome arm between the centromere and the spore color marker gene will generate second division segregation (SDS). One octad shows evidence of gene conversion, with a 2:6 pattern. (Courtesy of Denise Zickler, Université Paris Sud.)

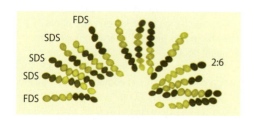

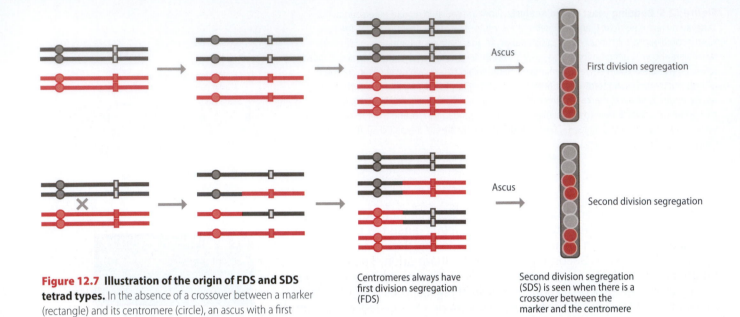

Centromeres always have first division segregation (FDS)

Second division segregation (SDS) is seen when there is a crossover between the marker and the centromere

Figure 12.7 Illustration of the origin of FDS and SDS tetrad types. In the absence of a crossover between a marker (rectangle) and its centromere (circle), an ascus with a first division segregation (FDS) pattern is obtained, whereas when there is a crossover between the marker and its centromere a second division segregation (SDS) pattern is generated.

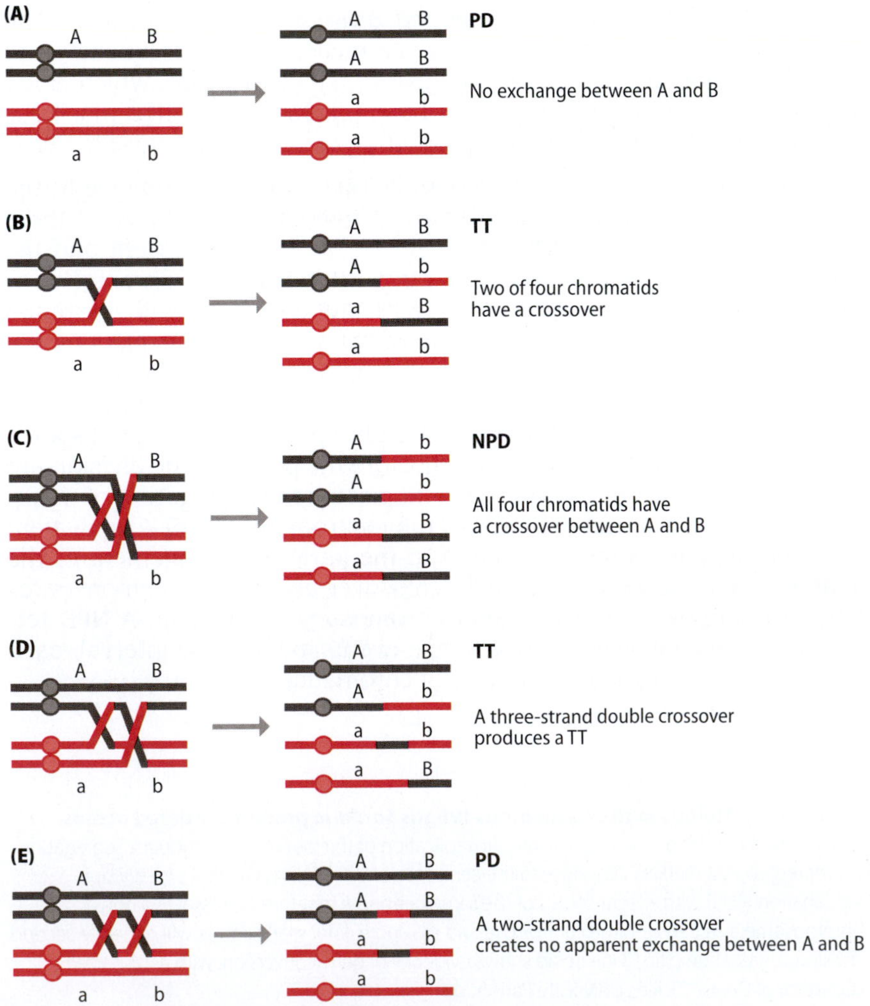

(A)

PD

No exchange between A and B

(B)

TT

Two of four chromatids have a crossover

(C)

NPD

All four chromatids have a crossover between A and B

(D)

TT

A three-strand double crossover produces a TT

(E)

PD

A two-strand double crossover creates no apparent exchange between A and B

Figure 12.8 Tetrad types in budding yeast involving linked genetic markers. In the absence of a crossover between markers A and B, the tetrad will have a parental ditype (A). A single exchange, involving only two of the four chromatids, yields a tetratype (B), while less frequent double crossovers (DCO) in the interval will—one-quarter of the time—generate a nonparental ditype (C). In addition to a four-strand DCO, a three-strand DCO will create a TT (D). A two-strand DCO will yield a PD (E).

Figure 12.9 A colorful way to score segregation of two markers. (A) In budding yeast, the integration of red and blue (*cyan*) fluorescent protein genes at two locations on chromosome 8 will produce spores that could be red, blue, purple, or colorless; hence tetrad types can be scored without dissection. (B) In *Arabidopsis* carrying three linked fluorescent markers one can identify different patterns of crossing over. (A, adapted from Thacker D, Lam I, Knop M & Keeney S [2011] *Genetics* 189:423–439. With permission from the Genetics Society of America. B, adapted from Francis K, Lam Y, Harrison B et al. [2007] *Proc Natl Acad Sci USA* 104:3913–3918. With permission from the National Academy of Sciences.)

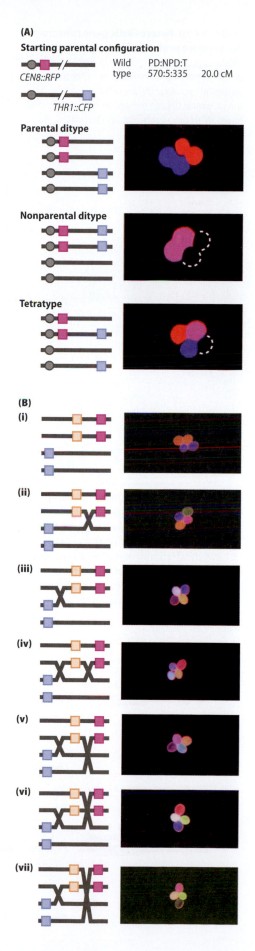

Recently the use of fluorescent fusion proteins that bind to arrays of bacterial operator sequences (LacO or TetR) inserted at defined chromosomal sites has made it possible to follow marker segregation without dissecting tetrads, both in budding yeast and in a "tetrad" mutant of *Arabidopsis* that keeps the four pollen products of meiosis together (**Figure 12.9**).

The frequencies of PD, TT, and NPD can be used to calculate map distances, in the same way as the percentage of recombinant offspring in *Drosophila* is used. For very short intervals, the fraction of crossing over is simply TT/2 divided by the total number of tetrads. When the chance of having a crossover is small, then the probability of finding a NPD becomes vanishingly small and can be ignored. The number of TT is divided by 2 because only half of the chromatids in the tetrad have a crossover. This correction makes it possible to compare results with organisms such as *Drosophila* where only one of the four products of meiosis is recovered and therefore recombination is based on individual chromatids and not tetrads.

When markers are farther apart, where there might be more than one crossover in an interval, one must take into account NPDs. But the situation is a bit more complicated: only one-quarter of the double crossovers between chromatids will yield a NPD, that is, those in which all four strands are involved (Figure 12.8). Another one-quarter are invisible two-strand double crossovers (DCO) that are indistinguishable from PD with no crossovers. The remaining half are three-strand DCOs that are phenotypically TT, despite having two exchanges. So one of the two exchanges of a three-strand DCO has already been counted among the TT. To correct for all these complications, the generally used map distance formula is

$$d = [TT/2 + 3\ NPD]/\text{total} \quad \text{or} \quad d = [TT + 6\ NPD]/[2 \times \text{total}].$$

The percent crossover is reported in centimorgans (cM).

In flies, to see double crossovers you need a diploid heterozygous for three linked markers, whereas in fungi the presence of all four spore products allows one to see some of these events with a single pair of markers. In both cases, the *expected* number of double crossovers, calculated by simply multiplying the probabilities of obtaining each single crossover event, is much greater than observed. The paucity of observed NPDs is a reflection of *crossover interference*, the tendency of crossover events to be well separated from each other. This is a subject we will examine in more detail in Section 12.15.

12.4 GENE CONVERSIONS ARISE FREQUENTLY IN MEIOSIS

Meiotic gene conversions occur 100 to 1000 times more frequently than in mitotic cells. Hence it is easy to recover prototrophic recombinants from diploids with two different auxotrophic alleles within the same gene. As we discussed in Chapter 6, few of these wild-type recombinants come from simple reciprocal crossovers; rather they are gene

Figure 12.10 Heteroallelic gene conversion can produce prototrophic recombinants. Most often, wild-type segregants are the result of a short patch gene conversion rather than a reciprocal exchange between the two alleles, as the reciprocal double mutant recombinant is rarely found to accompany prototroph formation.

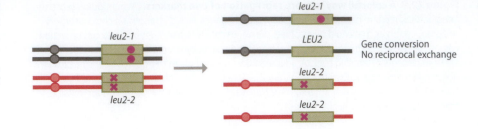

conversions, because none of the remaining three spores have the double mutant genotype expected from a reciprocal exchange between the two markers (**Figure 12.10**). The genotypes of the auxotrophic colonies in the other three spores can easily be determined by both genetic and molecular techniques. In the twenty-first century, it is easy to use PCR and DNA sequencing, or—better—allele-specific PCR primers that will only amplify one of the two mutations. In the decades before PCR, the genotype of these segregants had to be determined by genetic means, by test crosses to distinguish between the two alleles (**Figure 12.11**). For example in a cross between two *leu2* alleles, every Leu⁻ segregant would be crossed with two strains of opposite mating type, each carrying one or the other parental allele (*leu2-1* or *leu2-2*). These new diploids could then be sporulated and the mass of tetrads plated as a patch on medium lacking leucine. If the segregant being tested had *leu2-1*, then the patch would produce many Leu⁺ recombinants when crossed with *leu2-2*, but no prototrophs when crossed with *leu2-1* because the test diploid would be homozygous for *leu2-1*. If a segregant carried both *leu2-1* and *leu2-2*, then neither test cross would yield Leu⁺ recombinants. Allele testing can also be done by stimulating mitotic recombination by the test-cross diploids with ultraviolet (UV) light and plating them on medium lacking leucine (called drop-out medium), as shown in Figure 6.2.

There is a large range of gene conversion frequencies among different genes, with some genes being 10 times "hotter" than others. Genes in chromosomal regions with high levels of initiation of recombination also show an expansion of the genetic map relative to their genomic DNA distances (**Figure 12.12**). The molecular nature of hot spots will be presented in the following chapter.

Figure 12.11 Allele testing of auxotrophic segregants. Segregants are crossed with cells of the opposite mating type and carrying either *leu2-1* or *leu2-2*. If the resulting diploid can generate *LEU2* recombinants after being irradiated or sent through meiosis, then it must not carry the same allele as the tester. A *leu2-1,2* double mutant will fail to yield prototrophs when crossed with both testers.

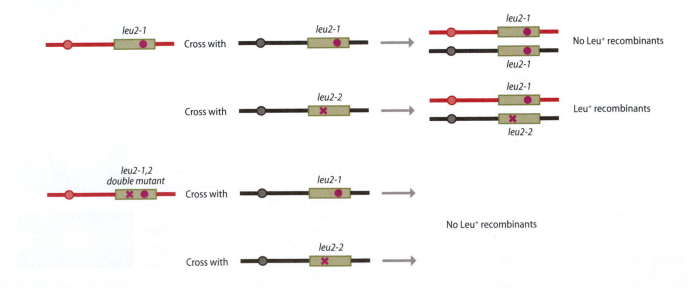

Figure 12.12 Differences in genetic and physical maps of chromosome III in budding yeast. Some regions such as the one surrounding *thr4* are genetically "hot," whereas the region around the centromere is "cold." Note that the order of *glk1* and *cha1* was incorrect on the genetic map, most likely because map distances were based on two-factor crosses and the markers were far apart. (Adapted from Oliver SG, van der Aart QJ, Agostoni-Carbone ML et al. [1992] *Nature* 357:38–46. With permission from Macmillan Publishers, Ltd.)

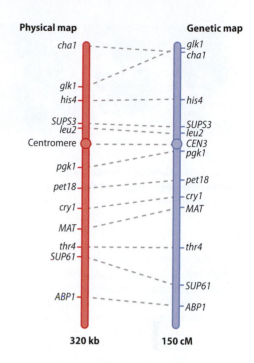

When there are a number of different alleles of a gene that can each be tested for gene conversion, it is generally possible to make a good map of the positions of different alleles *within* a gene by the frequency with which pairs of alleles produce prototrophs: the closer together, the fewer the number of recombinants, just as in constructing maps between genes (**Figure 12.13**; another example was shown in Figure 6.4). In budding yeast and other organisms where one can collect enough data, we also often observe a strong *polarity of gene conversion*, where alleles near the 5′ end of the gene have significantly higher levels of gene conversion than alleles at the 3′ end (**Figure 12.14**). These differences can be greater than fivefold from the hot and cold ends of the gene. The polarity gradient is consistent with finding that the sites of initiation of most meiotic recombination in budding yeast and some other organisms are found in or near promoter regions, at the 5′ end of genes.

12.5 POST-MEIOTIC SEGREGATION IS SEEN IN THE ABSENCE OF MISMATCH REPAIR

Gene conversions are examples of non-Mendelian segregation, an exception to the expected 2:2 segregation of alleles. There are additional exceptional patterns of inheritance in which one or more spores give rise to a *post-meiotic segregation* event. In fungi with a post-meiotic mitotic division, it is easy to see these exceptional events. Several examples are shown for asci in *Ascobolus* heterozygous for both a centromere-linked spore-shape allele and the red or white alleles of the *b2* gene. For example,

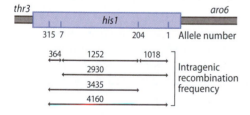

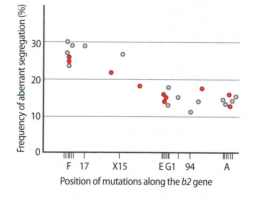

Figure 12.13 A meiotic fine-structure map of alleles within a budding yeast gene. Genetic map distances reflected by the frequency of His⁺ prototroph formation per 10⁶ spores plated when different pairs of *his1* alleles are crossed. (Adapted from Fogel S & Hurst DD [1967] *Genetics* 57:455–481. With permission from the Genetics Society of America.)

Figure 12.14 Polarity of gene conversion in the *b2* locus of *Ascobolus*. The frequencies of aberrant segregation (that is, those not displaying normal 4:4 segregation) were determined from many markers, both those that gave predominantly 6:2 and 2:6 outcomes (*gray circles*) and those that displayed frequent post-meiotic segregation (5:3 or 3:5) (*red circles*). (Adapted from Nicolas A & Petes TD [1994] *Experientia* 50:242–252. With permission from Springer Science and Business Media.)

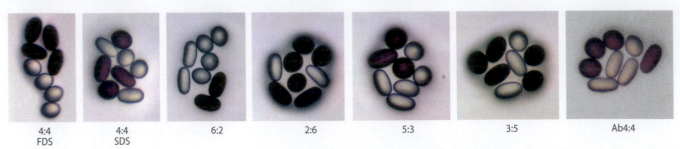

| 4:4 | 4:4 | 6:2 | 2:6 | 5:3 | 3:5 | Ab4:4 |
| FDS | SDS | | | | | |

Figure 12.15 Gene conversions and post-meiotic segregations in the *b2* locus of *Ascobolus*. A diploid heterozygous for the *b2* locus gives red-brown (wild-type) and white (*b2*) spores. Another centromere-linked marker is segregating that yields four oval and four round spores that arise from a post-meiotic mitotic division. Normal segregation should yield four white spores in pairs (that is, two oval and two round) as seen in the examples of 4:4 segregation. First division segregation (FDS) of the *b2* alleles yields four round spores of one color and four oval spores of the other color. Second division segregation (SDS) yields two round and two oval spores of each color. Gene conversions yielding 6:2 and 2:6 outcomes are shown, again with each color/shape combination in pairs. Two examples of post-meiotic segregation are shown, one 5 white to 3 brown and one 5 brown to 3 white. In the 5:3 case, one of the oval pairs has one brown and one white spore; in the 3:5 case, one of the round pairs has one brown, one white. An example of an aberrant 4:4 segregation (Ab4:4) is also shown; here, an oval pair and a round pair have a brown and a white spore. (Courtesy of Alain Nicolas, Institut Curie.)

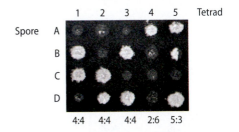

	1	2	3	4	5	Tetrad
Spore A						
B						
C						
D						
	4:4	4:4	4:4	2:6	5:3	

Figure 12.16 Post-meiotic segregation as seen by replica plating spores in budding yeast. A budding yeast diploid heterozygous for *trp1* was sporulated and tetrads dissected on rich medium. Tetrad 4 shows a non-Mendelian segregation pattern (2:6). Tetrad 5 illustrates a 5:3 pattern in which spore 5B is a sectored colony in which the left half is able to grow in the absence of tryptophan in the medium, while the right half cannot. The sectored colony reflects a post-meiotic segregation (PMS) event that is indicative of a segregant whose DNA carried a region of heteroduplex DNA in which a mismatch was not corrected. (Courtesy of Qiuqin Wu, Brandeis University.)

one can see five dark and three white spores (5:3), or the reverse 3:5 pattern as well as full gene conversions (6:2 and 2:6) (**Figure 12.15**).

Post-meiotic segregation (PMS) can also be seen by carefully replica plating the dissected spore colonies of budding yeast from rich medium onto selective medium. For example, in a cross of *TRP1/trp1-1*, one might obtain two colonies that were able to grow without tryptophan, one that could not grow, and a fourth that was sectored, with half growing without tryptophan and the other half failing to grow (**Figure 12.16**). In these instances the spores have inherited heteroduplex DNA, where "Watson" is *TRP1* and "Crick" encodes *trp1* or vice versa. This heteroduplex DNA must not have undergone mismatch repair. When this DNA undergoes the first post-meiotic mitotic DNA replication, one of the two chromatids of mitotic cell (and all of its subsequent progeny) will be fully *TRP1*, while the opposite chromatid is copied into DNA carrying *trp1*. To discuss such exceptional segregations, we keep track of the eight DNA strands of the four meiotic segregants, so that the event in budding yeast is also termed 5:3. Normal segregations are often termed 4:4 rather than 2:2, to be consistent.

Another important exception is termed an aberrant 4:4 (Ab4:4) segregation, where the pattern is best explained by unrepaired heteroduplex DNA in two of the four spores (Figure 12.15) (instead of a pattern such as ++++−−−−, the pattern is ++ +−+ −−). Ab4:4 tetrads only appear if mismatches escape mismatch repair and if heteroduplex DNA forms on two of the four chromatids. Robin Holliday's original model for recombination envisioned a reciprocal swap of strands between two homologous chromosomes to generate a tetrad with *symmetric hDNA*, that is, with hDNA on two chromatids (**Figure 12.17A**). Heteroduplex on a pair of chromatids can also arise by branch migration of a Holliday junction (HJ) (**Figure 12.17B**).

The likelihood of initiating more than one recombination event at a given location is extremely rare, so that other possible exceptional segregation patterns such as 8:0, 0:8 (that is, two gene conversions), or 7:1 or 1:7 (a gene conversion and a PMS) are indeed very rare and need not concern us.

The analysis of *post-meiotic segregation* (PMS) provided key discoveries about the way meiotic recombination is both initiated and processed. As already mentioned above, PMS depends on the packaging of heteroduplex DNA into a spore; hence it must not have undergone mismatch repair (MMR). Seymour Fogel's lab took advantage of this idea to look for mutations in budding yeast that would exhibit elevated levels of PMS; the first gene that they identified, named *PMS1*, proved to be the homolog of the MutL mismatch repair gene first identified in bacteria. Other PMS mutations were found to be other homologs of both bacterial MutL and MutS. There are actually six MutS homologous (Msh) genes in budding yeast, of which Msh1 functions in mitochondria, and Msh2, Msh3, and Msh6 are involved in mismatch repair. Msh2–Msh6 heterodimers recognize single base pair substitutions while Msh2–Msh3 act to correct small insertion/deletion heterologies. As discussed in Chapter 13, Msh4 and

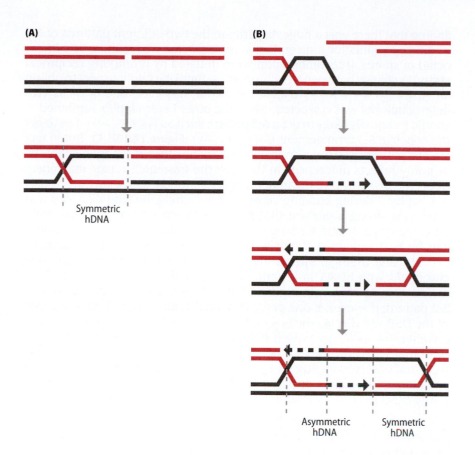

(A)

Symmetric
hDNA

(B)

Asymmetric
hDNA

Symmetric
hDNA

Figure 12.17 Two ways to generate symmetric hDNA. (A) Robin Holliday's model for recombination presumed that two strands of the same polarity were nicked and exchanged, thus generating two regions of hDNA. Taking into account the other two chromatids, this exchange generates an Ab4:4 outcome unless there is mismatch repair. (B) In the dHJ mechanism of DSB repair, strand invasion produces asymmetric hDNA on either side of the break, but branch migration can produce a region of symmetric hDNA.

Msh5 play key roles in regulating meiotic crossovers but are not involved in MMR. Similarly, there are four MutL homologous (Mlh) proteins, one of which has kept its historical name, Pms1, while the others are called Mlh1, Mlh2, and Mlh3. The principal MMR role is carried out by Pms1 and Mlh1, though Mlh1–Mlh2 and Mlh1–Mlh3 dimers also appear to have minor roles. Mlh1 and Mlh3 also are involved in regulating crossing over. In mammals, the genes have similar names, except that yeast Pms1 is the homolog of mammalian Pms2 and yeast Mlh2 is known as mammalian Pms1 (don't ask!). The Msh2–Msh6 dimer is named MutSα whereas the Msh2–Msh3 dimer is also called MutSβ.

Even in wild-type cells some mismatches escape MMR in heteroduplex DNA (hDNA) because not all mismatches are equally repairable. In most organisms, C:C mismatches are quite poorly repaired and give rise to 5:3 or 3:5 PMS tetrads. Frameshifts and small insertion/deletion mismatches in hDNA are usually efficiently repaired by Msh2–Msh3 and Mlh1–Pms1. Insertion/deletions of more than a few nucleotides require not only Msh2–Msh3 but also the flap endonuclease Rad1-Rad10 (XPF-ERCC1). However, insertions of a short palindromic sequence that can form a hairpin structure in hDNA are poorly corrected. Such hairpin insertions have been particularly informative as markers to examine hDNA formation in the absence of any mutations affecting MMR, as we discuss more in Chapter 13.

12.6 MEIOTIC SEGREGATION PATTERNS REVEAL HOW RECOMBINATION IS INITIATED

A key conclusion arising from the analysis of non-Mendelian segregation in fungi is that the initiating events in meiotic recombination are confined to only one of the four chromatids. This conclusion stemmed from the finding that there was a huge disparity in the two different patterns of 5:3 or 3:5 segregants for a given locus. In *Ascobolus* where there is a linear octad of spores, if recombination were initiated by promoting reciprocal strand exchange between two chromatids, then one might expect to obtain an Ab4:4 (++ +− + −−) (**Figure 12.18A**). If, however, one of these two heteroduplexes was corrected, while the other heteroduplex remained, it should be equally likely to get a 5:3 pattern such as (++ ++ − +−−) as to get the opposite 5:3 arrangement (++ +− ++ −−) (**Figure 12.18B, C**). But in fact the first pattern is 50 times more likely than the other. It is possible that the disparity reflects differences in the way the mismatch repair machinery treats the invading and the intact recipient strands, as we saw previously in Chapter 7 during budding yeast *MAT* switching, but the disparity was used as a strong argument that strand invasion was asymmetric, with only one single strand forming heteroduplex. The (++ ++ −+ −−) pattern can be best explained if only a single chromatid suffers DNA damage (a DSB) and a single strand invasion event occurs, leaving only one chromatid with hDNA in a particular region (**Figure 12.19A**). The SDSA mechanism as shown in Figure 12.19A does indeed result in only one 5:3 pattern; however, if one considers the dHJ mechanism, then one side of the DSB will display the (++ ++ −+ −−) pattern but the other side will have the (++ +− ++ −−) result (**Figure 12.19B**). Indeed, a marker lying on the other side of a meiotic hotspot would have the opposite disparity than that observed by Stadler and Towe. But the key point is that in each situation will be a clear disparity between type 1 and type 2, instead of the parity expected from the Holliday model.

Another important genetic finding came from fission yeast where one particular allele of the *ade6* gene had exceptional behavior. First, the *ade6-M26* allele was very "hot"; the frequency of gene conversion was 10 times higher than with any other *ade6* allele, even though M26 appeared to be in the middle of the gene, surrounded by many other much colder allelic markers (**Figure 12.20**). Second, almost all the gene conversion outcomes of the cross *ADE6/ade6-M26* were three wild type and one *ade6-M26* segregant (a 6:2 tetrad) and not 2:6; that is, when M26 was involved in a gene

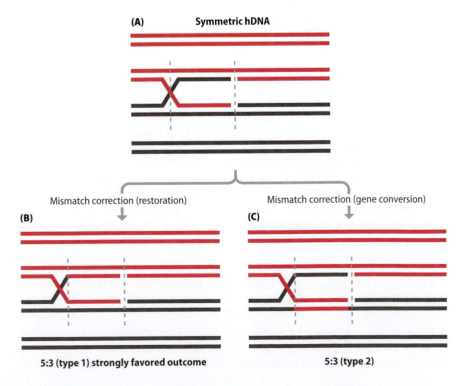

Figure 12.18 Mismatch correction of symmetric DNA can generate 5:3 or 3:5 patterns. Symmetric hDNA (A) can be mismatch corrected either to give a gene conversion or a restoration of the original genotype (defined by the intact strand in the heteroduplex). (B) One of the hDNA regions is restored and one left unrepaired, yielding a 5:3 pattern designated 5:3 (type 1). (C) One of the hDNA regions is left unrepaired and the second is mismatch corrected in favor of the invading strand (gene conversion), yielding a different 5:3 pattern (type 2). The type 1 5:3 outcome was strongly favored in an experiment by Stadler and Towe. (Adapted from Stadler DR & Towe AM [1971] *Genetics* 68:401–413. With permission from the Genetics Society of America.)

(A)   **Symmetric hDNA**

Mismatch correction (restoration)   Mismatch correction (gene conversion)

(B)   (C)

5:3 (type 1) strongly favored outcome   **5:3 (type 2)**

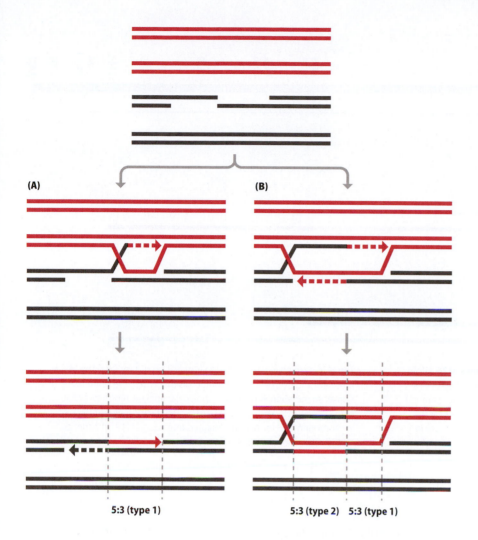

Figure 12.19 Fate of hDNA in SDSA and dHJ repair mechanisms. (A) Repair by the SDSA mechanism will generate asymmetric hDNA consistent with the 5:3 (type 1) outcome illustrated in Figure 12.18. In this instance the primer-extended invading strand is displaced and anneals to the opposite side of the DSB, leaving the donor intact. (B) The dHJ mechanism yields two regions of asymmetric hDNA ,one of which displays the 5:3 (type 1) pattern and the other the 5:3 (type 2) pattern. If the marker being studied lies to the right of the hotspot where DSBs are generated, then type 1 events will predominate. If a different marker had been selected, lying to the left of the hotspot, then type 2 events would be prevalent. In either case there is a large disparity between type 1 and type 2 outcomes.

conversion it was almost always lost! The M26 allele apparently acts as an *initiator* of recombination; that it was self-destructive suggested that M26 was the *recipient* rather than the donor of genetic information. M26 could be a recipient if it were the site of a double-strand break, whose repair would eliminate the very sequences that were the basis for creating the lesion. Though we haven't yet encountered the data, it is well known that essentially all meiotic recombination is induced by DSBs. Interestingly, despite being "hot" M26 did not lead to 4:0 outcomes, so in both budding and fission yeasts there is a strong inhibition of creating DSBs on both sister chromatids.

12.7 BRANCH MIGRATION CAN LEAD TO AB4:4 SEGREGATION

Although Ab4:4 tetrads are generally rare, certain alleles give rise to a significant proportion of Ab4:4 among non-Mendelian segregations. Most of the data concerning these events came from the study of *Ascobolus*, by the lab of Jean-Luc Rossignol. Using alleles of a spore color gene, *b2*, they could examine large numbers of tetrads visually. Examination of many alleles showed that there was a strong polarity of gene conversion, decreasing from high frequencies of gene conversion near a hot spot at one end of the gene (Figure 12.14). Well-corrected alleles gave 6:2 and 2:6 segregation and poorly corrected alleles displayed 5:3 and 3:5 PMS. However, at the low end of the polarity gradient, PMS alleles

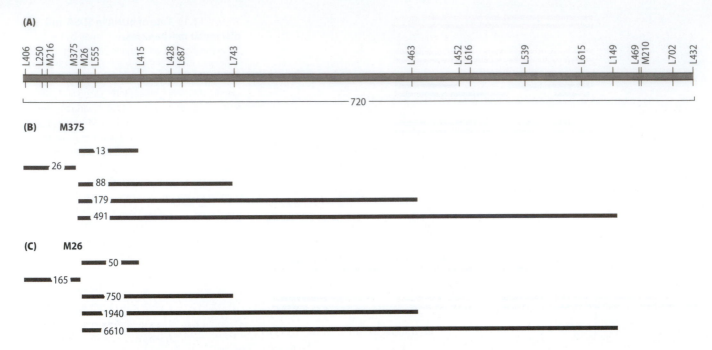

Figure 12.20 Map expansion caused by the *ade6-M26* allele in *Schizosaccharomyces pombe*. (A) A map of many *ade6* alleles determined by the frequencies of prototrophs per 10⁶ spores among various heteroallelic pairs. (B) Representative map distances for the *ade6-M375* allele and several other alleles. (C) A tenfold increase in prototroph formation for the same intervals when *ade6-M26* was used.

The M26 single base pair mutation is only 3 bp distant from the M375 mutation and both are G → T mutations that create the same TGA nonsense mutation. Subsequent molecular analysis has shown that M26 creates a binding site for a transcription factor that is implicated in promoting DSB formation. (Adapted from Gutz H [1971] *Genetics* 69:317–337. With permission from the Genetics Society of America.)

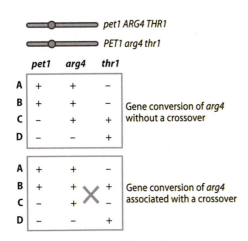

Figure 12.21 Gene conversions can occur with and without an accompanying crossover. Crossovers accompanying gene conversion can be assessed by the configuration of nearby flanking markers. A diploid resulting from a cross of a *pet1 ARG4 THR1* strain with a *PET1 arg4 thr1* strain was analyzed. Gene conversion unaccompanied by a crossover leaves the *pet1* and *thr1* markers in a parental configuration. A crossover accompanying gene conversion yields a tetratype arrangement for *pet1* and *thr1*. (Adapted from Fogel S, Mortimer R, Lusnak K & Tavares F [1979] *Cold Spring Harb Symp Quant Biol* 43:1325–1341. With permission from Cold Spring Harbor Laboratory Press.)

often exhibited Ab4:4 segregation. Rossignol argued that these events might have arisen by a transition from a single region of heteroduplex associated with a Holliday junction to a *symmetrical* heteroduplex structure that could then be extended by branch migration (Figure 12.17B). An elegant experiment that supported this idea was that placing a large heterology upstream of the allele that showed Ab4:4 behavior did not prevent the appearance of 5:3 or 3:5 outcomes but blocked the Ab4:4 types. Rossignol argued that a large heterology would prevent branch migration of the symmetric HJ from some transition point where it was formed down to the marker that would show Ab4:4. The study of the *b2* locus had a profound impact on the development of the molecular model of recombination proposed by Matthew Meselson and Charles Radding, (see Appendix for the historical account). However, there have been no other well-defined examples of this sort of transition in other fungi.

12.8 GENE CONVERSION AND CROSSING OVER HAVE A STRONG CORRELATION

Gene conversions and crossovers are intimately related. As we have already seen, both single and double HJs can be resolved (or dissolved) to produce both crossovers and noncrossovers. In budding yeast, somewhat less than half of all meiotic gene conversions are accompanied by a crossover, which can be seen in tetrads by looking at flanking markers (**Figure 12.21**). The proportion of gene conversions with a crossover is much higher among meiotic events than seen in mitotic recombination.

This increase depends on a number of meiosis-specific proteins that direct resolution of the HJ toward reciprocal exchange, as described in the following sections.

12.9 MAJOR EVENTS IN MEIOSIS CAN BE OBSERVED CYTOLOGICALLY

Before we turn to the molecular events of meiotic recombination, we need to relate these events to cytological stages that have been used to discuss the progression of the nucleus through meiosis. The meiotic prophase is divided into five stages prior to the completion of the first meiotic division: *leptotene, zygotene, pachytene, diplotene*, and *diakinesis* (**Figure 12.22**). During *leptotene* the DNA replication occurs and replicated chromosomes begin to condense. Protein components of an *axial element* form between the sister chromatids that are connected by meiosis-specific cohesin complexes. At *zygotene*, the condensed chromosome pairs begin to *synapse*, so that pairs of homologous chromosomes are aligned. As we will discuss in great detail in Chapter 13, this process is driven in most organisms by the action of a specialized topoisomerase called Spo11 which creates double-strand breaks that initiate meiotic recombination. During this time, in most eukaryotes, an elaborate *synaptonemal complex* (SC) begins to form between the paired homologs (**Figure 12.23**). The SC regulates the distribution of crossovers during meiosis. At *pachytene*, the homologs are fully synapsed and recombination has produced *chiasmata*, crossovers between homologous chromatids. At *diplotene*, the SC dissolves and homologs are held together by the chiasmata and sister chromatid cohesion. At *diakinesis*, the sister chromatid cohesion distal to chiasmata is lost and homologous chromosomes are ready for disjunction at Meiosis I.

Figure 12.22 Cytological stages of meiosis. The five stages of meiotic prophase and the first meiotic division are shown. The initial stages of homologous chromosome pairing are facilitated by the clustering of chromosome ends (telomeres) at the nuclear membrane. Homologous chromosomes are synapsed by formation of the synaptonemal complex. Crossovers are viewed as chiasmata. Bivalents are aligned on the meiotic spindle and homologous pairs separate in a reductional division.

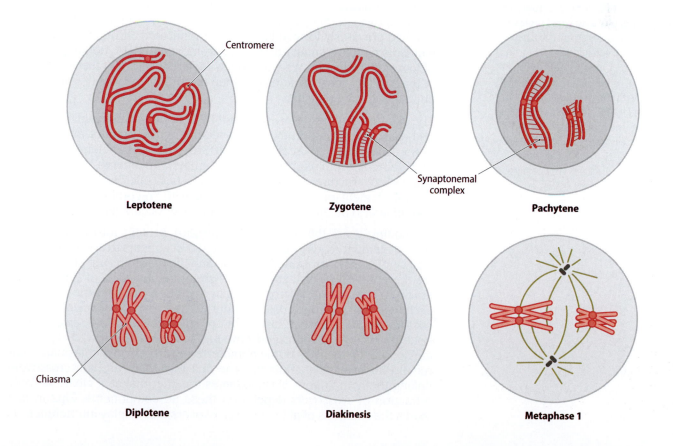

Leptotene   Zygotene   Pachytene

Diplotene   Diakinesis   Metaphase 1

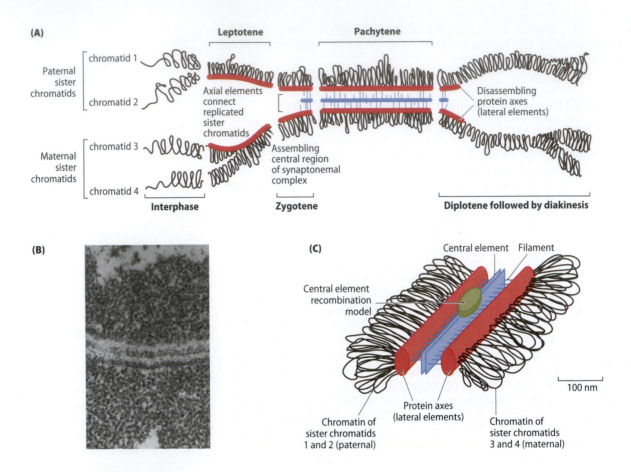

Figure 12.23 A timeline of cytological events during budding yeast meiosis.
(A) Formation of axial elements (*red*) between replicated sister chromatids. These protein structures become the lateral elements of the synaptonemal complex (SC). Assembly of the central elements (*blue*) occurs at the end of zygotene and the SC dissolves at diplotene. (B) An electron micrograph showing the SC. (C) The general architecture of the SC. (Adapted from Alberts B, Johnson A, Lewis J et al. [2008] *Molecular Biology of the Cell.* With permission from Garland Science.)

The SC consists of central element proteins that bridge the distance between two lateral elements, which are derived from the axial elements that had previously formed between the replicated chromatids. The axial elements bind to a small fraction of chromosomal DNA, holding the ends of loops of chromatin. We will discuss the architecture of the SC in terms of the budding yeast proteins and then discuss the similarities and differences in other organisms.

The axial element proteins Hop1 and Red1 apparently organize the DNA and also play an important role in regulating the amount and location of Spo11 cleavage to create meiotic DSBs. The central element consists of coiled-coil proteins that separate the lateral elements, arranged in a zipper-like structure (Figure 12.23). In electron microscope images, the central element of the SC is associated with "recombination nodules" that appear to be the loci where crossing over occurs (Figure 12.23B).

In budding yeast the assembly of the Zip1 protein into a continuous central element requires the Zip2 protein that appears to associate specifically with sites of crossing over. Zip3 and Zip4 proteins are also required for the creation of an extended SC structure. Deletion of Zip1 causes about 50% reduction in crossovers in budding yeast and the remaining exchanges fail to show crossover interference. In the absence of Zip1, the continuous SC is absent, but there are still a few *axial associations* that hold the homologous chromosome pairs together (**Figure12.24A**). These are apparently the sites of the noninterfering crossovers (chiasmata) that remain in these "Zip" mutants. The fact that there are still crossovers linking the homologous chromosomes together makes it clear that only a fraction of crossovers depend on these SC components. Zip2 in fact marks the locations of at least the crossovers that display interference, as

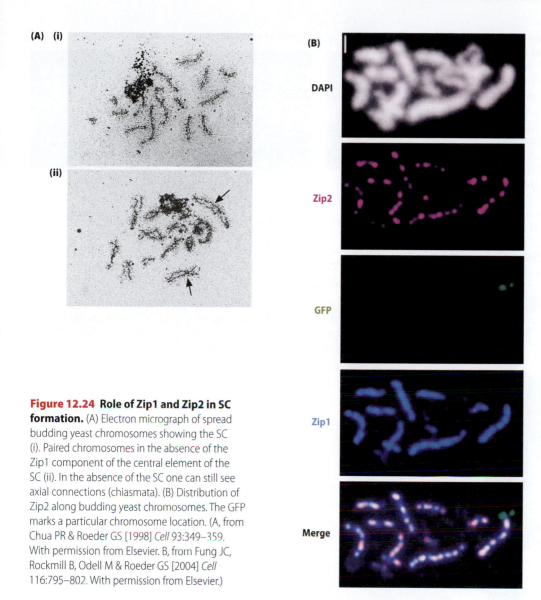

Figure 12.24 Role of Zip1 and Zip2 in SC formation. (A) Electron micrograph of spread budding yeast chromosomes showing the SC (i). Paired chromosomes in the absence of the Zip1 component of the central element of the SC (ii). In the absence of the SC one can still see axial connections (chiasmata). (B) Distribution of Zip2 along budding yeast chromosomes. The GFP marks a particular chromosome location. (A, from Chua PR & Roeder GS [1998] *Cell* 93:349–359. With permission from Elsevier. B, from Fung JC, Rockmill B, Odell M & Roeder GS [2004] *Cell* 116:795–802. With permission from Elsevier.)

shown by immunostaining of chromosome spreads from budding yeast meiosis (**Figure 12.24B**).

Zip3 is a SUMO E3 ligase that adds the ubiquitin-like SUMO peptide to meiotically important proteins such as the axial element protein Red1. Zip1 binds preferentially to SUMOylated targets and thus Zip3, as one of the synapsis initiating complex (SIC) proteins, is required for Zip1 spreading.

Zip4 is also known as Spo22. It has clear homologs in *Arabidopsis* and other plants, as well as in mice. Its precise function has not been established but its absence impairs the spreading of Zip1 in yeast and impairs crossing over. In another plant, rice, the absence of Zip4 impairs the recruitment of another key meiotic regulator, Mer3, to sites of recombination.

Although Zip protein's functions are clearly conserved, the sequences of analogous proteins in other organisms are so diverged (or converged) that one cannot identify homologs in other organisms by sequence gazing. In the case of Zip1, the best understood of the yeast proteins, it is possible to find structural homologs in many eukaryotes, where they are known respectively as C(3)g (*Drosophila*); SYP1, SYP2, and SYP3 (*Caenorhabditis*); ZYP1 (*Arabidopsis*); and SYCP1 (mammals). All of these

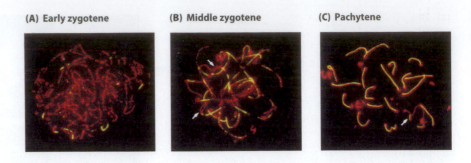

(A) Early zygotene   **(B) Middle zygotene**   **(C) Pachytene**

Figure 12.25 SC formation in mouse spermatocytes at different stages of meiosis. Axial/lateral elements (AE) were stained with anti-Scp3 (*red*). Central elements were stained with anti-Scp1 (*green*). (A) AE (*red*) formation is not yet complete, but short fragments of SC (appear *yellow* from overlapping of Scp3 and Scp1 signals) have begun to form. (B) AE formation is essentially complete and synapsis is extensive. The AEs of homologous chromosomes are aligned in regions that are not yet synapsed (*arrows*). (C) Nineteen fully synapsed bivalents are visible in addition to the X–Y pair—in which only the pseudoautosomal region is synapsed (*arrow*). Human CREST serum was used to visualize centromeres (*red*). (From Baudat F, Manova K, Yuen JP et al. [2000] *Mol Cell* 6:989–998. With permission from Elsevier.)

proteins appear to function as transverse fibers in the central region of the SC, holding homologs together. Each of these proteins has a very long coiled-coil domain that defines the width of the SC. In budding yeast Roeder's group engineered shorter and longer coiled-coils and showed that the SC width changed accordingly.

Analogous staining of mouse meiosis during spermatogenesis shows the assembly of SC during prophase (**Figure 12.25**). Sites of crossover in mammalian cells appear to correlate with foci of the Mlh1 protein (**Figure 12.26**). This also shows that the inhibition of assembly of the central element of the SC leaves each pair of homologous chromosomes linked by one or more exchanges, but showing two separated axial elements.

12.10 SC ASSEMBLY CAN PROCEED IN TWO DISTINCT WAYS

In many organisms, SC formation is driven by the recombination process itself. In budding yeast, *Arabidopsis*, or mammals, mutations that fail to create the DSBs that promote recombination also fail to form SC. Indeed, strand invasion is required to bring homologous chromosomes together so that SC can be elaborated. Thus, in the absence of Spo11, which creates meiotic DSBs, there is no SC formation in yeast or mice. In contrast, in both *Drosophila* and *Caenorhabditis*, a complete SC can form even in *spo11* mutants that cannot create meiotic DSBs. These organisms can build the SC because homologous chromosomes carry chromosome-specific "pairing sites" to which sequence-specific proteins bind and hold homologs in register, thus allowing SC to form. In *Caenorhabditis elegans* a set of four related C2H2 zinc-finger proteins, alone or in combination, facilitate the pairing of specific chromosomes. For example, in the absence of the *him-8* protein, all chromosomes are paired except for the X chromosomes (**Figure 12.27**). The pairing sites are located near the extremities of the chromosomes. These pairing proteins are themselves localized at the nuclear envelope, where proper synapsis of homologs seems to be mediated, as discussed below. Of course, the creation of Spo11-mediated DSBs is still needed for recombination in flies and worms.

Syce3⁺/⁺   pachytene

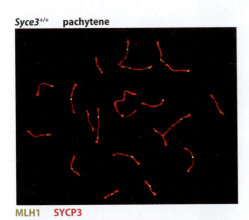

MLH1   SYCP3

Syce3⁻/⁻   pachytene-like

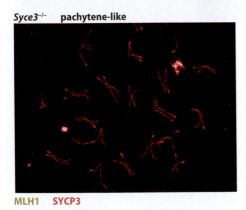

MLH1   SYCP3

Figure 12.26 Failure to recruit Mlh1 in the absence of SC. In mouse meiosis, defective SC formation in the absence of the synaptonemal complex protein SYCE3 leads to the failure to recruit the Mlh1 protein, which marks sites of crossing over. Loss of SYCE3 does not impair the assembly of the similarly named, but distinct, SYCP3 protein that is located along the axial elements that form between sister chromatids. (From Schramm S, Fraune J, Naumann R et al. [2011] *PLoS Genet* 7:e1002088.)

12.11 CHROMOSOMES EXHIBIT DYNAMIC MOVEMENTS PRIOR TO ZYGOTENE

In most organisms there is a telomere-mediated association of chromosomes prior to synapsis and SC formation. In this "bouquet" structure, the telomeres are clustered at the nuclear periphery (**Figure 12.28**). The bouquet arrangement apparently facilitates the homology searching and sorting out of chromosome pairs. Recently it has become clear that, associated with bouquet formation, there is a dynamic back and forth motion imparted to the telomere ends of chromosomes either by cytoplasmic

(A) Wild type

(B) *him-8* mutant

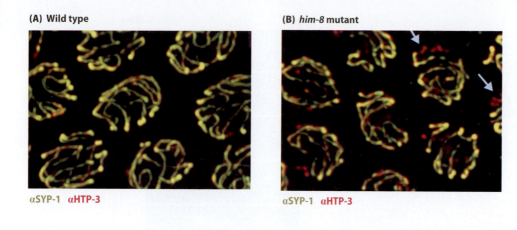

αSYP-1 αHTP-3   αSYP-1 αHTP-3

Figure 12.27 Role of pairing sites in worm meiosis. In *C. elegans* meiosis, nuclei are stained with antibodies against the SC component SYP-1 (*green*) and a component of the axial element, HTP-3 (*red*). (A) In wild type, all chromosomes pair normally. A field of eight nuclei is shown. (B) Absence of the pairing site protein Him-8 shows a failure to pair X chromosomes, whereas all other homologs pair properly. Arrows point to the unsynapsed X chromosomes. (From Phillips CM, Wong C, Bhalla N et al. [2005] *Cell* 123:1051–1063. With permission from Elsevier.)

actin cables (in budding yeast) or by cytoplasmic microtubules (in fission yeast and in worms) that are attached to the nuclear envelope. These dynamic movements are most evident in *S. pombe*, where a "horsetail"-shaped nucleus exhibits large telomere-led movements (**Figure 12.29**), but it is now evident that a similar phenomenon happens in budding

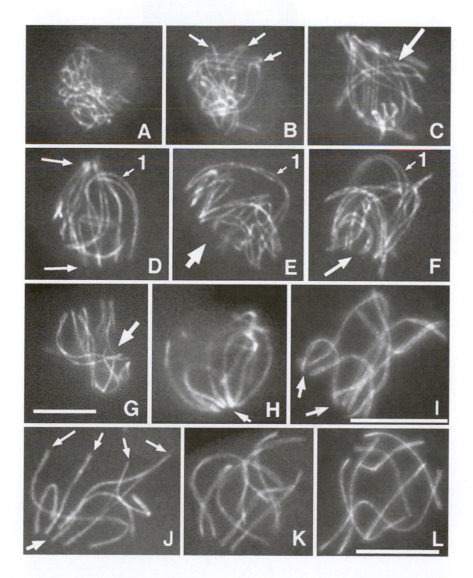

Figure 12.28 Bouquet stage of meiosis facilitates the sorting out of homologous chromosome pairs, as seen in *Sordaria*. Leptotene to late pachytene nuclei of *Sordaria*. Chromosome axes are stained by a GFP derivative of Spo76/Pds5, a cohesin-associated protein. (A) Early leptotene. (B) At mid-leptotene telomere regions start pairing (*arrows*). (C) At the end of leptotene all homologs are aligned. Arrow points to two pairs of homologous chromosomes that are interlocked. How such interlocks are resolved is not known. (D) Early bouquet with two clusters of telomeres (*arrows*). The arrow numbered "1" points to the pair of chromosomes 1. (E) Loose bouquet (*arrow*). (F) Tight bouquet with almost all telomeres grouped (*arrow*). (G) Zygotene bouquet; the arrow points to two interlocked pairs. (H) At early pachytene a very tight bouquet is visible (*arrow*). (I) As pachytene proceeds a looser bouquet is seen (*arrows*). (J) At mid-pachytene release from the bouquet starts mostly at one telomere of each pair (*arrows*). (K, L) Toward the end of pachytene chromosome ends disperse completely. White bars indicate 5 μm. (From Zickler D [2006] *Chromosoma* 115:158–174. With permission from Springer Science and Business Media.)

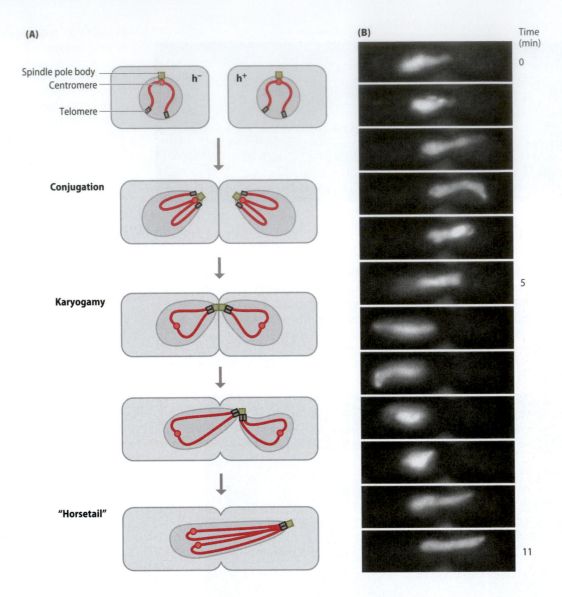

(A)

Spindle pole body
Centromere
Telomere
h⁻   h⁺

Conjugation

Karyogamy

"Horsetail"

(B)   Time (min)

0

5

11

Figure 12.29 Horsetail nuclear morphology in meiosis in S. pombe. (A) Prior to mating, vegetative fission yeast cells of opposite mating types, h⁺ (indicating *mat1P*, for plus) and h⁻ (indicating *mat1M*, for minus) have their centromeres (*red circle*) clustered at the spindle pole body (*green box*). When haploid cells mate, their telomeres (*gray boxes*) as well as their centromeres become clustered together at the spindle pole body. The formation of a diploid nucleus (karyogamy) results in centromere detachment and the elongation of the nucleus, leading for the formation of a "horsetail" where the movement of the spindle pole body elongates the chromosomes from their telomere attachments. Pre-meiotic DNA replication (not shown) begins as horsetail movement initiates and recombination takes place in this extended alignment. (B) The horsetail nucleus exhibits dramatic dynamic behavior that is needed for proper chromosome alignment and recombination. (A, adapted from Chikashige Y, Ding DQ, Imai Y et al. [1997] *EMBO J* 16:193–202. With permission from Macmillan Publishers, Ltd. B, from Davis L & Smith GR [2001] *Proc Natl Acad Sci USA* 98:8395–8402. With permission from the National Academy of Sciences.)

yeast, in worms, and probably in mammals. The movement imparted by cytoplasmic actin or microtubules is transmitted through the membrane by KASH and SUN domain proteins to which other, telomere-associated proteins are bound.

The consequences of disrupting telomere-led chromosome movements vary in severity in different organisms. In fission yeast, horsetail movements are driven by dynein motors attached to microtubules. Deletion of Bqt1 and Bqt2, which associate with telomeres, or deletion of the membrane-associated Sad1 protein, severely disrupts meiotic recombination. In budding yeast, deletion of the Csm4 or Mps3 membrane proteins severely compromises meiosis, although deletion of the telomere-asso-ciated Ndj1 protein has a less profound effect. Without Ndj1, the initial steps of homologous recombination appear normal, but there is a delay in SC formation and in the appearance of molecular intermediates, namely dHJs. While these chromosome movements apparently aid in the search for homology by the recombination machinery, Nancy Kleckner has argued that chromosome mobility may be designed to help chromosomes avoid entanglements as different regions of homologous chromosomes undergo pairing. In worms, inhibition of dynein that mediates microtubule action at the nuclear membrane results in the loss of synapsis between homologs while depletion of the SUN domain protein results in promiscu-ous pairing of nonhomologs and a severe defect in meiosis.

12.12 A STRONG BIAS FAVORS INTERHOMOLOG OVER INTERSISTER RECOMBINATION

A critical distinction between mitotic and meiotic recombination is that DSBs in meiosis are, in most organisms, much more frequently repaired by gene conversion with the homolog than with the sister chromatid. This difference is most evident in events accompanied by a crossover. A clever experiment was designed in 1933 by L. V. Morgan, the talented wife of T. H. Morgan, who examined the meiotic outcomes in flies carry-ing a circular X chromosome and a normal linear chromosome (**Figure 12.30**). A crossover between the ring and rod chromosome produces a dicentric linear product that apparently cannot be recovered in eggs; sim-ilarly, a crossover between two sister chromatids produces a dicentric ring chromosome that is also lost. Morgan argued that ring × rod cross-overs would eliminate an equal number of rings and linears while a ring × ring event would reduce only rings. In contrast, crossovers between two

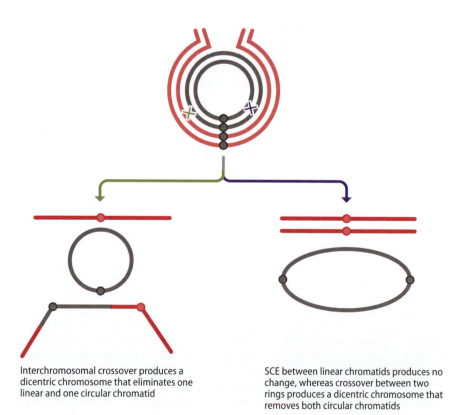

Interchromosomal crossover produces a dicentric chromosome that eliminates one linear and one circular chromatid

SCE between linear chromatids produces no change, whereas crossover between two rings produces a dicentric chromosome that removes both circular chromatids

Figure 12.30 Measurement of sister chromatid exchange in meiosis. A single interhomolog meiotic crossover between a ring chromosome and its homologous linear (rod) results in a dicentric linear chromosome that contains all of the sequences of the ring and rod. Failure to recover the dicentric among meiotic progeny eliminates an equal number of ring and rod chromosomes. Sister chromatid exchange (SCE) between ring chromatids will produce a dicentric ring, eliminating both, whereas SCE between rods has no genetic consequence. Thus, the deficit in recovering ring chromosomes provides an estimate of the frequency of SCE. (Adapted from Morgan LV [1933] *Genetics* 18:250–283. With permission from the Genetics Society of America.)

linear sister chromatids would not cause any loss. Thus there should be a deficiency of ring chromosomes among meiotic products proportional to the incidence of sister chromatid exchange (SCE). Morgan estimated that sister chromatid crossovers were rare compared to interhomolog events. A similar experiment 50 years later by Jim Haber examined ring × rod exchanges in budding yeast and found that interhomolog crossovers were about 25 times more frequent than between sisters. This conclusion was supported by molecular analysis examining double HJ intermediates in budding yeast meiosis, where it is again evident that intersister events are much rarer than interhomolog exchanges (see Chapter 13).

The constraints on intersister repair of a DSB are enforced by the synaptonemal complex, particularly by the axial elements that form before synapsis. Mutations that eliminate the components of the axial element, Red1 and Hop1, or the Mek1 kinase that phosphorylates these proteins, produce a highly significant increase in intersister repair. Moreover, *hop1* or *red1* mutations result in rescuing the viability of mutations such as *dmc1Δ* that preferentially interfere with interhomolog recombination, by permitting sister chromatid repair.

Another protein complex that plays a role in constraining intersister recombination is the 9-1-1 DNA damage checkpoint "clamp" consisting of Ddc1–Rad17–Mec3 and the clamp loader component, Rad24. These proteins are key in activating the Mec1(ATR) checkpoint kinase whose actions are discussed in detail in Chapter 16. Recent studies have shown that after DSB induction, Hop1 is phosphorylated by Mec1. When the Mec1-dependent phosphorylation sites of Hop1 are mutated, there is a shift to intersister recombination that is independent of the meiosis-specific Dmc1 recombinase. How Dmc1 is excluded from intersister recombination remains one of the most important mysteries in understanding meiotic recombination.

12.13 SOME MEIOTIC MUTANTS CAN BE ANALYZED BY BYPASSING MEIOSIS I

In budding yeast, it is reasonably simple to isolate mutations that fail to complete meiosis and produce inviable spores. Some mutants have severe defects in the repair of DSBs; others fail to initiate recombination at all. Without recombination there will be no homologous chromosome pairing and—even if pairing is accomplished—no tension between homolog pairs at Meiosis I; consequently there will be rampant nondisjunction.

spo13Δ rescue of recombination-defective diploid yeast cells

The inviability associated with some recombination-defective mutations can be overcome by exploiting the *spo13Δ* mutation, which has the convenient phenotype of effectively bypassing Meiosis I and carrying out a single Meiosis II division, resembling a mitotic division. Thus *spo13Δ/spo13Δ* diploids yield two diploid spores. In otherwise wild-type cells these diploid segregants still provide evidence of normal recombination because the diploid segregants often become homozygous for markers distant from their centromeres (**Figure 12.31A**). Using *spo13Δ* has revealed important distinctions among mutations that fail to produce viable spores. A *rad50Δ/rad50Δ spo13Δ/spo13Δ* diploid (lacking any meiotic recombination), produces two diploid spores each having the genotype of the original parent, without crossovers (**Figure 12.31B**). In contrast, a *rad52Δ/rad52Δ spo13Δ/spo13Δ* diploid produces only dead spores. These results lead to the conclusion that Rad50 must be needed to create the DSBs required for meiotic recombination, whereas Rad52 is needed for DSB repair. This assignment of different functions to two members of the

Genotype	DNA synthesis	Pairing and exchange	Meiosis I division reductional	Meiosis II division equational	Products

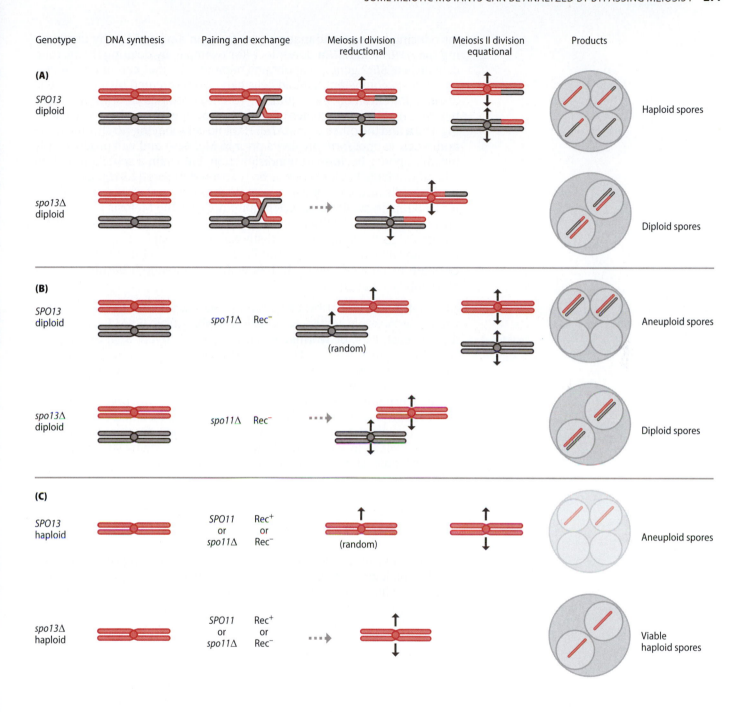

(A)

SPO13 diploid — Haploid spores

spo13Δ diploid — Diploid spores

(B)

SPO13 diploid — *spo11Δ* Rec⁻ (random) — Aneuploid spores

spo13Δ diploid — *spo11Δ* Rec⁻ — Diploid spores

(C)

SPO13 haploid — *SPO11* or *spo11Δ* / Rec⁺ or Rec⁻ (random) — Aneuploid spores

spo13Δ haploid — *SPO11* or *spo11Δ* / Rec⁺ or Rec⁻ — Viable haploid spores

Figure 12.31 Budding yeast meiosis in a *spo13Δ* diploid bypasses the first meiotic division and produces diploid spores.
(A) If cells are recombination proficient then one can recover diploids homozygous for markers far from the centromere. (B) In a *spo11Δ* diploid, there is no recombination and in *SPO13* cells there is massive nondisjunction. However, in *spo13Δ* diploids, two diploid spores, identical to the original diploid genotype are produced. (C) In haploid meiosis, *spo13Δ* creates two viable spores identical to the original haploid genotype both in wild-type and *spo11Δ* backgrounds. (Adapted from Rutkowski LH & Esposito RE [2000] *Genetics* 155:1607–1621. With permission from the Genetics Society of America.)

Rad52 epistasis group can be confirmed by creating a diploid homozygous for *rad52Δ*, *spo13Δ*, and *rad50Δ*, which produces two viable (diploid) spores in which there has been no recombination. This type of analysis revealed that the Mre11-Rad50-Xrs2 (MRX) proteins, in addition to being important in DSB repair, have a crucial role in meiosis to create DSBs, along with Spo11 and other proteins discussed below.

spo13Δ rescue of haploid meiosis

Deleting Spo13 can also rescue spore inviability when haploid budding yeast cells are tricked to enter meiosis. Normally haploid yeast do not turn on the meiotic program because it requires the co-expression of both *MAT**a*** and *MATα* to inhibit expression of *RME1*, a repressor of meiosis. By introducing the opposite mating-type gene on a plasmid or by unsilencing the *HMLα* and *HMRa* donor loci (for example, by deleting the histone deacetylase *SIR2* gene), one obtains haploid cells that can initiate meiosis. Meiosis-competent haploid cells can also be created by making cells disomic for chromosome 3 (that is, 15 chromosomes exist in one copy, whereas there are two copies of chromosome 3, with one homolog carrying *MATa* and the other with *MATα*). Haploid cells, having no chromosome homologs, cannot form bivalents prior to Meiosis I and will produce only inviable spores, because of nondisjunction, but again a *spo13Δ* mutation will allow viable haploid cells to be recovered (**Figure 12.31C**). Of course, haploid cells also cannot use a homologous chromosome as a template for DSB repair; so viable spores require not only the Meiosis I bypass but also the elimination of barriers that inhibit repair of DSBs using a sister chromatid as the template. Interestingly, *spo13Δ* haploids are efficient in repairing DSBs by sister chromatid recombination, while *SPO13* haploid cells are much less able to do this without removing a barrier to sister chromatid exchange.

Although no Meiosis I bypass strategy has been found in other organisms, it is possible to recover viable offspring from Spo11$^-$ cells. With fission yeast, flies, or worms, which have respectively 3, 4, and 6 pairs of chromosomes instead of the 16 in budding yeast, one can recover rare euploid, nonrecombined segregants from meiosis in spite of nondisjunction. For example, one in eight fission yeast meioses should, just by chance, partition all three chromosomes to the same pole in Meiosis I and then produce viable spores after Meiosis II. In fact, defects in recombination in worms were found by "high incidence of males" mutations (such as *him-8*) in which nondisjunction leads to many XO (male) offspring because of nondisjunction of the X chromosome.

12.14 RETURN-TO-GROWTH EXPERIMENTS REVEAL A PERIOD OF COMMITMENT TO MEIOTIC LEVELS OF RECOMBINATION

One other useful strategy in analyzing meiosis in budding yeast has been to allow diploids to initiate the process but then pull them back into vegetative conditions that block progression to the later stages of meiosis. A return-to-growth (RTG) experiment is shown in **Figure 12.32**. Diploids carrying the heteroalleles *lys2-1* and *lys2-2* were placed in sporulation medium for the time indicated and then washed and transferred back to growth medium. Whereas even after 12 hr incubation in sporulation medium the diploids do not form asci, by 8 hr they exhibit meiotic levels of Lys2$^+$ prototroph formation. In such return-to-growth protocols the cells end up as diploids as there is no reductional division (Figure 12.32B). Genetic analysis confirms that the Lys2$^+$ prototrophs that are recovered early in the time course are still diploid and usually heterozygous for other genetic markers (*leu1/LEU1* in this example). Thus cells are "committed" to undergo a high level of recombination before they are irrevocably channeled into meiosis and spore formation.

Although Dmc1 is required to complete interchromosomal meiotic recombination, in return-to-growth assays *dmc1Δ/dmc1Δ* diploids can be recovered with a high level of recombination, because in mitotic cells (and returned-to-growth cells) Rad51 is sufficient to repair the DSBs. In keeping with DSB repair following the rules of mitotic rather than meiotic cells, most RTG recombination intermediates are resolved as noncrossovers.

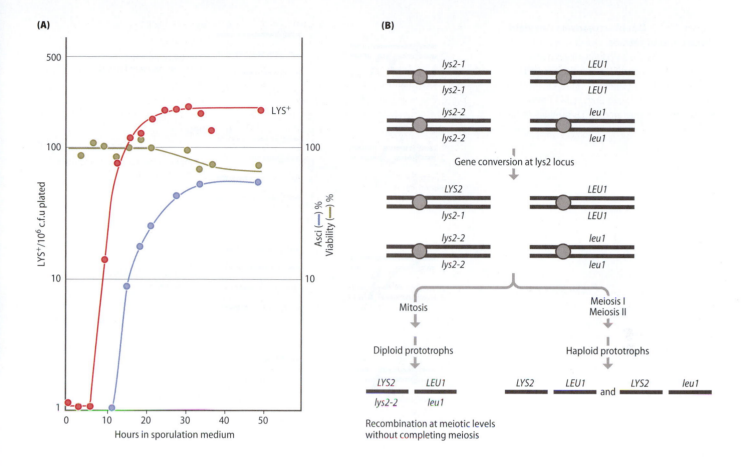

Figure 12.32 Return-to-growth (RTG) assay in budding yeast. (A) Diploid cells carrying heteroalleles of *lys2* and heterozygous for *leu1/ LEU1* were placed in sporulation medium and at intervals were plated onto plates lacking lysine and the number of Lys$^+$ colony-forming units (c.f.u.) scored. Cells at each time point were also examined for ascus formation. This RTG assay shows that cells become committed to a high level of recombination even when they exit meiosis before chromosome segregation and spore formation. (B) Illustration of *LYS2* prototroph formation before and after commitment to meiotic chromosome segregation. Because *leu1* is very tightly linked to a centromere, there is almost no crossing over between the marker and the centromere and hence diploids will nearly always be Leu1$^+$, whereas if cells progress through meiosis half of the Lys2$^+$ recombinants will be *leu1*. (Adapted from Esposito RE & Esposito MS [1974] *Proc Natl Acad Sci USA* 71:3172–3176. With permission from the National Academy of Sciences.)

12.15 INTERFERENCE REGULATES THE DISTRIBUTION OF CROSSOVERS

One of the most perplexing aspects of meiotic recombination is the phenomenon of crossover interference, which prevents a crossover from being located close to another crossover. This mechanism ensures that most crossovers will be well spaced from each other and not distributed at random. In *Drosophila*, where each chromosome arm usually has one or two exchanges, a single exchange is most likely to appear in the middle of the arm, whereas two exchanges will be located away from the center. Interference is readily measured by the fact that the observed number of double crossovers (DCO) in a three-factor cross is markedly lower than the expected number, based on the individual map distances (see Figure 12.3). Interference is often expressed as a *coefficient of coincidence* (c.o.c.), which is defined as the observed number of DCO divided by the expected number based on map distances. Interference is then defined as 1 − c.o.c.

In budding yeast, there are often more than two exchanges along a chromosome arm, but again they are well spaced from each other. Interestingly, in yeast, where one can recover all four products of meiosis, there is no *chromatid interference*; that is, in a region where there are two crossovers, there is no bias as to which chromatids will be involved. For example, if one crossover has occurred between homologs 2 and 3 in **Figure 12.33**, a second crossover is just as likely to occur between chromatids 1 and 4 as it is for a second exchange to involve the original

Figure 12.33 Double crossovers can yield three types of tetrads. Two-strand double crossovers will yield a PD for flanking markers, while a four-strand DCO will produce an NPD tetrad. Two different three-strand DCOs, involving different chromatids, will both yield TT tetrads. Hence only one-quarter of DCOs produce an NPD outcome.

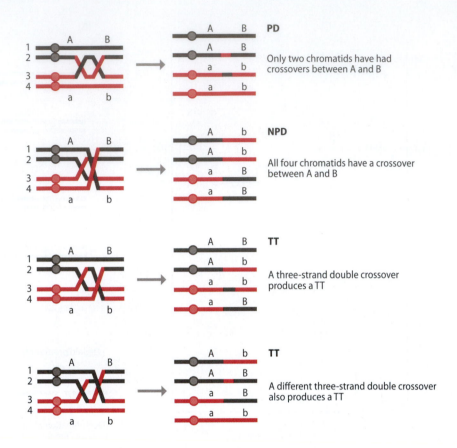

Only two chromatids have had crossovers between A and B

All four chromatids have a crossover between A and B

A three-strand double crossover produces a TT

A different three-strand double crossover also produces a TT

pair of chromatids. Since each possible double crossover arrangement is equally likely, there should be as many PD as NPD for the flanking markers and twice as many outcomes exhibiting a tetratype pattern as either parental or nonparental ditypes.

In yeast one can estimate crossover interference in several different ways. First, as with flies, one can show that the expected number of double crossovers in three-factor crosses exceeds the observed number, based on known map distances. Second, in a diploid carrying two linked markers, the number of DCO can be assessed by assessing the observed number of NPD tetrads (which can only arise by four-strand double crossovers) compared to the expected number based on the size of the interval and the proportion of observed TT.

Haig Papazian showed that the expected number of NPD double crossovers (in the absence of interference) can best be determined by the formula

$$f\mathrm{NPD_{exp}} = \tfrac{1}{2}[1 - f\mathrm{TT_{obs}} - (1 - 3/2\, f\mathrm{TT_{obs}})^{2/3}]$$

so long as there are not more than two crossovers in an interval (that is, when the proportion of TT is not close to two-thirds of all events). This expected number can then be compared with the observed number by a chi-squared test. An example of this analysis is shown in for an interval along a budding yeast chromosome in **Figure 12.34A**. In each of the three intervals the expected number of NPDs significantly exceeds the observed number (**Figure 12.34B**).

Finally, interference between crossovers in two adjacent intervals (A–B and B–C) can be determined by comparing the distribution of the three tetrad-types (PD, NPD, and TT) for one interval (B–C) in the cases when

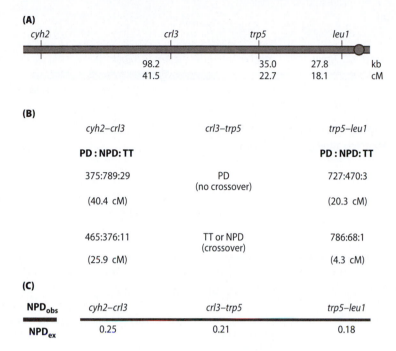

(A)

(B)

	cyh2–crl3	crl3–trp5	trp5–leu1
	PD : NPD : TT		**PD : NPD : TT**
	375:789:29	PD (no crossover)	727:470:3
	(40.4 cM)		(20.3 cM)
	465:376:11	TT or NPD (crossover)	786:68:1
	(25.9 cM)		(4.3 cM)

(C)

$\dfrac{\text{NPD}_{obs}}{\text{NPD}_{ex}}$	cyh2–crl3	crl3–trp5	trp5–leu1
	0.25	0.21	0.18

Figure 12.34 Crossover interference seen in budding yeast tetrads. (A) A genetic and physical map of part of the left arm of chromosome 7. (B) Tetrad data for two intervals flanking *crl3–trp5*. Among tetrads in which *crl3–trp5* exhibited the parental arrangement of markers (PD), the distribution of PD:NPD:TT in the two adjacent intervals shows a much longer map distance than among those tetrads in which there is evidence of at least one crossover in the *crl3–tp5* interval (TT or NPD). Calculated map distances are given in parentheses. (C) Another evidence of interference is the reduced number of NPDs in three genetic intervals relative to the expected number of NPDs, as calculated by the method of Papazian. (From data in Malkova A, Swanson J, German M et al. [2004] *Genetics* 168:49–63. With permission from the Genetics Society of America.)

the adjacent interval (A–B) either does or does not exhibit a crossover (**Figure 12.34C**). First one identifies all the tetrads in which the A–B interval contains either a TT or NPD pattern (indicative of crossing over) and determines the distribution of PD, NPD and TT in the adjacent B–C interval. Then this distribution is compared to the distribution of PD, NPD and TT in interval B–C for all the tetrads in which the A–B interval exhibits a PD pattern (taken as indicative of no crossing over). Significant interference can be shown by suitable statistical test, such as a contingency chi-squared test. In Figure 12.34C, the two intervals adjacent to *crl3–trp5* are shown. Among crossovers in that interval, the map distance for both *cyh2–crl3* and *trp5–leu1* are much smaller than when there was no crossover in the *crl3–trp5* region.

This last method of measuring interference substantiated earlier evidence that gene conversions accompanied by crossing over are interfering whereas gene conversions without an associated exchange are not interfering. That NCO gene conversions do not interfere suggests that they might arise from different intermediates from gene conversions associated with crossovers. Molecular evidence supporting this idea is presented in Chapter 13.

12.16 HOMEOSTASIS ASSURES THAT SMALL CHROMOSOMES USUALLY GET AT LEAST ONE CROSSOVER DURING MEIOSIS

If one plots the relationship between genetic map distance (cM) and physical chromosome length in kilobases, the 16 chromosomes of budding yeast lie on a straight line, but importantly the intercept of this line does not go through the origin; rather, it intercepts the *y*-axis (**Figure 12.35A**). These data indicate that smaller chromosomes have a higher crossover density, which seems to reflect the necessity of ensuring that small chromosomes will have an exchange, which is needed to assure proper meiotic chromosome disjunction. A similar correlation is seen among human chromosomes.

(A)

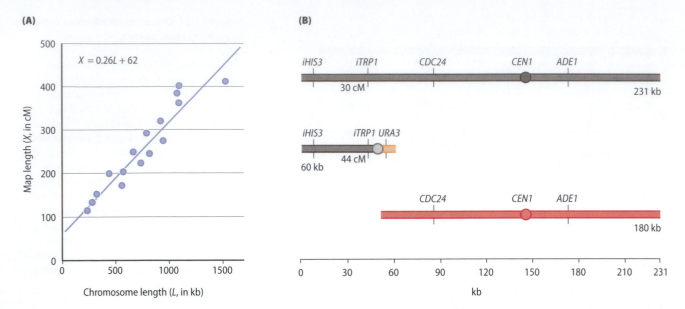

(B)

Figure 12.35 The relation of genetic map length to chromosome length. (A) Budding yeast chromosomes exhibit a genetic map proportional to their length but there is more crossing over per unit length in short versus long chromosomes. (B) If the 231-kb chromosome I (the smallest yeast chromosome) is bisected by the addition of a second centromere and telomeres to produce 60- and 180-kb derivatives , then the genetic distance between markers on the very small chromosome segment (for example, the inserted *HIS3* and *TRP1* genes) increases by 40%, supporting the idea that larger size enhances interference. (A, adapted from Stahl FW, Foss HM, Young LS et al. [2004] *Genetics* 168:35–48. With permission from the Genetics Society of America. B, from Kaback DB, Barber D, Mahon J et al. [1999] *Genetics* 152:1475–1486. With permission from the Genetics Society of America.)

David Kaback carried out a very informative experiment that involved bisecting a chromosome arm, by adding a centromere and new telomeres (Figure 12.35B). One could then compare the same interval as a segment of a long chromosome arm or as a major part of a short chromosome arm. Whereas the *his3–trp1* interval in its normal 231 kb chromosome context has a map distance of about 30 cM, this distance dramatically increases to about 44 cM when the arm is only 60 kb. Thus when there are no adjacent crossovers, the level of exchange is much higher.

In *Caenorhabditis* there is another manifestation of interference. Each pair of homologs has on average, and almost always, a single crossover. Remember that one crossover per meiosis translates to a map distance of 50 cM, because a single crossover will yield two parental and two recombined chromosomes. This control appears to be exerted at the level of crossovers, as *C. elegans* chromosomes experience more DSBs than result in crossovers. But the most striking evidence of interference as a form of crossover control came from the analysis of crossovers in a situation when two chromosomes were fused together and made homozygous. Despite their combined length, there was only a single crossover (**Figure 12.36**). Thus in *C. elegans* there must be a mechanism

Figure 12.36 Evidence of crossover interference in *C. elegans*. The measured map distance for each of two normal chromosomes is nearly 50 cM each, which is the equivalent of one crossover per meiosis. Note that not all crossovers are recovered because the genetic markers do not cover the entire chromosome length. These two chromosomes were fused by X-irradiation and a diploid was constructed marked with heterozygous markers. There was only one crossover for this double-length chromosome instead of the expected two. (Adapted from Hillers KJ & Villeneuve AM [2003] *Curr Biol* 13:1641–1647. With permission from Elsevier.)

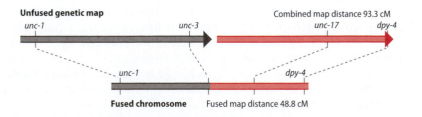

to enforce a single exchange per chromosome, regardless of its length. Crossover control not surprisingly depends on the SC, as partial depletions of one of the central element SYP proteins results in an increased number of crossovers.

12.17 THERE ARE SEVERAL MODELS OF INTERFERENCE

Three models, none of them yet with a sound molecular basis, have emerged to account for crossover interference: a stress-relief model, a polymerization model, and a counting model (**Figure 12.37**). The stress-relief mechanism, advocated by Kleckner, is predicated on the assumption that meiotic chromosomes are constrained and under torsional or mechanical stress (Figure 12.37A). Nancy Kleckner analogizes the situation to a rigid, but flexible, beam with a coating of a more rigid polymer. As the beam is twisted, the stresses result in a cracking of the polymer at some point; this relieves stress on the polymer coating in that vicinity, so that further torque will create a second crack that will be far away from the first, in a region that is still under stress. In molecular terms, it is not yet evident how stress is established or relieved, especially since the point where interference seems to be established is early after the induction of DSBs, which one might imagine have by themselves relieved stress. The roles of the so-called ZMM proteins in establishing interference will be addressed in the following chapter.

Figure 12.37 Models of crossover interference. (A) The stress model uses the analogy of a twisted beam with a coating that can crack. This crack absorbs stress so that further twisting of the beam will result in a second crack far away from the first. (B) A polymerization model envisions the propagation of an inhibitor of crossing over that moves away from the site of a crossover (CO) and displaces a component required from crossing over. (C) The counting model suggests that each CO is followed by a fixed number of encounters with recombination intermediates that will be resolved as noncrossovers (NCO). (Adapted From Berchowitz LE & Copenhaver GP [2010] *Curr Genomics* 11:91–102. With permission from Bentham Science Publishers.)

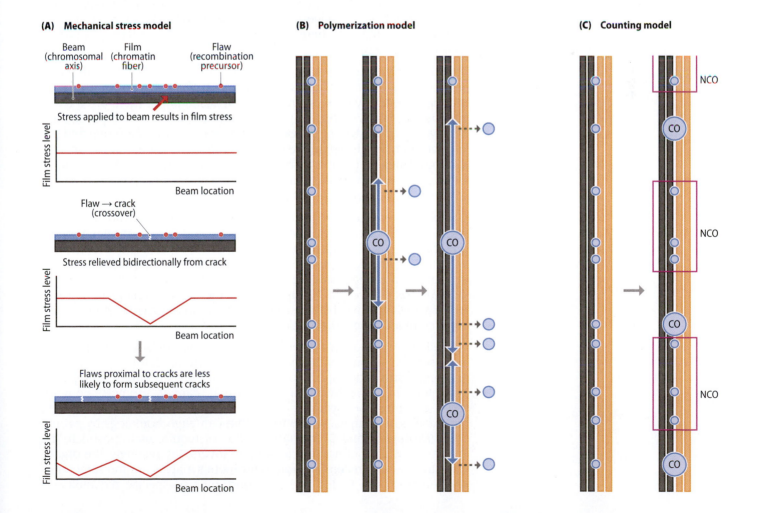

(A) Mechanical stress model

Beam (chromosomal axis)   Film (chromatin fiber)   Flaw (recombination precursor)

Stress applied to beam results in film stress

Film stress level

Beam location

Flaw → crack (crossover)

Stress relieved bidirectionally from crack

Film stress level

Beam location

Flaws proximal to cracks are less likely to form subsequent cracks

Film stress level

Beam location

(B) Polymerization model

CO

CO

CO

(C) Counting model

NCO

CO

NCO

CO

NCO

CO

A different possible mechanism would be the polymerization of a crossover inhibitor from a crossover site, as proposed by Jeff King and Robert Mortimer (Figure 12.37B). In this model, a zone around the crossover would ensure that any nearby recombination intermediates would be resolved as noncrossovers. Inhibition could travel down the SC, where crossover resolution appears to occur. This is an attractive feature of this model because it can account for the huge difference in genomic distance over which interference acts in yeast versus flies versus mammals. Although the sizes of the genomes varies tremendously, the lengths of SC in all organisms are much more similar. The molecular basis for such an inhibitor has not been specified.

An alternative model, advanced by Frank Stahl, begins with the well-documented assertion that there are considerably more DSB repair events than are resolved as crossovers (Figure 12.37C). Thus some fraction of DSB repair intermediates must be resolved without an exchange. This can occur by an early decision, channeling some events to be noncrossovers by an inherently noncrossover pathway such as SDSA or by resolving common dHJ intermediates to yield noncrossover outcomes. The model is predicated on the simplifying assumption (which is not supported by molecular data) that both crossovers and noncrossovers are derived from a common intermediate. For simplicity, we imagine that there is a common HJ intermediate that can be resolved to give either a crossover or a noncrossover. The counting model suggests that there is an apparently processive counting "machine" moving along a chromosome arm that keeps track of the way one or more previous intermediates was resolved. If there was a crossover (CO), then the next several intermediates will be obligately resolved as noncrossovers (NCO). Analysis of the data in flies and in budding yeast suggests that there are two NCOs between each CO in *Drosophila*, but that there will be three such NCOs between a pair of COs in budding yeast.

The counting model provides a good description of crossover distribution in *Drosophila*, but the fit in budding yeast or *Arabidopsis* can be made substantially better by making an additional assumption: there are a small number of crossovers that do not exert interference. For example, in budding yeast, about 10% of crossovers appear to be noninterfering and the proportion of noninterfering exchanges would be higher on the smaller chromosomes. These extra crossovers can account for the higher level of crossing over in smaller chromosomes. These extra crossovers would also account for the chiasmata seen in Zip1 mutants.

Evidence against a simple counting model (when there are no extra, noninterfering chromosomes) came from using budding yeast hypomorphic *spo11* alleles that initiate fewer DSBs. If there are fewer total events then, maintaining the same one CO, three NCO rule, there should be a significant drop in map distances. But in fact these strains show *crossover homeostasis* (**Figure 12.38**), implying that there is a mechanism that will ensure that there will be at least one interhomolog CO per chromosome arm to ensure proper disjunction.

A credible explanation for these noninterfering crossovers is that they represent a second pathway of recombination. Stahl observed that the organisms that appear to have a small, but important, number of noninterfering crossovers are those that cannot promote homologous chromosome synapsis without recombination (*Saccharomyces*, *Arabidopsis*, mammals), while those that can align homologs by pairing sites (*Drosophila* and *Caenorhabditis*) do not require us to postulate the existence of a set of noninterfering crossovers. Interestingly, the organisms that require recombination to promote synapsis are the same ones that employ both Rad51 and a second, meiosis-specific recombinase,

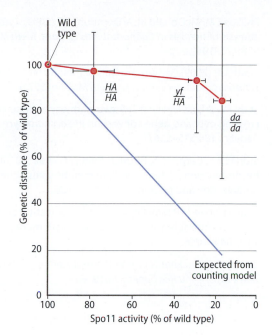

Figure 12.38 A test of the counting model of interference. Fewer DSBs are generated in strains carrying different mutants of the Spo11 protein. An HA tag on the end Spo11 reduces its activity, while a Y316F (*yf*) mutation or a D290A (*da*) mutation at the active site eliminates its activity. If crossovers are spaced by a strict counting model, genetic distance should decrease proportionally to their measured Spo11 activity to create DSBs, but in fact the mutant strains exhibit crossover homeostasis. (Adapted from Martini E, Diaz RL, Hunter N & Keeney S [2006] *Cell* 126:285–295. With permission from Elsevier.)

Dmc1, as well as several proteins that aid the function of Dmc1, such as Hop2 and Mnd1. In these organisms, the absence of Dmc1 abolishes all interhomolog crossing over. This finding suggests that no interhomolog recombination can happen unless the homologs are first aligned by Dmc1-specific events. However, at least in *Saccharomyces*, Dmc1 seems to be directly involved in all exchange events, whereas Rad51 is surprisingly much less important, as we will see in Chapter 13.

SUMMARY

Recombination in meiosis is a highly choreographed event, making sure that there is at least one exchange per chromosome to provide the tension between homologs needed to assure proper segregation. The formation of the synaptonemal complex is critical in maintaining synapsis and in controlling crossing over. SC formation between homologous chromosomes in yeast and mammals depends on functional homologous recombination to pair homologs, but in worms and flies its formation is independent of creating DSBs or strand invasion. Mismatch repair is efficient for most single base-pair mismatches but C:C mismatches and poorly repaired palindromic sequences show a high level of post-meiotic segregation that helps to define the extent of strand invasion and the connection to crossing over. The least understood aspect of meiosis is how interference is enforced.

SUGGESTED READING

Andersen SL & Sekelsky J (2010) Meiotic versus mitotic recombination: two different routes for double-strand break repair: the different functions of meiotic versus mitotic DSB repair are reflected in different pathway usage and different outcomes. *BioEssays* 32:1058–1066.

Haber JE, Thorburn PC & Rogers D (1984) Meiotic and mitotic behavior of dicentric chromosomes in *Saccharomyces cerevisiae*. *Genetics* 106:185–205.

Hillers KJ & Villeneuve AM (2003) Chromosome-wide control of meiotic crossing over in *C. elegans*. *Curr Biol* 13:1641–1647.

Housworth EA & Stahl FW (2009) Is there variation in crossover interference levels among chromosomes from human males? *Genetics* 183:403–405.

Hunt PA (2006) Meiosis in mammals: recombination, non-disjunction and the environment. *Biochem Soc Trans* 34:574–577.

King JS & Mortimer RK (1991) A mathematical model of interference for use in constructing linkage maps from tetrad data. *Genetics* 129:597–602.

Kleckner N, Zickle, D, Jones GH et al. (2004) A mechanical basis for chromosome function. *Proc Natl Acad Sci USA* 101:12592–12597.

Malkova A, Swanson J, German M et al. (2004) Gene conversion and crossing-over along the 405-kb left arm of *Saccharomyces cerevisiae* chromosome VII. *Genetics* 168:49–63.

Martinez-Perez E & Colaiacovo MP (2009) Distribution of meiotic recombination events: talking to your neighbors. *Curr Opin Genet Dev* 19:105–112.

Niu H, Li X, Job E et al. (2007) Mek1 kinase is regulated to suppress double-strand break repair between sister chromatids during budding yeast meiosis. *Mol Cell Biol* 27:5456–5467.

Phillips CM, McDonald KL & Dernburg AF (2009) Cytological analysis of meiosis in *Caenorhabditis elegans*. *Meth Mol Biol* 558:171–195.

Smith GR (2009) Genetic analysis of meiotic recombination in *Schizosaccharomyces pombe*. *Meth Mol Biol* 557:65–76.

Stahl FW & Foss HM (2010) A two-pathway analysis of meiotic crossing over and gene conversion in *Saccharomyces cerevisiae*. *Genetics* 186:515–536.

Storlazzi A, Gargano S, Ruprich-Robert G et al. (2010) Recombination proteins mediate meiotic spatial chromosome organization and pairing. *Cell* 141:94–106.

Zhao H, Speed TP & McPeek MS (1995) Statistical analysis of crossover interference using the chi-square model. *Genetics* 139:1045–1056.

Zickler D & Kleckner N (1998) The leptotene-zygotene transition of meiosis. *Annu Rev Genet* 32:619–697.

CHAPTER 13
MOLECULAR EVENTS DURING MEIOTIC RECOMBINATION

Meiotic recombination uses the same basic homologous recombination machinery as somatic cells, but it has become overlaid by a number of additional features that result in the significant distinctions between meiotic and mitotic recombination. First, meiotic recombination is initiated by the creation of breaks by a topoisomerase-like protein, Spo11, that acts preferentially at "hot spots." Second, in many organisms, strand invasion depends not only on Rad51 but also on a meiosis-specific homolog, Dmc1, and on a number of additional proteins. Third, meiotic gene conversions are associated with a very high level of reciprocal crossing over. This chapter will examine each of these meiosis-specific features in detail.

13.1 SPO11 CREATES DSBs TO INITIATE MEIOTIC RECOMBINATION

In all eukaryotes studied, meiotic recombination depends on a homolog of the yeast *SPO11* gene. Spo11 is itself homologous to the topoisomerase VI family in bacteria and archaea. Spo11 creates a DSB with a covalent linkage of a tyrosine residue to overlapping 5′ ends (**Figure 13.1**). The presence of Spo11 blocks 5′-to-3′ resection of the DSB ends. Budding yeast Spo11 is removed by the Mre11 endonuclease (in collaboration with Rad50, Xrs2, and Sae2) that leaves a small tail of ssDNA attached to the protein and creates 3′-ended ssDNA that can be further enlarged by 5′-to-3′ exonucleases or helicase/endonucleases. A single amino acid mutation, Spo11-Y135F, alters the tyrosine by which Spo11 is covalently attached to DNA, and abolishes its activity.

In budding yeast, the creation of DSBs occurs nonrandomly at "hot spots" that correlate with the sites where gene conversions are most frequent. Spo11-mediated cleavage is tightly regulated. It requires the participation of a surprising number of other factors, whose precise functions are not yet fully understood. Among these factors are Rec102, Rec104, Rec114,

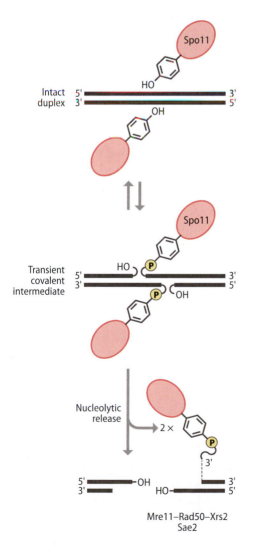

Figure 13.1 Spo11, a modified topoisomerase, creates meiotic DSBs. Spo11 acts by creating a covalent link to the 5′ end of the DSB via a tyrosine. Removal of Spo11 is needed to allow 5′-to-3′ resection of the DSB ends. Excision of Spo11 requires the Mre11 endonuclease complex as well as Sae2, by clipping off short oligonucleotides connected to the enzyme. (Adapted from Keeney S, Giroux CN & Kleckner N [1997] *Cell* 88:375–384. With permission from Elsevier.)

Mei4, Mer2, and Ski8. Spo11 will act at novel sites when the protein is fused to a *GAL4* binding domain (Gal4BD) and the fusion protein is then directed to *GAL4* binding sites. However, even in these new locations, Spo11-Gal4BD cannot create DSBs unless all of the other factors listed above are expressed. No one has succeeded in getting Spo11 to create DSBs in vegetative cells, likely because not all necessary co-activators are present.

In budding yeast, MRX proteins play at least two distinct roles in meiotic recombination. First, MRX proteins are required to facilitate the creation of DSBs by Spo11. This function is distinct from its role in cleaving Spo11 from the DSB end, as nuclease-defective *mre11-H125N* or *rad50-S* mutations both enable creation of DSBs but cannot process them. The removal of Spo11 through Mre11's endonuclease activity yields oligonucleotides bound to Spo11. Presumably MRX also plays its normal end-binding roles in initiating 5'-to-3' resection, though this is hard to assess in view of its prior requirements. In other organisms, MRX (or MRN) does not appear to be required for Spo11 cleavage of DNA but does play a role in processing.

13.2 DSB FORMATION IS ASSOCIATED WITH REPLICATION AND CHROMATIN MODIFICATIONS

DSB formation in budding yeast is also strongly correlated with meiotic DNA replication. A seminal experiment by Michael Lichten's group examined DSB formation in a strain in which all the origins of replication along one chromosome arm were deleted (**Figure 13.2A**), so that a replication fork must traverse a long region before reaching the segment containing a DSB hot spot. Remarkably, the timing of DSB formation was delayed, proportional to the time it took for the region to replicate (**Figure 13.2B**). The time of appearance of DSBs occurs several hours after DNA replication is complete. When pre-meiotic replication is blocked by hydroxyurea or by deleting both Clb5 and Clb6 cyclins, DSB formation is greatly reduced. The same *clb5Δ clb6Δ* double mutant does not prevent mitotic replication, probably because other B-type cyclins such as Clb2 are expressed.

Figure 13.2 Delayed replication results in delayed chromatin remodeling and delayed appearance of DSBs. (A) When three origins of replication (ARS sequences) on the left arm of yeast chromosome III are deleted, DNA replication at sites such as YCL49c is delayed. (B) Delayed replication results in a subsequent delay in the appearance of DSBs at YCL49c. (C) The delay in replication also results in a delay in the appearance of microccal nuclease hypersensitive sites (MNase HS) in chromatin that precedes the appearance of Spo11-induced DSBs. (Adapted from Murakami H, Borde V, Shibata T et al. [2003] *Nucleic Acids Res* 31:4085–4090. With permission from Oxford University Press.)

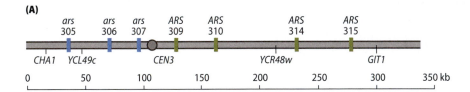

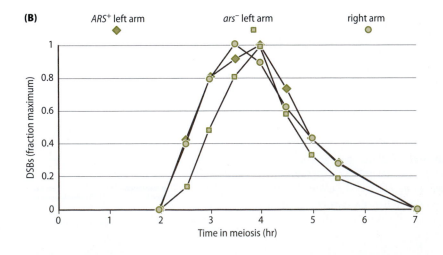

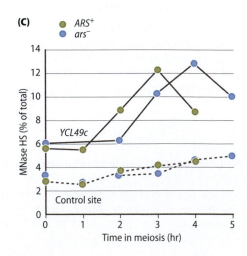

Replication fork passage is not absolutely required, as DSBs can be created in unreplicated chromosomes.

Still, under normal circumstances, the passage of the replication fork appears to set up conditions to initiate DSB formation some time later. First, replication fork passage is coupled to the assembly of cohesins that surround sister chromatids. The assembly of cohesins in meiosis is distinctive because a meiosis-specific cohesin protein, Rec8, replaces its mitotic homolog, Scc1 (also called Mcd1). When Rec8 is absent, DSB formation is greatly reduced in some chromosome regions but hardly affected in others. The absence of Rec8 also severely diminishes the ability of DSBs to be repaired. The connection to cohesins is poorly understood, but one possible link is the finding that Spo11 is initially localized around centromere regions and only later distributes to discrete sites along chromosome arms as DNA replication proceeds. This initial concentration may reflect the fact that Rec8-containing cohesins are also concentrated at centromeres as well as being located along chromosome arms.

A second replication-associated factor is the Cdc7–Dbf4 kinase (sometimes called the Dbf4-dependent kinase or DDK) that is associated with proteins moving with the replication fork. DDK phosphorylates Mer2, which in turn allows the recruitment of Rec114 and Mei4, which are some of the proteins required for Spo11 to create DSBs.

There is also a connection between DSB formation and chromatin modifications. In budding yeast, prior to DSB formation there are significant changes in chromatin structure that can be revealed by increased digestion of chromatin in the hot spot area by micrococcal nuclease (MNase) (**Figure 13.2C**). DSB formation selectively occurs in regions in which histone H3 lysine 4 is trimethylated by the Set1 histone methyltransferase prior to meiosis.

How these chromatin modifications and protein bindings are choreographed to create DSBs in budding yeast is beginning to be clear. Mer2 and cohesin proteins aggregate to create a chromosome axis (**Figure 13.3i**). This axis is elaborated by the binding of additional meiosis-specific proteins such as Hop1 and Red1. Meiotic hot spots, marked by trimethylation of histone H3 on lysine 4 (H3-K4)—predominantly at promoter regions—are located in the loops created by axis formation (Figure 13.3ii). The Spp1 protein, part of the COMPASS complex that includes the histone methyltransferase Set1, has two important affinities: it binds H3-K4-Me3 but it also binds to Mer2. Thus the hot spot is dragged to the axis, where the DSB is created by Spo11, which is associated with Mer2, Rec114, Mei4 (Figure 13.3iii).

How "hot" a given hot spot is depends not only on the specific sequences where cleavage occurs but also on the surrounding larger chromosomal context. When plasmids containing each of two *leu2* alleles were integrated at a number of different sites in the budding yeast genome, there was at least a 20-fold difference in allelic recombination frequencies (producing Leu$^+$ recombinants) among the different sites of insertion, even though the alleles were embedded in a common plasmid backbone (**Figure 13.4**). Insofar as one can judge, the inserted sequences adopt the hot or cold character of the surrounding sequences. Thus, there are unknown regional influences on the frequency of initiating recombination. These factors could include the density of Rec8 or Mer2 or other axis proteins. This principle seems to extend to fission yeast and perhaps beyond.

One other constraint needs to be noted. Even very hot hot-spots do not result in Spo11 cleavage of most chromatids; in fact fewer than 1 in 4 molecules is cleaved even at prominent hotpots. Tetrad analysis also confirms that usually only 1 of 4 chromatids is cleaved in any region, as most non-Mendelian events are 2:6 and 6:2 events, with few, if any, 0:8 or 8:0. These results imply that both sisters are rarely cleaved in the same region. If one sister in each homolog was cleaved it is possible that sequential repair of the DSBs could result in what would seem to be a single gene conversion event. How cleavage is restricted to one of two sister chromatids is not known. As described later, induction of the HO or VDE endonucleases in meiosis does not suffer from this constraint: both sisters can be readily cut.

13.3 MEIOTIC HOT SPOTS CAN BE IDENTIFIED IN MANY DIFFERENT ORGANISMS

In budding yeast, increasingly high-resolution methods have been used to identify the locations of Spo11-induced DSBs. As mentioned above, mutations such as *rad50-S* allow DSBs to be created but fail to remove Spo11.

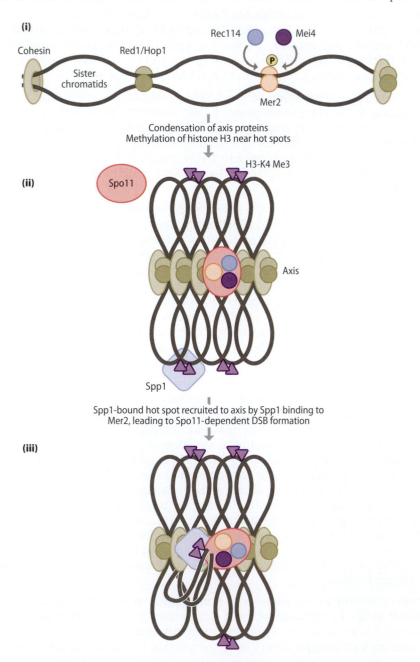

Figure 13.3 Creation of DSBs by recruitment of hot spot DNA to the sister chromatid axis. (i) Cohesin and axial element proteins Red1 and Hop1 bind at dispersed sites, linking sister chromatids, along with the Mer2 protein that is phosphorylated (P) by the passage of the replication fork. (ii) Aggregation of the axial element proteins forms an axis, while hot spots are located in chromatin loops. Hot spots are marked by histone H3-K4 trimethylation. (iii) Spo11 assembles with Mer2 and other partner proteins to cleave a hot spot that is dragged to the axis by the Spp1 protein that interacts both with trimethylated H3-K4 and Mer2. (Adapted from Panizza S, Mendoza MA, Berlinger M et al. [2011] *Cell* 146:372–383. With permission from Elsevier.)

(A)

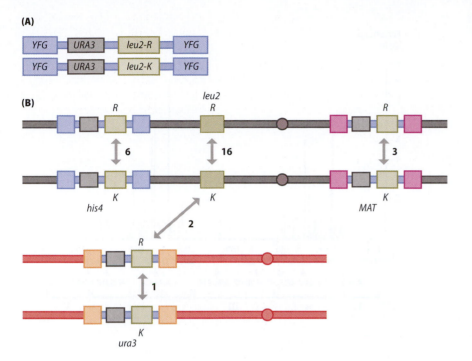

Figure 13.4 Chromosome location governs recombination activity. (A) A circular plasmid carrying *URA3* and one of two *leu2* alleles and a segment of chromosome homology can be integrated at "your favorite gene" (YFG). (B) Allelic and ectopic recombination can be scored as *LEU2* prototrophs arising by meiotic recombination between *leu2-K* and *leu2-R* in their different locations. Each measurement was made in a diploid having only one pair of *leu2* alleles at the locations indicated by arrows. The relative rates of recombination are shown, normalized to that when the alleles are integrated at *URA3* (that is, the rate at that locus = 1). At *leu2*, the two alleles were tested in the absence of surrounding plasmid sequences. One ectopic recombination event was scored when only *leu2-R* was integrated at *ura3* and it could recombine with *leu2-K* at the normal *leu2* site when the second allele was the *leu2-K,R* double mutant. (Adapted from Lichten M, Borts RH & Haber JE [1987] *Genetics* 115:233–246. With permission from the Genetics Society of America.)

The presence of Spo11 prevents the DSB ends from being resected or to engage in recombination (or re-ligation). Separation of entire chromosomes by pulse field gel electrophoresis, isolated during meiotic prophase in a *rad50-S* mutant, revealed that there were a few prominent fragments when the gel was probed with sequences from one end of a chromosome (**Figure 13.5A**). These are sites of frequent DSB formation (that is, meiotic hot spots). A higher resolution map was obtained by fragmenting DNA, isolating DNA to which Spo11 was covalently attached and probing tiled microarrays (**Figure 13.5B**). A similar approach examined DNA from cells that had progressed beyond the release of Spo11 but was prevented from subsequent recombination by deleting the Dmc1 protein. Thus the ends were resected, leaving long-enough ssDNA ends to bind to BND cellulose, which does not bind dsDNA. The purified resected fragments adjacent to DSB cleavage sites could then be used to probe a microarray to identify hot spots (**Figure 13.6**). These methods have shown that there is no specific DNA sequence that specifies Spo11 cleavage; rather, a given hot spot is made up of a number of different nearby cleavage sites that are used in different cells.

There are important differences between the cleavage maps produced by isolating DNA from *rad50-S* (or *sae2Δ*) cells and that taken from *dmc1Δ* cells, most notably that regions that seemed particularly "cold" in the *rad50-S* maps—subtelomeric regions and centromeric regions—are not nearly so extreme in the *dmc1Δ* maps (Figure 13.6). It is possible that the *rad50-S* mutant, by "trapping" Spo11 on ends, leads to the depletion of a factor needed to initiate recombination in late-replicating regions such as those near telomeres and thus DSBs in these regions are underrepresented. Genetic experiments have long indicated that there are frequent exchanges—even ectopic exchanges—among subtelomeric regions, so it is likely that the *dmc1Δ*-based maps are more complete. Another feature of these hot spot maps is that the peaks do not emerge out of a "floor" that is completely cold; there is a substantial elevation of DSBs across the entire genome that underlies the peaks of the hot spots. The great majority of DSB sites lie in promoter regions of genes, but what makes some genes hot is not yet evident.

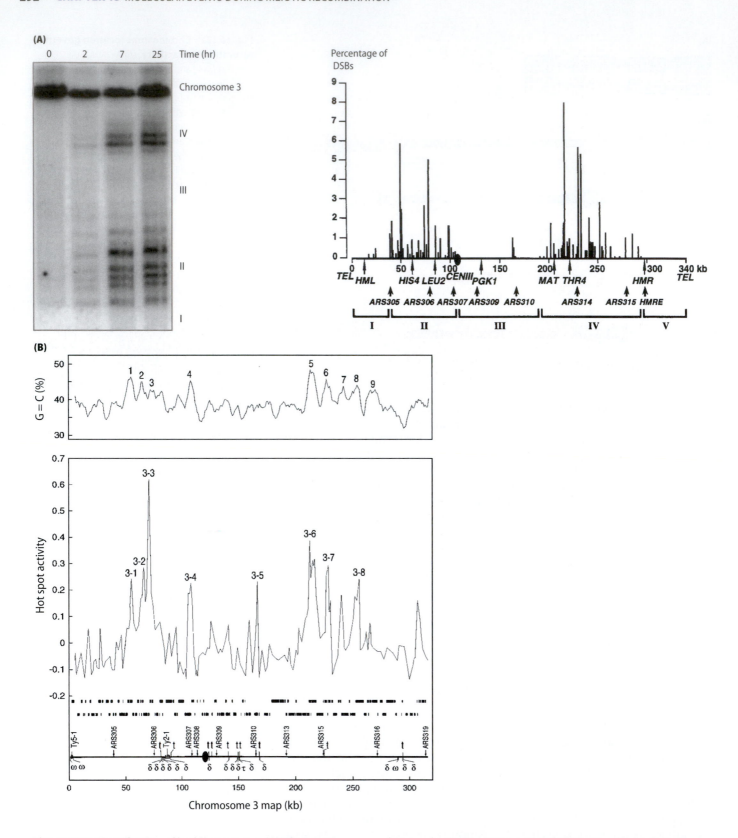

Figure 13.5 Identification of budding yeast meiotic hot spots.
(A) Meiotic DSBs cannot be processed in a *rad50-S* mutation. By probing a chromosome-separating gel with a probe from the left end of chromosome 3, it is possible to map the most frequent sites of breakage. (B) DNA fragments covalently attached to Spo11 in a *rad50-S* mutant can be hybridized to a microarray of chromosome 3. The positions of transcripts on the top and bottom strands are shown in two lines, as are the locations of origins of replication (ARS sequences), ttRNAs (t) and Ty retrotransposon elements, and solo LTR sequences (δ). (A, from Baudat F & Nicolas A [1997] *Proc Natl Acad Sci USA* 94:5213–5218. With permission from the National Academy of Sciences. B, from Gerton JL, DeRisi J, Shroff R et al. [2000] *Proc Natl Acad Sci USA* 97:11383–11390. With permission from the National Academy of Sciences.)

An even more precise way to map DSB sites takes advantage of the fact that short oligonucleotides are still covalently attached to Spo11 when Mre11 excises Spo11 from the DSB ends (**Figure 13.7A**). Removal of Spo11 leaves equal amounts of two distinct families of DSB ends (**Figure 13.7B**). One set of ends has 10–15 nt of ssDNA at the 3' end, whereas the other set has 3' ssDNA of 24–40 nt that can be purified and then amplified and sequenced to determine precisely where the DSBs have formed, and in

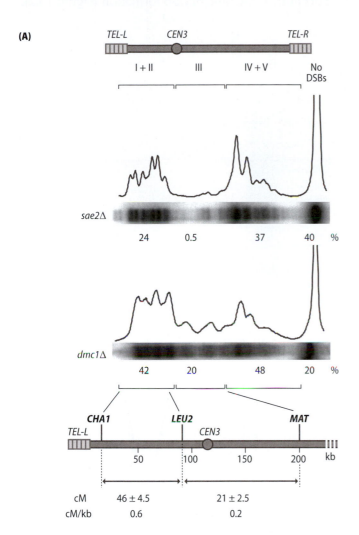

(A)

Figure 13.6 Comparison of DSB mapping methods. Comparison of mapping meiotic DSBs by mapping fragments of DNA attached to Spo11 in a *sae2Δ* (A) or *rad50-S* diploid (B) with mapping ssDNA located at resected DSBs in a *dmc1Δ* strain. (A) Separation of Spo11-cut chromosome 3 as in Figure 13.5. The locations of prominent hot spots are compared to the genetic map of part of chromosome 3. The distribution of DSBs recovered from the *dmc1Δ* experiment agrees more closely than that found with *sae2Δ*. (B) Microarray results for the two methods. (Adapted from Buhler C, Borde V & Lichten M [2007] *PLoS Biol* 5:e324.)

(B)

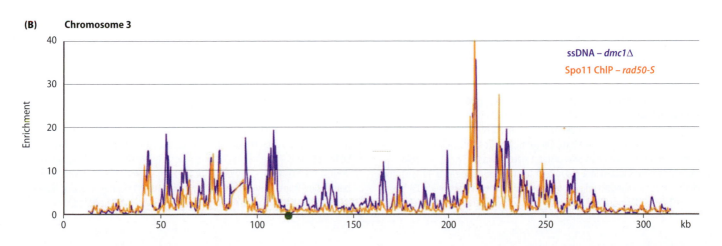

what abundance. The results of this analysis made it clear that DSBs appear in nucleosome-deficient regions (NDR) in transcription start sites (TSS) (**Figure 13.7C**). One example is shown in **Figure 13.7D**. Another key demonstration from this work is that the changes in chromatin structure that are needed for DSB formation are already established when cells enter meiosis; nucleosomes are already absent in the regions where DSBs will form (Figure 13.7D). Note, however, that measurement of MNase hypersensitive sites does show that there are additional changes in chromatin structure that occur near the time of DSB formation (Figure 13.2).

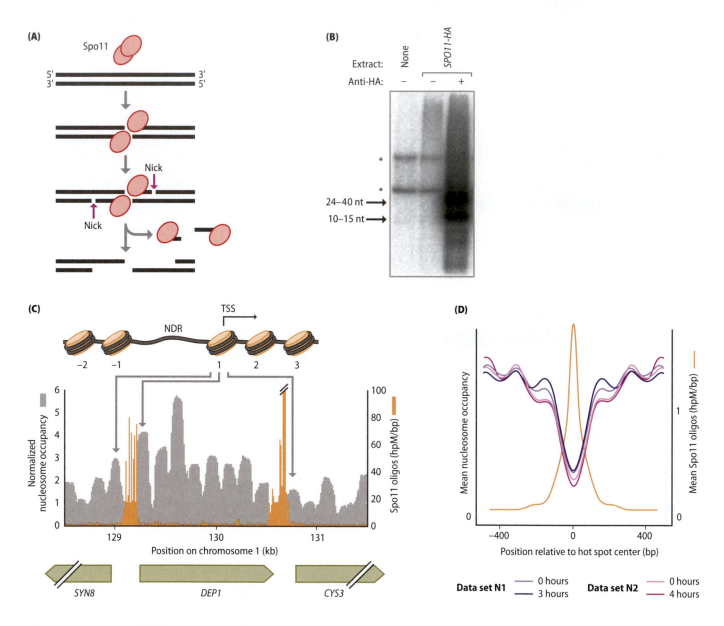

Figure 13.7 Mapping of DSBs by sequencing Mre11-liberated oligonucleotides. (A) Spo11 is excised from DSBs by Mre11-dependent nicks that liberate different sized oligonucleotides (abbreviated oligos) covalently attached to Spo11. (B) Oligonucleotides attached to Spo11-HA can be immunopurified as two distinct size groups (*arrows*). Asterisks represent bands found in a *spo11Δ* strain. (C) DSBs appear in nucleosome-deficient regions (NDR) adjacent to the transcription start site (TSS). Nucleosome positions are shown in *gray*, where the average nucleosome

density = 1. Arrows point to +1 nucleosomes in three transcribed genes. DSBs (*orange*) are evident in the NDRs of each gene and are scored as "hits per million per base pair" (hpM/bp) based on deep sequencing of the Spo11 oligos. (D) The absence of nucleosomes in the region where DSBs appear is evident even before meiosis is initiated. (A, C, D, adapted from Pan J, Sasaki M, Kniewel R et al. [2011] Cell 144:719–731. With permission from Elsevier. B, from Neale MJ, Pan J & Keeney S [2005] *Nature* 436:1053–1057. With permission from Macmillan Publishers, Ltd.)

Fission yeast

In fission yeast, the distribution of DSBs can also be assessed by examining DNA covalently bound to the Spo11 homolog, Rec12, in a *rad50-S* mutant. Strikingly, the distribution of prominent hot spots over the genome is very different from budding yeast. There are major regions of cleavage each separated by about 65 kb of "cold" DNA. Many of these are in large intergenic regions, but the best characterized hot spot, *ade6-M26,* lies within the ORF of the *ade6* gene. As we saw in Figure 12.20, the *ade6-M26* allele is much hotter than a similar mutation, *ade6-M375*, lying only a few nucleotides away. The mutation in *ade6-M26* creates a cyclic AMP-responsive element (CRE) to which the Atf1/Pcr1 transcription factor can bind. Other normal genomic sites containing the ATGACGT CRE motif proved to be hot spots as well. A recent survey has suggested that there are several other small sequence motifs that can also act as meiotic hot spots when they are introduced into *ade6*. In some of these cases, the binding of transcription factors causes an open chromatin conformation that may be required for DSB induction; in some other cases, the CRE-containing region is "open" even in the absence of Atf1. Some *S. pombe* sites are much hotter when placed in the *ade6* gene than they are in their normal chromosome context. What remains unsolved in *S. pombe* is the contradiction between very widely spaced sites of DSBs and a more uniform distribution of crossovers. One idea is that a significant fraction of initiating events in fission yeast are ssDNA nicks rather than DSBs—a proposal that finds some resonance in the observation that the detected intermediates of meiotic recombination in *S. pombe* are not double Holliday junctions but single HJs, as discussed below.

Mammals

In mammals, notably mice and primates, the distribution of meiotic recombination hot spots has been analyzed both in terms of crossover frequencies along a chromosome and from the spectrum of DSB cleavages. Crossovers can be inferred from an analysis of linkage disequilibrium, in which specific alleles of adjacent genes have remained historically linked. These conserved blocks are adjacent to other regions that are in equilibrium, that is, where crossing over has occurred often enough so that certain haplotypes are not favored. In mouse and in humans, PCR analysis of thousands of single sperm has allowed the identification of short, 1–2 kb regions, in which crossing over is hundreds of times higher than in adjacent regions. In addition a high-resolution analysis of crossing over along mouse chromosomes from large numbers of mouse progeny has added unprecedented detail. By these approaches it is possible to estimate that there is a meiotic hot spot every 50–100 kb but that hot spots themselves are confined to very small regions.

Recently, Galina Petukhova and Daniel Camerini-Ortero's labs have surveyed the landscape of DSBs in mouse meiosis by the same chromatin-immunoprecipitation/high throughput sequencing (ChIP-Seq) that were used in budding yeast. The sites of both Dmc1 and Rad51 binding to Spo11-dependent DSBs are shown for chromosome 1 in **Figure 13.8Ai**. Individual hot spots exhibit a roughly symmetrical pattern that spans several kilobases (**Figure 13.8Aii**). Overall, the hot-spot map correlates well with classic genetic maps for this chromosome. The special case of the hot spot(s) that assure there will be a crossover in the pseudoautosomal region (PAR) shared by X and Y chromosomes is shown in Figure 13.8B. Here there seem to be a cluster of hot spots while the rest of the X and Y chromosome have very low DSB activity. Unlike budding yeast, hot spots are not principally found in 5′ promoter regions.

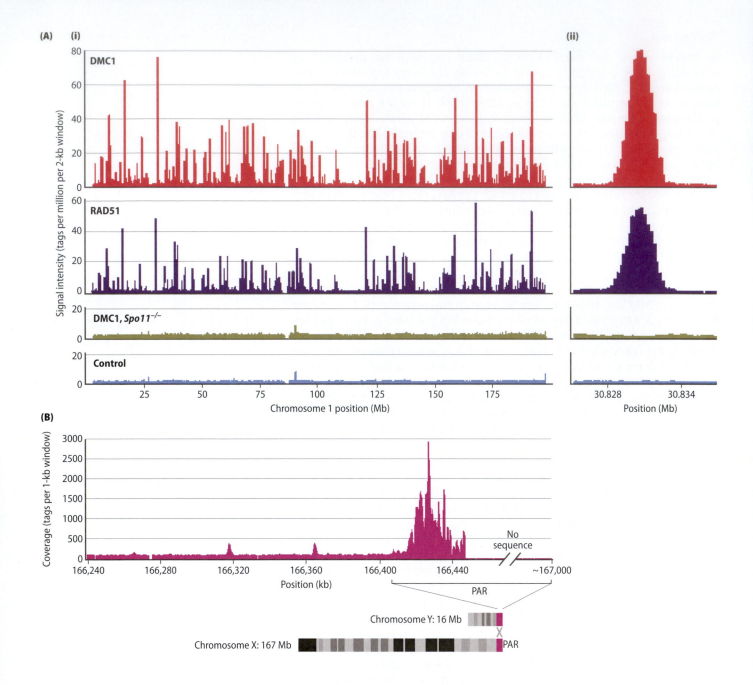

Figure 13.8 Mapping meiotic DSBs in mouse by ChIP-Seq. (A) Chromatin immunoprecipitation using anti-Rad51 and anti-Dmc1 bound to ssDNA created by Spo11 at hot spots was used to determine the locations of the meiotic hot spots. DSBs distribute roughly symmetrically around the center of a representative hot spot. (B) DSBs are confined to the pseudoautosomal region (PAR) of the Y chromosome that recombines with a homologous region of the X chromosome. The hot spot(s) distribute over a large region compared with a single hot spot. (Adapted from Smagulova F, Gregoretti IV, Brick K et al. [2011] *Nature* 472:375–378. With permission from Macmillan Publishers, Ltd.)

13.4 MEIOTIC HOT SPOTS IN MAMMALS CORRELATE STRONGLY WITH PRDM9 HISTONE METHYLTRANSFERASE

A key breakthrough in understanding the regulation of mammalian meiotic hot spots was the discovery that in mice and in humans, the *PRDM9* gene exerts effects on meiotic recombination on all the chromosomes. Prdm9 encodes a zinc-finger DNA binding protein with histone methyltransferase activity, directed at histone H3-K4. The structure of zinc-finger proteins was shown in Figure 10.11. Each 28 amino acid domain of the protein recognizes a three-nucleotide sequence; in combination, the zinc fingers of Prdm9 recognize a 13-bp motif. Strikingly, two different mouse species have significantly diverged zinc-finger motifs of Prdm9 that recognize different sequences; these differences can account for their different recombination profiles. Indeed there is remarkable variation in the sequences of the Prdm9 proteins in different mammals (**Figure 13.9**).

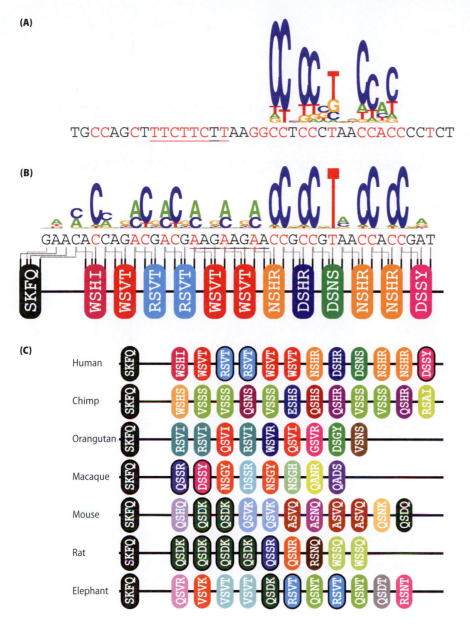

Figure 13.9 Comparison of PRDM9 consensus sequences. (A) A13-bp PRDM9 consensus motif in humans, represented by a logo plot that shows by relative letter height the estimated probability of hot spot activity and by total letter height the degree of base specificity, as well as an extended ~39-bp motif shown in *red*. (B) *In silico* prediction of the binding consensus for PRDM9, aligned with the 13-nucleotide oligomer, with more influential positions shown in *red*. Below the text is the sequence of four predicted DNA-contacting amino acids for the 13 successive human PRDM9 zinc fingers (one oval per finger, differing colors for differing fingers, and their predicted base contacts within the motif). An N-terminal zinc finger is separated from the others. (C) Sequence of four predicted DNA-contacting amino acids for the PRDM9 zinc fingers in seven mammalian species, presented as in (B). Distinct fingers are given different colors. (Adapted from Myers S, Bowden R, Tumian A et al. [2010] *Science* 327:876–879. With permission from the American Association for the Advancement of Science.)

A similar allelic variation in the human Prdm9 protein alters the DNA binding motifs recognized by the human protein and is likely to be responsible for more than 50% of the variation in hot spot activity in some human populations. In a recent study of more than 2 million crossovers in 30,000 African-Americans, a 17-bp Prdm9 recognition sequence could account for nearly all the hot spot activity.

A detailed analysis of all of the hot spots identified by Petukhova and Camerini-Ortero (Figure 13.8) revealed that they share a motif at the center of the hot spot (**Figure 13.10A**). This motif identifies the consensus (strain-specific) binding site for the mouse Prdm9 protein, confirming that most DSBs created in mouse testis are indeed determined by this zinc-finger protein. Unlike budding yeast, where DSBs form in nucleosome-free regions, the centers of the DSBs coincide with the position of the center of a nucleosome (**Figure 13.10B**). As with budding yeast, DSB activity is associated with testis-specific trimethylation of lysine 4 on histone H3. However, unlike yeast, the sites of DSB cleavage are far

Figure 13.10 Relation of human meiotic hot spots to the PRDM9 consensus sequence. (A) PRDM9 consensus sequence from mouse hot spot analysis. A three-letter amino acid recognition code for adjacent zinc fingers specifying a preferred DNA sequence is shown below. (B) Both predicted (*blue*) and experimentally determined (*red*) nucleosome occupancy profiles peak at the centers of DSB hot spots as determined by ChIP-sequencing and shown in Figure 13.8. (Adapted from Smagulova F, Gregoretti IV, Brick K et al. [2011] *Nature* 472:375–378. With permission from Macmillan Publishers, Ltd.)

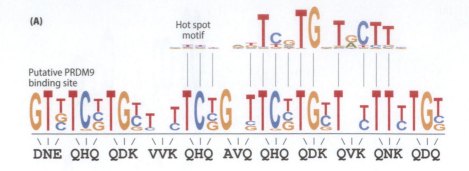

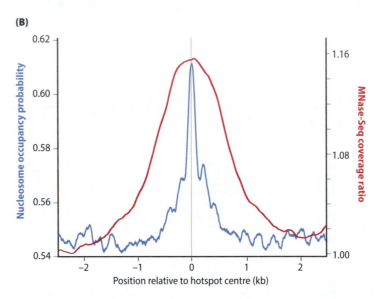

from transcription start sites, even though these regions also have H3-K4 trimethylation. Somehow the presence of Prdm9—perhaps by recruiting other proteins in the way that Spp1 facilitates DSB formation in yeast— assures that DSBs will be made at sites distant from promoter regions. When *PRDM9* is deleted then mice still show a high level of DSB formation, but now the sites of cleavage are indeed at promoter regions, as they are in yeast! Again the center of DSB activity coincides with the position of the +1 nucleosome in the transcribed region. Why yeast favors nucleosome-free regions and mammals center the activity over a nucleosome is not understood.

The variation among Prdm9 proteins also explains the apparent lack of evolutionary conservation in hot spots when the same chromosomal regions have been compared between chimpanzees and humans. Even though both primates have comparable levels of crossing over across a large region, the positions of hot spots in humans have little or no resemblance to the set in chimpanzees. It should be remembered that in fungi, when a "hot" allele is crossed to a "cold" one, the resulting gene conversion events occur with the *loss* of the hot spot, as expected if a DSB were created at the hot locus and repaired using a colder one. So hot spots will constantly be eliminated in heterogeneous populations; but there must also be the evolution of new ones to keep exchange levels sufficiently high. A rapidly evolving Prdm9 protein provides a simple explanation. Recent studies show that the zinc-finger repeat structure of Prdm9 is indeed rapidly evolving even in the human population, so new hot spots are continually revealed.

Beyond the distribution of Prdm9 binding sites, very little is yet known in mammals—or for that matter in budding yeast—about the global and regional control of crossing over. As mentioned above, in budding yeast, the same recombination reporter region, integrated at different locations shows a 20-fold variation in activity, but there are no specific factors yet identified that could control these differences. The same region may show significant differences when tested in strains of different backgrounds. It is also known that the presence of one strong hot spot will decrease hot spot activity at sites along the same pair of sister chromatids, though how this is effected is not yet known.

In many metazoans, the most obvious global regulator of meiotic recombination is gender. Male fruit flies lack meiotic crossing over entirely. In many mammals the rates of crossing over in females is statistically significantly higher than in males, although these effects vary by chromosomal region. As with yeasts, the level of crossing over for a given region can vary substantially.

In mouse, the level of Spo11 cleavage turns out to be feedback-regulated by the DNA damage checkpoint, specifically by the ATM protein kinase. The level of Spo11 cleavage is much higher in an *ATM* mutant, leading to male infertility, but when one of the Spo11 alleles is ablated, the level of cleavage diminishes and fertility is restored.

13.5 HOMOLOGOUS CHROMOSOME PAIRING OFTEN REQUIRES Spo11-INDUCED RECOMBINATION

In budding and fission yeasts, and in mammals, the pairing of homologous chromosomes at the pachytene stage of meiosis depends on Spo11 and on homologous recombination. In these cells there is no other mechanism that will align homologs. This can be seen in the mouse by fluorescently staining spermatocytes or oocytes in meiotic prophase with an antibody that reacts with the SYCP3 axial element protein and with an antibody that recognizes SYCP1, a protein of the central element of the SC (see Figure 12.25). In Spo11$^{+/+}$ cells, at late zygotene, each pair of chromosomes (bivalents) is aligned and an extensive SC has formed between the 20 pairs of mouse chromosomes (except that the X and Y chromosomes are only synapsed in a small, pseudoautosomal region). In Spo11$^{-/-}$ many nuclei show aberrant synapsis, sometimes with parts of two different chromosomes paired together. In some cases, there is no SCYP1 staining at all—just 40 unsynapsed pairs of chromosomes.

As discussed in Chapter 12, both flies and worms appear to have chromosome-specific "pairing centers" and specific pairing proteins that result in the synapsis of homologs even in the absence of Spo11. Interestingly, in flies and worms, the number of DSBs is on the order of a few dozen, while in budding yeast or mice, where synapsis of homologs depends on homologous recombination, there are more than 100. In lilies, the number of DSBs in meiosis appears to be in the thousands.

13.6 RECOMBINATION CANNOT BEGIN UNTIL Spo11 IS RELEASED FROM THE DSB ENDS

The covalent binding of Spo11 to 5′ ends of the DSB must be released before 5′- to -3′ resection of the ends can occur. In budding yeast removal of Spo11 occurs by an endonucleolytic cleavage of the 5′ strand in a reaction that requires not only the MRX complex (which was also needed for Spo11 to cleave), but also the Sae2 protein. More specifically, the endonuclease activity of Mre11 is required for this cleavage, although a nuclease-dead mutation of Mre11 (for example, Mre11-H125N) is normal in the creation of DSBs by Spo11. Rad50 is also required, as a set of *rad50-S* mutations that alter Rad50's ATP binding also have the phenotype whereby Spo11 remains covalently attached to DSB ends. As Sae2 also has endonuclease activity, it is not clear why both proteins are needed, but *sae2Δ* also prevents Spo11 removal and resection. The unprocessed DSB ends found in these mutants have been invaluable in mapping the sites of DSBs and in characterizing the nature of the cleavage event, as illustrated before in Figure 13.6. It may be possible to gain some insight by inducing Sae2 after all the DSBs have formed, if it can trigger Spo11 removal when expressed at a later time.

As noted in Section 13.3, the removal of Spo11 leaves equal amounts of two distinct families of long and short DSB oligonucleotides attached to Spo11 (Figure 13.7). It is not yet certain that each DSB is asymmetrically processed in this way or if the two types of cleavage products occur stochastically. If each DSB is initially processed to create one long and one short ssDNA end, this asymmetry may strongly bias subsequent resection and strand invasion events, as discussed below. A recent study has suggested that Mre11 may initially cleave the DNA much further from the end and that the nicked DNA is processed bidirectionally, primarily by Exo1, to leave a long ssDNA region.

13.7 IN MEIOSIS, IN MANY ORGANISMS, Rad51 IS NOT THE ONLY RecA-LIKE STRAND EXCHANGE PROTEIN

Even though Rad51 is a highly efficient recombinase, it is insufficient to carry out meiotic recombination in many organisms. In yeasts, mammals, and plants, a second meiosis-specific RecA-like protein, Dmc1 (disrupted meiotic cDNA), is required (**Figure 13.11**). However, neither flies nor worms have a Dmc1 homolog. Frank Stahl has speculated that Dmc1 might play a special role in early recombination events—the ones that would promote the initial synapsis between homologs—and that Dmc1 would therefore not be required in flies and worms, which achieve synapsis and SC formation through the use of pairing sites. However, it seems that Dmc1 is normally responsible for almost all meiotic recombination.

Figure 13.11 Comparison of yeast Rad51 and Dmc1 amino acid sequences. Alignment of Rad51 and Dmc1 is shown using the one-letter amino acid abbreviations. Identical amino acids are shown in between the aligned sequences by the shared letter. Related amino acid residues are indicated by +. The Walker A and Walker B ATPase domains are noted by *blue* lines. Protein sequences are deposited in the National Center for Biotechnology Information (NCBI) as accession numbers P25453 (Dmc1) and P25454 (Rad51).

```
Dmc1    9  DSDTAKNILSVDELQNYGINASDLQKLKSGGIYTVNTVLSTTRRHLCKIKGLSEVKVEKI   68
           D    + + +++LQ  GI +D++KL+  G++T    V    R+ L +IKG+SE K +K+
Rad51  73  DEAALGSFVPIEKLQVNGITMADVKKLRESGLHTAEAVAYAPRKDLLEIKGISEAKADKL  132

       69  KEAAGKIIQVGFIPATVQLDIRQRVYSLSTGSKQLDSILGGGIMTMSITEVFGEFRCGKT  128
             A +++ +GF+ A      R  +  L+TGSK LD++LGGG+ T SITE+FGEFR GK+
      133  LNEAARLVPMGFVTAADFHMRRSELICLTTGSKNLDTLLGGGVETGSITELFGEFRTGKS  192

      129  QMSHTLCVTTQLPREMGGGEGKVAYIDTEGTFRPERIKQIAEGYELDPESCLANVSYARA  188
           Q+ HTL VT Q+P ++GGGEGK  YIDTEGTFRP R+  IA+ + LDP+  L NV+YARA
      193  QLCHTLAVTCQIPLDIGGGEGKCLYIDTEGTFRPVRLVSIAQRFGLDPDDALNNVAYARA  252

      189  LNSEHQMELVEQLGEELSSGDYRLIVVDSIMANFRVDYCGRGELSERQQKLNQHLFKLNR  248
           +N++HQ+ L++    + +S   + LIVVDS+MA +R D+ GRGELS RQ  L + + L R
      253  YNADHQLRLLDAAAQMMSESRFSLIVVDSVMALYRTDFSGRGELSARQMHLAKFMRALQR  312

      249  LAEEFNVAVFLTNQVQSDPGASALFASADGRKPIGGHVLAHASATRILLRKGRGDERVAK  308
           LA++F VAV +TNQV +        F + D +KPIGG+++AH+S TR+   +KG+G +R+ K
      313  LADQFGVAVVVTNQVVAQVDGGMAF-NPDPKKPIGGNIMAHSSTTRLGFKKGKGCQRLCK  371

      309  LQDSPDMPEKECVYVIGEKGITD      331
           + DSP +PE ECV+ I E G+ D
      372  VVDSPCLPEAECVFAIYEDGVGD      394
```

Most surprising is the fact that in budding yeast Dmc1 is essential for spore viability, whereas deleting Rad51 reduces spore viability only to about 30–40%. Molecular analysis of strand invasion and the production of crossovers support the conclusion that meiotic recombination can get by even in the absence of Rad51, Rad55, and Rad57, whereas Dmc1 is essential. Spore viability also strongly depends on Rad52.

As we saw in Chapter 12, a valuable way to measure the effect of recombination proteins in budding yeast is by "return-to-growth" assays in which cells are placed in sporulation medium for a period of time and then plated on selective medium to score prototrophs arising from heteroallelic recombination. In this situation, recombination can be completed by the mitotic apparatus, so long as nothing irreversible has happened to DSBs in the meantime. In wild-type cells, prototroph formation rises >100-fold if cells are returned to growth at the time that DSBs are generated. In what seems like a paradox compared with what happens in an unperturbed meiosis, recombination is almost eliminated in the absence of Rad51. In contrast, *dmc1Δ* strains produce 10–20% of the wild-type level of prototrophs, with kinetics only a bit slower. These results mean that the approximately 150 DSBs generated in these cells were completely repaired in a significant fraction of the cells when they were returned to growth conditions. This repair is carried out by Rad51 in what is in essence a mitotic cell (**Figure 13.12**).

Remarkably *dmc1Δ* diploids retain the ability to generate a significant level of recombination in a return-to-growth experiment even when the sporulation culture is maintained for 24 hr. Retaining the ability to produce viable recombinants in cells that cannot produce viable spores could only result if cells have been blocked from executing meiotic chromosome segregation and spore formation, and indeed a *dmc1Δ* diploid is blocked from entering chromosome segregation by a DNA damage checkpoint. When the cells are finally returned to selective growth medium they can repair the DSBs using the Rad51-dependent mitotic machinery. Diploids lacking Rad51 also succeed in producing prototrophs, but these only arise after a delay that reflects the time it takes for cells to complete meiosis using Dmc1.

One important finding from return-to-growth assays reveals that, although heteroallelic recombination leading to prototrophs is high, the accompanying level of crossing over is relatively low, similar to events initiated entirely in mitotic cells. Molecular analysis of recombination intermediates suggests that—as in normal mitotic cells—the Sgs1–Top3–Rmi1 complex dissolves the great majority of dHJs that were created by repair of Spo11 DSBs so that they are recovered as noncrossovers.

In fission yeast, the roles of Dmc1 and Rad51 are different, but unlike budding yeast, spore viability is very significantly reduced in *rad51Δ* (*rhp51Δ*) and not in *dmc1Δ*. Both proteins contribute to crossovers. Both Rad55 and Rad57 appear to act, as mediators, in the Rad51 pathway. Another RecA homolog, Rlp1, which most resembles the Rad51 paralog Xrcc2 in mammalian cells, appears to inhibit intragenic recombination by either Rad51 or Dmc1.

13.8 Dmc1 WORKS WITH SEVERAL MEIOSIS-SPECIFIC AUXILIARY FACTORS

The analysis of budding yeast revealed a number of other meiosis-specific genes whose deletion phenotype resembled that of *dmc1Δ* (**Figure 13.13**). Michael Dresser identified Tid1 (two-hybrid interactor with Dmc1), which proved to be homologous to Rad54 (and is also known as Rdh54).

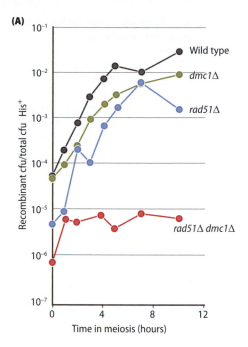

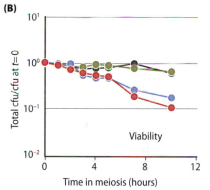

Figure 13.12 Return-to-growth assay of *rad51Δ* and *dmc1Δ* diploids, measuring *his4* heteroallelic recombination. (A) Cells lacking Dmc1 can complete recombination that is initiated in meiosis but completed in mitosis, after cells are returned to growth medium. In contrast, most *rad51Δ* diploids can only produce viable His+ recombinants when they complete meiotic recombination. Colony forming units = cfu. (B) The effect on cell viability after return-to-growth medium from sporulation medium is plotted showing a more severe effect of deleting *RAD51* when cells complete repair in a mitotic environment. (Adapted from Shinohara A, Gasior S, Ogawa T et al. [1997] *Genes Cells* 2:615–629. With permission from John Wiley & Sons, Inc.)

Figure 13.13 Auxiliary factors for Rad51 and Dmc1 in meiotic recombination. In addition to the Rad51 loading mediators, Rad51 is inhibited by Hed1. Loading of Dmc1 requires Mei5-Sae3. Dmc1's activity is facilitated by Hop2-Mnd1. Later steps in Rad51-mediated recombination are facilitated by Rad54, whereas Dmc1 interacts with the Rad54 paralog, Tid1 (Rdh54). Rad51 also acts as a co-factor for Dmc1 action.

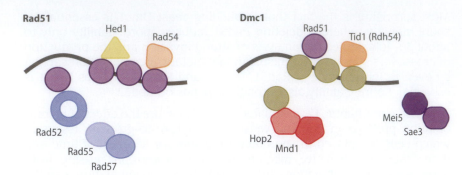

Tid1 is not meiosis specific, although its roles in mitotic recombination are relatively minor, and seen only in the absence of Rad54. But in meiosis, Tid1 is strongly required. A combination of *in vivo* and *in vitro* studies argues that Tid1 has two important roles. First, *in vitro* it stimulates strand exchange of Dmc1 (but it also stimulates Rad51). Second, Tid1/Rdh54 displaces Dmc1 from binding nonspecifically to dsDNA and thus frees up Dmc1 to bind to ssDNA at sites of recombination. As noted before in mitotic cells Rad54, Rdh54, and Uls1 all seem to act to displace Rad51 from dsDNA. *In vitro*, Tid1/Rdh54 also can displace nucleosomes, similar to the activity described for its cousin Rad54.

Another key complex required for Dmc1 action comprises the budding yeast proteins Mei5 and Sae3. Although these proteins do not resemble the Rad51-like paralogs (Rad55–Rad57 or the PCSS proteins) they appear to have a related function in loading Dmc1 onto ssDNA. In fact, Mei5 and Sae3 share homology, respectively, with fission yeast Sfr1 and Swi5 proteins, which in mitotic cells act as alternative mediators for some Rad51-mediated recombination events. In meiosis *S. pombe* Sfr1–Swi5 promote both Rad51- and Dmc1-dependent recombination. The budding yeast proteins are only expressed in meiosis and seem to interact principally with Dmc1. Mei5–Sae3 interact not only with Dmc1 but also with RPA and appear to play an important role in the displacement of RPA and the binding of Dmc1 to ssDNA. Homologs of Sfr1 and Swi5 are also found in mammals but have not been identified in worms or *Arabidopsis*. Sae3, but not Mei5, has been found in *Drosophila*.

Yet another pair of proteins—Hop2 and Mnd1—are indispensable for meiotic recombination. These proteins are evolutionarily conserved and their mutants have similar meiosis-defective phenotypes in budding and fission yeast, *Arabidopsis*, and mouse. Although Rad51 and Dmc1 can load onto resected DSB ends in these mutants, DSB repair fails and there are no crossover products (**Figure 13.14**), nor does one find earlier recombination intermediates. These proteins form a stable heterodimeric complex that binds DNA and stimulates the recombinase activity of both Rad51 and Dmc1 *in vitro*. Hop2 binds to DNA while Mnd1 interacts with Rad51 or Dmc1. Hop2–Mnd1 stabilize Rad51- and Dmc1-ssDNA nucleoprotein filaments. The Hop2–Mnd1 heterodimer enhances the ability of the Rad51- and Dmc1-ssDNA nucleoprotein filaments to capture duplex DNA, which should facilitate homology searching and joint molecule formation. Curiously—and still unexplained—yeast *mnd1Δ rad51Δ* and *hop2Δ rad51Δ* double mutants are able to complete crossover recombination, but noncrossover outcomes seem to be absent.

13.9 Rad51 PLAYS ONLY A SUPPORTING ROLE IN YEAST MEIOSIS

Indirect immunofluorescent staining of prophase budding yeast cells undergoing recombination show that a significant fraction of foci of

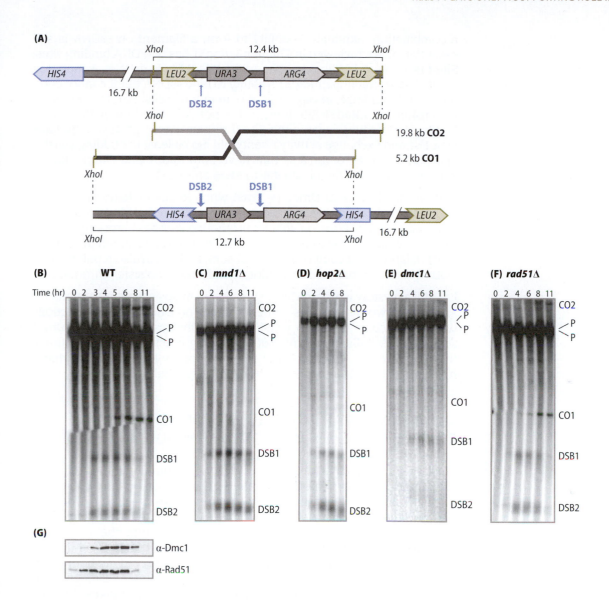

Figure 13.14 Role of Dmc1, Rad51, Hop2, and Mnd1 in completing recombination in budding yeast. (A) An ectopic recombination assay allows the detection of restriction fragments diagnostic of a crossover. (B) Kinetics of the appearance of DSBs and crossovers (CO) distinct from the parental bands (P) are shown in a Southern blot probed for *ARG4* sequences. COs are absent in diploids homozygous for deletions of *MND1* (C), *HOP2* (D), and *DMC1* (E), but are clearly visible in a *rad51Δ* strain (F). (G) Western blot analysis using anti-Dmc1 and anti-Rad51 antibodies (α-Dmc1 and α-Rad51) shows that Rad51 is present in both vegetative and sporulating cells whereas Dmc1 appears only after inducing meiosis. (Adapted from Henry JM, Camahort R, Rice DA et al. [2006] *Mol Cell Biol* 26:2913–2923. With permission from the American Society for Microbiology.)

Dmc1 and Rad51 overlap, which might suggest that they work together. From the molecular analysis of the defects of *rad51Δ* and *dmc1Δ* it seems that there cannot be an obligate participation of both proteins at each DSB, since *rad51Δ* cells are not nearly as defective in completing recombination as *dmc1Δ*.

The failure to repair meiotic DSBs in the absence of Dmc1 or Tid1 can be overcome in several ways. Overexpressing Rad51—or even overexpressing Rad54—suppresses *dmc1Δ* defects to yield viable tetrads. But recombination is not entirely normal. In particular, there is an absence of crossover interference. This result suggests that Rad51 is fully able to accomplish interhomolog meiotic recombination but is normally restrained from doing so.

The idea that Rad51 is inhibited in meiosis was substantiated by discovery of yet another meiosis-specific protein, Hed1, whose deletion also suppressed the defects of *dmc1Δ*. Hed1 does not impair the loading of Rad51 onto ssDNA but blocks its ability to interact with Rad54. It is still not clear what role this regulation plays in normal meiosis because deleting Rad54 has very little effect on budding yeast meiosis.

Recently, Doug Bishop made the remarkable discovery that budding yeast's Rad51 strand exchange activity per se is not required for meiotic recombination, although its ability to form a filament on ssDNA is still essential. As we reviewed in Chapter 3, Rad51 has two DNA binding sites: Site I is a ssDNA binding domain, while Site II binds dsDNA. A mutation in Site II does not impair Rad51's ability to bind ssDNA but it cannot carry out strand exchange, *in vitro* or *in vivo*, in mitotic cells. But, to everyone's astonishment, a Rad51 Site II mutant is not defective in meiotic recombination! Moreover, *in vitro*, the Site II mutant Rad51 protein stimulates Dmc1 strand exchange activity twentyfold. So, at least in budding yeast—and perhaps in all organisms that have both Rad51 and Dmc1—Rad51 acts as an auxiliary factor, similar to Mei5 and Sae3.

Still, in the absence of Dmc1 (at least when Rad51 or Rad54 is overexpressed or when Hed1 is deleted) wild type Rad51 is able to load at all DSBs and rescue spore viability by promoting sufficient crossing over to assure proper chromosome segregation. In the situation when Red1 or Hop1 axial element components are absent, Rad51 can also repair meiotic DSBs, but here repair occurs predominantly between sister chromatids.

It is still a mystery why Rad51 is insufficient in meiosis in many organisms, although in worms and flies—where there is no Dmc1—Rad51 functions without difficulty. In fission yeast and in *Arabidopsis*, Rad51 and Dmc1 seem to be equally important. But in budding yeast, Dmc1 has assumed the dominant role. Whether Dmc1 is more important than Rad51 in mammals is harder to assess, because Rad51 is essential.

13.10 MOLECULAR INTERMEDIATES OF MEIOTIC RECOMBINATION ARE WELL-STUDIED IN BUDDING YEAST

Most of the molecular analyses of meiotic recombination in budding yeast have been carried out in strain SK1, which enters and completes meiosis more rapidly than many other strains. It takes about 5–7 hr from the time that cells are placed in sporulation medium, consisting of 1% potassium acetate lacking a nitrogen source, to the time that cells complete Meiosis I (which is assayed by the appearance of two DAPI-stained nuclear masses). Pre-meiotic DNA replication is observed at about 2.5 hr and DSBs appear shortly thereafter (Figure 13.14). Completed recombination events, crossovers, can be seen at about 4 hr. Thus it takes about an hour for DSBs to be repaired, similar to the time that has been observed for HO-induced DSBs in mitotic cells.

Two particular hot spots have been the focus of the majority of analyses of DNA intermediates during recombination. Both of these lie in the same region of chromosome 3. The first was created by inserting *LEU2* sequences adjacent to the *HIS4* gene, which was known to have high levels of gene conversion (**Figure 13.15**). Most of the original *LEU2* sequences on the same chromosome were deleted. The junction between the *HIS4* and *LEU2* regions is very hot—hotter than for either *HIS4* or *LEU2*. Why these sequences are so hot has not been established, but it is likely that other, competing hot spots were removed. Perhaps this unusual hot spot can form a distinctive secondary structure, or exclude nucleosome sequences. Interestingly, bacterial pBR322 plasmid sequences also prove to have distinct hot spots. For example, in a second useful assay system, pBR322 sequences carrying *arg4* and *URA3* sequences, were inserted either at *leu2* or, 22 kb more distal, at *his4* (**Figure 13.16A**). Thus recombination between *arg4* sequences is ectopic. The different chromosome contexts assures that any branched structures will not migrate away if the DNA is cut by restriction enzymes so that the relevant fragments include nonhomologous flanking yeast chromosome sequences. Despite the fact that *ARG4* in its normal location on chromosome 8 has a significant DSB

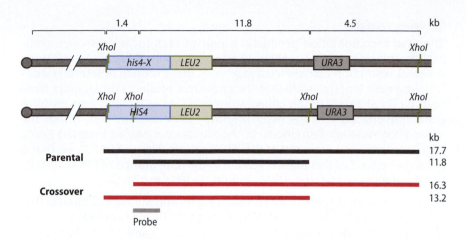

Figure 13.15 The well-studied *HIS4–LEU2* meiotic recombination assay system. Meiotic recombination can be monitored around the budding yeast hot spot created by the fusion of the nearby *HIS4* and *LEU2* genes. (Adapted from Cao L, Alani E & Kleckner N [1990] *Cell* 61:1089–1101. With permission from Elsevier.)

near the 5′ end, the two most prominent DSB sites in this construct are in pBR322 sequences, which did not co-evolve with histones and have a more accessible structure. Moreover, as a further striking demonstration that chromosome context influences Spo11 cleavage, the cleavages of the same sequences inserted at *his4* are much stronger than those in the *leu2* insert.

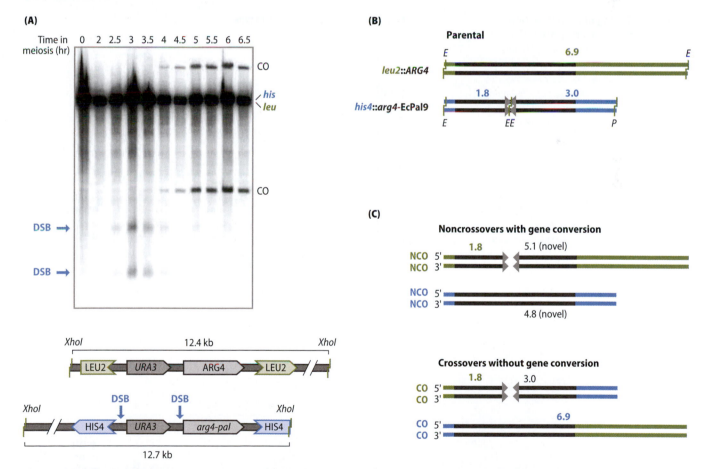

Figure 13.16 Detection of crossovers in the *arg4* system. (A) *ARG4* alleles were inserted by plasmid integration at the *leu2* and *his4* loci of the left arm of chromosome 3, about 25 kb apart. Appearance of meiotic DSBs and crossover products is shown on a Southern blot. The positions of flanking *Xho*I sites makes it possible to see crossover products with novel restriction fragment lengths. (B) Modification of the *ARG4* regions inside the *leu2* and *his4* duplications allows a more detailed analysis of recombination intermediates. (C) Sizes and diagnostic restriction sites for noncrossover outcomes and for two of the possible crossover products in the *arg4* regions when DNA digested with *Eco*RI and *Pvu*II (*P*). A mismatch repair-resistant palindromic site (an inverted pair of *Eco*RI sites) in the *arg4-ECPal9* allele (*EE*) is used both as a restriction fragment marker and as a mismatch repair-resistant palindrome when present in heteroduplex DNA. Crossovers are shown for the case when the exchange occurs in the region to the left of the palindrome. (Adapted from Allers T & Lichten M [2001] *Cell* 106:47–57. With permission from Elsevier.)

Crossover products

The final recombination products, a pair of reciprocal crossover chromatids, can be seen by Southern blots using a pair of homologs carrying different restriction enzyme cleavage sites that produce different restriction fragment lengths, such that the crossover products are distinct from both parental bands (often called Mom and Dad). One can also monitor gene conversions when a heterozygous restriction site lies within the shared homologous sequences. In this instance a pair of inverted EcoRI sites were inserted into one of the *ARG4* genes (**Figure 13.16B**), so that a gene conversion of this marker results in the appearance of novel restriction fragments containing *ARG4* sequences (Figure 13.16C).

A very important discovery by Thorsten Allers and Michael Lichten is that the kinetics of the appearance of noncrossover (NCO) products precedes the appearance of CO outcomes by about 1 hr (**Figure 13.17**). Moreover, in an *ndt80Δ* mutant that cannot complete meiosis, noncrossover recombinants still appear at nearly normal levels, whereas CO products are dramatically reduced. The absence of CO products is accompanied by the persistence of dHJ intermediates. These data have reinforced the idea

Figure 13.17 Timing of recombination events in budding yeast meiosis. (A) The appearance of DSBs, joint molecules (JM) and crossovers (CO) in the ectopic *arg4* system. Recombinants containing heteroduplex DNA (hDNA) are also shown. Noncrossovers containing heteroduplex DNA (hNCO) appear earlier than crossovers and COs containing hDNA (hCO). Heteroduplex DNA-containing intermediates are discussed in Section 13.11. Note that noncrossovers without hDNA are indistinguishable from parental molecules and cannot be identified. (B) Similar analysis in an *ndt80* deletion mutant that arrests in pachytene and fails to produce most COs but allow NCOs. (Adapted from Allers T & Lichten M [2001] *Cell* 106:47–57. With permission from Elsevier.)

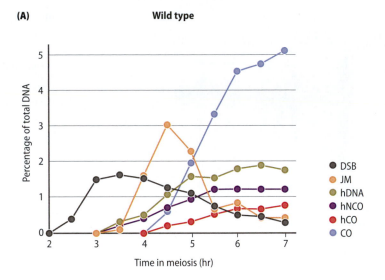

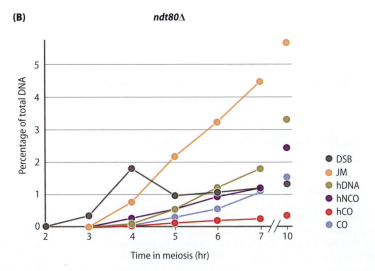

that most NCO products arise from a pathway that is distinct from the one leading to crossovers. Presumably the pathway leading to most NCOs never forms dHJs, an outcome consistent with an SDSA mechanism. In contrast, the CO pathway involves the formation of dHJs, which apparently are almost always resolved as crossover outcomes.

DNA end resection in meiosis is less extensive than in mitotic cells

5′-to-3′ resection can be followed on Southern blots after native gel electrophoresis as an increasing smear below the size of the restriction fragment containing the Spo11 cleavage site. On denaturing gels, the strand with the 3′ end at the DSB remains at full length while the 5′-ended strand shows progressive loss. Resection is more readily seen in a *dmc1Δ* strain, where recombination is prevented. Even in a *dmc1Δ* strain, the rate of resection is significantly slower than in mitotic cells with an HO cut. In mitotic cells, the Sgs1–Dna2 helicase nuclease degrades DNA about 4 kb/hr (1 nt/sec), although without Sgs1, Exo1 is responsible for degradation at about 1 kb/hr. In meiotic cells, Exo1 appears to be the principal exonuclease, although there is still enough resection in its absence to generate recombination intermediates. Presumably MRX-Sae2 accounts for the remaining resection.

Why Sgs1 does not play an important role in resection (though it clearly is involved in other aspects of meiotic recombination) is not known. The slower rate of resection in meiosis is consistent with the shorter gene conversion tract lengths that are seen in meiosis compared with mitotic cells, even when the same HO endonuclease is used to create DSBs in both cases. It makes sense that resection should be more restrained, as cells have a roughly 150 DSBs, so that chewing away large segments could create complications. We cannot rule out the possibility that the initial rate of resection is as fast as, or faster than, in mitotic cells, but instead there are severe constraints on continuing degradation.

Binding of Dmc1 and Rad51 to DSB ends

The same ChIP assays that have been used to follow Rad51 loading onto ssDNA at HO-induced DSBs have been applied in budding yeast meiosis. The lower level of cleavage at most DSB sites in meiosis, compared with nearly complete cleavage in mitotic cells, has made this a formidable problem; nevertheless, recruitment of Dmc1 and Rad51 to a Spo11-induced DSB can be documented. Analysis of a site-specific DSB, created by the VDE endonuclease (discussed extensively in Section 13.20) has provided more detail. First, *rad52Δ*, *rad55Δ*, or *rad57Δ* block Rad51 loading as was shown in mitotic cells for an HO break, but they do not block Dmc1 loading. Conversely, *mei5Δ* and *sae2Δ* prevent Dmc1 loading but have no effect on Rad51. Dmc1 loads in the absence of Rad51 and vice versa. Neither *hop2Δ*, nor *mnd1Δ*, nor *rad54Δ* prevent loading of either recombinase, though they are needed for Dmc1 and Rad51 recombination activities at a later step. Finally, in a *tid1Δ* mutant there is a general nonspecific increase in Dmc1 at sites where there is no DSB. This result is consistent with the idea that Tid1 releases nonspecific Dmc1 binding so that it can function at DSBs.

The binding of both recombinases has also been analyzed by immunofluorescence analysis of meiotic chromosomes. In budding yeast it seems that two recombinases are not interspersed in a mixed filament, but occupy distinct regions. Recent studies in *Saccharomyces* argue against the idea that the two proteins bind to opposite sides of the DSB, but one study of *Arabidopsis* meiosis argues that Rad51 and Dmc1 do bind to opposite sides of the break.

13.11 TRANSIENT STRAND INVASION INTERMEDIATES CAN BE IDENTIFIED BY 2-D GEL ELECTROPHORESIS

Strand invasion creates a branched DNA structure, as does the formation of double Holliday junctions. These structures are unstable and can be unwound or dissolved by branch migration; hence to study these structures it is necessary to stabilize them. The traditional method for holding unstable intermediates together is to cross-link DNA strands by UV irradiation after treatment of spheroplasted cells with psoralen (4,5',6-trimethylpsoralen). More recently, Allers and Lichten found that branch migration could also be suppressed by isolating DNA in the presence of high concentrations of divalent cations such as Mg^{++} or CTAB.

Recombination intermediates, like branched structures arising during replication, can be identified by using two-dimensional gel electrophoresis in which the electrophoresis conditions in the first dimension (no ethidium bromide and low voltage) separate molecules principally on the basis of total molecular weight, whereas the electrophoresis conditions of the second dimension (ethidium bromide and higher voltage) separate molecules on the basis of their shape as well as mass (**Figure 13.18**). Strand invasion intermediates will migrate similar to Y-shaped replication intermediates, while Holliday junctions and double Holliday junctions migrate as X-shaped molecules that form a "spike." Below the parental-sized spots are smears resulting from DSBs and resection. The gels are transferred to membranes for Southern blot analysis.

A more detailed identification of what DNA strands are included in any structure can be accomplished either by probing the Southern blot with strand-specific probes or by separating the individual strands by subjecting the DNA in a given spot to electrophoresis on another gel under denaturing conditions. If the DNA had been cross-linked, it must first be treated to reverse the cross-links, which happily is not a difficult procedure.

Single-end invasion

Following recombinase recruitment to ssDNA, the next step in DSB repair is strand invasion by one of the two ends of a DSB. Neil Hunter and Nancy Kleckner could identify two distinct intermediates, representing strand invasion of DSBs on each of the different-sized parental

Figure 13.18 Appearance of joint molecules in budding yeast meiosis as displayed in two-dimensional gels. In addition to restriction fragments identifying the parental-sized *arg4* genes contained within the constructs integrated at *leu2* and *his4* (*leu* and *his*), one can see the transient appearance of joint molecules (JM) that migrate much more slowly. Note that one does not see an arc of replicating molecules because pre-meiotic replication is completed before the 3 hr time point. (Adapted from Allers T & Lichten M [2001] *Cell* 106:47–57. With permission from Elsevier.)

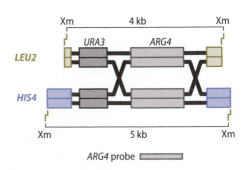

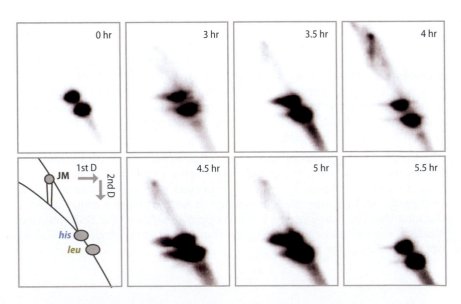

restriction fragments (**Figure 13.19**). *Single-end invasion* (SEI) intermediates disappear, as expected, with the appearance of dHJs and with the appearance of noncrossover outcomes. SEIs depend on the presence of Dmc1. A very important observation made by Hunter and Kleckner is that SEIs are strongly reduced in a *red1Δ* strain, in which there is also a dramatic reduction of interhomolog recombination, but an increase in sister chromatid repair. This finding has strongly influenced subsequent thinking about the distinctions between recombination leading to crossovers versus noncrossovers. Strand invasion intermediates stable enough to be recovered as SEIs lead predominantly to crossover outcomes. Certainly there must be strand invasion intermediates leading to noncrossovers, likely via an SDSA pathway, but these are apparently not stable enough to be detected on two-dimensional gels.

dHJ formation

HJ-containing molecules in meiosis were first identified by Leslie Bell and Breck Byers in 1983, but it was the insightful studies of Tony Schwacha and Kleckner that firmly established that, in budding yeast, meiotic recombination leads to the formation of dHJs rather than single HJ intermediates. The strains used had restriction-site polymorphisms such that the two parental-sized restriction fragments around the *HIS4::LEU2* hot spot were different in the Mom and Dad parental chromosomes and the two reciprocal crossover products were different from both parental bands and from each other. After two-dimensional gel electrophoresis one could identify the appearance of three branched molecule species that corresponded to the Mom × Dad *interhomolog* recombination

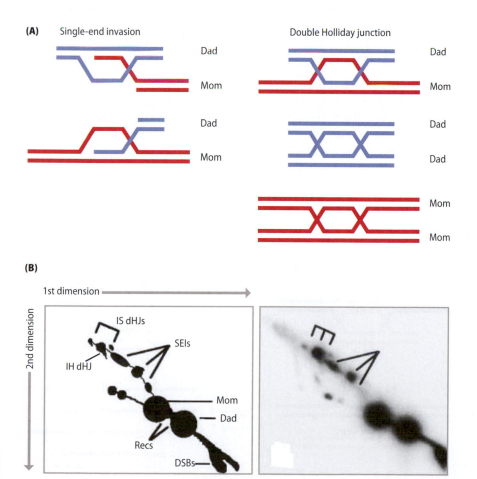

(A) Single-end invasion

Dad
Mom

Dad
Mom

Double Holliday junction

Dad
Mom

Dad
Dad

Mom
Mom

(B)

1st dimension

2nd dimension

IS dHJs

SEIs

IH dHJ

Mom
Dad

Recs

DSBs

Figure 13.19 Identification of SEIs and dHJs. Recombination at the *HIS4:LEU2* hot spot in budding yeast between strains whose chromosomes (Mom and Dad) have restriction fragment length differences in this region. (A) Structures of two alternative single-end invasion (SEI) intermediates and three different dHJs resulting from interhomolog (Mom × Dad) and intersister (Mom × Mom or Dad × Dad) recombination events. (B) Two-dimensional gel analysis separating SEIs and dHJs. A Southern blot probed with sequences common to both Mom and Dad sequences is diagrammed on the left. Note that the abundance of the two intersister (IS) dHJs is much less than the interhomolog (IH) event. The parental-sized fragments are indicated as Mom and Dad, between which are the much less abundant reciprocal crossover recombination products (Recs) of intermediate size. Fragments containing DSBs and their resected ends are shown at the bottom right. (Adapted from Hunter N & Kleckner N [2001] *Cell* 106:59–70. With permission from Elsevier.)

intermediate flanked by equivalent Mom × Mom and Dad × Dad *intersister* recombination structures (Figure 13.19). There is a strong bias in favor of interhomolog (Mom × Dad) dHJs.

The demonstration that these joint molecule bands contained dHJs involved cutting out the gel containing the Mom × Dad intermediate, reversing their psoralen cross-links and separating the strands on a denaturing gel. This analysis revealed several important aspects of these structures. First, all the strands were full length; there were no hidden single-strand nicks or gaps in these intermediates. Second, most of the bands were either Mom- or Dad-length; there were no strands with the lengths expected if there had been an exchange. Schwacha and Kleckner argued that if the intermediates contained a single HJ, there should be four DNA strands each of a distinctive length (**Figure 13.20**)—one Mom, one Dad, and two crossover strands—but this was not the case. In contrast a dHJ should only have strands of parental length. The dHJ intermediate can be resolved either to have NCO or CO outcomes by treating the DNA with purified bacterial RuvC HJ resolvase. After such treatment there is a roughly equal mix of all four expected restriction fragments.

These findings were confirmed and extended by Allers and Lichten, who showed that these dHJ intermediates contained heteroduplex DNA, as expected if they arose from homologous recombination between two different parental chromatids. Their analysis took advantage of the observation that heteroduplex DNA in which one strand had a short palindromic structure escapes mismatch repair at least half the time. Into one parental molecule they inserted into *ARG4* a pair of inverted EcoRI sites (see Figure 13.16) such that strand invasion and heteroduplex DNA formation would produce a hairpin containing an EcoRI site (*arg4-EcPal9*) (**Figure 13.21**). The parental band is cut into two fragments by EcoRI digestion whereas a heteroduplex-containing intermediate will be cut in the palindrome stem, lopping off the hairpin end. On native gels the band will still migrate as the size defined by the flanking restriction sites, but on denaturing gels one strand will be revealed as having been cut by *Eco*RI (Figure 13.21). The results of this analysis revealed a wealth of new information. First, NCO fragments were recovered that contained heteroduplex DNA (hDNA). Clearly there had been strand exchange and the formation of hDNA that would give rise to a PMS event if the recombinants had been recovered as spores. Indeed, tetrad analysis of the non-Mendelian outcomes from this cross showed that there were an equal number of 6:2 and 5:3 events (the latter being the PMS events expected if hDNA between *ARG4* and *arg4-EcPal9* was not mismatch corrected).

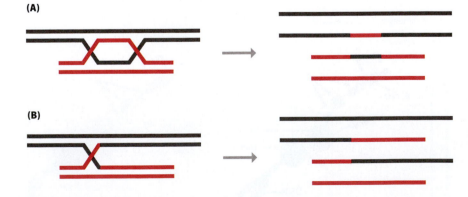

Figure 13.20 Demonstration that meiotic JMs are predominantly dHJs. Random cleavage by the bacterial RuvC resolvase of each of the two HJs in a dHJ (A) will generate only parental-sized restriction fragments, whereas random cleavage of a single HJ (B) will generate both parental- and recombinant-sized bands. (Adapted from Schwacha A & Kleckner N [1995] *Cell* 83:783–791. With permission from Elsevier.)

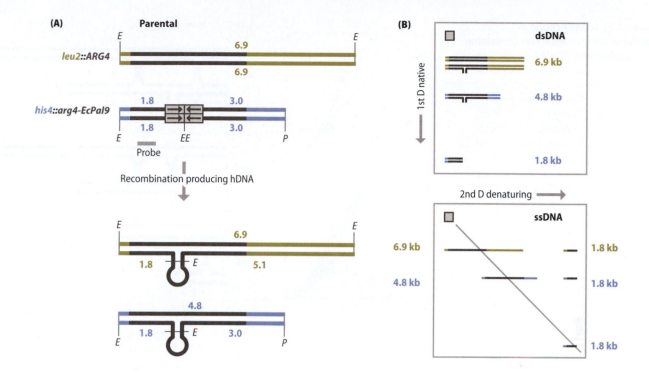

This work also confirmed that interhomolog joint molecules indeed contained dHJs, but made the important additional finding that these intermediates often contained hDNA (Figure 13.21). As with the NCO products, hDNA could be recognized because one of the DNA strands contained the palindromic *Eco*RI site and thus when molecules were run on denaturing gels (Figure 13.21B) a novel restriction fragment was seen.

However, not all of the joint molecules were of the same size; there were at least three distinct dHJ configurations (**Figure 13.22**). The canonical dHJ mechanism of recombination would predict that the two HJs should lie on either side of the DSB sites, which lie to the left of the *arg4-EcPal9* site; but in some cases, this could not be true. Allers and Lichten proposed two alternative ways in which the steps following strand invasion would lead to a dHJ that would not flank the initial DSB site (**Figure 13.23A, B, C**). Another alternative is shown in **Figure 13.23D**, in which strand invasion is initiated from the opposite DSB end and proceeds in an SDSA-like manner until the migrating D-loop is "captured" beyond the *Eco*RI site. In this case, to obtain the structure of JM3, it is necessary to invoke mismatch correction of the hDNA, as happens in about 50% of cases. All of these mechanisms will lead to crossovers downstream of the initiating DSB site and of the *Eco*RI site.

Initiation of new DNA synthesis

As with HO-mediated events in mitotic yeast cells (described in Chapter 8), it is possible to detect the appearance of recombinant strands at a known initiation site for meiotic recombination. Using PCR primers that are unique for each parental chromosome, and spanning the site of prominent DSB formation at a hot spot, Dean Dawson's lab detected new DNA synthesis from the 3′ end of an invading strand (**Figure 13.24**). This method has not been exploited yet to investigate in detail whether there is a bias in the initiation of primer extension from one or the other end of a specific DSB.

Figure 13.21 Identification of noncrossovers containing heteroduplex DNA. (A) In heteroduplex DNA the pair of *Eco*RI sites form a short palindrome (designated *EE*) that frequently escapes mismatch correction but can be cleaved in the stem of the hairpin as well as part of duplex DNA. (B) Cleavage of the *Eco*RI hairpin in an intermediate containing hDNA thus generates a single-stranded nick that is revealed when bands separated in one-dimensional electrophoresis are subjected to a second electrophoretic step in denaturing medium. (Adapted from Allers T & Lichten M [2001] *Cell* 106:47–57. With permission from Elsevier.)

Figure 13.22 Identification of recombination intermediates containing heteroduplex DNA (hDNA). (A) After two-dimensional gel electrophoresis, three distinct sizes of JMs can be identified, along with an unidentified additional intermediate (*). A faint arc of Y-shaped replication intermediates or recombination intermediates (R) is also observed. (B) Double Holliday junctions containing *Eco*RI in hDNA can be identified by subjecting JMs to denaturing gel electrophoresis. Verification of the structure of the JM1 intermediate is shown. (Adapted from Allers T & Lichten M [2001] *Cell* 106:47–57. With permission from Elsevier.)

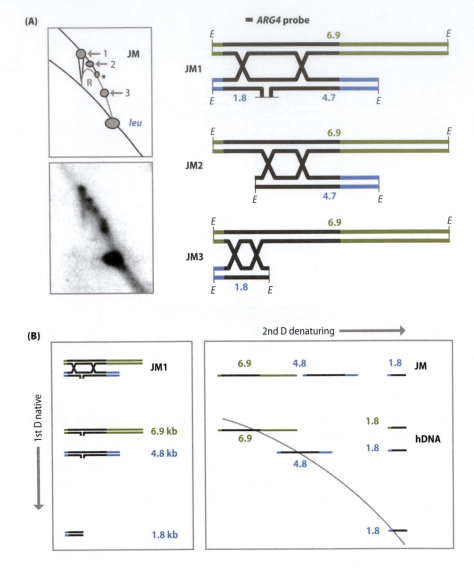

HJ formation in S. pombe

A similar analysis of joint molecules has been carried out in fission yeast. There are many differences between the two yeasts in how they carry out meiosis, the most striking of which is that the intermediates in *S. pombe* are single HJs (**Figure 13.25**). Resolution of the HJs in fission yeast depends almost completely on the Mus81–Eme1 endonuclease. Mus81 plays only a minor role in resolving the dHJs in *S. cerevisiae*.

13.12 TO COMPLETE GENE CONVERSION THE SECOND DSB END MUST BE CAPTURED

In all of these mechanisms there is a step in which the second end of the DSB becomes engaged. In SDSA, this step appears to be an annealing between a displaced ssDNA being copied from the template and a second end that had not participated in strand invasion. In events leading to dHJs there must also be engagement of the second end, pairing with a ssDNA region generated by D-loop formation. Whether the biochemical requirements to carry out strand annealing in SDSA are different from second

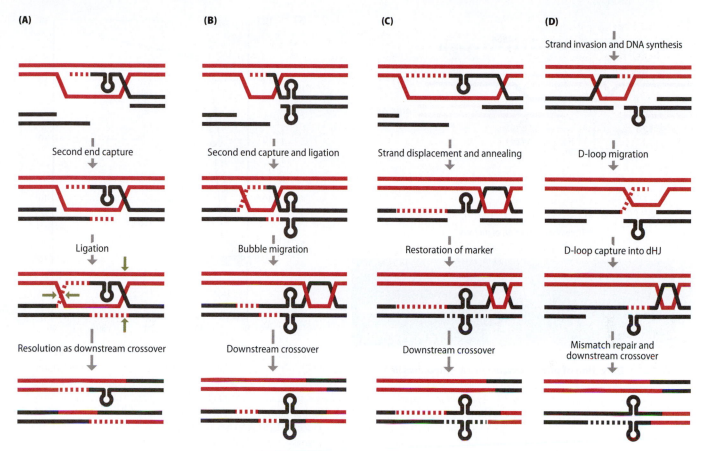

Figure 13.23 Formation of alternative crossover outcomes from intermediates containing dHJs and hDNA. Four possible alternatives of meiotic DSB repair are illustrated, each producing a crossover downstream of the hDNA. (A) Conventional dHJ mechanism with a HJ on either side of the DSB. One of the chromatids contains hDNA that is refractory to mismatch correction. (B). Limited resection and/or strand invasion leaves the *Eco*RI site as a homoduplex, but branch migration can place the site of crossing over far from the initial DSB. (C) An SDSA mechanism in which the second DSB end anneals and captures a displaced newly copied strand can also lead to a downstream crossover. (D) An alternative SDSA mechanism initiated from the other DSB end can also yield an outcome similar to that in (C), if there is mismatch correction of the hDNA. (Adapted from Allers T & Lichten M [2001] *Mol Cell* 8:225–231. With permission from Elsevier.)

end capture in a dHJ is not yet clear. In budding yeast, Rad52 should play an important role in this process; but it has been difficult to learn about the need for Rad52 at later steps because it is essential for the initial steps of Rad51 loading, at least in mitotic cells. In meiosis, however, Rad51 has been relegated to a less prominent role as only Dmc1 is essential for interhomolog homologous recombination.

In *rad52* diploids, there is a surprisingly high level of meiotic crossing-over—as much as 40–50% of the wild type level—although spores are dead. Similar to what was described in mitotic cells, these *RAD52*-independent crossovers are not reciprocal: they appear to be *half-crossovers* in which the reciprocal product fails to be completed. Spore viability is greatly improved in a diploid carrying a C-terminal truncation of Rad52 (*rad52-327*) that has lost its Rad51-interaction domain but which can still promote strand annealing. This has led to the conjecture by Akira Shinohara and Neil Hunter that the annealing function of Rad52 is required to convert a SEI into a dHJ. The N-terminal part of Rad52 is also needed for the annealing step of SDSA. These conjectures are supported by two-dimensional gel analysis of joint molecules in the absence of Rad52. SEIs are formed at a high level, but dHJs are greatly reduced. In

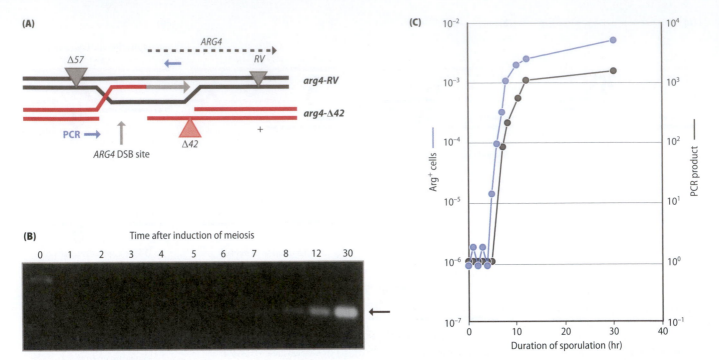

Figure 13.24 Detection of primer extension intermediates in yeast meiosis. (A) Two Arg⁻ alleles are distinguished by small deletions so that each allele can hybridize to only one PCR primer (*blue*). Strand invasion and primer extension will be detected by the appearance of the PCR product (B). The strain used in these experiments proceeds through meiosis more slowly than SK1 shown in previous figures. At later time points PCR will amplify the final product as well as the primer extension intermediate. (C) Time of appearance of the PCR product relative to the time of commitment to recombination (that is, initiation of DSBs) as monitored by return-to-growth plating of cells on medium lacking arginine. (Adapted from Bascom-Slack CA & Dawson D [1998] *Mol Gen Genet* 258:512–520. With permission from Springer Science and Business Media.)

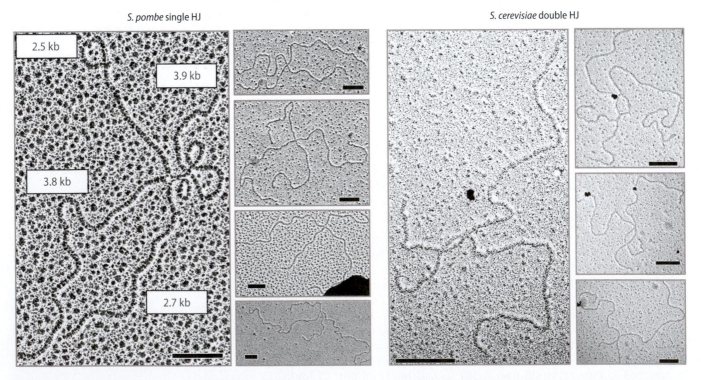

Figure 13.25 Intermediates of recombination in *S. pombe* have single HJs compared with dHJs in *S. cerevisiae*. JMs purified from two-dimensional gel analysis were visualized by electron microscopy. Whereas budding yeast intermediates appear to be dHJs separated by a pair of strands, the branched intermediates from meiosis in fission yeast appear to be single HJs. (From Cromie GA, Hyppa RW, Taylor AF et al. [2006] *Cell* 127:1167–1178. With permission from Elsevier.)

rad52-327, dHJs are prominent, though delayed in appearance, and SEIs are less evident, as they are transient.

Hunter and Shinohara suggest that the nonreciprocal crossovers that arise in the absence of Rad52 come from Dmc1-promoted SEIs in which the initial D-loop is stabilized by branch migration that proceeds away from the invading end, to produce a dHJ but where one chromatid arm is still broken—a half-crossover as we have seen earlier (**Figure 13.26**). HJ resolution of this intermediate yields a half-crossover, with one intact, viable chromatid and one still-broken crossover fragment. This idea is supported by their finding that the production of half crossovers also depends on the Msh5 protein, which apparently stabilizes dHJs and is part of an elaborate mechanism devoted to regulating which DSB events will become crossovers and which will be repaired as NCOs.

13.13 AXIAL ELEMENT COMPONENTS ARE IMPORTANT IN THE CONTROL OF THE INTERHOMOLOG BIAS

A striking aspect of meiotic recombination is that it must overcome the strong tendency for a DSB to be repaired by the sister chromatid that is seen in mitotic cells. In budding yeast, deletions of the axial element proteins Hop1 and Red1 promote sister chromatid repair in meiosis. This is especially evident in a *dmc1Δ* mutant that cannot repair DSBs, but in *red1Δ dmc1Δ* or *hop1Δ dmc1Δ* mutants, repair does occur—by intersister recombination. An important recent finding from Nancy Hollingsworth's lab is that Red1 and Hop1 are required to recruit a dimeric form of the serine/threonine protein kinase Mek1, which in turn enforces the strong choice of interhomolog donors to repair the DSB. Inhibition of an ATP analog-sensitive allele of Mek1, *mek1-as1*, results in a loss of the "barrier to sister chromatid repair," allowing DSBs in a *dmc1Δ* strain to be repaired. Mek1 enforces this block by phosphorylating a specific site (T132) on the Rad54 protein; consequently a *rad54-T132A* mutation can also rescue *dmc1Δ*. Biochemical evidence suggests that the phosphorylation of Rad54 results in a decreased ability to interact with Rad51; *in vitro* assays show that Rad54-T132D is also defective in Rad51-mediated D-loop formation. Moreover, even though normally Rad54 is not phosphorylated in mitotic cells, the Rad54-T132D mutation causes mitotic cells to exhibit increased sensitivity to radiomimetic drugs.

But this regulation of Rad54 is clearly not the entire story of preventing intersister repair and promoting interhomolog repair. Even when Rad54-T132A is combined with a deletion of Hed1, the other meiotic inhibitor of Rad51, the viability of wild-type *DMC1* strains is unaffected (that is, recombination is not predominantly with sisters), suggesting that Dmc1, working presumably with other factors, is able to direct most events to homologs. Thus other factors must actively promote interhomolog repair besides the inhibition of Rad51's interaction with Rad54. Moreover, since Rad51 or Rad54 overexpression can largely bypass *dmc1Δ*, there must be an "escape route" to allow interhomolog recombination even when Mek1 is active.

Recent molecular models from the Kleckner, Hunter, Klein, and Bishop labs have attempted to explain interhomolog bias by suggesting that one of the two ends of a DSB initiates an incomplete strand exchange with its sister and becomes "tied up" at this point (**Figure 13.27**). This idea is entirely consistent with mitotic data that strand invasion at the *MAT* locus in mitotic cells is accomplished in the absence of Rad54, but the intermediate is in some way blocked from progressing to the next step of initiating new DNA synthesis. While one end is therefore "occupied", the

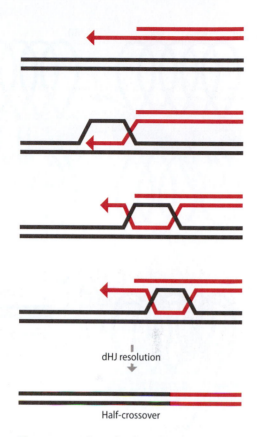

dHJ resolution

Half-crossover

Figure 13.26 Branch migration can produce a dHJ from a single-end invasion. A strand invasion intermediate can be branch-migrated to form a dHJ. Resolution of the dHJ yields a half crossover, as the second product will be incomplete and inviable. (Adapted from Lao JP, Oh SD, Shinohara M et al. [2008] *Mol Cell* 29:517–524. With permission from Elsevier.)

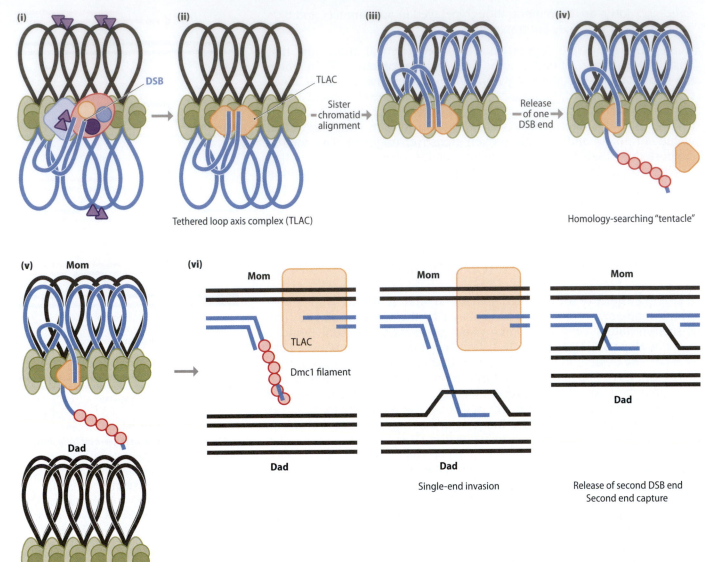

Figure 13.27 Creation of a DSB and tethering of one end enhances interhomolog recombination. (i) As shown in Figure 13.3, a DSB is created at a hot spot after it is brought into the chromosome axis (as marked by the axial element). (ii) The ends are held by a postulated tethered loop axis complex (TLAC). (iii) Sister chromatids align. (iv) One end remains sequestered and tethered while the second resected end, coated with Dmc1, is liberated to search for homology with non-sister sequences (v). (vi) Single-end invasion subsequently leads to the formation of dHJ-containing intermediate that holds homologous chromosomes together. (Adapted from Kim KP, Weiner BM, Zhang L et al. [2010] *Cell* 143:924–937. With permission from Elsevier.)

second end is free to explore the rest of the genomic space for a homologous donor sequence. How only one end is sequestered is not at all clear, but the findings that the removal of Spo11 from two sides of a DSB may leave a long and a short 3'-ended tail might be part of the explanation (though one imagines that both Dmc1- and Rad51-mediated strand invasions need substantial ssDNA ends). One might have imagined that DSB ends might have Dmc1 on one side and Rad51 on the other and that one end would be treated differently; but this now does not seem to be the case: both Rad51 and Dmc1 appear to occupy each end. So at this point it is unclear how one end would be sequestered and how the other can carry out a Dmc1-mediated strand invasion that will synapse the two homologs. Kleckner's lab has suggested that cohesins that would link sister chromatids together might play a role in this step, based on the observation that deleting the meiosis-specific cohesin subunit Rec8 restores an interhomolog bias in *red1Δ* cells.

If one DSB end is sequestered, there must then be a second step in which this first end is released so that it can join the second end in forming a covalently closed dHJ. The Sgs1 helicase could conceivably play a role in this step, given that Sgs1-defective meiotic cells frequently form multiple joint molecules, with one link between the two homologs and another between two sisters. If interhomolog recombination fails, then the sequestered end might be in a position to facilitate intersister repair.

13.14 ZMM PROTEINS PLAY KEY ROLES IN REGULATING CROSSOVERS AND IMPLEMENTING INTERFERENCE

As we have noted before, the meiotic recombination pathways leading to noncrossovers and to crossovers are kinetically different and mutations such as *ndt80Δ* affect only the crossover pathway. However, from the data concerning gene conversions in mitotic cells one would expect that a proportion of the dHJ intermediates would give rise to noncrossovers, by the action of the "dissolvasome" (Sgs1–Top3–Rmi1). But apparently such unwindings of dHJs are largely prevented in meiotic recombination. This difference can be explained by the participation in meiotic recombination of a set of meiosis-specific proteins that apparently prevent dHJ dissolution. Collectively, they are referred to as the ZMM proteins and involve the synaptonemal complex Zip1, Zip2, Zip3, Zip4, and Spo16 proteins, mismatch repair homologs Msh4, Msh5, and the helicase Mer3. Two other mismatch repair proteins, Mlh1 and Mlh3, are also key elements in crossover control. However, the roles of these proteins, and of Sgs1, appear in fact to be much more complicated than simply an antagonism over dissolving dHJs. The ZMM mutants are grouped together because they share a common general phenotype in regulating crossovers in budding yeast—functions that seem evolutionarily conserved up to mice—but in fact they have distinct phenotypes. We will examine them in smaller groupings.

Msh4–Msh5

The proteins that would seem most directly to prevent dissolution of dHJs are the meiosis-specific homologs of the mismatch repair MutS homologs, Msh4 and Msh5. These proteins play no role in mismatch repair but instead have been shown *in vitro* to bind strongly to Holliday junctions; thus it would seem reasonable to imagine that by binding they block the action of Sgs1 to unwind dHJs. In budding yeast, *msh4Δ* or *msh5Δ* mutations result in a roughly 50% reduction in crossovers (**Figure 13.28**).

As noted above, the reduction in crossovers in these ZMM mutants results in a dramatic alteration in spore viability (**Figure 13.29**). The loss of viability is not random; instead there is a large increase in tetrads with two or no viable spores. This is the pattern expected if there are one or more chromosomes that fail to disjoin properly at the first meiotic division. Paired homologous chromosomes require at least one exchange event to provide the tension needed to assure Meiosis I disjunction.

The actual numbers show some significant differences in different chromosomal regions, an observation that we will discuss below, in thinking about noninterfering and interfering crossovers. For future reference we note that the remaining reciprocal exchanges in *msh4Δ* or *msh5Δ* yeast tetrads fail to exhibit crossover interference. This defect is most readily seen by comparing the proportion of observed versus expected NPDs based on map distance. While wild-type diploids have only about 25% as many NPDs as expected, the crossovers remaining in *msh4Δ* or *msh5Δ* show an equal number of expected and observed NPDs.

Consistent with the idea that Msh4–Msh5 bind to dHJs and prevent their being unwound is the satisfying result that deleting Sgs1 restores crossovers to a normal level (**Figure 13.30**). This result suggests that if one prevents Sgs1-mediated dHJ dissolution, then crossovers will increase without the stabilizing effect of the Msh4–Msh5 proteins. However, analysis of the level of COs and NCOs in Southern blots does not support the idea that the reduction of COs in *msh4Δ* or *msh5Δ* is simply caused by the removal of dHJs and conversion of them into NCOs, because the level of NCOs remains the same in *msh4Δ* and *msh4Δ sgs1Δ* strains. Instead, it

Figure 13.28 Effects of *msh4Δ*, *mlh1Δ*, and *mms4Δ* on meiotic recombination in budding yeast. (A) A series of intervals along one chromosome was created by insertion of genetic markers and ablation of resident genes. These markers span approximately 100 cM genetic distance. (B) Spore viability (SV) is given for each genotype. (C) Total recombination from all intervals was measured from complete tetrads (T) and from random spores (S). Each interval is shown for the colors in (A). (Adapted from Argueso JL, Wanat J, Gemici Z & Alani E [2004] *Genetics* 168:1805–1816. With permission from the Genetics Society of America.)

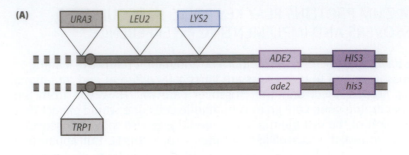

(A)

(B)

Strain	Wild type	*mms4Δ*	*mlh1Δ*	*msh5Δ*	*mlh1Δ* *msh5Δ*	*mlh1Δ* *mms4Δ*	*mms4Δ* *msh5Δ*	*mlh1Δ* *mms4Δ* *msh5Δ*
Spore viability (%)	97	53	68	36	37	42	19	18

(C)

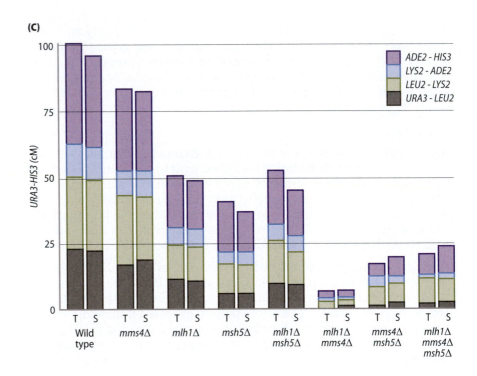

seems that the elevated crossovers are generated by a *different* CO pathway involving the Mus81–Mms4 endonuclease; this pathway is discussed below. Among many other questions, we would like to know if *sgs1Δ msh4Δ* strains have not only regained a high level of crossovers but have also reestablished crossover interference. The idea that ZMM proteins act very early in the crossover/noncrossover decision pathway, long before formation of covalently closed dHJs, has gained traction because of results such as these.

Mlh1/Mlh3

Yeast Mlh1, with Pms1, plays a central role in mismatch repair, but this activity is entirely different from the function of Mlh1 in controlling meiotic crossovers. In meiosis Mlh1 partners with Mlh3, which has a minor role in mismatch repair. In their absence, crossovers are reduced by about half and—according to some observers—there is a loss of interference among the remaining exchanges. *mlh1Δ* or *mlh3Δ* behave epistatically to *msh4Δ* or *msh5Δ*, a result that implies that the four Msh and Mlh proteins work at a common step (Figures 13.28 and 13.29); however, in some

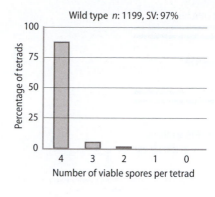

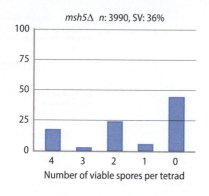

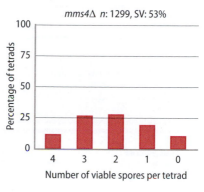

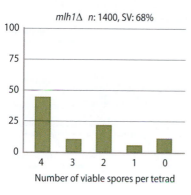

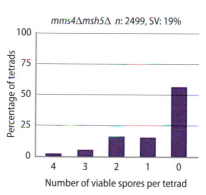

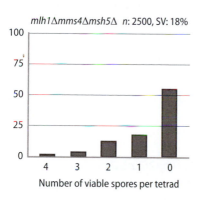

Figure 13.29 Effects of *msh5Δ*, *mms4Δ*, and *mlh1Δ* on spore viability (SV) in budding yeast. Tetrads for each genotype were dissected and the numbers of viable spores in each tetrad are shown. Patterns with many more tetrads with 2 and 0 viable spores are indicative of Meiosis I nondisjunction. SV = spore viability. (Adapted from Argueso JL, Wanat J, Gemici Z & Alani E [2004] *Genetics* 168:1805–1816. With permission from the Genetics Society of America.)

assays *mlh1Δ* has a more severe effect than *msh4Δ* or *msh5Δ*, suggesting it could play additional roles. Recent evidence points to the possibility that Mlh1–Mlh3 can act as a HJ resolvase, as Mlh3 contains endonuclease motifs that—in some bacterial MutL homologs—have been shown to encode an active endonuclease.

Mer3

Mer3 is a meiosis-specific helicase that unwinds DNA in a 3′-to-5′ direction. *mer3Δ* has a phenotype that resembles *msh4Δ* and the double mutant of *mer3Δ* and a deletion of the Msh or Mlh proteins resembles either single mutation. *In vitro*, Mer3 promotes the Rad51-dependent formation of a longer region of heteroduplex DNA, assimilating ssDNA in a 3′-to-5′ direction (for example, from the 3′ end of a DSB) (**Figure 13.31**). Mer3 is imagined to facilitate the formation of a more stable strand invasion structure by creating a longer region of heteroduplex DNA.

Mer3 has also been analyzed in *Arabidopsis*, where its absence causes a phenotype similar to the lack of Msh5, with a 75% reduction in meiotic crossovers.

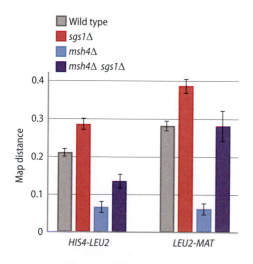

Figure 13.30 Inactivation of Sgs1 restores crossovers in *msh4Δ* diploids. Recombination in two intervals on chromosome 3 in budding yeast was measured in wild-type and mutant cells. (Adapted from Jessop L, Rockmill B, Roeder GS & Lichten M [2006] *PLoS Genet* 2:e155.)

Figure 13.31 Role of Mer3 helicase in extending hDNA after strand invasion. The action of Mer3 is suggested to stabilize the strand-invasion intermediate and direct recombination toward dHJ formation and away from SDSA and noncrossovers. (Adapted from Mazina OM, Mazin AV, Nakagawa T et al. [2004] *Cell* 117:47–56. With permission from Elsevier.)

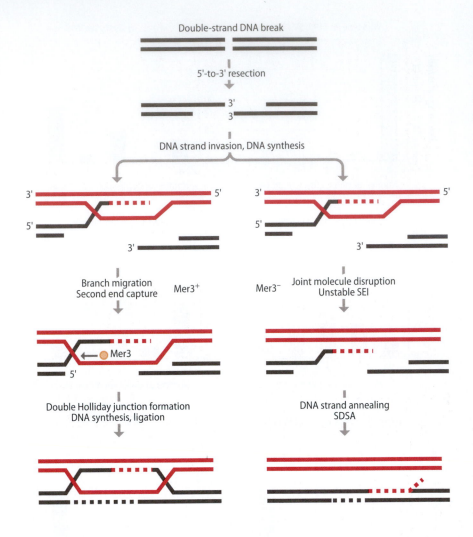

Zip1/Zip2/Zip3/Zip4/Spo16

This is a heterogeneous group of proteins that have distinct functions in budding yeast meiosis related to the formation and extension of the synaptonemal complex (SC). Their roles in SC formation have been described in Chapter 12. Here we will simply say that knockouts of these proteins behave like other ZMM mutants: reduced COs and a lack of interference among the remaining exchanges. The one exception appears to be *spo16Δ*, which has reduced CO, but apparently normal interference, as measured by the reduction in the expected number of NPDs among tetrads. Consistent with the idea that crossover interference requires Msh4–Msh5, *spo16Δ* also does not affect their localization to apparent sites of DSB repair.

The ZMM proteins have also been studied in the fungus *Sordaria*, where there is excellent cytology to complement a good genetic system. Recent studies by Denise Zickler have argued that the ZMM proteins may play important roles even before their roles in crossover resolution. A *mer3* mutation, for example, resulted in greater entanglement, or trapping, of nonhomologous replicated chromatid pairs as co-alignment and synapsis progressed. Quite striking are images in which the Mer3-GFP protein, which generally co-localizes with Rad51, can be shown to be in pairs on either side of the SC. One interpretation would be that Mer3 helicase facilitates strand invasion such that one end of DSB is occupied with its sister while the second end has located its homolog. In *Sordaria*, however, we cannot correlate these images with the proportion of intersister and interhomolog SEIs, just as in yeast we cannot visualize Mer3-GFP with such precision. *S. pombe*, which uses Dmc1 but has no crossover interference, also lacks Msh4 and Msh5.

13.15 ZMM PROTEINS ARE EQUALLY IMPORTANT IN MEIOSIS OF METAZOANS AND PLANTS

The functions of Msh4–Msh5 and Mlh1–Mlh3 are well conserved among most eukaryotes. In *Arabidopsis*, the absence of AtMsh5 results in a 75% reduction in chiasmata. In *Caenorhabditis elegans* loss of the Msh5 homolog virtually eliminates crossovers, suggesting that worms (which also lack Dmc1 and its many partners) have only one crossover-generating pathway (**Figure 13.32**). It is not known if knocking out the BLM helicase in worms would suppress *msh-5* defects in meiosis, but by itself loss of BLM reduces meiotic crossovers.

The situation is different in mice, as both Msh4 and Msh5 homologs are required for normal chromosome synapsis (without which there are no crossovers). *MLH3*$^{-/-}$ mice have a 90% reduction in crossovers; the remaining exchanges may represent a separate pathway. The positions of crossovers in mouse meiosis can be assayed cytologically by immunostaining with an antibody against Mlh1, as shown in Figure 12.26.

Indeed, mice lacking the homologs of yeast Zip3 (RNF212) and Zip4 (TEX11) also have more extreme phenotypes than their absence in fungi. For example RNF212 is required for crossing over in general. It appears that RNF212 selectively localizes to a subset of recombination sites and that this is a key early step in the crossover designation process. RNF212 acts to recruit or stabilize MSH4–MSH5. A very important discovery by Hunter's lab is that a RNF212$^{+/-}$ heterozygote exhibits *haploinsufficiency*; this finding indicates that RNF212 is a limiting factor for crossover control. They raise the possibility that human alleles could alter the amount or stability of RNF212 and be risk factors for aneuploid conditions.

Of course there needs to be an outlier in any evolutionary story and here it is *Drosophila* that lacks Msh4 and Msh5. Similarly, although flies have an Mlh1 homolog that is required for mismatch repair, it seems to be dispensable for crossover control. But flies apparently do have an alternative way to stabilize dHJs without Msh4–Msh5. Their role seems to be taken over by Mcm8 and Mcm9, a pair of helicases related to the Mcm2-7 proteins that are part of the GCM helicase during replication. Crossovers are significantly reduced without Mcm8 or Mcm9 (called MEI-217 and -218); but this problem can be suppressed by removing the BLM helicase that is presumably engaged in dissolving unstabilized dHJs. Mcm8 and Mcm9 also appear to play important, but undefined, roles in homologous recombination in mammals as well.

13.16 THE CROSSOVER/NONCROSSOVER DECISION APPEARS TO BE MADE EARLY IN DSB REPAIR

In the vision of Robin Holliday, whose 1964 model of recombination has left an indelible mark on everyone interested in recombination, all meiotic recombination proceeded through a common intermediate containing a single HJ and crossovers and noncrossovers resulted from alternative cleavage of the HJ. Certainly for the majority of crossovers, this does not appear to be the case. The kinetics of appearance of crossovers and noncrossovers are not the same, whereas resolution of a

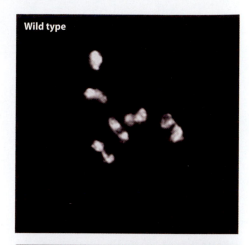

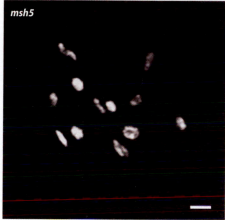

Figure 13.32 Msh-5 is required for all crossovers in *C. elegans*. In wild-type meiotic cells, the formation of crossovers (chiasmata) results in synapsis between the six pairs of homologous chromosomes at pachytene. In *msh-5* mutant cells, one sees 12 univalents because of the lack of crossovers holding homologous pairs together. (From Kelly KO, Dernburg AF, Stanfield GM & Villeneuve AM [2000] *Genetics* 156:617–630. With permission from the Genetics Society of America.)

common intermediate would demand that they appear simultaneously. In an *ndt80Δ* mutant, crossovers are absent but noncrossovers are recovered at an undiminished (but also not higher) level. Finally when *msh4Δ/ msh5Δ* or other ZMM mutants impair crossovers, and when *sgs1Δ* then restores them, the levels of NCOs remains constant. All of these observations suggest that the ZMM mutants control the level of crossing over at an early stage. Further support for this idea comes from the analysis of single-end invasion intermediates, which also are strongly reduced in the ZMM mutants. This finding suggests that the process that channels intermediates into CO or NCO happens very early, at or around the time of strand exchange. The choice is between an NCO SDSA mechanism and a CO-generating dHJ process.

One way to rationalize all these data is to imagine that it is the Mer3 protein that is the key decision maker. Mer3's helicase activity can establish a more stable heteroduplex which could in turn provide the opportunity to form a dHJ (Figure 13.31). Possibly Msh4–Msh5 also bind, not to the HJ but to an earlier branched intermediate. Recall that Msh2 and Msh3 bind not only small heterologies but also to branched intermediates in which a 3'-ended nonhomologous tail is then cleaved, as seen in SSA events. Possibly Msh4 and Msh5 have a similar role in stabilizing an initial strand invasion.

What role Zip1 and other proteins play in this process is harder to specify, but the central region of the SC appears to harbor the recombination nodules that may contain the other ZMM proteins. Zip1 and its cousins may also underlie the establishment of torsional stress that Kleckner has invoked in her stress-relief model of crossover interference. But in this regard it is striking to discover that the spacing of Zip2 or Zip3 foci along yeast chromosomes displays a nonrandom pattern similar to that expected for the interference-based distribution of crossovers (**Figure 13.33**). Thus the SC proteins are already distributed nonrandomly even in *msh4Δ* or *zip1Δ* mutants. These results argue that establishment of SC along the chromosome axis is not required to establish the meiotic pattern of COs. It would be most interesting to know if *mer3Δ* still allows the nonrandom distribution of Zip2 foci.

13.17 AT LEAST ONE MORE CROSSOVER SYSTEM EXISTS, IN ADDITION TO THAT CONTROLLED BY ZMM PROTEINS

The fact that the ZMM mutants only eliminate between 30 and 60% of crossovers in most intervals implies that there is a second crossover-generating pathway. Indeed there is, involving the Mus81–Mms4 endonuclease. Knocking out Mus81 or Mms4 alone causes only a small

Figure 13.33 Zip2 foci exhibit cytological interference. (A) Spreads of synapsed yeast chromosomes after lysing nuclei on microscope slides were stained with the DNA stain DAPI and with antibodies against Zip1, Zip2, and GFP. Chromosome 15 is marked near one end with an array of LacO sequences to which LacI-GFP binds. A merged image of the three antibody stains is shown on the right. (B) The positions of Zip2 foci were located in 10 intervals along chromosome 15. Based on this distribution, the probability of finding Zip2 foci in adjacent intervals was calculated and compared with the observed frequencies. In each interval there is clear evidence of interference, preventing Zip2 foci in adjacent regions. (Adapted from Fung JC, Rockmill B, Odell M & Roeder GS [2004] *Cell* 116:795–802. With permission from Elsevier.)

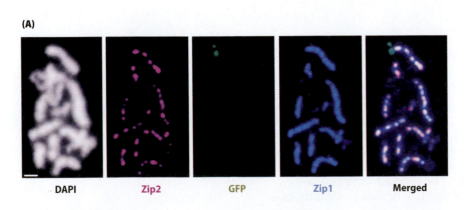

(A)

DAPI   Zip2   GFP   Zip1   Merged

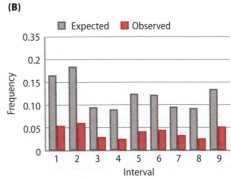

(B)

reduction in crossing over, but there is a significant loss of spore viability. In contrast to *msh5Δ* or *mlh1Δ*, spore inviability of *mus81Δ* or *mms4Δ* does not seem to arise primarily from Meiosis I nondisjunction as shown in Figure 13.29. In ZMM mutants tetrads often contain either two viable spores or no viable spores, but *mms4Δ* and *mus81Δ* diploids exhibit a more random pattern of spore death. Moreover, among compete tetrads the crossovers in *mus81Δ* or *mms4Δ* strains still display crossover interference. When crossovers and crossover intermediates are monitored in *mms4Δ* they exhibit a delay. There is a greater reduction in interhomolog joint molecules than in intersister JMs. One other interesting finding is that deleting the axial element Red1 protein suppresses the barriers to completing Meiosis I and Meiosis II divisions. These data imply that removing impediments to intersister DSB repair allows *mus81Δ* cells to survive much better; in turn this interpretation suggests that the block to completing meiosis rests with problems in completing interhomolog DSB repair.

That Mus81/Mms4 operate in a different pathway from Msh4/Msh5/Mlh1/Mlh3 is evident by examining double mutants such as *mus81Δ msh5Δ* or *mms4Δ msh5Δ* (Figure 13.28). These strains show a more than 10-fold reduction in crossovers, although there are still some crossovers remaining. Despite these reductions in exchange, spore viability (which demands that almost all of the 16 chromosome pairs have a crossover to ensure proper segregation) is surprisingly robust (20–40%). The reasonably efficient chromosome segregation in *mus81Δ msh5Δ* or *mms4Δ msh5Δ* could reflect the persistence of a special class of initial crossovers needed to promote synapsis. These residual exchanges are not eliminated by deleting Mlh1, so the search for new crossover factors will continue.

As we will discuss in more detail in Chapter 14, Mus81–Mms4 can weakly cleave HJs *in vitro* but the complex is much more active on nicked HJs and other branched DNA structures. Whether Mus81–Mms4 represents the HJ resolvase in an alternative CO pathway or instead acts like Msh4/Msh5 in stabilizing some as-yet-uncharacterized recombination intermediate has yet to be sorted out. In *S. pombe* meiosis, Mus81–Eme1 is responsible for nearly all crossovers and overexpression of a bacterial HJ resolvase (RusA) can largely compensate for its absence. But in budding yeast, the defect in meiotic COs in *mus81Δ* is not complemented by RusA.

The relationship between the two CO pathways is not simple. Analysis of the distribution of COs led to the idea that there are both interfering and noninterfering exchanges and that ZMM proteins controlled the interfering COs. That the noninterfering COs rely on Mus81–Mms4 is inferred from (a) the fact that the remaining crossovers in the absence of Mus81–Mms4 continue to show interference, and (b) total exchanges and spore viability are strongly reduced in *msh5Δ mms4Δ* compared with either single mutant. So there seem to be two crossover pathways, although no one has accumulated enough data to demonstrate directly that the noninterfering COs have diminished in a *mus81Δ* strain.

Paradoxically, in both *msh4Δ* and *mms4Δ* the rate of gene conversion goes up, at least when measured by 3:1 and 1:3 segregation among the relatively rare complete tetrads. This observation differs from other studies that found no difference compared with wild type when gene conversions were measured as the rate of prototroph formation by recombination between auxotrophic heteroalleles. Possibly the two assays of gene conversion reveal a subtle difference in the mechanisms of the recombination events that can be recovered in these mutant strains. A longer region of heteroduplex would more often cover a single allelic site in a heterozygote (increasing 3:1 or 1:3) but would lead to more frequent co-conversion of a pair of heteroalleles and thus no net increase in prototrophs.

A different demonstration of the differences between the ZMM-regulated CO pathway and the Mus81 pathway comes from exploiting an important observation by David Kaback that smaller yeast (and human) chromosomes have a higher level of crossing over, as discussed in Chapter 12 (Figure 12.35). When map distances are scored for same intervals in the original and bisected chromosomes, there is a significant increase in map distance on the shortened version. Stahl and colleagues have now shown that much of the increased level of crossing over in the smaller chromosomes results from Msh4-independent exchanges. This result supports the idea that the Mus81-dependent pathway produces non-interfering crossovers that are preferentially distributed on the smaller chromosomes.

13.18 DISTRIBUTION OF CROSSOVERS AND NONCROSSOVERS CAN BE ANALYZED GENOMEWIDE IN BUDDING YEAST

Until recently, our understanding of the distribution of crossovers and noncrossovers came predominantly from examining meiotic events along a well-marked chromosome arm, where one could detect all of the crossover events and a fraction of noncrossover gene conversions. For example, one study looked at 10 markers spread over a 405-kb arm of budding yeast chromosome VII. There were about 3.3 crossovers per meiosis along this arm. Remembering that 1 crossover = 50 cM (that is, half the chromatids at one site have a crossover), these numbers show that in budding yeast there is about 2.3 kb/cM. The distribution of tetrads with 1, 2, 3, and so forth, crossovers per meiosis is not well described by a simple Poisson distribution; instead the distribution was better fit by an interference model in which there were three noncrossover recombination events for every crossover (the so-called "counting" model described in Figure 12.37). An even better fit was achieved by postulating that there is an additional small number of crossovers that are randomly distributed and, moreover, noninterfering. However, other predictions of the counting model (a lack of crossover homeostasis) have not been supported by experiment, as shown in Figure 12.38.

Another important consideration is that from these same tetrads not nearly all the postulated crossover- or noncrossover-associated gene conversions are detected. Only about 22% of the tetrads had evident gene conversions (1:3 or 3:1 segregation) at one of the 10 loci, half of which were accompanied by crossing over. So, although there were 3.3 crossovers along this 400-kb interval, one detects a very small fraction (half of 22%) of all the noncrossover events with the 10 markers along this part of one chromosome arm. So most crossovers and most (invisible) noncrossovers must have occurred in regions where there were no genetic markers.

A radical improvement in being able to detect all (or at least most) genetic exchanges, with and without crossover, was accomplished by studying the meiosis of a diploid heterozygous for thousands of single nucleotide polymorphisms (SNPs) (**Figure 13.34**). Lars Steinmetz' group used high-density microarrays to map the location of both crossovers and gene conversions in all four DNA strands of 56 tetrads; one example is shown in **Figure 13.35**. On average they found about 90 CO and 46 NCOs per meiosis; but since 30% of the COs occurred *between* the SNP markers,

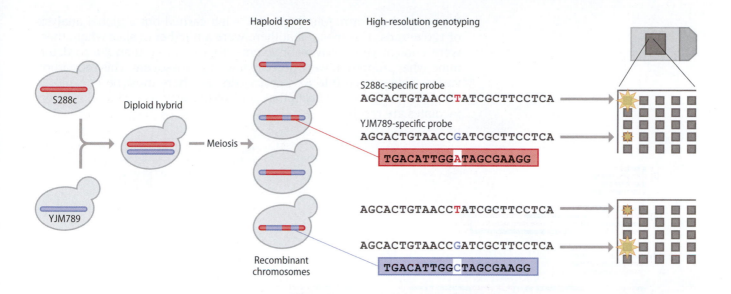

Figure 13.34 Microarray detection of single nucleotide polymorphisms (SNPs) among spores derived from a diploid. Single-nucleotide differences between pairs of alleles can be detected by hybridization on microarrays, as each sequence hybridizes strongly to the oligonucleotide that perfectly matches the sequence in a given meiotic segregant (large star), though it weakly hybridizes to the other. (Adapted from Mancera E, Bourgon R, Brozzi A et al. [2008] *Nature* 454:479–485. With permission from Macmillan Publishers, Ltd.)

the authors corrected the totals to reflect NCOs that would have occurred without an accompanying gene conversion. Thus, on average, there were 90 COs and 66 NCOs. The number of NCOs is still quite low compared with the analysis of individual markers, where only 30–50% of gene conversions at a particular locus are accompanied by CO. The proportion of NCOs is also markedly lower than the estimates based on "counting" models of interference, which estimate three NCO for every CO.

One explanation for the reduced number of NCOs would be that gene conversion tracts for NCOs are much shorter than for COs. A significant difference in the lengths of gene conversions of COs and NCOs has been seen in mouse, but in the yeast data the difference is minimal (2.0 kb versus 1.8 kb, respectively). It is unlikely that some of the "missing" NCOs have been removed by nonhomologous end-joining because NHEJ is turned off in *MAT**a**/MATα* diploids by repression of the *NEJ1* gene. A more likely explanation is that heteroduplex DNA is more often restored, instead of converted, in the SDSA events that apparently account for most NCOs compared to the dHJ pathway leading to crossovers. In this case a fraction of NCOs would lack the necessary conversion event to be scored. The ratio of NCO to CO is much closer to 1 in a *msh2* diploid lacking most MMR.

One should also remember that the global analysis of gene conversions and crossing over is carried out in strains having thousands of heterologies, where the mismatch repair machinery may be overtaxed, so that the outcomes may be different from what is seen in near-isogenic diploids carrying a few mismatches. This may explain the finding that 10% of the non-Mendelian segregation events in the highly polymorphic diploid used in microarray analysis were PMS events. This analysis confirmed that C:C mismatches are globally refractory to mismatch repair, but some of the other PMS events may be a reflection of too many heteroduplexes for the mismatch repair machinery to deal with. Consistent with the observations in several organisms that a poorly corrected mutation is more likely to be corrected if it lies in hDNA with a well-corrected mismatch, the PMS events involving C:C mismatches tended to be further away from the next polymorphic site.

Global analysis of gene conversion and crossing over has also revealed that the proportion of gene conversions accompanied by crossing over is not at all uniform. John McCusker's lab carried out a global analysis of 120 tetrads. This meant that there were a number of sites where there were enough gene conversion events (that is, more than 20) to determine what proportion were accompanied by crossovers. This proportion varied widely, from 0 to 52% crossovers. So there must be unexplored chromosome context effects that control the way recombination intermediates are resolved.

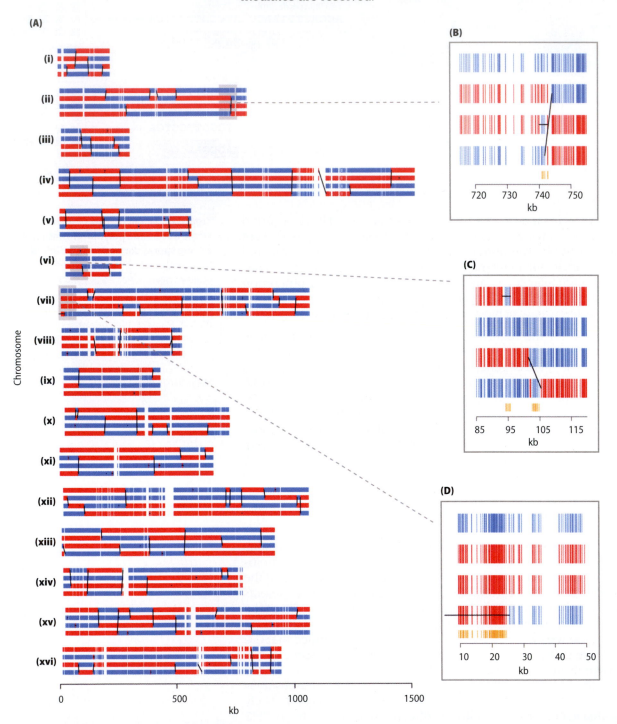

Figure 13.35 Detection of all recombination events from a single budding yeast tetrad. Recombination was determined by analyzing microarray data for approximately 52,000 markers (A). In the close-ups of three regions, diagonal or vertical lines show crossovers while a horizontal line indicates a gene conversion. Yellow bars below show the lengths of gene conversion tracts. In (B), a CO overlapped by an independent NCO in a third spore; (C) a CO with complex gene conversion tract, and a nearby, independent NCO; and (D) a long NCO at the end of the chromosome. (Adapted from Mancera E, Bourgon R, Brozzi A et al. [2008] *Nature* 454:479–485. With permission from Macmillan Publishers, Ltd.)

13.19 GLOBAL ANALYSIS OF MEIOTIC EVENTS HAS BEEN EXTENDED TO METAZOANS AND PLANTS

Recently, similar microarray and high-throughput DNA sequencing methods have been used to obtain similar chromosome- or genome-wide pictures of crossing over and gene conversion in *Drosophila*, *Arabidopsis*, and mouse. In these organisms we discover more complexities. For example, there are highly significant differences in the patterns of meiotic events in males and females (ignoring *Drosophila*, where there is no crossing over in males). Genotyping of sperm has also enabled us to see how these events happen in human males.

Drosophila

Two experiments have examined gene conversion and crossing over on all or part of the X chromosome. Scott Hawley's lab sequenced 30 males coming from a heterozygous female with >than 93,000 SNPs and small insertion/deletions to identify 15 crossovers and five gene conversions without an associated crossover. Based on the assumption that only a fraction of NCOs were found, and correcting for the nonuniform distribution of markers, Hawley estimated that there would be 1 CO and 1.6 NCO on the X chromosome per meiosis. Single crossovers tended to be in the middle of the arm. The mean gene conversion tract length was estimated to be 476 bp. Interestingly a motif GTGGAAA was identified near crossovers but not noncrossovers, but whether it is important in specifying crossover-associated recombination events is not yet evident.

A second approach, from Andrew Clark, was to select specific crossovers genetically and then sequence the intervening region to determine where crossovers had occurred. His lab analyzed 2500 flies that had a single crossover in a 2.1 Mb region of the X chromosome. They found a 90-fold variation in the rates of crossing over when the region was broken down into 5-kb units. Again a motif (here ATGGAAA) showed up as a possible factor in the location of crossovers. There was no obvious correlation with the GC content of the regions or any correlation with the location of genes. There were no obvious hotpots at promoter regions.

Arabidopsis

There is very substantial sexual dimorphic behavior in meiosis. The total genetic map in males is 575 cM versus only 332 cM in females. Christine Mézard's analysis of crossovers in a genomewide study showed further that the higher male CO rates were elevated at both ends of each chromosome. There was also a much stronger correlation of COs in males in regions with a high CpG content (which can be sites of DNA methylation). COs in females were also higher in regions containing genes. Interestingly, the difference in genetic map lengths is also seen in the lengths of synaptonemal complex in males and females. This study did not use enough SNPs to identify gene conversion tracts, but other studies have suggested that there is a surprising excess of crossovers over noncrossovers, a result that might mean that NCO gene conversions are very short and often do not include a SNP, or that a significant proportion of DSBs could be repaired by sister chromatid recombination.

Mouse

A great deal has been learned from deep-sequencing studies of mouse meiosis in both males and females. Maria Jasin and Scott Keeney focused their attention on a specific meiotic hot spot known as A3 to look at gene conversions and crossovers with a SNP on average every 30 bp. In this hybrid, there are apparently many more DSBs initiated on one parental chromosome than the other. In keeping with estimates of the number of DSBs seen cytologically (for example, Rad51 foci) and the number of crossovers (for example, Mlh1 foci) this sequencing study showed that 90% of the gene conversions were NCO. The gene conversion tract lengths of COs averaged about 500 bp, while the NCOs were only about 100 bp. The positions of NCOs also argued that the A3 hot spot did not suffer DSBs at one specific sequence but rather represented a set of possible DSB locations. Thus, while a single Prdm9 binding site lies at the center of the hot spot and recruits Spo11 and other proteins, the cleavage away from this central location and the short span of NCO gene conversions means that the "hotter" sequence will not always be lost.

Humans

Similar analyses have been done to study recombination in human sperm, as well as by examining children from well-defined families. From various studies it emerges that at least half, and probably more, of the hot spots harbor a 13-bp Prdm9 consensus site (which is different in humans than in mouse). Unlike budding yeast, the locations of crossovers are generally tens or hundreds of kilobases from the nearest transcription start site (**Figure 13.36**). Different hot spots appear to show differences in male and female meiosis, while others appear to be gender independent.

13.20 MEIOTIC RECOMBINATION CAN BE INDUCED BY MEGANUCLEASE-INDUCED DSBs

A direct comparison between DSB-initiated meiotic and mitotic recombination events is difficult because the events are normally initiated in different ways. However, several approaches to address the similarity of mitotic and meiotic recombination have been undertaken using site-specific endonucleases. In some cases the same nuclease can be expressed in both stages of the life cycle; in other instances one can compare Spo11-initiated meiotic events with those initiated by a different nuclease.

HO endonuclease

By expressing HO endonuclease under the control of a meiosis-specific (*SPO13*) promoter, it has been possible to make such comparisons (**Figure 13.37**). Among tetrads from *SPO13::HO* diploids about half show gene conversion between a *leu2* gene carrying an HO cleavage site (*leu2-cs*) and a *LEU2* donor. In half of these cases there was a single gene

Figure 13.36 Meiotic crossovers in humans are located far from transcription start sites (TSS). Locations of crossovers from 728 children of related families are mapped relative to the closest TSS. (Adapted from Coop G, Wen X, Ober C et al. [2008] *Science* 319:1395–1398. With permission from the the American Association for the Advancement of Science.)

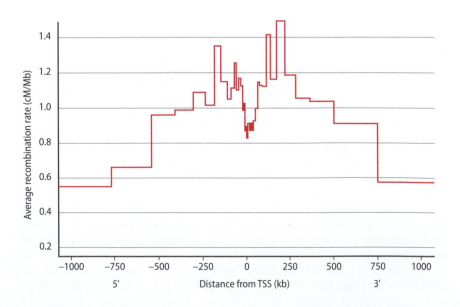

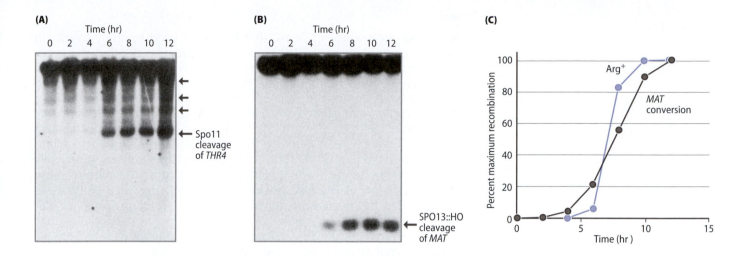

Figure 13.37 Comparison of meiotic events induced by Spo11 and SPO13::HO. Spo11 induces recombination at *THR4* (A) whereas *SPO13::HO* creates a meiotic-specific DSB at *MAT* (B). (C) Gene conversions at *MAT* compared with Spo11-induced heteroallelic recombination at *arg4*. (Adapted from Malkova A, Ross L, Dawson D et al. [1996] *Genetics* 143:741–754. With permission from the Genetics Society of America.)

conversion (yielding 3 *LEU2* : 1 *leu2-cs* spores), while the rest have 4 *LEU2* where both sister chromatids must have been cut and gene-converted. Cells with a single gene conversion behave in most respects like Spo11-induced events. First, about 50% are associated with crossing over, similar to other meiotic gene conversions. Second, the percentage of crossovers is reduced threefold by deleting Msh4, similar to the reductions seen for the ZMM mutations in Spo11-driven crossovers. Together these results show that the DSBs induced by Spo11 are not treated intrinsically differently from those induced by HO (for example, in the formation of the necessary intermediates or in the recruitment of factors required to produce a high level of crossovers).

In several respects the HO-induced DSBs are different from Spo11 events. First, neither cleavage nor 5′-to-3′ resection necessary to promote gene conversion is prevented by deleting Rad50, whereas both DSB formation by Spo11 and subsequent 5′ and 3′ resection depend on Rad50, Mre11, and Xrs2. Second, HO is not prevented from cleaving both sister chromatids at the same site, whereas cases where such events occur with Spo11 are at best rare, even for hot spots. When both sisters are cleaved there is a striking reduction in the proportion of tetrads that exhibit even a single crossover (that is, a TT), from 52 to 23%. Thus, cleaving both sisters disrupts some critical feature of the architecture/tension surrounding the site of repair and results in a reduction in crossovers. In addition, when both sisters are cleaved the frequency of co-converting a marker 1 kb away from the DSB rose dramatically, from 6 to almost 40% (the level of co-conversion seen in mitotic cells expressing HO). Thus, some constraints on 5′-to-3′ resection may depend on the integrity of the axial element or SC that in turn depends on one sister chromatid being intact.

In the absence of Spo11, in *spo13Δ spo11Δ* diploids that produce two diploid spores, *SPO13::HO* is able to induce a high level of gene conversions. These events were still meiotic-like, with 27% associated crossovers compared with about 5% in mitotic cells. Still, this is half the level seen for a control *spo13Δ SPO11* diploid. The difference between crossovers with and without Spo11-induced events might be attributed to the formation of SC that is absent in *spo11Δ*. Surprisingly, a significant proportion of apparent crossovers in these *spo11Δ spo13Δ SPO13::HO* cells actually seem to be BIR events, in which sequences distal to the DSB have been copied all the way to the end of the chromosome. This is the first demonstration that BIR can occur in meiotic cells and suggests that the presence of Spo11 or the many DSBs produced by Spo11 play a very important role in ensuring the normal outcomes of meiosis.

VDE

A second site-specific nuclease that has been used to study events in meiotic recombination is VDE, a nuclease related to HO that promotes the insertion of a DNA element at a specific cleavage site. VDE is encoded by the *VMA1* gene. VDE-induced DSBs are very efficient in meiosis, so that nearly all targets are cleaved. In a diploid heterozygous for *arg4-BglII* and an *arg4* allele carrying a VDE cleavage site and surrounded by *URA3* sequences, repair can either yield gene conversions of the VDE site or deletions by SSA between the *URA3* sequences. In large fraction of the tetrads, repair of the DSBs on the two sisters yielded genetically distinct results, so each was repaired independently. Resection of these Spo11-independent DSBs is slowed down by deleting Sae2 or Exo1, as well as with the nuclease-dead Mre11-H125N. Deleting the axial element component, Hop1, results in a higher proportion of SSA. Evidence of extensive resection was also found in a *spo11* mutant diploid; here we presume axial elements are intact but no SC will form. So, the action of Spo11 is necessary to accomplish the normal, meiotic-like recombination of a meganuclease-induced DSB. Most likely a full level of crossovers and short gene conversion tracts requires Spo11-mediated assembly of SC.

I-SceI cleavage in S. pombe meiosis

One unsolved problem in *S. pombe* meiosis is the contradiction between very widely spaced sites of DSBs and a more uniform distribution of crossovers. The idea that the recombination intermediates branch migrated from the site of the DSB to locations 20 kb or more away was tested by Gerald Smith's lab, who deleted *S. pombe's* Spo11 homolog (Rec12) and placed the I-SceI nuclease under the control of the *REC12* promoter. Because *S. pombe* has only three chromosomes, the lack of crossing over still allows the recovery of viable spores even in the face of chromosome nondisjunction. I-SceI-induced DSBs at one site was accompanied by a very high level of crossing over in that local interval but there was no increase 25 kb away. This finding has led to the suggestion that many of the crossover-producing lesions in *S. pombe* could be ssDNA nicks and not DSBs, which are seen only at widely separated sites. Another benefit of using I-SceI was finding that a very high proportion of the gene conversions were accompanied by crossing over, just as had been seen for normal DSBs. This finding means that there is no special mechanism of ensuring crossovers linked to Rec12-generated DSBs. Why gene conversions should be accompanied by crossing over more than two-thirds of the time (at some loci more than 85%) might reflect the fact that fission yeast does not exhibit crossover interference.

Mos1 transposon

In *C. elegans* a *Mos1* transposon was found that disrupted the *unc-5* gene. Excision of *Mos1* in meiotic cells leaves DSBs with ssDNA ends (similar to *Drosophila* P-element excision that we discussed in Section 6.19). *Mos1* excision can be induced by heat-shock, resulting in recombination with a different *unc-5* allele to produce unc5+ recombinants. These recombinants arose in as many as 7% of the progeny and both CO and NCO outcomes were obtained, with COs representing ~11% of the recombinants. A detailed analysis led to a number of important conclusions. First, DSBs induced after SC assembly were competent to become COs. This result suggests that the decision to form a CO or NCO is not necessarily an early event, as the analysis of ZMM mutants suggested for budding yeast; but recall that in worms SC formation is not driven by homologous recombination. Second, a *Mos1*-initiated crossover exerted interference on other crossovers. On chromosomes with an NCO event at the *Mos1* site, nearly 40% of the chromosomes showed a CO elsewhere on the chromosome, in agreement with expectations based on the genetic

Wild type　　　　　　　　*spo11*　　　　　　　*spo11* + X-rays

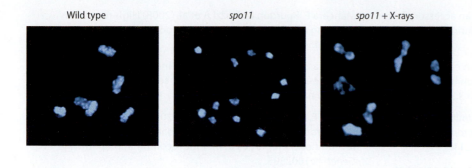

Figure 13.38 X-irradiation can partially rescue Spo11-deficient meiosis in *C. elegans*. Whereas wild-type pachytene cells exhibit six synapsed pairs of chromosomes, a Spo11⁻ cell (*spo11*) shows 12 unpaired homologs. In Spo11⁻ cells, which have progressed to the point for forming synaptonemal complex, X-irradiation partially restores homologous chromosome pairing, although one pair have failed to synapse and some paired chromosomes show morphological abnormalities. (From Dernburg AF, McDonald K, Moulder G et al. [1998] *Cell* 94:387–398. With permission from Elsevier.)

map and the finding that there is only one CO per bivalent. In contrast, on chromosomes with a CO event at the *Mos1* site, no additional COs were detected. Again, this result suggests that a different kind of DSB can compete with Spo11-induced DSBs to form a crossover.

Gamma irradiation

In worms lacking Spo11, there are no crossovers and hence homologous chromosomes are not linked by chiasmata (**Figure 13.38**). However, X-irradiation partially restores crossing over to such animals, as seen by the formation of synapsed homologous chromosome pairs. So, again, an exogenous way of generating DSBs can lead to meiotic levels of crossing over.

SUMMARY

Meiosis is initiated by Spo11-mediated cleavage, which exhibits regional preference but not sequence specificity. In mammals the zinc-finger binding protein Prdm9 plays a key role in determining the location of DSBs. Many of the important molecular steps in meiotic recombination—strand-invasion, primer extension, second-end capture, and double Holliday junction formation and resolution—have been visualized by Southern blots. In many organisms, including both yeast and mammals, recombination depends on both Rad51 and its paralog, Dmc1. Dmc1 acts with a number of other factors that are not involved in Rad51-mediated mitotic recombination. A large proportion of crossovers depend on the action of ZMM proteins, but there is at least one alternative crossover-generating process.

SUGGESTED READING

Allers T & Lichten M (2001) Differential timing and control of noncrossover and crossover recombination during meiosis. *Cell* 106:47–57.

Allers T & Lichten M (2001) Intermediates of yeast meiotic recombination contain heteroduplex DNA. *Mol Cell* 8:225–231.

Andersen SL & Sekelsky J (2010) Meiotic versus mitotic recombination: two different routes for double-strand break repair: the different functions of meiotic versus mitotic DSB repair are reflected in different pathway usage and different outcomes. *BioEssays* 32:1058–1066.

Baudat F, Buard J, Grey C et al. (2010) PRDM9 is a major determinant of meiotic recombination hot spots in humans and mice. *Science* 327:836–840.

Borde V & de Massy B (2013) Programmed induction of DNA double strand breaks during meiosis: setting up communication between DNA and the chromosome structure. *Curr Opin Genet Dev.* 23:147-155.

Borner GV, Kleckner N & Hunter N (2004) Crossover/noncrossover differentiation, synaptonemal complex formation, and regulatory surveillance at the leptotene/zygotene transition of meiosis. *Cell* 117:29–45.

Cloud V, Chan YL, Grubb J et al. (2012) Rad51 is an accessory factor for Dmc1-mediated joint molecule formation during meiosis. *Science* 337:1222–1225.

Ehmsen KT & Heyer WD (2008) Biochemistry of meiotic recombination: Formation, processing, and resolution of recombination intermediates. *Genome Dyn Stab* 3:91–164.

Esberg A, Muller LA & McCusker JH (2011) Genomic structure of and genome-wide recombination in the *Saccharomyces cerevisiae* S288C progenitor isolate EM93. *PLoS One* 6:e25211.

Keeney S & Neale MJ (2006) Initiation of meiotic recombination by formation of DNA double-strand breaks: mechanism and regulation. *Biochem Soc Trans* 34:523–525.

Kohl KP & Sekelsky J (2013) Meiotic and mitotic recombination in meiosis. *Genetics* 194:327–334.

Lichten M & de Massy B (2011) The impressionistic landscape of meiotic recombination. *Cell* 147:267–270.

Mancera E, Bourgon R, Brozzi A et al. (2008) High-resolution mapping of meiotic crossovers and non-crossovers in yeast. *Nature* 454:479–485.

Mancera E, Bourgon R, Huber W & Steinmetz LM (2011) Genome-wide survey of post-meiotic segregation during yeast recombination. *Genome Biol* 12:R36.

Martini E, Borde V, Legendre M et al. (2011). Genome-wide analysis of heteroduplex DNA in mismatch repair-deficient yeast cells reveals novel properties of meiotic recombination pathways. *PLoS Genet* 7:e1002305.

Oh SD, Jessop L, Lao JP et al. (2009) Stabilization and electrophoretic analysis of meiotic recombination intermediates in *Saccharomyces cerevisiae*. *Meth Mol Biol* 557:209–234.

Schwacha A & Kleckner N (1995) Identification of double Holliday junctions as intermediates in meiotic recombination. *Cell* 83:783–791.

CHAPTER 14

HOLLIDAY JUNCTION RESOLVASES AND CROSSING OVER

Crossovers are generated by Holliday junction (HJ) resolvases. Our understanding of this important group of enzymes has dramatically improved over the past decade, with the characterization of two new enzymes. There are still some profound mysteries associated with HJ resolution, most notably in the connections to mismatch repair. Much of what we know molecularly and structurally about HJ resolvases comes from work on enzymes from bacteriophage, bacteria, and mitochondria. Our knowledge about eukaryotic HJ resolvases comes primarily from studies of meiotic recombination in budding yeast. Only recently has this analysis extended to the role of these enzymes in meiosis of other organisms and in the resolution of HJs in mitotic cells. There are both "true" resolvases that cleave symmetrically at HJs and other endonucleases that cleave nicked HJs and other related structures.

14.1 HOLLIDAY JUNCTIONS CAN ADOPT ALTERNATIVE CONFIGURATIONS

First, a word about HJ structures. In the absence of divalent cations, HJs form an open, square-planar structure with approximately fourfold symmetry (**Figure 14.1**). In the presence of Mg^{2+}, however, HJs in solution have a stacked, twofold symmetrical structure with two antiparallel "continuous" strands and two "crossing strands" that show a change in direction (Figure 14.1). In the stacked arrangement the strands that we image to be exchanged appear to be the continuous strands. This structure seems to be at variance with the way we imagine HJs to form between aligned, homologous chromatids and it is quite likely that the arrangement of

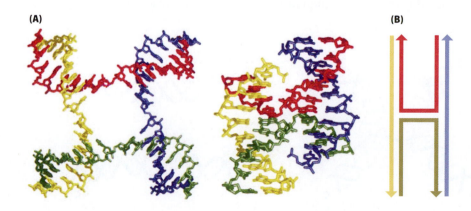

(A) **(B)**

Figure 14.1 Holliday junction structure.
(A) Open (square-planar) and stacked-X configurations of a Holliday junction.
(B) A flattened cartoon of the four strands is shown.
(Adapted from Sharples GJ [2001] *Mol Microbiol* 39:823–834. With permission from John Wiley & Sons, Inc.)

these molecules *in vivo* is very differently constrained by the global chromosome structure, cohesins, and many other proteins bound to DNA; however, there may be significant distortions of the parallel alignment of homologous DNA molecules at the point where resolvases act.

There are no structural studies of double HJs to show their relative orientation or whether there is a minimum separation between the HJs. We do not understand how dHJs can be resolved in a biased fashion (recall that most dHJs in yeast meiosis appear to be resolved as crossovers and that there is strong evidence against there being closely spaced double crossovers). These observations make it unlikely that each HJ is resolved independently. Electron microscope studies of DNA containing dHJs has indicated that they are, on average, about 260 bp apart (see Figure 13.24), but of course proteins could move them to a different, more optimal distance for resolution.

14.2 CANONICAL HJ RESOLVASES MAKE SYMMETRICAL CUTS THAT CAN BE RE-LIGATED

The best studied of the "true" HJ resolvases are bacteriophage T7 endonuclease I, T4 endonuclease VII, the lambdoid bacteriophage RusA, *E. coli* RuvC, the mitochondrial resolvases Cce1 of budding yeast and Ydc2 of fission yeast, and Gen1 in mammalian cells. All of these enzymes act as dimers, cleaving the same phosphodiester bond in symmetrical sites across the HJ. Thus the crossover products have no missing or overhanging bases that would prevent re-ligation of the ends (**Figure 14.2**). Although these resolvases all yield similar outcomes, they do not all act in the same way, as discussed below.

HJ resolution can be assayed in supercoiled plasmids containing short inverted repeats that will adopt a cruciform structure to relieve the stress of supercoiling (**Figure 14.3**). This clever assay shows that both sides of the cruciform must be cleaved simultaneously (or at least before the proteins become unbound) because cutting a single strand will yield a nicked circle in which the cruciform is no longer extruded and thus the second strand can no longer be cut. Nicked circles are a very minor product in this assay.

Another convenient assay in budding yeast is the resolution of HJ-containing joint-molecules arising from sister chromatid repair in the absence of Sgs1, after cells are treated with a low dose of MMS (**Figure 14.4**). The absence of Sgs1 prevents dissolving of dHJs and thus they accumulate. The bacterial RusA expressed in yeast can resolve these structures, as can expression of the mammalian GEN1 enzyme. Curiously, the success of dHJ resolution after expressing these non-yeast enzymes implies that the endogenous yeast enzymes are either not expressed sufficiently or are prevented from resolving these structures.

Figure 14.2 Resolution of a Holliday junction by the RuvC enzyme. (i) Configuration of a HJ in the absence of protein binding. Binding of a RuvC dimer (ii) results in cleavage at the same sequence on two strands of the HJ (iii). HJ resolution yields two products whose single-strand nicks can be easily re-ligated. (Adapted from Shah R, Cosstick R & West SC [1997] *EMBO J* 16:1464–1472. With permission from Macmillan Publishers, Ltd.)

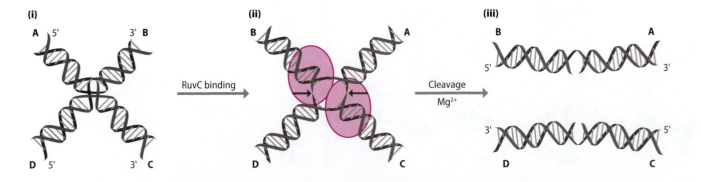

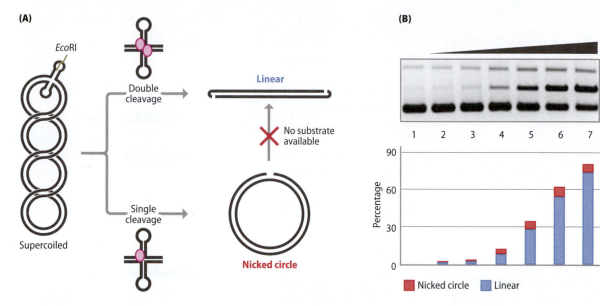

Figure 14.3 Demonstration that HJ resolution involves concerted cutting of two strands. An assay for HJ resolution using a supercoiled plasmid with an inverted repeat (palindromic) sequence that is extruded as a cruciform in a supercoiled plasmid (A). Simultaneous cleavage of two opposite strands yields a linear molecule—the major product when the GEN1 enzyme is added (B). If only one strand is cut before the other, then the plasmid relaxes into a nicked circle and the cruciform structure is lost and cannot be further cleaved. Analysis of the products on an agarose gel shows that nearly all the product is that expected from cleavage of both strands. (Adapted from Rass U, Compton SA, Matos J et al. [2010] *Genes Dev* 24:1559–1569. With permission from Cold Spring Harbor Laboratory Press.)

Some of these enzymes show sequence preference; for example, RuvC cleaves preferentially at the sequence A/T TT ⇓ G/C, whereas its homolog Cce1 prefers ACT ⇓ A. RusA cleaves after CC, and in general the cleavage follows a pyrimidine. The T4 and T7 enzymes, as well as RuvC, do not entirely act on HJs, as they have all been shown also to cleave Y-shaped molecules that might somewhat mimic stalled replication forks.

The strong sequence preference for the A/T TTG sequence in RuvC is illustrated by its failure to generate linear products if the site is missing, even in a hybrid HJ in which one arm has the preferred site (**Figure 14.5**). RuvC is perhaps the most fastidious of the resolvases and requires homology between the arms being cut; most resolvases will cut symmetrically at any four-way junction.

The different sequence requirements are only one indication that these evolutionarily distant enzymes do not all work in the same way. Enzymes such as Cce1 bind to the HJ and "flatten it out" into an open X structure, where it can cleave either pair of strands of the same polarity (**Figure 14.6A**). In the case of phage T4's endonuclease VII, the enzyme cleaves the exchanging strands. A more detailed image of the way T4

Figure 14.4 Resolution of HJ structures generated in budding yeast by replication stalling. HJ-containing joint molecules accumulate in *sgs1Δ* budding yeast treated with MMS. Under these conditions the Sgs1–Top3–Rmi1 complex cannot dissolve HJs and they must be eliminated by HJ resolvases. The X-spike (*red arrow*) seen in two-dimensional gels represents a family of HJ-containing molecules that are removed when cells overexpress *E. coli* RusA or the active fragment of human GEN1 containing amino acids 1–527. (From Mankouri HW, Ashton TM & Hickson ID [2011] *Proc Natl Acad Sci USA* 108:4944–4949. With permission from the National Academy of Sciences.)

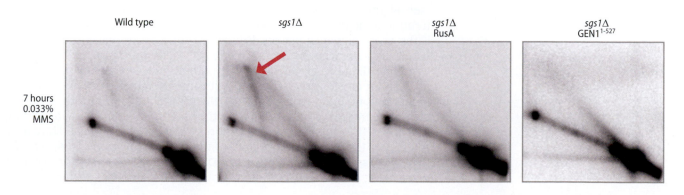

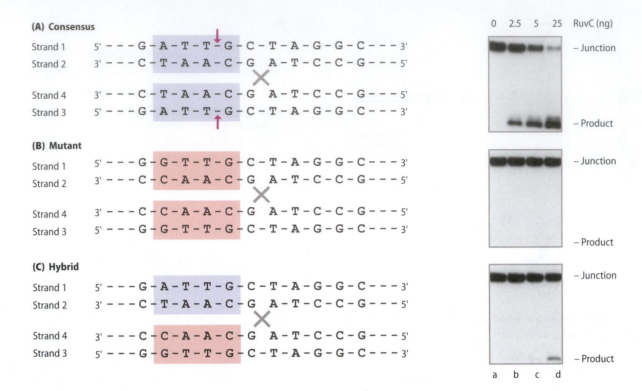

Figure 14.5 Requirement of RuvC for a consensus cleavage site. The strong preference of RuvC for a specific cleavage site is illustrated in (A), whereby a sequence containing ATTG is readily cleaved into linear products whereas HJs modified to remove this site are not cleaved (B). A very small amount of cleavage for the hybrid HJ shows that both sites must be recognized (C). (Adapted from Shah R, Cosstick R & West SC [1997] *EMBO J* 16:1464–1472. With permission from Macmillan Publishers, Ltd.)

endonuclease VII binds is shown in **Figure 14.6B**. In contrast T7's endonuclease I binds to the stacked form and orients the pairs of arms in a 90° angle. T7 endonuclease I cleaves the continuous (noncrossover) strands. Although RuvC binds *in vitro* to a stacked arrangement of strands and cuts the noncrossing strands, this configuration may not reflect its action *in vivo*, where it interacts with the RuvA and RuvB proteins, which themselves hold HJs in a square, planar configuration.

Another resolvase that appears to share these general properties is the metazoan and plant Gen1 protein and its yeast homolog, Yen1; however, there is yet neither a structure nor detailed information about preferential sites of cleavage. An example of the symmetrical cleavage of a synthetic HJ (one lacking homology between the arms) by the eukaryotic Gen1 enzyme is shown in **Figure 14.7**. Note that in this case the DNA sequences themselves are not symmetrical even though the cuts are in symmetrical positions. Because of this symmetrical cleavage, the products of cleavage and exchange can be readily ligated *in vitro*.

If there are special properties of *double HJs*, we have yet to learn about them and how any of these resolvases would deal with them *in vivo*. We do know that bacterial RuvC (without RuvA and RuvB) will cleave unnicked dHJs isolated during yeast meiosis into all four possible arrangements (see Figure 13.21, showing that denaturing dHJs without HJ resolvase treatment yields only parental strands). However, it remains to be seen if the yeast enzymes act on dHJs *in vivo* in a more biased fashion that would explain why apparently nearly all dHJs are resolved as crossovers (one HJ cut to give a CO and the other cut to give a NCO). From the analysis of crossovers associated with intermediates of meiotic recombination containing heteroduplex DNA (in a 5:3 or 3:5 tetrad), it seems clear that crossovers can occur on either side of the mismatch, assuming that the dHJ forms around the region of hDNA. However, as shown by Allers and Lichten this is not always the case (see Figure 13.24), so that finding crossovers on either side of hDNA might not mean that resolution occurred at one or the other of a pair of flanking HJs.

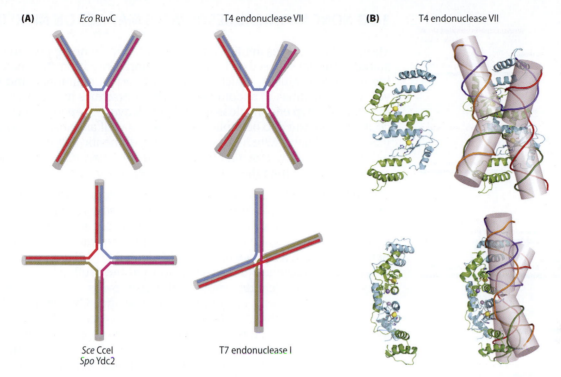

Figure 14.6 Structure of the HJ bound by different HJ resolvases, as revealed by crystal structures. (A) Distortion of a single HJ structure by different resolving enzymes. *E. coli* (*Eco*) RuvC, *S. cerevisiae* (*Sce*) Cce1, *S. pombe* (*Spo*) Ydc2, and bacteriophage T4 and T7 endonucleases. (B) X-ray crystal structure of the T4 endonuclease VII resolvase alone and bound to a HJ. (Adapted from Declais AC & Lilley DM [2008] *Curr Opin Struct Biol* 18:86–95. With permission from Elsevier.)

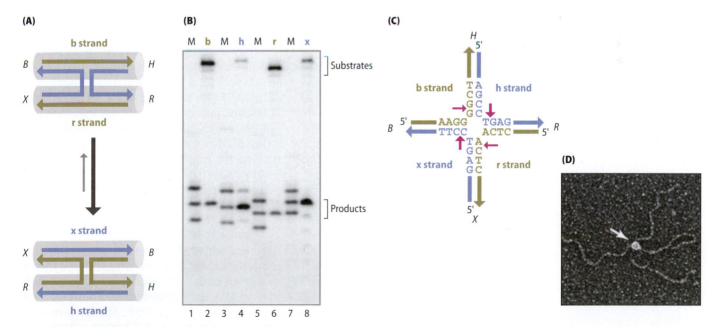

Figure 14.7 GEN1 cleavage of a synthetic HJ. (A) Four strands of DNA (b, h, x, and r) anneal into a cruciform structure, termed J3, that strongly adopts one of two possible configurations. The resulting four arms are designated B, H, X, and R. (B) In four separate experiments, one of the four strands of J3 (b, h, x, or r) was labeled at the 5′ end with ^{32}P and then was incubated with GEN1[1-527]. DNA products were analyzed by denaturing PAGE and autoradiography. Lanes marked M contain marker oligonucleotides of 15, 16, and 17 nt, which are the sizes of the 5′ ends of the b, h, r, and x strands, respectively. Their different migrations reflect sequence-specific differences. (C) The HJ is cleaved symmetrically as indicated by the pairs of large (predominant products) and small, purple arrows. Incision occurs on the continuous strands of the HJ. Note that the four arms of this synthetic HJ have different DNA sequences. GEN1 has less sequence preference than RuvC (Figure 14.5) and thus can cleave this heterologous HJ. (D) An electron microscopic image of GEN1 bound to a HJ. (Adapted from Rass U, Compton SA, Matos J et al. [2010] *Genes Dev* 24:1559–1569. With permission from Cold Spring Harbor Laboratory Press.)

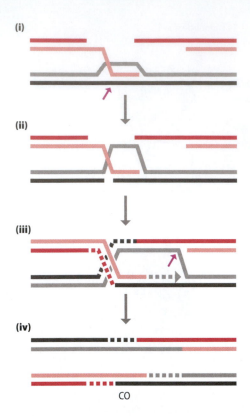

Figure 14.8 Nicking of incomplete HJs can lead to crossovers without dHJs. (i) Strand invasion generates a D-loop that can be nicked (*arrow*) (ii). Initiation of DNA synthesis could complete the formation of a crossover and extension of a branched structure (iii) that can be cleaved (*arrow*). These cleavages result in a crossover outcome (iv). (Adapted from Osman F, Dixon J, Doe CL & Whitby MC [2003] *Mol Cell* 12:761–774. With permission from Elsevier.)

14.3 NONCANONICAL RESOLVASES MAY ACT ON NICKED HJs

Many of the enzymes implicated in HJ resolution do not appear to be well suited to this task, based on their *in vitro* characterization. The most notable example is Mus81, which pairs with Eme1 in mammals, and fission yeast or with Mms4 in budding yeast. These enzymes appear to be much less specific "flap endonucleases," able to cleave Y-shaped molecules that could be construed to be replication forks as well as substrates with single-stranded flaps. When confronted with authentic HJs these enzymes either do not show a symmetrical pattern of cleavage and/or they are much less active than they are on nicked HJ substrates.

The apparent preference of Mus81 for such nicked substrates has had a profound influence on current thinking about how recombination intermediates may be processed, as we have discussed in previous chapters. Crossovers can be generated by nicking intermediates in recombination before they have a chance to form fully ligated dHJs (**Figure 14.8**). In fact, *in vivo*, these enzymes, much more so than the "authentic" Yen1/Gen1 enzyme, seem to generate most of the crossovers in both meiotic and mitotic budding yeast cells as well as resolving other branched intermediates of recombination and replication. Whether they work with as-yet-unspecified factors that influence their activity (remembering RuvA and RuvB act with RuvC) remains to be seen. In mammals Mus81 interacts not only with Eme1 but also with Eme2. Perhaps Mus81–Eme2 will have more "canonical" properties than Mus81–Eme1.

14.4 THERE ARE AT LEAST FOUR HJ RESOLVASES IN EUKARYOTES

Understanding which resolvase does what is still in flux and is made all the more complicated by the differences among different eukaryotes. As a first approximation, we can say that most eukaryotes have at least three major endonucleases capable of cleaving HJs and related structures: Gen1/Yen1, Mus81–Eme1/Mms4 (and possibly Mus81–Eme2), and Slx4–Slx1 (or in some cases Slx4–XPF). In budding yeast, and perhaps elsewhere, there is another major contribution by a complex that includes Mlh1–Mlh3 and Exo1. We do not yet understand how this complex works. An additional complication is that Slx4 not only interacts with Slx1, but also appears to be a scaffold on which a number of other flap endonucleases assemble. Below we review how each of the resolvase families were identified and discuss their biochemical properties. Then we will return to see what each one does in different organisms.

Mus81 complex

Mus81 was identified in two different ways. Mus81 was first identified as clone 81 in a yeast two-hybrid interaction with Rad54 and showed that it had homology with the XPF family of endonucleases. Subsequent work showed that *mus81Δ* is sensitive to UV- and MMS-induced damage, but surprisingly not to γ-irradiation or an HO-induced DSB. Mus81–Mms4 in budding yeast and Mus81–Eme1 in fission yeast and higher eukaryotes exhibit flap endonuclease activity on branched substrates with 3′ ends as well as cleavage of nicked Holliday junctions. This activity has been reported to be stimulated by Rad54 in a species-specific fashion (*S. cerevisiae* Rad54 does not stimulate mammalian Mus81–Eme1). Compared to its robust cleavage of nicked HJs *in vitro*, Mus81 has limited activity on intact HJs. Subsequently budding yeast *mus81Δ* was identified as one of a number of recombination-associated proteins that are synthetically lethal (slx) with deletion of Sgs1, including *slx1Δ*, *slx4Δ* and *mms4Δ*. The list of mutations that are very sick or lethal with *sgs1Δ* (or *top3Δ*) also includes *srs2Δ*, *rdh54Δ*, *rad50Δ*, *mre11Δ*, *xrs2Δ*, and *rad1Δ*. Exactly why all of these double mutant combinations are lethal in the absence of the BLM helicase and its partners remains a mystery, as further genetic analysis revealed that they had distinct genetic interactions. For example, *sgs1Δ srs2Δ* is suppressed by deleting Rad52, Rad51, Rad55, and Rad57—that is, all the proteins needed to allow Rad51 to form a filament that might engage in some sort of out-of-control recombination event in *sgs1Δ* cells. However, this double mutant is not suppressed by *rad54Δ*, consistent with the idea that Rad54 predominantly acts later in the process of recombination. But *sgs1Δ mus81Δ* or *sgs1Δ mms4Δ* double mutants are different; they are suppressed not only by *rad51Δ* and *rad52Δ* but also by *rad54Δ*. It is not obvious what *rad54Δ* fails to do.

As noted above, Mus81–Eme1/Mms4 cleaves nicked HJs much more avidly than intact HJs. There is a crystal structure for human Mus81–Eme1—the only structure so far of any of the nonmitochondrial eukaryotic resolvases (**Figure 14.9**). A nicked HJ is shown modeled into this structure.

Recent work has revealed how the Mus81 complex is regulated during the cell cycle and in meiosis. In both cases, it is activated by hyperphosphorylation of Mms4/Eme1 by the Cdc5/Plk1 polo kinase only after S phase. We will return to this result when we look more closely at the phenotypes of *mus81Δ* in budding yeast.

Yen1/Gen1

Yeast Yen1 was initially recognized by sequence motifs as belonging to the Rad2 family of endonucleases that includes Fen1 (Rad27) and Exo1. However, its role as a HJ resolvase came from arduous biochemical fractionation by Stephen West's lab that identified both Yen1 and mammalian Gen1 as canonical HJ resolvases that, like RuvC, cleave symmetrically to produce nicks that can be easily ligated. A truncation of mammalian Gen1 is active when expressed in budding yeast.

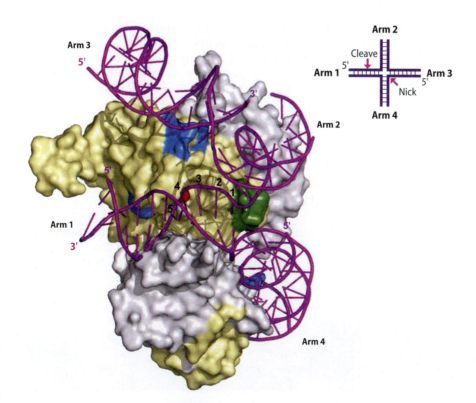

Figure 14.9 Crystal structure of the Mus81–Eme1 complex bound to a nicked HJ. The configuration of the four arms of the nicked HJ is shown in an inset. (Adapted from Chang JH, Kim JJ, Choi JM et al. [2008] *Genes Dev* 22:1093–1106. With permission from Cold Spring Harbor Laboratory Press.)

Surprisingly, Gen1 appears to be mostly cytoplasmically located in mammalian cells, so that it only gains access to DNA at the initiation of mitosis, at the time of nuclear envelope breakdown. Again, this means that Gen1 is active only during mitosis and not during S phase.

Slx1–Slx4

As already noted, *slx1Δ* and *slx4Δ* were initially identified as synthetic lethal partners of *sgs1Δ*. In budding yeast, the two proteins have both common and separate functions. Slx4 is required to excise the 3'-ended flap generated during SSA and in strand invasions where there is a long 3' nonhomologous tail, but surprisingly Slx1 is not required for this function. Slx4 is only active in this process when it is phosphorylated at one site, at least, by Mec1 or Tel1 checkpoint kinases. Why this function should be inactive until there is damage is not understood, but perhaps it has something to do with whether Slx4 acts with the tail-clipping proteins (Saw1, Msh2, Msh3, Rad1, and Rad10) or instead partners with Slx1.

One of the remarkable findings of the past few years is that Slx4 (known as BTBD12 in mammals and Mus312 in *Drosophila*) is a very large protein that acts as a "tool belt" or scaffold that binds a surprisingly large number of proteins involved in HJ resolution or 3' flap removal, including Rad1–Rad10 (mammalian XPF-Ercc1), Msh2–Msh3, Mus81–Eme1, and Slx1, as well as the telomere-protection protein Trf2 and the Plk1 Polo-like protein kinase (**Figure 14.10**)!

Slx1–Slx4 has been reported to be both a flap endonuclease and a bona fide HJ resolvase. Recent work from Stephen West's lab has discovered that Slx1–Slx4 and Mus81–Eme1 mutually activate each other. It now appears that Slx1–Slx4 makes a single nick in a HJ that Mus81–Eme1 then takes over to make the second cut. These two cleavages are concerted, because using the cruciform assay described in Figure 14.3, there is almost no nicked circle product; rather the two enzymes cooperate to create a bona fide HJ resolvase. In *Drosophila* and *Caenorhabditis* the homologs of Slx1–Slx4 appear to have full HJ resolving activity.

Exo1–Mlh1–Mlh3

In both budding yeast and mouse meiosis the Mlh1–Mlh3 proteins (and unknown co-factors) are responsible for the majority of crossovers. As noted in Chapter 12, the localization of Mlh1 in mouse meiosis correlates very strongly with sites of crossing over. We await a definitive biochemical experiment to confirm the suspicion that Mlh3 has the necessary endonuclease activity to act as a HJ (or dHJ) resolvase. In budding yeast, Exo1 plays a key role even when its endonuclease activity is mutated. This situation may be clarified by the time this book is published.

14.5 DIFFERENT HJ RESOLVASES PLAY DIFFERENT ROLES IN DIFFERENT ORGANISMS

Whether it is advantageous to resolve recombination intermediates as crossovers depends on the types of cells or the time in the cell cycle. Consequently HJ resolvases are subjected to tight regulation that varies in different cell types and in different organisms.

Figure 14.10 Mammalian Slx4 (BTBD12) is a scaffold or "tool belt" for many other DNA repair proteins. Slx4 binds three endonucleases (Mus81–Eme1, Rad1–Rad10, and Slx1) in addition to the polo-like kinase Plk1, the telomere protection protein Trf2, and the mismatch repair proteins Msh2–Msh3. (From data in Svendsen JM, Smogorzewska A, Sowa ME et al. [2009] *Cell* 138:63–77. With permission from Elsevier.)

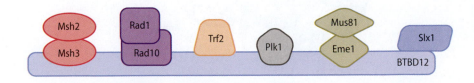

Budding yeast mitotic cells

In budding yeast, two resolvases have been implicated in the creation of crossovers in dividing cells: Mus81–Mms4 and Yen1. Lorraine Symington devised a clever diploid strain in which it was possible to examine many kinds of DSB repair from an I-SceI site (**Figure 14.11**). Most gene conversions co-converted the *ade2-n* allele and remained Ade⁻. About 25% of the events were associated with a reciprocal crossover that resulted in LOH of the distal NAT and HPH markers that are placed in allelic locations. Deleting Yen1 alone had no effect whereas a *mus81Δ* derivative showed a 50% reduction in exchanges. Crossovers were nearly eliminated in the *mus81Δ yen1Δ* double mutant. When conversion tracts were apparently short (or when heteroduplex DNA was restored by mismatch repair), leading to Ade2⁺ colonies, the total level of crossing over was much lower—about 11% instead of 25%—and both *mus81Δ* and *yen1Δ* single mutants had only about 5% crossovers. Crossovers vanished in the double mutant, and there was also a marked reduction in red/white sectored colonies that depend on crossing-over. Why *yen1Δ* had a much more profound effect in the Ade2⁺ (white) cells is not evident. Its action might be tied to the presence of a short heteroduplex DNA, or possibly the intermediates are different—nicked versus complete dHJs, for example. Alternatively there may be differences in timing of different types of recombination events, when Yen1 is active.

In mitotic cells, yeast Yen1 activity is tightly regulated, by dephosphorylation, and data from synchronized cells show that Yen1 is active only at the end of mitosis, apparently to take care of any otherwise unresolved HJs. A mutation that removes several phosphorylation sites on Yen1 makes a constitutively active protein that can suppress the MMS sensitivity of *mus81Δ*; thus, at least in some circumstances, the two activities can substitute for each other.

There is also evidence that Slx1–Slx4 plays a role in resolving intermediates arising in mitotic replication, at least in the absence of BLM helicase. By using a temperature-sensitive Sgs1 mutation, Stephen Brill concluded that both *slx1Δ* and *slx4Δ* are specifically required to complete rDNA replication, since the rDNA-containing chromosome became trapped in the well of a chromosome-separating pulse-field gel, whereas all the other chromosomes completed replication. rDNA replication is unusual in that the binding of the Fob1 protein to its site in rDNA blocks the replication fork so that replication is largely unidirectional. It is not yet clear whether the role of Slx1–Slx4 is related to this fork-blocking or what is the nature of the rDNA intermediates arising without Sgs1.

Another assay in which all of the yeast resolvases can be examined is the use of a truncated HO cleavage site that generates single-strand nicks in G1 that are converted to a DSB on one sister chromatid after replication (see Chapter 9). Deleting Yen1 reduces the proportion of SCE (exchange events that lead to the formation of a dimer circular plasmid) about 50% and *mus81Δ* causes an even more severe defect. The double mutant almost eliminates SCE. Overexpression of Mus81 or of Yen1 suppresses the defects of the other mutant. But most strikingly *slx4Δ*, by itself, eliminates almost all SCE. However, *slx1Δ* has no effect and the role of Slx4 appears to be related to a requirement for a flap endonuclease, since *rad1Δ* has a similar impact.

There are still other factors influencing crossing over. In the particular case of crossovers accompanying the DSB repair of a plasmid recombining with a homologous chromosomal site, crossovers are impaired by deleting the Rad1–Rad10 endonuclease or Msh2–Msh3, all of which are needed to clip off nonhomologous tails. Symington has suggested that Rad1–Rad10 may act to clip a D-loop. Whether this activity also requires

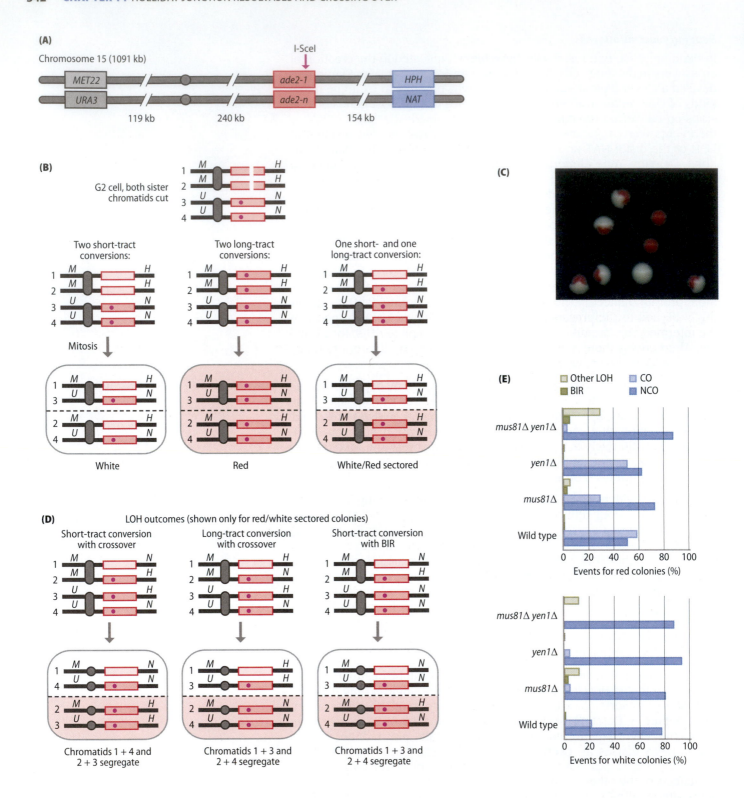

Figure 14.11 A budding yeast assay of crossing over and gene conversions. (A) An I-SceI cleavage within the *ade2* gene can be repaired by a short-patch gene conversion to create white Ade⁺ recombinants or by longer co-conversion that leaves cells red and Ade⁻ (homozygous for *ade2-n*). (B) Repair events that can occur in G2 cells where both sister chromatids are cleaved. Long- and short-patch gene conversions and both crossovers and BIR events can be distinguished as shown. (C) Different outcomes can be scored visually by the appearance of red, white, and red–white sectored colonies. (D) A detailed analysis of sectored colonies reveals three different outcomes displaying loss of heterozygosity for one or both of the distal drug-resistance markers, *HPH* (hygromycin B resistance) and *NAT* (nourseothrisin resistance). Not all versions of similar outcomes are shown. (E) Effect of deleting Mus81 and/or Yen1 on outcomes for short- and long-patch gene conversions. (Adapted from Ho CK, Mazon G, Lam AF & Symington LS [2010] *Mol Cell* 40:988–1000. With permission from Elsevier.)

Slx4 is not known. We also do not know if there is something special about plasmid × chromosome recombination, but an important role for Rad1–Rad10 is not evident when there is recombination between two chromosomal sites.

Budding yeast meiosis

The most intensive study of resolvases has been in *Saccharomyces cerevisiae* meiosis. Here, both Mus81–Mms4 and Yen1 are active, but again Yen1 has an apparently minor role. In fact at least half of yeast meiotic crossovers depend on a fourth HJ resolving activity that requires Exo1 (but not its exonuclease motif), Mlh1 and Mlh3. How this Exo1-Mlh complex acts is a matter of conjecture; it has not been purified. In some bacteria that do not use MutH to nick DNA at a distance from a mismatch, they use an endonuclease encoded by MutL. Both yeast Mlh3 and its mammalian homolog Pms2 have endonuclease activity and mutations that presumably ablate the endonuclease domain have defects in meiotic recombination similar to the complete deletion of Mlh3.

There are at least two major pathways that generate meiotic crossovers in budding yeast. One route depends on the ZMM proteins apparently to stabilize dHJs. Presumably this pathway uses Exo1–Mlh1–Mlh3 as its resolvase. In cells lacking any of the ZMM proteins, recombination is reduced by about 50% and most of the remaining crossovers depend on Mus81–Mms4. Yen1 seems to play a back-up role and only is seen to have an effect when these other pathways are impaired.

The activities of both Mus8–Mms4 and Yen1 are tightly regulated in budding yeast mitosis and meiosis (**Figure 14.12**). Yen1 is inactive and hyperphosphorylated throughout most of the cell cycle and through the first meiotic division, but then is activated by dephosphorylation only at the point where the last joint molecules must be resolved prior to chromosome separation. Thus Yen1 only becomes active late in meiosis, when most recombination should be completed. Yen1's meiotic activity is confined to Meiosis II where presumably it acts to remove any persistent dHJs (between sister chromatids?). These findings are consistent with genetic analysis that suggests yeast Yen1 has a surprisingly minor "back up" role in meiotic and mitotic joint molecule resolution. Similarly, Mus81–Mms4 (in yeast) or Mus81–Eme1 (in mammals) only becomes activated late in the cell cycle when Mms4/Eme1 is phosphorylated by polo-like kinase (Cdc5 in yeast, Plk1 in mammals). Stephen West argues this late activation allows most dHJs in mitotic cells to be dissolved rather than resolved, thus preventing crossovers and loss of heterozygosity.

As discussed in Chapter 13, in a *zmm* mutant such as *msh4Δ*, crossovers can be restored to a normal level by deleting Sgs1. Presumably loss of Sgs1 prevents the dissolution of dHJs. This would leave dHJs available as substrates for the resolvases; however, the situation is much more complex. Specifically, in the absence of Sgs1 one discovers many branched molecules in which the two ends of the DSB have invaded different chromatids, sometimes all four chromatids are found in a set of interconnected structures that probably are all dHJs (**Figure 14.13**). Apparently Sgs1 plays a role in dismantling one-ended strand invasion structures (or else it helps stabilize structures where both ends are engaged in a common joint molecule). Under these not-quite-normal circumstances, noncrossovers appear coincidentally with crossovers and NCOs are no longer independent of Ndt80 expression, as is the case in wild-type cells. Ndt80 controls the expression of a number of late meiotic genes but most notably the Cdc5 kinase. Crossovers appear in a *ndt80Δ* background if Cdc5 is expressed from a different promoter. In wild-type cells the difference in timing between the appearance of NCOs and COs suggested that they arose by different pathways not necessarily sharing a common

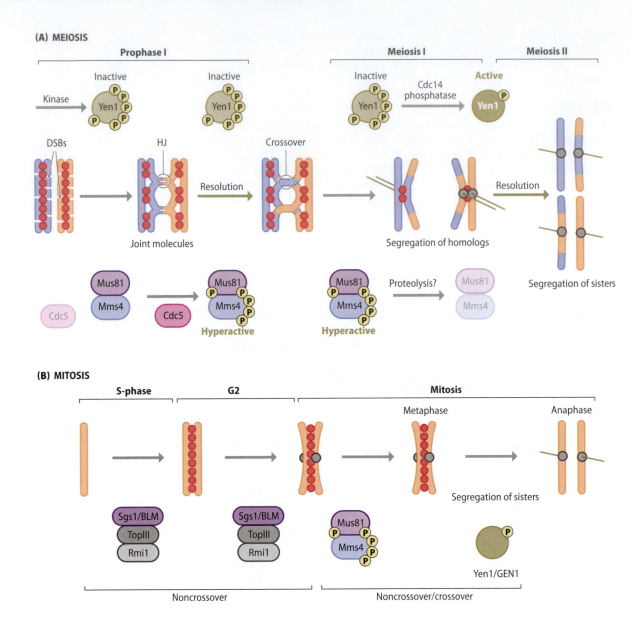

Figure 14.12 Regulation of crossing over by Mms4 and Yen1 in yeast meiosis and mitosis. Mms4–Mus81 becomes active late in meiosis (A) and in mitosis (B) by Cdc5-dependent phosphorylation, while Yen1 gains activity late in the cell cycle by dephosphorylation. Recently Mark Hall, Purdue University, has shown that the key mitotic regulator, Cdc14, dephosphorylates Yen1. (Adapted from Matos J, Blanco MG, Maslen S et al. [2011] *Cell* 147:158–172. With permission from Elsevier.)

intermediate, but in *sgs1*Δ the simultaneous appearance of COs and NCOs suggests that they might be produced by resolution of common intermediates. Crossovers depend primarily on Mus81, with Yen1 having a small role and Slx4 now showing a small effect (**Figure 14.14**). Quite possibly some of the crossovers now arising did not come from authentic dHJs. In any case, it is not at all clear that inhibiting Sgs1 leads to a "restoration" of crossovers so much as it opens up a new crossover-producing pathway.

Fission yeast

In fission yeast, Mus81–Eme1 seems to be the only essential resolvase in meiosis. As noted in Figure 13.25, there is strong evidence that most intermediates are single HJs. There is no Yen1 homolog, but expressing human GEN1 in *S. pombe* will take over for Mus81.

Mus81 also plays a critical role in sister chromatid repair when mating type switching is initiated in the unusual situation where the replication-associated DSB at *mat1* cannot be repaired by using one of the adjacent *mat2* or *mat3* donors. Instead, because only one of two sister chromatids is cleaved, the DSB can be repaired by sister chromatid recombination

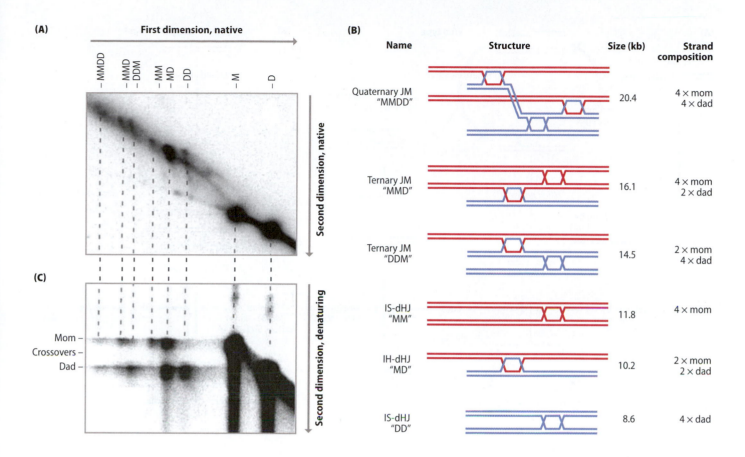

with its sister. Why this does not happen normally, and a *cis*-acting *mat2* or *mat3* is used instead, has not been worked out. Why Mus81 is essential in this type of sister chromatid repair has not been explained.

Similar to what has been reported for budding yeast, the Slx1–Slx4 pair has been implicated as a structure-specific nuclease that initiates Rad52-dependent recombination at stalled rDNA replication forks. Again, what intermediates arise in this situation remains to be learned.

Drosophila

Fruit flies have homologs of Mus81–Eme1 and Gen1. Similar to budding yeast, *Drosophila*'s Slx4 (known as Mus312) has two partners: a Slx1 homolog and XPF, the homolog of yeast's Rad1. In meiosis, Mus312-XPF is responsible for the great majority of crossovers. This remains an exceptional situation; in no other case so far examined do Slx4 and a Rad1–Rad10 nuclease act as the major meiotic HJ resolvase. We noted in Figure 5.8 that mitotic budding yeast Slx4 (without Slx1), plus Rad1 and Rad10, are critical in cleaving nonhomologous 3'-ended flaps in strand invasion and SSA intermediates. In flies, the most important somatic role of Slx4 is in conjunction with Slx1. The roles of the various *Drosophila* resolvases have been assessed in the absence of the BLM helicase, here called Mus309. As in yeast, the absence of BLM is synthetically lethal during development with mutations of Mus81, Slx4, or Slx1 and, again similar to yeast, the lethality of *mus-309 mus-81* is suppressed by deleting the fly Rad51 homolog, Spn-A. Deleting Rad51 does not strongly suppress the absence of the BLM helicase and Slx1 or Slx4, again similar to what has been found in yeast. Unlike budding yeast, however, *mus-81 gen-1* double mutants are also (embryonically) lethal.

Figure 14.13 Sgs1 prevents multichromatid joint molecules in budding yeast meiosis. Identification of multichromatid joint molecules on two-dimensional gels of cells lacking Sgs1 activity, collected 8 hr after initiation of meiosis. The Southern blot (A) is probed with a radioactive fragment that lights up all strands. The region analyzed contains different-sized restriction fragments for the two parental chromosomes, called Mom (M) and Dad (D). The sets of interhomolog (IH) and intersister (IS) intermediates are deduced (B). Note that the absence of crossover-sized strands on denaturing gels (C) indicates that most of the connections are dHJs rather than single HJs. (Adapted from Oh SD, Lao JP, Hwang PY et al. [2007] *Cell* 130:259–272. With permission from Elsevier.)

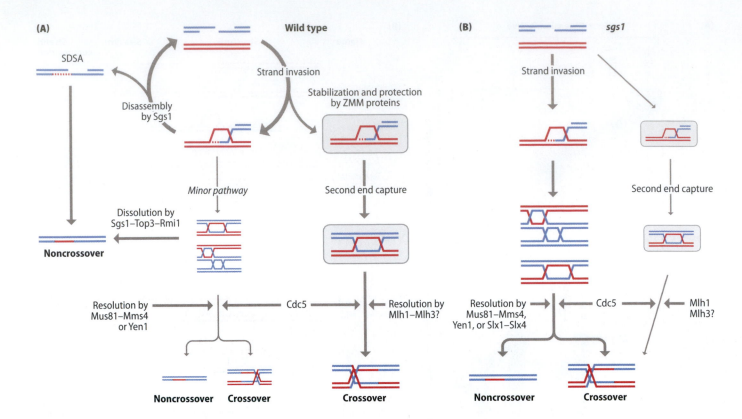

Figure 14.14 Control of crossovers in budding yeast meiosis. Resolution of dHJs by different resolvases in wild type (A) and in cells lacking Sgs1 helicase (B). Different pathways to resolve meiotic recombination intermediates are shown, with the thickness of the arrows in each pathway proportional to its importance. Cdc5 kinase plays a key role in completing meiosis in cells lacking the Ndt80 transcription factor. Cdc5 phosphorylates and activates Mus81-Mms4 for HJ resolution. In wild-type cells different pathways lead to CO and NCO outcomes. (B) In the absence of Sgs1, dHJs are intrinsically stable and dHJ resolution leads to both CO and NCO outcomes from a common intermediate. (Adapted from De Muyt A, Jessop L, Kolar E [2012] *Mol Cell* 46:43–53. With permission from Elsevier.)

Caenorhabditis

In worms, the Gen1 homolog, GEN-1 is not essential for DSB repair during meiotic recombination, though *gen1* mutants show DNA damage sensitivity. However, GEN-1 does seem to play a novel role in DNA damage signaling. GEN-1 promotes germ-cell cycle arrest and apoptosis in a pathway that acts in parallel to the conserved DNA damage response pathway mediated by RPA loading, Chk1 activation, and induction of apoptosis. Moreover, GEN-1 has been shown to act redundantly with the 9–1–1 complex to ensure genome stability.

Worms also have an ortholog of MUS312/Slx4, called HIM-18. Worm Slx4 physically interacts both with Slx1 and XPF, and the *him-18* mutant is synthetically lethal with the absence of BLM helicase (*him-6*). Mutants show somatic DNA damage sensitivity and an accumulation of Rad51 foci that could indicate Slx4 has a late role in HR-mediated repair at stalled replication forks. In meiosis, inhibition of HIM-18 causes a notable reduction in crossover recombination, which is accompanied by an increase in Rad51 foci, germ-cell apoptosis, unstable bivalent attachments, and subsequent chromosome nondisjunction (indicating a lack of crossing-over—but apparently not the persistence of interhomolog HJs).

Worms have an apparent ortholog of Mus81, but by the time of this book's printing, nothing has been published about its roles.

Mammals

Very similar to the results reported in yeast, worms, and flies, mammalian cells lacking BLM are sensitive to the absence of the several endonucleases. Depletion of MUS81 and GEN1, or SLX4 and GEN1 results in severe chromosome abnormalities, in which sister chromatids remain interlinked in a side-by-side arrangement and the chromosomes are elongated and segmented (**Figure 14.15A**). These results indicate that normally replicating human cells require Holliday junction processing activities to prevent

(A)

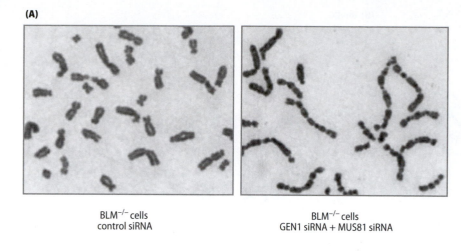

BLM$^{-/-}$ cells
control siRNA

BLM$^{-/-}$ cells
GEN1 siRNA + MUS81 siRNA

(B)

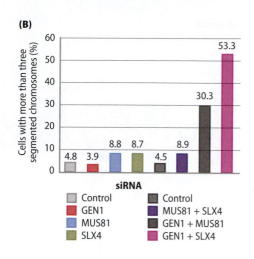

Figure 14.15 Role of several mammalian HJ resolvases in the formation of SCEs in the absence of BLM helicase. (A) Chromosome segmentation caused by abnormal chromosome condensation in mammalian cells lacking BLM helicase and impaired by siRNA inhibition of GEN1 and MUS81. Note that the chromosomes are not labeled with BrdU to show SCE; here, one sees the hypercondensation and segmentation of chromosomes when both the expression of GEN1 and MUS81 is impaired. (B) Effect of different combinations of siRNA against different HJ resolvases. (Adapted from Wechsler T, Newman S & West SC [2011] *Nature* 471:642–646. With permission from Macmillan Publishers, Ltd.)

sister chromatid entanglements that must be removed to allow accurate chromosome condensation. In contrast, depletion of both MUS81 and SLX4 in BLM cells had little effect, suggesting that GEN1 is able to compensate (**Figure 14.15B**). Depletion of MUS81 or SLX4 reduces the high frequency of SCEs in Bloom's syndrome cells, indicating that MUS81 and SLX4 promote SCE. A similar reduction in SCEs is not accomplished by knocking down GEN1.

In meiosis, Gen1 also seems to have a minor role. In contrast, inhibiting Slx4 (BTBD12) impairs mouse meiosis and leads to the persistence of Mlh1–Mlh3 staining. As we noted in Chapter 13, sites of crossovers appear to be marked by immunofluorescent staining of Mlh1, so it is quite possible that the predominant resolvase activity is similar to the Exo1-MutLγ activity in budding yeast. However, in cells lacking BTBD12, Mlh1–Mlh3 is not sufficient to resolve the problems in recombination. There is also a clear role for Mus81 in generating Mlh1–Mlh3-independent crossovers, though male mice are fertile in its absence. Although some DSBs are apparently not repaired, there is an apparent compensatory increase in Mlh1 and Mlh3, but now the constraints of interference have apparently been lost (Mlh1–Mlh3 associates with sites that normally would be dealt with by Mus81).

14.6 BRANCH MIGRATION ENZYMES CAN INFLUENCE HJ CLEAVAGE

One topic that we have not yet discussed in detail concerns the role of other HJ-interacting proteins on directing the resolution of the junctions. Virtually everything we know about this subject comes from work on the *E. coli* RuvA, RuvB, and RuvC proteins. RuvA binds to cruciform structures and recruits a pair of hexameric RuvB helicase motors that orient oppositely so that their action is to pull a pair of strands through the helicases and move the HJ (**Figure 14.16**). A pair of RuvC HJ resolvases binds to this complex, apparently displacing two RuvA subunits and cleaving the HJ in symmetrical positions. Interestingly, cleavage is directed to the strands that pass through the RuvB rings 3′ toward the Holliday junction, so that on any one HJ the cleavage is strongly biased.

Gareth Cromie and David Leach have suggested that this bias may explain why double HJs would be resolved preferentially as crossovers (**Figure 14.17**). If, as in the *E. coli* RuvABC case, the orientation of the 3′ ends of the two HJs directs resolution, one of the two junctions will

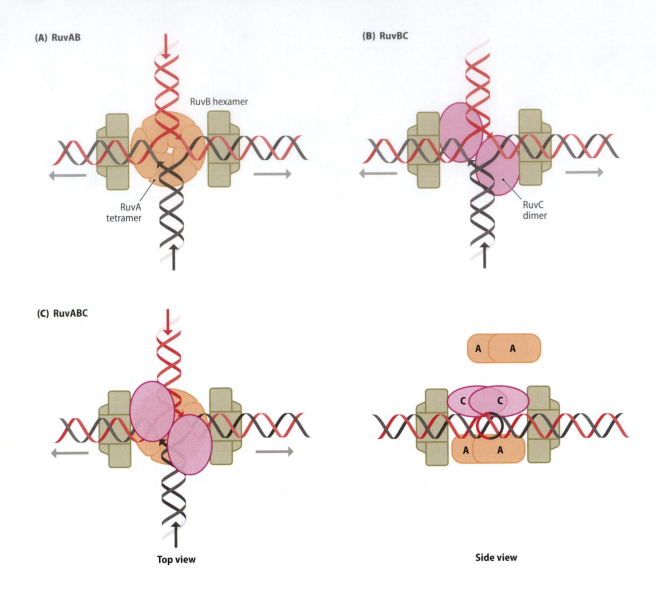

(A) RuvAB

RuvB hexamer

RuvA
tetramer

(B) RuvBC

RuvC
dimer

(C) RuvABC

Top view

Side view

Figure 14.16 RuvA and RuvB assist RuvC in HJ resolution. (A) RuvA binds to cruciform structures and assembles RuvB helicases that can catalyze branch migration. (B) Binding of RuvC which is thought to displace two RuvA subunits (C) then cleave the HJ at preferred nucleotide sequences. RuvA–RuvB therefore plays a key role in branch migrating the HJ until a suitable RuvC site is exposed. (Adapted from van Gool AJ, Hajibagheri NM, Stasiak A & West SC [1999] *Genes Dev* 13:1861–1870. With permission from Cold Spring Harbor Laboratory Press.)

be resolved as a crossover and the other will be a noncrossover. This outcome is consistent with the observations in budding yeast that NCOs precede COs and that the time of resolving dHJs into COs is not accompanied by a further increase in NCOs (that is the dHJs are resolved only as COs and are not a random mixture of COs and NCOs).

One set of proteins that might act similarly to RuvAB are some of the ZMM proteins, such as Msh4–Msh5, that can bind to HJs and might recruit other proteins to yield a biased resolution of dHJs.

14.7 MISMATCH REPAIR ALSO INFLUENCES CROSSOVER REGULATION

To conclude our discussion of the way crossovers are generated, we must return to one of the most confusing aspects of meiosis: the role that mismatch correction and the mismatch repair (MMR) machinery play in producing crossing over and the role that crossing over plays in mismatch repair. It should be clear already that Mlh1 is a central actor in MMR as well as in the formation of some crossovers, but the problem apparently goes much deeper.

A striking indication of the complex interplay between crossing over and MMR has come from the findings of Frank Stahl's lab that tetrads with a 5:3 or 3:5 segregation (that is, where the mismatch is not corrected) behave differently from 6:2 and 2:6 gene conversions or from tetrads in which there was no apparent gene conversion. In these experiments one of the parents carries a small palindromic insertion so that heteroduplex includes a short palindrome on one strand; such mismatches are only repaired one-third to one-half of the time. Crossovers associated with unrepaired heteroduplex do not cause interference on adjacent intervals (as measured by the formation of NPDs). In contrast, in tetrads with full conversions of the same palindromic marker or in 2:2 tetrads—where, in some cases, the heteroduplex should have been restored to the parental configuration—crossovers show the expected interference effect. Moreover, in a *msh4Δ* mutant that eliminates the ZMM pathway of crossing over, there was an approximately 50% reduction in crossovers both in tetrads in which there was no associated gene conversion or in tetrads with 6:2 or 2:6 full gene conversions, but *msh4Δ* had much less effect on tetrads with 5:3 or 3:5 post-meiotic segregation. These data argue that recombination intermediates containing heteroduplex DNA that is not mismatch repaired are dealt with in an entirely different way from cases where heteroduplex is corrected. And we do not understand this. One idea that could have explained such data would be that 6:2 and 2:6 events arise from gap-repair events while the 5:3 and 3:5 events must have involved heteroduplex DNA formation. But this explanation is unlikely because one can examine well-repaired and poorly repaired mutations at the same site, which are likely to have common intermediates.

The question is, then, (1) does mismatch repair establish the way HJs are resolved; or (2) are mismatches handled differently at different stages of a single recombination process; or (3) are mismatches handled differently by different recombination pathways? It is possible that nicks associated with MMR create a bias in how dHJs are resolved, but so far there is no experimental way to answer this conjecture. We can imagine that such nicks, coupled to branch migration, could create nicked HJs that would be preferentially acted upon by Mus81. Mus81 seems to generate non-interfering COs. But clearly not all mismatches are shifted to the Mus81 pathway.

The alternative—that mismatches are handled differently at different steps in the repair process—is more likely. How MMR knows which strand to correct is clearly different from what happens just after replication, where there is some cue as to which is the older strand. But in meiosis, recombination between homologs can take place with either of two sisters, one of which has an old Watson and the other an old Crick, so that in one case strand invasion would anneal with an old strand while in the second case it would be with a new strand. So there must be different cues. Nicks associated with strand invasion could provide such cues, as illustrated in Figure 14.17. Several experiments in mitotic cells suggest that MMR repairs mismatches in favor of the base on the unbroken strand. Moreover, there are two different times when MMR might occur. A study of MMR during *MAT* switching in mitotic yeast argued that MMR could occur rapidly, repairing a mismatch at the time of strand invasion, before the beginning of new DNA synthesis. But some mismatches are only formed at later steps of recombination, at the capture of the second end in SDSA or in dHJ intermediates, and after branch migration. Here the rules may be different; for example in SDSA, there are nicks on both strands near a mismatch. Depending on the pathway of recombination, this bias may be reflected in an excess of restorations or of conversions. Thus, as noted in Chapter 13, in a survey of recombination by microarray analysis of thousands of SNPs, the ratio of NCO to CO is about 1 in

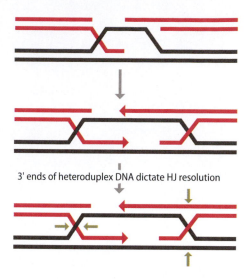

3' ends of heteroduplex DNA dictate HJ resolution

Figure 14.17 Biased resolution of dHJs can be determined by the orientation of the 3' ends in each HJ. The strands ending 3' to the gaps in dHJ formation are proposed to be the targets of HJ resolvases (*green arrows*). Thus in an intermediate on the pathway to form a complete dHJ, one HJ will be cleaved in the crossover mode while the second will be resolved as a noncrossover. (Adapted from Cromie GA & Leach DR [2000] *Mol Cell* 6:815–826. With permission from Elsevier.)

an *msh2Δ* diploid that lacks most MMR. MMR may produce an excess of restorations in NCO events, which render them invisible in this assay.

Alternatively, perhaps the (at least) two different crossover-generating pathways have different rules about mismatch correction. As we discussed in Section 14.5, there appear to be at least two distinct recombination pathways. Both Mus81 and one of the ZMM components must be eliminated to remove most meiotic crossovers. One possibility that has been raised in this regard is that the Mus81 pathway might operate predominantly on transient, nicked HJs whereas the ZMM pathway would involve covalently closed dHJs. Mismatches might be less efficiently repaired in the Mus81 pathway.

Another argument for two different recombination pathways comes from the analysis of the position of meiotic crossovers. Getting a good fit of the crossover data in budding yeast and in other organisms requires that there be both a crossover pathway that displays interference and a scattering of noninterfering crossovers. The existence of two pathways is especially evident in organisms that cannot form a synaptonemal complex without recombination-driven pairing, including yeast and mammals; but it is not evident in either worms or flies, which employ pairing sites. Stahl has noted that the organisms that require initial pairing by recombination have Dmc1 and all of its associated actors (Hop2, Mnd1, Mei5, and Sae3) whereas they are absent in flies and worms. These two pathways might be mechanistically very different, and they might treat mismatches differently. This idea is also consistent with the idea that in worms all crossovers are Msh5-dependent (there is no pairing pathway that would be Mus81-dependent).

Thus there would be an initial pairing pathway followed by a "disjunction" pathway, the latter manifesting interference and being required to establish chiasmata to hold homologous pairs of chromosomes together. Frank Stahl and Jette Foss suggest that in the disjunction (dHJ) pathway mismatches are corrected only after the formation of fully ligated dHJs and that these mismatches will be resolved half of the time as full gene conversions and the other half as restorations, depending on which strands of the HJ are cleaved. In this conception, all dHJs are resolved as CO and all COs depend on ZMM and Mlh1.

SUMMARY

Our understanding of how HJs are resolved is rapidly evolving as two new resolvases (Gen1/Yen1 and Slx1–Slx4) have only recently joined the Mus81–Mms4/Eme1 in the arsenal of structure-specific endonucleases. In budding yeast the principle resolvase still awaits purification, though it appears to need Mlh1–Mlh3 and a noncatalytic role for Exo1. In budding yeast meiosis, the prevalent joint molecule is a double HJ, and at least four resolvase activities have been implicated in assuring proper chromosome segregation. In contrast, in fission yeast the key intermediate is a single HJ requiring only Mus81–Eme1 for resolution. What happens in mammalian cells still awaits "resolution," but we already know that all of these activities play important roles in resolving branched intermediates in mammalian cells, for example in BLM$^{-/-}$ cells. In *E. coli*, the way that dHJs are resolved is also influenced by the RuvAB proteins interacting with RuvC. Very little is yet known about such branch-migration activities in eukaryotic cells.

SUGGESTED READING

Cromie GA & Leach DR (2000) Control of crossing over. *Mol Cell* 6:815–826.

Declais AC & Lilley DM (2008) New insight into the recognition of branched DNA structure by junction-resolving enzymes. *Curr Opin Struct Biol* 18:86–95.

Matos J, Blanco MG, Maslen S, Skehel JM & West SC (2011) Regulatory control of the resolution of DNA recombination intermediates during meiosis and mitosis. *Cell* 147:158–172.

Schwartz EK & Heyer WD (2011) Processing of joint molecule intermediates by structure-selective endonucleases during homologous recombination in eukaryotes. *Chromosoma* 120:109–127.

Sharples GJ (2001) The X philes: structure-specific endonucleases that resolve Holliday junctions. *Mol Microbiol* 39:823–834.

Wechsler T, Newman S & West SC (2011) Aberrant chromosome morphology in human cells defective for Holliday junction resolution. *Nature* 471:642–646.

West SC (1996) The RuvABC proteins and Holliday junction processing in *Escherichia coli*. *J Bacteriol* 178:1237–1241.

CHAPTER 15

NONHOMOLOGOUS END-JOINING

The most straightforward way to repair a broken chromosome is simply to re-join the ends. End-joining is indeed an efficient repair route in bacteria, archaea, and eukaryotes. In mammals, end-joining appears to be the predominant way of repairing breaks outside of S phase; breaks made during replication are apparently most often repaired by sister chromatid recombination. Even in budding yeast, which seems to be the champion of efficient homologous recombination mechanisms, re-ligation of broken ends occurs in a significant proportion of DSB repair events. Since the joinings can occur between blunt ends or in situations where only a few base pairs could stabilize the junction, these mechanisms were often referred to as illegitimate recombination, but are now generally called nonhomologous end-joining (NHEJ). In many cases, end-joining is a mutagenic process, leading to the deletion or even the addition of some base pairs at the junction. We now distinguish between "classical" NHEJ and an alternative pathway, called microhomology-mediated end-joining (MMEJ) or simply "alternative" NHEJ.

15.1 "CLASSICAL" NHEJ IS ESSENTIAL FOR THE MAMMALIAN IMMUNE SYSTEM

Eukaryotes have a number of different pathways of end-joining. This is one aspect of DSB repair where much of what we understand first came from the analysis of humans with defects in the immune system. The creation of a rich diversity of immunoglobulin genes relies on NHEJ to join alternative V, D, and J segments together to create thousands of immunoglobulin genes and T-cell receptor genes (**Figure 15.1**). NHEJ is also required for the further diversification of immunoglobulin molecules by class switching. Many, but not all, components of the NHEJ system(s) in mammals are evolutionarily conserved in yeasts. Both in mammals and in yeast, there are several distinct NHEJ mechanisms.

In pre-B immune cells, the Rag1 and Rag2 proteins bind and cleave recognition sequences adjacent to V, D, and J sequences in such a way that the two recognition sequences are blunt-end ligated while the two coding ends are left as hairpin ends (**Figure 15.2**). The hairpin ends are opened by a specialized endonuclease, Artemis, which leaves single-stranded ends which are held together by a pair of Ku proteins (Ku70 and Ku80) along with their associated DNA protein kinase (DNA-PKcs). Ligation is accomplished by a specialized DNA ligase, DNA ligase 4, in a complex with Xrcc4 and XLF (also called Cernunnos) (**Figure 15.3**).

The Ku proteins form a basket-like structure that can hold two ssDNA ends together (**Figure 15.4**); there seems to be a preference for the molecule to sit at the "inner" end of the ssDNA, adjacent to dsDNA. Thus Ku can present the two ends to the ligase complex for joining. Exactly how this process depends on the protein kinase activity of DNA-PKcs remains unclear. DNA-PKcs probably phosphorylates and activates Artemis and also regulates the stability of various NHEJ proteins at the DSB ends.

Figure 15.1 NHEJ promotes immunoglobulin gene rearrangements.
(A) The heavy chain gene of immunoglobulins consists of multiple copies of dispersed V, D, and J segments adjacent to a single constant region. V–D–J joinings by NHEJ result in a functional immunoglobulin heavy chain gene, which combines with a light chain to recognize a specific epitope. The large number of combinations of V–D–J joinings results in a wide variety of different antibody specificities. (B) Initially each V–D–J region is expressed as an IgM (membrane bound molecule) with a Cμ constant region; however, class-switch recombination between various "switch" regions (S) makes it possible for a particular antibody specificity to be present in several other antibody classes such as IgA, IgG, and IgE. Similar rearrangements occur to construct both immunoglobulin light chains and T-cell receptors. (Adapted from Murphy K [2011] Janeway's Immunobiology, 8th ed. With permission from Garland Science.)

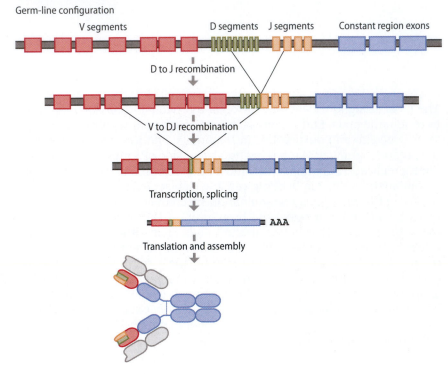

(A) VDJ recombination

Germ-line configuration

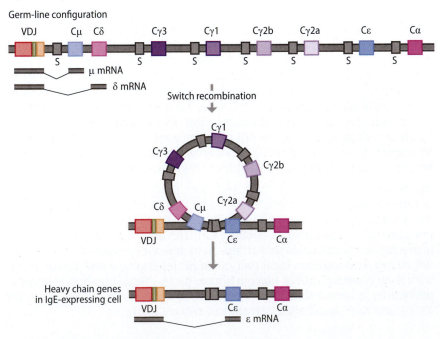

(B) Antibody class-switch recombination

Germ-line configuration

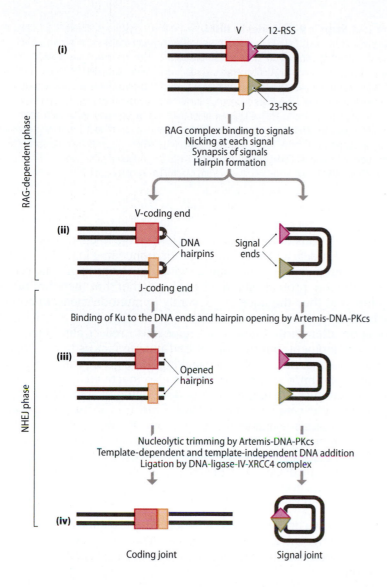

RAG-dependent phase

(i)

V   12-RSS

J   23-RSS

RAG complex binding to signals
Nicking at each signal
Synapsis of signals
Hairpin formation

(ii)   V-coding end

DNA
hairpins

Signal
ends

J-coding end

NHEJ phase

Binding of Ku to the DNA ends and hairpin opening by Artemis-DNA-PKcs

(iii)

Opened
hairpins

Nucleolytic trimming by Artemis-DNA-PKcs
Template-dependent and template-independent DNA addition
Ligation by DNA-ligase-IV-XRCC4 complex

(iv)

Coding joint

Signal joint

Figure 15.2 Steps in RAG-mediated cleavage and processing of coding and signal joint ends in VDJ recombination. (i) In this example from light chain immunoglobulin gene rearrangements (where there is no D region), RAG proteins recognize and cleave next to a 12-bp and a 23-bp recombination signal sequence (RSS) adjacent to V and J segments, respectively. (ii) Cleavage results in hairpin-ended V and J segments and to blunt-ended RSS sequences (signal ends). (iii) Hairpin ends are opened by the Artemis endonuclease in concert with the DNA-PKcs kinase. (iv) The coding ends are joined by NHEJ to create a V–J junction while the signal ends are also joined. The coding ends may be modified before joining by the addition of nucleotides that increases the variability between the V and J segments. (Adapted from Lieber MR, Ma Y, Pannicke U & Schwarz K [2004] *DNA Repair* 3:817–826. With permission from Elsevier.)

Immunoglobulin class switching does not require DNA-PKcs even though it uses the Ku proteins; moreover, in lower eukaryotes, there is no DNA-PKcs, so that the Ku proteins apparently act on their own. Recent studies have suggested that the lack of requirement for DNA-PKcs reflects a redundancy with the XLF protein associated with XRCC4 and DNA ligase 4. In the absence of XLF, DNA-PKcs becomes essential for NHEJ.

One function of the Ku proteins both in mammals and yeast is to block more extensive resection of the DSB ends. In budding yeast, deletion of Ku70 or Ku80 causes a twofold increase in the resection of DSB ends generated by HO endonuclease. Ku proteins are essential for the predominant pathway of NHEJ, but there is an alternative pathway that is Ku independent that will be discussed below. In mouse, the absence of Ku leads to more extensive deletions at the junctions created by this alternative NHEJ mechanism.

DNA ligase 4 acts in a complex with XRCC4 and XLF/Cernunnos (**Figure 15.5**). Both ligase 4 and Xrcc4 are highly conserved from human to yeasts (although curiously *Schizosaccharomyces pombe* appears to lack Xrcc4). In contrast only a few highly conserved residues can be aligned to argue that the XLF/Cernunnos protein is evolutionarily related to the Nej1 protein of budding or fission yeast, but they appear to carry out a similar function.

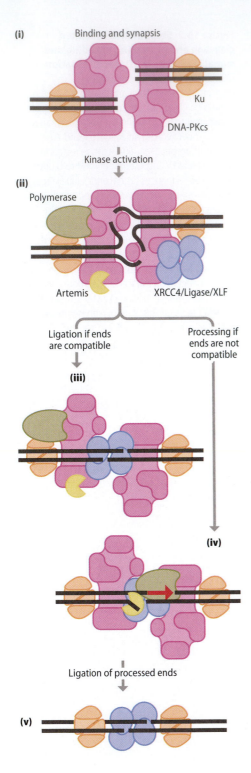

(i) Binding and synapsis

Ku

DNA-PKcs

Kinase activation

(ii) Polymerase

Artemis

XRCC4/Ligase/XLF

Ligation if ends are compatible

Processing if ends are not compatible

(iii)

(iv)

Ligation of processed ends

(v)

Figure 15.3 Steps in V–J joinings by NHEJ. V–D–J and V–J joinings require NHEJ. (i) Ku binds to DNA ends and recruits DNA-PKcs, which mediates synapsis of the ends. (ii) Artemis and perhaps another nuclease assemble at the synaptic complex, along with the DNA ligase 4/XRCC4/XLF proteins and one of several DNA polymerases. Threading of single-stranded DNA ends into cavities in the DNA-PKcs molecule activates the kinase. DNA-PKcs phosphorylates Artemis, activating its endonuclease activity. DNA-PKcs undergoes autophosphorylation and moves away from the DNA ends, allowing access to XRCC4/Ligase4. (iii) If the ends are compatible, ligation occurs immediately. (iv) If the ends are incompatible, DNA polymerase can fill in 3′ protruding ends while nuclease activities can remove single-stranded overhangs. When the ends are processed into a compatible substrate, XRCC4/Ligase4 completes NHEJ (v). (Adapted from Budman J & Chu G [2005] *EMBO J* 24:849–860. With permission from Macmillan Publishers, Ltd.)

Complete deletions of the Ku proteins yield viable mice, though they are quite small. Ku-deficient mice can still form noncoding joints between the signal sequence ends during immunoglobulin rearrangements, possibly because the Rag proteins still hold ends together, but there are no coding joints and thus the mice are severely immunodeficient. In contrast, mice homozygous for deletions of ligase 4, XRCC4, or XLF are all inviable at, or soon after, birth. Their death appears to reflect abnormal neurological development, but the role of end-joining factors in development is not understood. But, viable mice lacking these genes can be recovered when the p53 tumor suppressor gene is also inactivated. Such p53- and Lig4-deficient mice are at great risk of developing lymphoblastic tumors. Surprisingly, the characteristic chromosome translocations associated with tumorigenesis—such as those that join the immunoglobulin heavy chain gene region to c-Myc—are still found and still have translocation junctions that lack extensive homology (that is, they did not form by SSA between the many dispersed copies of Alu or other repetitive sequences). In fact, these junctions provided the first evidence of an efficient alternative, ligase 4- and Ku-independent NHEJ mechanism, discussed below.

Unexpectedly, Ku deficiency also rescues the embryonic lethality of DNA ligase 4-defective mice. This discovery can also be explained by the existence of a second end-joining pathway that is more efficient in the *absence* of Ku proteins.

15.2 VDJ JOININGS OFTEN EXHIBIT ADDITIONAL MODIFICATIONS AT THE JUNCTION

In immunoglobulin VDJ junctions, one can see evidence of additional enzymatic activities. Some junctions have P (palindromic) nucleotides that are generated when a hairpin end is cleaved away from the end and unfolded. This ssDNA can be joined to another such end usually at a point where at least one base pair can be formed that apparently stabilizes the junction long enough to be joined (**Figure 15.6**). In these cases there are small gaps that have to be filled in. In mammals the PolX family of polymerases (Polλ or Polμ) are both important in such fill-in processes. These polymerases can also add a base at the end of ssDNA tails, which may be used to pair with a second end. In addition, some IgH junctions have untemplated additional nucleotides that are the product of an enzyme called terminal deoxynucleotidal transferase (TDT). This enzyme can add several bases to the 3′ ends that can be incorporated into the NHEJ junctions; there is a strong preference for the addition of G or C. Examples of coding end junctions are shown in **Figure 15.7**. TDT does not appear to modify immunoglobulin light chain genes. Terminal transferase is another function lacking in lower eukaryotes.

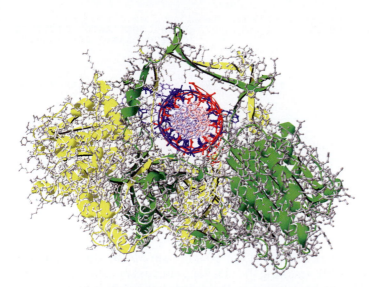
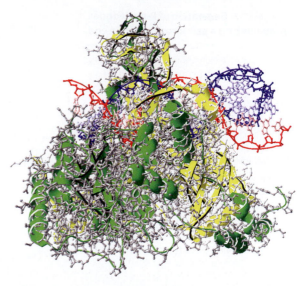

Figure 15.4 The Ku heterodimer surrounds the DNA ends. Two views of human Ku heterodimer protein attached to DNA. Ku70 is marked *green* and Ku80 *yellow*. The two chains of DNA are marked *red* and *blue*. Two arms of Ku surround the DNA. (Courtesy of Karol Głąb, "Crystal Structure of the Ku Heterodimer," Wikipedia, The Free Encyclopedia [accessed September 30, 2013].)

15.3 NHEJ CONTRIBUTES TO DSB REPAIR IN YEAST

Yeast cells lacking Ku proteins or other components of NHEJ are not noticeably radiation sensitive, whereas *rad52Δ* cells are markedly sensitive. However, one can see that NHEJ contributes to viability by comparing *rad52Δ* cells to *rad52Δ yku70Δ* cells, where it becomes evident that there is a Rad52-independent pathway of repairing X-irradiated cells (**Figure 15.8**). Another experiment that illustrates the relative use of NHEJ and HR used a yeast cell with *MAT***a** but where both *HML* and *HMR* donors have Yα sequences. When HO endonuclease is turned on for an hour,

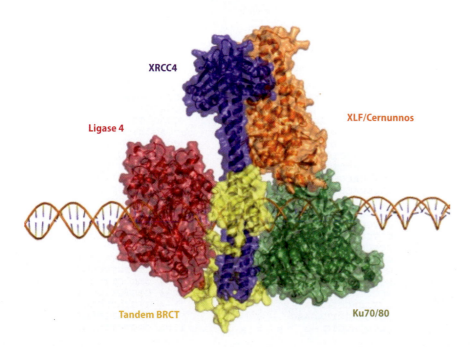

Figure 15.5 Assembly of the NHEJ proteins at a pair of DSBs, based on their individual protein structures. X-ray crystal structures of DNA ligase 4, XRCC4 (including its two BRCT domains), XLF, and the Ku70 and Ku80 proteins are shown. (Courtesy of Murray Junop, McMaster University.)

Figure 15.6 Generation of P-nucleotides by opening up a palindromic hairpin end. (A) A hairpin end created by excision of the maize Ds transposable element inserted in a budding yeast chromosome is cut on one strand away from the center. A similar opening occurred on the left end (not shown). (B) Opening this flap creates a single-strand region that can anneal in a variety of ways with a strand from a second end, one of which is shown with blue boxes. (C) After microhomology-mediated annealing, the overhanging 3′-ended nonhomologous strands are clipped off and the gap is filled in (D), resulting in a junction with palindromic (P) nucleotides. (Adapted from Yu J, Marshall K, Yamaguchi M et al. [2004] *Mol Cell Biol* 24:1351–1364. With permission from the American Society for Microbiology.)

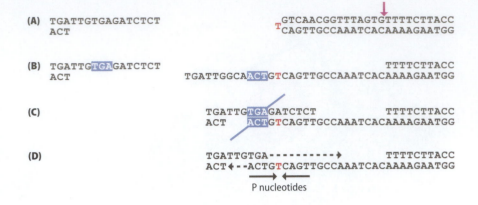

90% of the cells switch to *MAT*α but 10% remain *MAT***a**. The *MAT***a** cells are the result of NHEJ restoring the cleavage site and not simply cells that failed to be cleaved, because when the same experiment is performed in a *yku70*Δ derivative, all the cells switch to *MAT*α. Thus even in a system evolved to promote *MAT* switching, NHEJ is a competitive process. As a comparison, in mammals, an analogous experiment in which the outcomes are NHEJ or gene conversion suggests that NHEJ is more often the winner, with about half of the products showing end-joining.

The core components of NHEJ are found in budding yeast: Ku70–Ku80, DNA ligase 4, Xrcc4 (called Lif1), and Nej1. Deletion of any of these components causes a 20- to 100-fold reduction in ligation of the cohesive ends of a transformed, linearized plasmid carrying a selectable marker. The defects are even greater, on the order of 500-fold, when a DSB is created in a chromosome with HO or I-SceI, which leaves 4-bp 3′ overhanging ends. With these endonucleases, two different types of NHEJ can be explored. If HO is expressed for 20–60 min and then turned off, most end-joining restores the perfect HO cleavage site by simply ligating the ends back together (**Figure 15.9A**). Re-ligation requires all of the core end-joining components. This experiment is possible because the HO protein is rapidly degraded. But if HO is expressed continuously, then simple

Figure 15.7 Examples of coding joints in immunoglobulin end-joining. Deletions are shown in numbers next to the coding ends. The P nucleotides are shown adjacent to the coding ends in lower case (*blue*). Untemplated nucleotides (N) added by terminal transferase are shown in *green*.(Adapted from Lu H, Shimazaki N, Raval P et al. [2008] *Mol Cell* 31:485–497. With permission from Elsevier.)

12-RSS coding end	P	N	P	23-RSS coding end
GCTCGACCTCAGAAC				CGAGCTGTGTTCCGT
GCTCGACCTCAGAAC				CGAGCTGTGTTCCGT
GCTCGACCTCAGAAC	g			CGAGCTGTGTTCCGT
GCTCGACCTCAGAAC		C		CGAGCTGTGTTCCGT
GCTCGACCTCAGAAC	gt			CGAGCTGTGTTCCGT
GCTCGACCTCAGAAC			cg	CGAGCTGTGTTCCGT
GCTCGACCTCAGAAC	gt	CC		CGAGCTGTGTTCCGT
GCTCGACCTCAGAAC	gt	AG	g	CGAGCTGTGTTCCGT
GCTCGACCTCAGAAC		T		−1 GAGCTGTGTTCCGT
GCTCGACCTCAGAA −1		GC		CGAGCTGTGTTCCGT
GCTCGACCTCAGAAC		AAA		−1 GAGCTGTGTTCCGT
GCTCGACCTCAG −3		C		CGAGCTGTGTTCCGT
GCTCGACCTCAG −3		C	cg	CGAGCTGTGTTCCGT
GCTCGACCTCAGAAC	gt	AGG		−3 GCTGTGTTCCGT
GCTCGACCTCAG −3				CGAGCTGTGTTCCGT
GCTCGACCTCAG −3		C		CGAGCTGTGTTCCGT
GCTCGACCTCAGAA −1		G		−3 GCTGTGTTCCGT
GCTCGACCTCAGAAC		TA		−4 CTGTGTTCCGT
GCTCGACCTCAGAAC		CT		−4 CTGTGTTCCGT
GCTCGACCTCAGAA −1				−4 CTGTGTTCCGT
GCTCGACCTCAGA −2		GCC		−4 CTGTGTTCCGT
GCTCGACCTCAGAAC		G		−5 TGTGTTCCGT
GCTCGACCTC −5				−1 GAGCTGTGTTCCGT

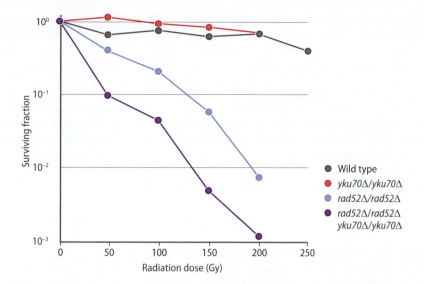

Figure 15.8 NHEJ contributes to DSB repair in yeast. Although a yeast strain lacking Yku70 is not by itself radiation sensitive, a mutant lacking both Rad52 and Yku70 is significantly more sensitive than one lacking Rad52 alone. (Adapted from Siede W, Friedl AA, Dianova I et al. [1996] *Genetics* 142:91–102. With permission from the Genetics Society of America.)

re-ligation is futile and the re-joined cut site is cut again. In this case, the only cells that can grow have altered the HO cut site.

The most common changes are small 2- and 3-bp insertions that arise by misalignment of the two ends, stabilized by a single base pair, followed by filling in the small gap by a DNA polymerase (**Figure 15.9B**). Another set of HO-resistant survivors have deleted the terminal 3 bp of the DSB, again apparently by mispairing and trimming of the ends (**Figure 15.9C**). Other HO-resistant survivors have small deletions, in rare cases as large as several hundred base pairs, and again usually with one or a few base pairs homology between the ends. A very important experiment compared the deletions arising in repairing HO cuts with those resulting from dicentric chromosome breakage that removed one centromere; the proportion of joinings with 0, 1, 2, and so forth, base pairs at the junction were remarkably similar, thus arguing that the NHEJ events involving HO were not special.

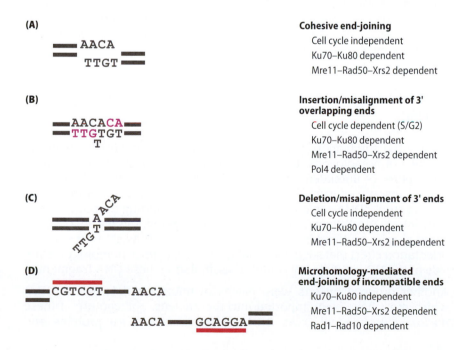

(A)   Cohesive end-joining

AACA
TTGT

Cell cycle independent
Ku70–Ku80 dependent
Mre11–Rad50–Xrs2 dependent

(B)   Insertion/misalignment of 3′ overlapping ends

AACACA
TTGTGT
T

Cell cycle dependent (S/G2)
Ku70–Ku80 dependent
Mre11–Rad50–Xrs2 dependent
Pol4 dependent

(C)   Deletion/misalignment of 3′ ends

AACA
A
T
TTGT

Cell cycle independent
Ku70–Ku80 dependent
Mre11–Rad50–Xrs2 independent

(D)   Microhomology-mediated end-joining of incompatible ends

CGTCCT   AACA

AACA   GCAGGA

Ku70–Ku80 independent
Mre11–Rad50–Xrs2 dependent
Rad1–Rad10 dependent

Figure 15.9 Different NHEJ outcomes in repairing a DSB. If HO endonuclease is expressed in a strain lacking *HML* or *HMR* then repair can only occur by NHEJ. The HO cut creates 3′-ended 4-bp overhanging ends. (A) If HO is expressed transiently, then most ends can be re-ligated to re-form the cleavage site. This event requires the full complement of classical NHEJ factors. If HO is continuously expressed then only events that mutate the cleavage site will be recovered. These outcomes have different genetic and cell cycle requirements, as indicated. (B) A frequent outcome is the templated addition of 2 bp, stabilized by the misalignment of the 3′ overhanging ends. Newly added bases are shown in magenta. (C) Alternatively, misalignment can lead to the loss of part of the 3′ overhang by clipping off the small tails and ligating the ends. (D) If the two ends of the DSB are incompatible for annealing (for example if there are two nearby HO cleavage sites in opposite orientation), then NHEJ is normally very inefficient in budding yeast. In the absence of Ku, however, alternative NHEJ (MMEJ) becomes efficient, removing larger regions from the DSB ends and using longer microhomologies (*red* underlining). (A–C, adapted from Moore JK & Haber JE [1996] *Mol Cell Biol* 16:2164–2173. With permission from the American Society for Microbiology. D, adapted from Ma JL, Kim EM, Haber JE & Lee SE [2003] *Mol Cell Biol* 23:8820–8828. With permission from the American Society for Microbiology.)

Many NHEJ events require processing of ends or filling in small gaps. Surprisingly there is evidence that both normal replicative polymerases, δ and ε, and one translesion polymerase, Pol4, may all be involved. The fill-in synthesis of misaligned ends such as those described above for HO-generated ends requires the DNA Pol X protein, Pol4. It appears that Pol4 and Polδ redundantly play an important role even with the re-joining of compatible plasmid ends. The end trimming when a few nucleotides are not paired at the end appears to be performed by the proofreading activity of DNA polymerase ε, as a mutation in this activity has a notable effect on imprecise end-joining of partially incompatible plasmid ends. The proofreading action of both Polδ and Polε also can trim off up to 20 nt of 3′-ended overhanging ends that arise as intermediates in SSA or in strand invasion.

In budding yeast there is an equivalent requirement for the MRX proteins as for Ku or Nej1. There is no obvious requirement for the MRN proteins in the main NHEJ pathway in mammalian cells, although since these proteins are essential, it has been difficult to assure that they have been sufficiently depleted to guarantee that MRN is not needed for NHEJ. In yeast, without MRX, the small fill-in events are absent but the deletion events are still recovered. Thus there must be several sub-pathways of NHEJ, with slightly different requirements. Indeed, in mammals, MRN proteins have been suggested to be quite important in an alternative NHEJ mechanism described below.

Among the differences in NHEJ in budding yeast and mammals is that blunt ends are very inefficiently joined in yeast. It is unclear why this should be, although most enzymatically induced DSBs, by site-specific nucleases such as HO or I-SceI, or by Spo11 in meiosis, do not leave blunt ends. Possibly yeast has no helicase that can open up blunt ends, but need 3′ or 5′ overhanging tails to get started.

It should be noted that *NEJ1* and to a lesser extent *LIF1* are transcriptionally regulated such that *MAT**a**/MATα* diploid cells repress NHEJ. Why this should be advantageous is not yet established; possibly this repression is designed to prevent *MAT**a**/MATα* diploids undergoing meiotic recombination from repairing some of the approximately 100 DSBs by NHEJ.

In *S. pombe*, in addition to the surprising absence of the highly conserved Xrcc4 homolog, there is an XLF homolog. Moreover, there is also no requirement for MRN for NHEJ as assayed in cycling cells; however, when G_0 cells are damaged and use NHEJ to repair chromosome breaks the Mre11 complex proteins are required.

15.4 NHEJ FACILITATES CAPTURE OF DNA FRAGMENTS AT DSBs

Among the survivors of continuous HO cutting of the *MAT* locus in the absence of homologous donors was an unexpected type of DSB repair, the capture of unrelated DNA fragments into the site of cleavage. One type of insert, occurring in about 1 in 50 NHEJ events, was the 95-bp "strong stop" cDNA copied as the first step in converting retrotransposon RNA into cDNA. When Ty RNA was overexpressed, the entire Ty cDNA could also be captured in a similar fashion. The cDNAs showed a strong orientation effect that suggested that a few base pairs of homology on one side of the HO cut were important in assimilating these DNA fragments.

Another group of inserts were pieces of mitochondrial DNA (mDNA) which must have been imported into the nucleus. The capture of these *nuclear mitochondrial DNAs* (NUMTs) requires both Ku proteins and

Mre11–Rad50, but little else is known about what other proteins might be needed apart from the standard NHEJ factors. NUMTs may have arisen during NHEJ repair of DSBs. There are many instances of mDNA insertions into the mammalian genome. It is also possible that the insertions of pseudogenes—when mRNA is copied into cDNA—into the mammalian genome also depend on NHEJ.

15.5 END-JOINING MAY ALSO OCCUR BY ALTERNATIVE NHEJ

Mice lacking DNA ligase 4 or XRCC4 still exhibit a high level of chromosomal translocations. Clearly, there is an alternative NHEJ pathway, possibly using DNA ligase 3 and other factors that can be seen when the standard NHEJ process is ablated. This also proves to be the case in budding yeast, in the absence of Ku70–Ku80. (Because there is no third DNA ligase in budding yeast, all events are DNA ligase 4 dependent.) When the overlapping ends of a DSB created by an endonuclease are complementary, the absence of Ku virtually eliminates NHEJ. However, if the ends are incompatible, for example by transforming in a plasmid with different restriction site cleavages at each end of a gap or by creating two HO cleavages in two nearby sites of opposite orientation (**Figure 15.9D** and **Figure 15.10**), *yku70Δ* cells exhibit a surprisingly high level of repair, similar to that seen in wild-type cells with compatible ends. DNA sequence analysis of the junctions revealed that these events contained more extensive deletions than the simple joining of complementary ends and require more extensive microhomology, on the order of 5–9 bp. These outcomes define a *microhomology-mediated end-joining* pathway, termed MMEJ. Note that this terminology does not mean that normal "classical" NHEJ does not benefit from base pairing at the junction; NHEJ often uses 1–3 bp of homology. But MMEJ demands considerably longer regions of base pairing. MMEJ is not the same as single-strand annealing, as it is Rad52 independent.

One reason that Ku-independent end-joining produces deletions that are larger than in wild-type cells has to do with the locations of microhomology, but another factor is that the absence of Ku promotes more rapid end resection, at least in budding yeast. Ku-independent MMEJ has distinct requirements not seen for normal NHEJ. Because there are usually longer 3'-ended tails to clip off at the junctions, Rad1-Rad10 are also required, as they are in SSA.

In mammals, alternative NHEJ also proves to use more extensive microhomology at the junctions. One difference that is seen between cells lacking XRCC4 or DNA ligase 4 and those without the Ku proteins, is that the latter have more extensive deletions, consistent with Ku restraining the rate of 5'-to-3' resection.

Figure 15.10 Microhomology-mediated end-joining. Cleavage of two HO recognition sites in opposite orientation results in deletions bounded by microhomology. Three examples are shown. The 4-bp single-stranded ends generated by HO endonuclease cutting are underlined. Microhomology at the junctions are indicated in *blue* boxes on the left side. Deleted nucleotides are shown in parentheses and the magenta sequence is the microhomology at the junction. The other copy of the microhomology is shown in a *blue* box. (Adapted from Ma JL, Kim EM, Haber JE & Lee SE [2003] *Mol Cell Biol* 23:8820–8828. With permission from the American Society for Microbiology.)

MATα HO cleavage sequence

TGTT | ACAA | URA3 | Yα | AACA | TTGT

MATa HO cleavage sequence

HO cut ends

```
TTTATAAAATTATACTGTT                    GTATAATTTTATAAACCCTGGTTTTGGTTTTGTAGA
AAATATTTTAATATG                        TTGTCATATTAAAATATTTGGGACCAAAACCAAAACATCT

TTTATAAAATTATA(CTGTT                    GTATAATTTTATA)AACCCTGGTTTTGGTTTTGTAGA
TTTATAAAATTATACTG(TT                    GTATAATTTTATAAACCCTG)GTTTTGGTTTTGTAGA
TTTATAAAATTAT(ACTGTT --------------------------------------TTGACGAATATTAT)GCTGAAG
```

53 bp

Alternative end-joining appears to depend on DNA ligase 3 and on poly-ADP ribose polymerase (PARP), though what the precise role of PARP might be has not been established. In mammalian cells and in *Xenopus* extracts that are proficient for end-joining, the MRN complex also plays an important role in resection-dependent MMEJ.

A very clever system for studying different end-joining events in mammals was devised by Bernard Lopez, using a pair of I-SceI cleavage sites in the same or different orientations in which one could recover either inversions of a fragment or a deletion of the fragment, which resulted in expression of two different surface antigens, CD8 and CD4 (**Figure 15.11**). As noted above, loss of Ku80 had a profound impact on cohesive end-joining but little effect on microhomology-mediated ligation. Why Ku is so precisely needed when the very ends are complementary is still a mystery. Interestingly, deletions were only two- to eightfold more frequent than inversions, where there had to be two NHEJ events and the ends had to remain in a local repair environment. In some experiments, a third I-SceI site in a *neo* gene on another chromosome could be repaired to

Figure 15.11 An assay system to study NHEJ and MMEJ in mammalian cells. (A) Cutting of two I-SceI sites in either direct or inverted orientation allows the formation of a deletion or inversion that expresses different cell surface markers (CD4 or CD8, respectively) that can be detected and isolated by fluorescence activated cell sorting (FACS) and recovered by magnetic activated cell sorting (MACS) using antibodies against CD4 or CD8 epitopes, coupled to magnetic beads. Repair junctions can then be analyzed by PCR amplification and DNA sequencing. (B) Sequence of two sets of ends cleaved by I-SceI in the substrate shown in (A). (C) Deletions arising in the inverted orientation (expressing CD8), and in the absence of Ku80. The magenta sequence is the 4-bp region at the I-SceI cleavage site that becomes single-stranded as shown in (A). Microhomologies ranging from one to six bases at different junctions are shown in blue letters, aligned to the intact sequence at the top. In three cases there are no overlapping base pairs. (Adapted from Guirouilh-Barbat J, Huck S, Bertrand P et al. [2004] *Mol Cell* 14:611–623. With permission from Elsevier.)

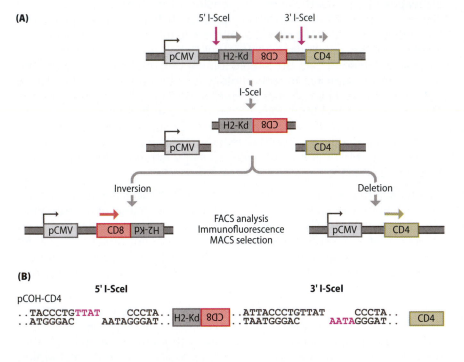

(C)

1 clone : accurate repair : high fidelity (HiFi)
TCTAGAGCAACACGGAAGGAATTACCCTGTTATCCCTATCTAGATATGAAATCACG

30 clones : deletions		Microhomology (nt)
TCTAGAGCAACACGGAAGGAATTACCCTG	CCCTATCTAGATATGAAATCACG	-
TCTAGAGCAACACGGAAGGAATTA	CCCTATCTAGATATGAAATCACG	4
TCTAGAGCAACACGGAAGGAATTA	TCCCTATCTAGATATGAAATCACG	3
TCTAGAGCAACACGGAAGGAATTACCCTG	TAGATATGAAATCACG	1
TCTAGAGCAACACGGAAGGAAT	CCTATCTAGATATGAAATCACG	-
TCTAGAGCAACAC	GATATGAAATCACG	1
TCTAGAGCAACACGGAAGGAAT	ACG	2
TCTAGAGC	AATCACG	2
TCTAGAGCAACACG	GATATGAAATCACG	2
TGGCTAGCGC	TCTAGATATGAAATCACG	5
TCTTGGCT	CCCTATCTAGATATGAAATCACG	-
CCACTGCTTA	CTATCTAGATATGAAATCACG	2
CGAAATTAAT	CGATTCCTTGCGGTCC	3
TTATCGAAATTAATA	CGATTCCTTGCGGTCC	3

Neo$^+$ by gene conversion from a nearby donor copy on the same chromosome. Only about one-third of the cells with a deletion on the first chromosome (by NHEJ) had a gene conversion (that is, Neo$^+$); presumably NHEJ restored the remainder of the I-SceI cuts. Even more striking, about 1–2% of the NHEJ events that resulted in a deletion (expressing CD4) also expressed CD8 because the liberated fragment was "captured" by cohesive end NHEJ at the I-SceI site on the second chromosome. Thus, the liberated fragment can participate in another event.

15.6 THE 53BP1 PROTEIN PLAYS MULTIPLE ROLES IN END-JOINING

In most model systems used to study end-joining, a single DSB, created on a chromosome by a site-specific endonuclease or introduced by transformation as a linearized DNA fragment, must be ligated; but there is a class of NHEJ events that require the conjunction of DSB ends generated at a distance. Immunoglobulin V(D)J joining events involve joinings between relatively close D and J segments as well as joining of DJ or J to much more distant V segments. Immunoglobulin class switching also joins regions that are many kilobases apart. Translocations and telomere end fusions require contact between sites on different chromosomes. A number of studies have suggested that the mammalian 53BP1 protein plays a striking role in NHEJ. In the absence of 53BP1, long-range NHEJ events are markedly reduced but there is little effect—in some cases there was an increase—on close-range joinings. For example, D–J joining between regions separated by 8 kb were normal, whereas V–DJ joinings of segments separated by 28 or 172 kb were reduced fourfold. Similarly, in immunoglobulin class-switch recombination, joinings between different switch (S) regions were impaired whereas there was an increase in intra-S region deletions that occur because there are multiple DSBs created within a single S region. These results support the idea that 53BP1 plays a role specifically in long-distance end-joining.

53BP1 is also required for fusions between unprotected mouse telomeres. When part of the shelterin complex that normally protects telomeres is disrupted by inactivating the TRF2 protein, there are rampant fusions of telomere ends (**Figure 15.12A**). These joinings are dependent on classic end-joining factors such as DNA ligase 4 or the Ku proteins, but they are also dependent on 53BP1 (**Figure 15.12B**). Titia de Lange's group made the remarkable discovery that the *movement* of chromosome ends (marked with a GFP fusion to a telomere protein, TRF1) was dramatically increased in the presence of 53BP1. Telomeres with 53BP1 explored about an eight times larger volume than when 53BP1 was deleted. The mobility of damaged ends without 53BP1 was estimated to be normal, compared to other chromosome regions; so it is the 53BP1-mediated increase that is exceptional. Increased motion does not depend on functional NHEJ but is dependent on the ATM checkpoint kinase. How 53BP1 can increase chromatin mobility is not yet understood; it does not depend on actin-mediated movements, which play a role in meiotic chromosome alignment, but appears to require cytoplasmic microtubules that may indirectly attach, via transmembrane proteins, to DNA bound at the nuclear periphery. de Lange has suggested that 53BP1 may normally act to discourage non-allelic recombination events, by allowing single ends that have invaded a non-allelic template to dissociate and synapse with the other end at a single template. Thus 53BP1 might be part of a recombination execution checkpoint, similar to that described in budding yeast.

In addition to its role in promoting access of DSB ends to distant DSBs, 53BP1 also plays a profound role in regulating the choice between end

Figure 15.12 Role of 53BP1 in promoting telomere fusions in mouse cells lacking *TRF2*. Mouse cells are treated with Cre recombinase to delete a floxed *TRF2* gene, resulting in telomeres that have lost their end protection. (A) In the presence of one functional allele of 53BP1, telomeres engage in end fusions, often linking many chromosomes together, as seen after 120 hr of Cre expression. (B) Without 53BP1, telomeres fail to end-join. (From Dimitrova N, Chen YC, Spector DL & de Lange T [2008] *Nature* 456:524–528. With permission from Macmillan Publishers, Ltd.)

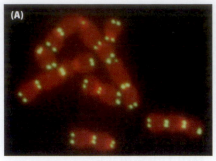
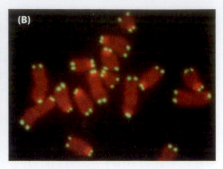

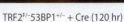

TRF2^F/–^53BP1^+/–^ + Cre (120 hr)   TRF2^F/–^53BP1^–/–^ + Cre (120 hr)

joining and homologous recombination. It does this by combining with the Rif1 protein to inhibit 5′-to-3′ resection carried out by CtIP, Exo1, and BLM. Rif1 is an ortholog of a telomere-associated protein (Rap1-interacting factor) in budding yeast, but in mammals Rif1 plays a key role in regulating DSB end resection. ATM-dependent phosphorylation of 53BP1 enhances its interaction with Rif1, resulting in an inhibition of resection. When 53BP1 or Rif1 is absent, resection increases in an ATM kinase-dependent fashion. Thus when 53BP1 is ablated, the proportion of DSB events repaired by homologous recombination increases because well-resected ends are not good substrates for end-joining. Rif1 accumulation in G1 and S phase cells is antagonized by BRCA1 that associates with, and regulates, CtIP. Rif1 also appears to play a second role, in facilitating the loading of BLM helicase that in turn facilitates resection.

An analysis of mutations of different domains of 53BP1 has led to the conclusion that the requirements for 53BP1 in mediating NHEJ at dysfunctional telomeres and in class-switch recombination are not identical. In particular, a mutant that prevents 53BP1 oligomerization is only modestly impaired in promoting NHEJ of dysfunctional telomeres and does not affect the repression of CtIP-dependent resection; however, this mutation is very defective in immunoglobulin class-switch recombination.

Changes in 5′-to-3′ resection also affect NHEJ in budding yeast, in a situation where there are simultaneous HO endonuclease-induced breaks on two chromosomes. In wild-type cells more than 90% of the survivors use NHEJ to re-ligate the breaks within each chromosome, although 10% have interchromosomal joinings that create reciprocal translocations. In strains deleted for the ATM homolog, Tel1, or the MRX-associated Sae2 protein, neither of which is required for NHEJ, there is a fivefold or greater increase in reciprocal translocations. Similar increases were seen with *mre11-H125N* and *rad50-S*, mutants, which are also not impaired in NHEJ of a single DSB. All of these mutants slow down 5′-to-3′ resection, perhaps enabling MRX proteins to persist longer at the DSB ends and to interact with distant partners.

15.7 GENE AMPLIFICATION CAN OCCUR VIA NHEJ AND BFB CYCLES

Many gene amplifications are almost certainly the consequence of the formation of a dicentric chromosome that then enters a sequence of breakage, fusion of broken sister chromatid ends and the re-formation of a stretched, dicentric chromatid at mitosis, a cycle first defined as the

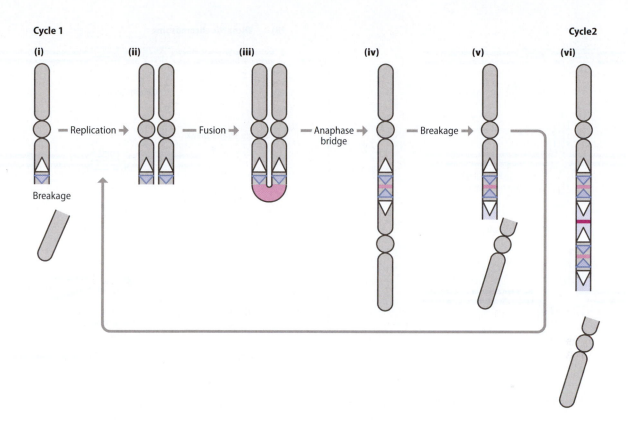

Cycle 1   **Cycle2**

(i)   (ii)   (iii)   (iv)   (v)   (vi)

— Replication →   — Fusion →   — Anaphase → — Breakage →
                                      bridge

Breakage

breakage–fusion–bridge (BFB) cycle by Barbara McClintock in the 1930s (**Figure 15.13**). A characteristic feature of gene amplifications by BFB mechanisms is that the amplified sequences are arranged in inverted orientation (Figure 15.13). This can be explained if there was a chromosome break prior to DNA replication, leading to the formation of a pair of broken sister molecules whose ends can fuse by NHEJ (or possibly by SSA between different repeated sequence copies). This fusion results in a dicentric chromosome that can break, leaving one chromatid with two copies of a region, in inverted orientation. This cycle can continue, producing nested, inverted copies of the amplified region. More recent analysis by Kirill Lobachev has suggested that another way that such BFB cycles can be initiated is by the cleavage of cruciform DNA structures formed at inverted repeated sequences, producing a hairpin-ended chromosome that will produce a dicentric chromosome after DNA replication (**Figure 15.14**). Alternatively a DSB could be at a distance from a pair of inverted repeats, such that resection of the broken end will lead to annealing of complementary sequences, creating a hairpin end. In these amplifications, it seems that broken chromosome ends will most likely be joined by nonhomologous end-joining mechanisms.

Figure 15.13 The breakage–fusion–bridge cycle. Replication of a broken chromosome (i) can result in end fusion, presumably by NHEJ (ii, iii). The *light blue* triangle designates a sequence near the break site and the *white* triangle a more distant sequence. (iv) The resulting dicentric chromosome can then break at mitosis, leaving an inverted repeated sequence at the longer end (v). This breakage–fusion–bridge (BFB) cycle can be repeated generating more complex head-to-head and tail-to-tail fusions (vi). Multiple iterations of this cycle can produce amplifications of these terminal sequences, in nested, inverted orientations. (Adapted from Kwei KA, Kung Y, Salari K et al. [2010] *Mol Oncol* 4:255–266. With permission from Elsevier.)

SUMMARY

There are both Ku-dependent and Ku-independent pathways of end-joining. When both pathways are functioning the Ku-dependent pathway predominates, but in its absence the Ku-independent MMEJ pathway is surprisingly efficient. The best-studied role for NHEJ is in creating the programmed rearrangements of the mammalian immune system. Most chromosomal rearrangements associated with cancer cells appear to have arisen by NHEJ.

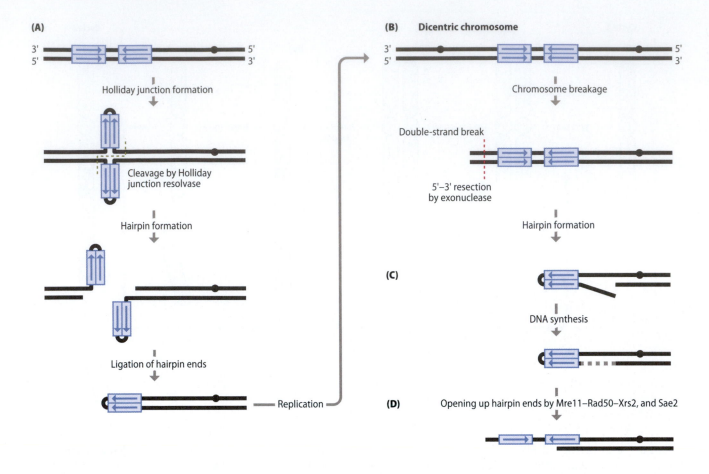

Figure 15.14 Formation of a hairpin-ended chromosome. (A) The presence of palindromic sequences can lead to the formation of a Holliday junction that can be cleaved and then ligated to form a hairpin-ended chromosome. Replication will produce a dicentric duplicated structure (B) that can then begin a BFB cycle. (C) 5′-to-3′ resection of a DSB adjacent to a pair of inverted repeats can lead to the formation of a hairpin end that, after removal of unpaired sequences and filling in the gap, can enter the BFB cycle or can be opened by the MRX complex and Sae2 (D) either to re-form a hairpin or to engage in recombination involving sequences further from the opened end. (Adapted from Haber JE & Debatisse M [2006] *Cell* 125:1237–1240. With permission from Elsevier.)

SUGGESTED READING

Brill SJ (2013) Linking the enzymes that unlink DNA. *Mol Cell* 52: 159-160.

Cahill D, Connor B & Carney JP (2006) Mechanisms of eukaryotic DNA double strand break repair. *Front Biosci* 11:1958–1976.

Daley JM, Palmbos PL, Wu D & Wilson TE (2005) Nonhomologous end joining in yeast. *Annu Rev Genet* 39:431–451.

Decottignies A (2013) Alternative end-joining mechanisms: a historical perspective. *Front Genet* 4:48.

Dimitrova N, Chen YC, Spector DL & de Lange T (2008) 53BP1 promotes non-homologous end joining of telomeres by increasing chromatin mobility. *Nature* 456:524–528.

Kasparek TR & Humphrey TC (2011) DNA double-strand break repair pathways, chromosomal rearrangements and cancer. *Semin Cell Dev Biol* 22:886–897.

Lieber MR, Gu J, Lu H et al. (2010) Nonhomologous DNA end joining (NHEJ) and chromosomal translocations in humans. *Subcell Biochem* 50:279–296.

CHAPTER 16

DNA DAMAGE CHECKPOINTS AND GENOME INSTABILITY

One cannot discuss the basis of genome stability without including the role of DNA damage checkpoints. References to the actions of checkpoint kinases have crept into many chapters, but it is worth consolidating some ideas in one chapter. This is a subject worthy of an entire book, and only some of the most salient points will be reviewed here. Cells detect DNA damage in several ways and activate both a G1-to-S and a G2-to-M checkpoint. These checkpoints delay cell cycle progression, presumably to give the cell more time to repair damage before entering S phase or to initiate mitosis, respectively. In this brief review we will not examine the DNA replication checkpoint (often called the S-phase response) in which different sensors feed into the same network of kinases described here for the DSB checkpoint response to stabilize DNA if replication itself is compromised. Checkpoint-mediated arrest of the cell cycle provides an opportunity for DNA repair processes to be completed before mitosis would segregate broken or covalently linked chromosomes. In many organisms, checkpoint-mediated apoptosis affords a way to destroy damaged cells before they undergo genome rearrangements that might promote cancer. The astonishing number of rearrangements in human cancer cells mostly appear in cells that have lost their ability to repair DSBs by homologous recombination, and, in addition, have lost their ability to restrain cell growth in the face of damage.

16.1 THE DNA DAMAGE CHECKPOINT PROVIDES A CELL CYCLE DELAY TO ALLOW DNA REPAIR

The first identification of a DNA damage checkpoint was accomplished by Ted Weinert and Leland Hartwell who studied the viability of budding yeast cells lacking the *RAD9* gene after exposure to ionizing radiation. Repair per se was not affected by deleting *RAD9* (in contrast to knocking out genes such as *RAD1* or *RAD10* that were active players in nucleotide excision repair); but *rad9Δ* cells were sensitive to X-rays or other DNA damaging agents because they failed to arrest prior to mitosis. Consequently the cells attempted to divide with damaged or broken chromosomes. Wild-type cells, with an intact G2–M checkpoint, arrest as large budded cells with a single nucleus for several hours, allowing time for repair of DSBs, while *rad9Δ* cells plow ahead and divide several times, often producing dead daughter cells (**Figure 16.1**). Importantly, the X-ray or UV sensitivity of *rad9Δ* can be suppressed by delaying mitosis, for example by adding a microtubule inhibitor, thus accomplishing by a mechanical arrest of mitosis what Rad9 normally enforces by arresting cells through the DNA damage checkpoint.

Figure 16.1 Yeast *RAD9* is required to prevent mitosis after γ-irradiation. Individual haploid yeast cells or small clusters plated on rich medium and then γ-irradiated. The same cells were then photographed 10 hr later. (A, B) In wild-type cells the majority of cells remain arrested as large dumbbells (*white arrows*). Note that growth per se is not inhibited and these cells have become enormous relative to the initial cells. Some cells have repaired the DNA damage and grown into microcolonies (*black arrow*). (C, D) In contrast, in the absence of *RAD9*, cells divide without delay but the daughter cells inherited DSBs that had not been repaired. Consequently many cells were unable to grow after a few divisions and so many microcolonies have ceased dividing. (From Weinert TA & Hartwell LH [1988] *Science* 241:317–322. With permission from the American Association for the Advancement of Science.)

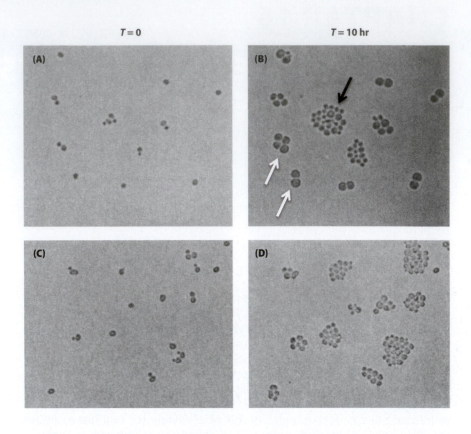

$T = 0$ $T = 10$ hr

Rad9 was the first of many proteins to be linked to the inhibition of cell division in response to DNA damage. There are in fact several overlapping checkpoints. Cells treated with agents that stall DNA replication (for example, hydroxyurea that depletes dNTP pools, or UV and MMS that block replication forks) engage the *S phase checkpoint*. DSBs and other damage triggers the *DNA damage checkpoint* that is our main focus. Agents such as nocodazole that impair microtubule function or mutations that interfere with normal kinetochore function also block attempts to complete mitosis through the *spindle assembly checkpoint* (SAC). Finally, cells in which mitosis has produced a daughter cell with broken chromosomes, or cells that suffer a chromosome break in G1, can be stalled and eliminated by a *G1-to-S checkpoint*.

16.2 PI3 KINASE-LIKE KINASES ARE AT THE APEX OF DNA DAMAGE SIGNALING

Three evolutionarily related protein kinases are the master regulators of DNA damage responses. ATM (ataxia telangiectasia mutated), ATR (ataxia telangiectasia and Rad3 related), and DNA-PKcs (DNA-dependent protein kinase, catalytic subunit) are all serine/threonine protein kinases. These very large proteins (more than 2500 amino acids) are related to phosphoinositide-3-kinase (PI3K) and are thus known as PI3K-like kinases (PI3KK). Another related kinase is mTOR (mammalian target of rapamycin) that regulates cell growth at many levels including regulating both autophagy and protein synthesis. Yeasts, flies, and worms lack DNA-PKcs, which principally plays an important role in NHEJ in mammals.

Both ATM and ATR phosphorylate SQ and TQ peptide sequences in a large number of targets (**Figure 16.2**). A detailed examination by Stephen

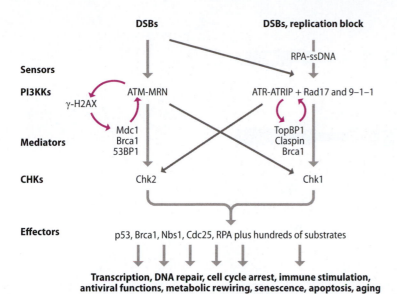

Figure 16.2 The role of ATM and ATR in the human DNA damage response. DSBs and other types of DNA damage trigger activation of the PI3KK proteins ATM and ATR. Full activation of these kinases depends on several mediator proteins that promote activation of downstream checkpoint kinases (Chk1 and Chk2). These in turn modify many downstream effectors. (Adapted from a drawing by Stephen Elledge, Harvard Medical School.)

Elledge's lab revealed phosphorylation of more than 900 sites in more than 700 human proteins in response to DNA damage. One notable target is the terminal SQ site in histone H2AX (S139 in mammals). This phosphorylated form is known as γ-H2AX and serves as marker of DSB damage.

ATM is known as Tel1 in budding yeast, Rad3 in fission yeast, dATM in *Drosophila*, and Atm-1 in *Caenorhabditis*. Although ATM plays very important roles in DSB repair, it is not essential, whereas knocking out BRCA1 is lethal in mammalian cells. Indeed cells lacking ATM are able to complete gene conversion, whereas viable truncation *BRCA1* mutations are severely deficient (**Figure 16.3**). In contrast ATM-deficient mouse cells show severe defects in end joining, though it is not yet clear why this should be so. Budding yeast *tel1Δ* cells are proficient for both homologous recombination and end joining.

Orthologs of ATR are called Mec1 in budding yeast, ATR in fission yeast, Mei-41 in flies, and Atl-1 in worms. In mammals and in most organisms, ATR is essential. Exactly what is its most essential role in mammals is not well established, but cells lacking ATR exhibit extreme chromosome instability including the appearance of many chromosome breaks during replication. In budding yeast a deletion of *MEC1* is viable if the level of dNTPs is raised by overexpressing ribonucleotide reductase (*RNR*) genes. This result suggests that Mec1's control of the transcriptional and post-transcriptional regulation of RNR proteins is its most important function.

16.3 DIFFERENT MECHANISMS ACTIVATE ATM AND ATR IN RESPONSE TO A DSB

ATM is activated in several distinct ways. In the best-understood pathway ATMs associate with the Nbs1 subunit of the MRN complex that bind to blunt or near-blunt DSB ends. But there is an MRN-independent pathway as well. The association with MRN is sufficient to activate ATM, which appears to act as a dimer. A large fraction of ATM in the cell can be activated, apparently by autocatalytic phosphorylation of the protein at an SQ site (S1981), followed by an exchange of activated dimers with other unmodified partners. The retention of ATM at sites of DNA damage depends on its interaction with the MDC1 protein that in turn associates

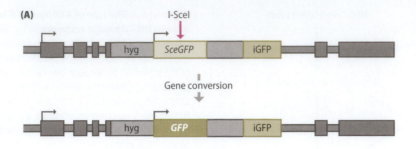

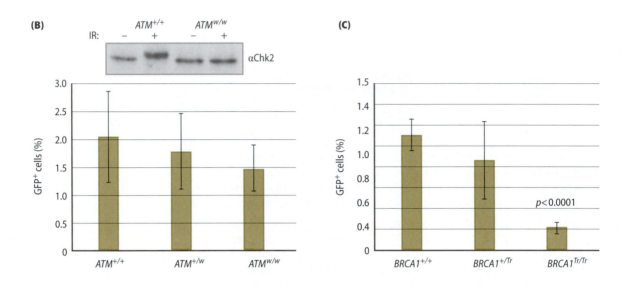

Figure 16.3 ATM is not required for homologous recombination. (A) An I-SceI-mediated cleavage of *SceGFP* gene can be repaired by gene conversion, using a homologous internal segment of the gene lacking both 5′ and 3′ ends (iGFP), resulting in *GFP* expression. (B) A cell line having a truncation of ATM (*ATM^w*) is unresponsive to γ-irradiation (IR) and cannot phosphorylate ATM's downstream target Chk2. Nevertheless, cells homozygous for *ATM^w* are recombination proficient. (C) A cell line homozygous for a truncation of BRCA1 (*BRCA1^Tr*) is severely impaired in I-SceI-mediated gene conversion. (Adapted from Kass EM, Helgadottir HR, Chen CC et al. [2013] *Proc Natl Acad Sci USA* 110:5564–5569. With permission from the National Academy of Sciences.)

with γ-H2AX (**Figure 16.4**). In budding yeast Nbs1 is replaced by Xrs2, which is required to recruit Tel1. Yeasts lack Mdc1, and the association of Tel1 with the DSB diminishes as MRX is displaced, as 5′-to-3′ resection proceeds.

In contrast, ATR does not bind to DSB ends but is activated by its association with single-stranded DNA that is created as the ends are resected. ATR binds to its partner ATRIP, which in turn binds to RPA, which avidly binds ssDNA. ATRIP is called Ddc2 in budding yeast. ssDNA adjacent to stalled replication forks is likely the principal trigger in activating ATR under conditions of replication stress.

In addition, ATR activation depends on its interaction with the so-called 9–1–1 clamp—a trimer of proteins that form a DNA clamp that is structurally similar to PCNA. But whereas PCNA has an affinity for the dsDNA/ssDNA junction where there is a 3′-recessed end (as there would be as DNA is replicating), the 9–1–1 clamp has a strong affinity for a junction with a 5′-recessed end (as there is when 5′-to-3′ resection removes one strand) (**Figure 16.5**). The name 9–1–1 resonates with the emergency telephone number in the USA, but is derived from the names of the three fission yeast proteins that make up the clamp: Rad9, Rad1, and Hus1 (hydroxyurea sensitive 1). The loading of this clamp requires four of the five RFC (replication factor C) proteins that load PCNA, but one PCNA-loading subunit, Rfc1, is replaced by a 9–1–1 loading subunit, Rad17. Confusingly, in budding yeast Rad17 is called Rad24 and one of the 9–1–1 subunits is called Rad17. Oh well! ATR in turn phosphorylates some of these clamp proteins as well as RPA.

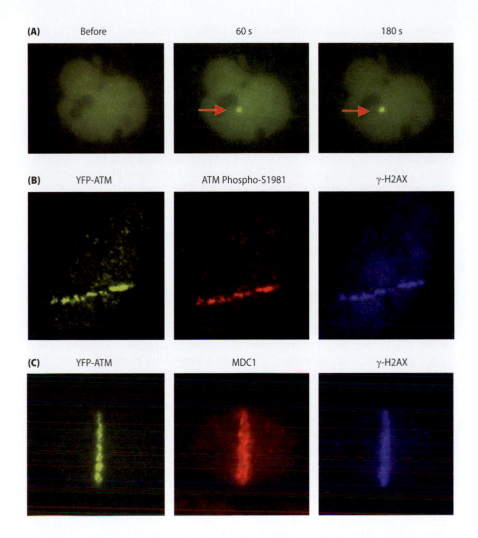

Figure 16.4 Localization of ATM to sites of laser-induced DNA damage. (A) Cells expressing a fluorescently tagged GFP-ATM were exposed to focused laser light that produces a small spot of DNA damage. ATM can be seen to associate with this damage within seconds. (B) After a laser stripe producing DNA damage is "etched" across a cell, YFP-ATM co-localizes with an antibody that recognizes ATM-S1981 phosphorylation and with γ-H2AX. (C) MDC1 co-localizes with both YFP-ATM and γ-H2AX. (From So S, Davis AJ & Chen DJ [2009] *J Cell Biol* 187:977–990. With permission from the Rockefeller University Press.)

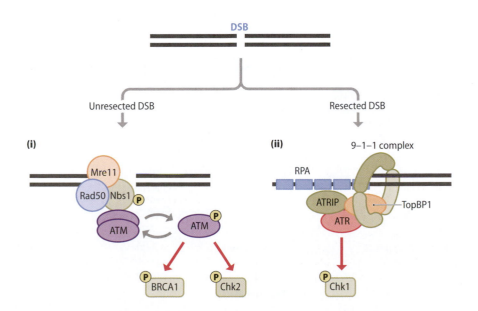

Figure 16.5 Activation of ATM and ATR. (i) The MRN complex binds to unresected DSB ends and recruits ATM to the break. Dimerization of ATM leads to its autophosphorylation of S1981 and the activated ATM monomer promotes both further autophosphorylation and phosphorylation of other targets (*yellow*). (ii) ATR is activated by the association of its binding partner ATRIP with RPA that is bound to ssDNA. Full activation of ATR requires also its interaction with TopBP1 or other co-activators as well as with the 9–1–1 clamp complex that preferentially binds to 5'-resected ends. (Adapted from Bakkenist CJ & Kasten MB [2004] *Cell* 118:9–17, and from Lopez-Contreras AJ & Fernando-Capitillo O [2012] in *Protein Phosphorylation and Human Health* [C Huang ed.]. With permission from Elsevier and the authors, respectively.)

Finally, ATR activation in response to DSBs depends on ATM. ATM and its associated MRN complex are needed to initiate 5′-to-3′ resection of the DSB ends. This resection is needed to recruit RPA, to recruit ATRIP, to activate ATR. But then ATR phosphorylates CtIP, which promotes further resection. ATM also phosphorylates two other MRN-associated proteins, BRCA1 and 53BP1. As noted in earlier chapters, BRCA1 promotes 5′-to-3′ resection whereas 53BP1 in conjunction with Rif1 retards it. We will not dwell more on these proteins here.

A very clever experiment by David Toczyski showed that one can activate a checkpoint response in the absence of a DSB simply by clustering together multiple copies of budding yeast's ATRIP and 9–1–1 complexes. In a yeast strain carrying an array of LacO binding sites, simultaneous expression of LacI fusion proteins of Ddc2 (yeast's ATRIP) and Ddc1 (one of the 9–1–1 subunits) is sufficient to cause activation of Mec1 and cell cycle arrest prior to mitosis.

16.4 ATM AND ATR INITIATE A PROTEIN KINASE CASCADE

By themselves ATM and ATR do not cause cell cycle arrest. Downstream of the PI3 kinase-like kinases are two other key protein kinases called CHK1 and CHK2 in mammals (Chk1 and Rad53 in budding yeast; and Chk1 and Cds1 in fission yeast). These protein kinases are activated by phosphorylation by the "apical" protein kinases ATM and ATR. Then Chk1 and Chk2 phosphorylate many downstream targets that regulate cell division as well as eliciting a transcriptional response to the damage. The division of labor in controlling the checkpoint response is different in vertebrates and yeasts. In mammals, ATM principally controls Chk2 (Figure 16.5) whereas ATR drives Chk1. In budding yeast Tel1 (ATM) plays a relatively minor role and the activation of both Chk1 and Rad53 (Chk2) in response to induction of a DSB is dictated by the actions of Mec1 (ATR) (**Figure 16.6**).

The full activation of the budding yeast Rad53 (Chk2) pathway depends on another key regulator, Rad9, which is phosphorylated by Mec1 and which acts as a scaffold to recruit multiple Rad53 molecules and to

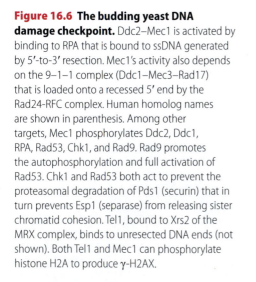

Figure 16.6 The budding yeast DNA damage checkpoint. Ddc2–Mec1 is activated by binding to RPA that is bound to ssDNA generated by 5′-to-3′ resection. Mec1's activity also depends on the 9–1–1 complex (Ddc1–Mec3–Rad17) that is loaded onto a recessed 5′ end by the Rad24-RFC complex. Human homolog names are shown in parenthesis. Among other targets, Mec1 phosphorylates Ddc2, Ddc1, RPA, Rad53, Chk1, and Rad9. Rad9 promotes the autophosphorylation and full activation of Rad53. Chk1 and Rad53 both act to prevent the proteasomal degradation of Pds1 (securin) that in turn prevents Esp1 (separase) from releasing sister chromatid cohesion. Tel1, bound to Xrs2 of the MRX complex, binds to unresected DNA ends (not shown). Both Tel1 and Mec1 can phosphorylate histone H2A to produce γ-H2AX.

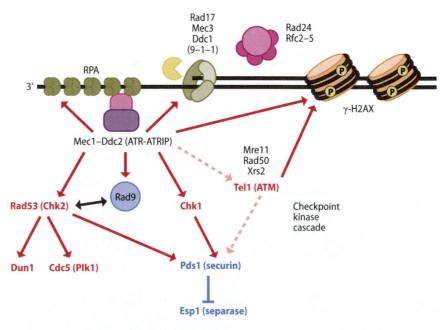

promote Rad53's full activation by autophosphorylation (Figure 16.6). Rad9 is not required to activate Chk1. In mammalian cells, 53BP1, which shares some protein motifs with yeast Rad9, is required for full activation of Chk2, but there is no evidence that this activation involves promoting the self-association of Chk2.

16.5 DSB-INDUCED CELL CYCLE ARREST IN MAMMALS OCCURS BEFORE S PHASE AND MITOSIS

For mammals, a major line of defense is to arrest damaged cells prior to DNA replication and—if repair fails—to eliminate damaged cells by apoptosis, so that they do not proliferate and accumulate chromosomal rearrangements. The central actor in this drama are the p53 tumor suppressor with its network of interacting factors that trigger both growth arrest and apoptosis (**Figure 16.7**). p53 can be activated by either ATM or ATR, as well as by downstream Chk1 or Chk2 kinases. p53 plays a central role in tumor suppression; indeed, p53 is mutated in all tumors, in many different kinds of cancer.

p53 induces anti-proliferative processes as a response to various tumor-promoting stresses. The abundance of p53 is regulated by the ubiquitin ligase Mdm2. The optimal function of Mdm2 requires Daxx, which stabilizes Mdm2 and also directly promotes Mdm2's ubiquitin ligase activity toward p53. Thus p53 is rapidly degraded in the absence of damage.

Figure 16.7 p53 regulation of DNA replication in response to ATM and ATR activation. In the absence of DNA damage Cdc25 dephosphorylates Cdk2, activating it to promote DNA replication. In addition, Daxx and Mdm2 collaborate to ubiquitylate p53, leading to its degradation. When ATM and ATR are activated, the interaction between Daxx and Mdm2 is disrupted and the stabilized, phosphorylated form of p53 accumulates. p53 induces the transcription of p21, which binds and inactivates Cdk2/cyclin E. In addition, ATM and ATR activation leads to the phosphorylation of Cdc25 that in turn triggers its degradation, so that Cdk2 remains in an inactive form. Thus, DNA damage triggers an inhibition of DNA replication in two ways: by inactivating Cdk2 through dephosphorylation and by the binding of p21. *Red arrows* indicate phosphorylations by different protein kinases and *red borders* indicate the active forms of Cdk2 and p53. (Courtesy of Koji Nakade, RIKEN Bioresource Center.)

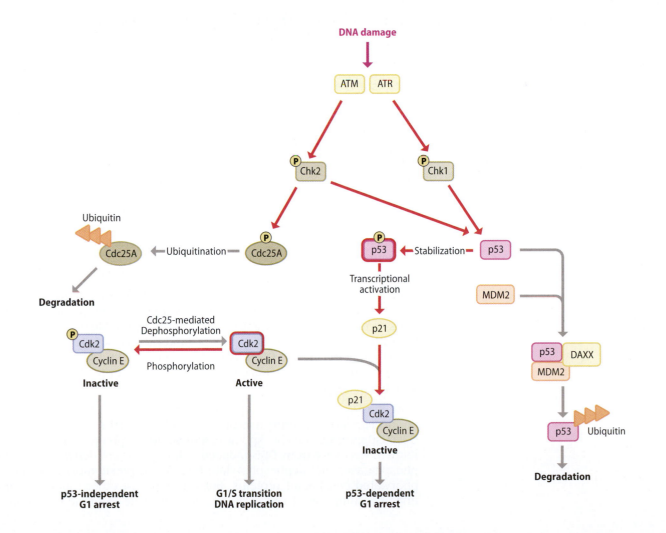

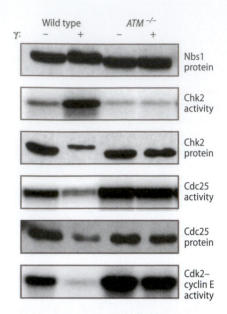

Figure 16.8 Activation of Chk2 and destruction of Cdc25 after γ-irradiation in human cells. In wild-type cells, γ-irradiation promotes the activation of Chk2, the phosphorylation of Nbs1 (causing it to migrate slightly slower in the western blot), and the destruction of Cdc25. In cells lacking ATM activity, Chk2 is not activated, Nbs1 is not phosphorylated and Cdc25A is not degraded, so that Cdk2–cyclin E activity remains high. (From Falck J, Petrini JH, Williams BR et al. [2002] *Nat Genet* 30:290–294. With permission from Macmillan Publishers, Ltd.)

The Daxx–Mdm2 interaction is disrupted upon DNA damage by ATM-dependent phosphorylation of Daxx; consequently the concentration of p53 rises and it then can trigger the many pathways under its control.

In mammals, DNA damage-induced arrest at the G1-to-S phase transition begins with ATM activation of Chk2, which in turn inactivates the key cell cycle regulator, the phosphatase Cdc25A (Figure 16.7 and **Figure 16.8**). Cdc25 is required to dephosphorylate Tyr15 in the Cdk2 cell cycle kinase; phosphorylated Cdk2 cannot trigger the initiation of new DNA synthesis. This initial arrest is maintained by p53's transcriptional activation of the p21 protein that binds to Cdk2/cyclin E and keeps it in an inactive state. p21 also inhibits Cdk4/cyclin D and prevents transcription of genes important for DNA replication.

One simple assay to show checkpoint defects monitors damage-independent DNA replication by the incorporation of BrdU into cells previously exposed to ionizing radiation. DNA synthesis that occurs despite the presence of DSBs—termed radio-resistant DNA synthesis—is thus a good indicator of checkpoint failure (**Figure 16.9**).

DSB-damaged cells also arrest prior to mitosis. The ATM–Chk2 kinase cascade again results in the inactivation of Cdc25A phosphatase, but here Cdc25A fails to dephosphorylate the inactive form of Cdk2/cyclin B and thus Cdk2 cannot activate mitosis. Two other isoforms of Cdc25 (Cdc25B and Cdc25C) also contribute to checkpoint arrest by regulating the level of phosphorylation of other proteins.

16.6 DSB-INDUCED CELL CYCLE ARREST IN BUDDING YEAST OCCURS PRIMARILY BEFORE ANAPHASE

Budding yeast lack p53 and have a very modest G1-to-S checkpoint response. Yeasts show some apoptotic-like responses but do not have the characteristic cleavage of DNA into fragments that is one of the hallmarks of mammalian apoptosis. Instead, yeasts have a strong G2–M checkpoint arrest (Figure 16.6). The block, however, is distinct from the entry into mitosis that is seen in mammalian cells. Instead, yeast arrest after chromosome condensation in metaphase, prior to the initiation of anaphase. The focus of attention of the checkpoint kinases is securin, known as Pds1 in budding yeast and Cut2 in fission yeast. Securin is both the chaperone and the inhibitor of separase, the protease that cleaves the kleisin subunit of cohesins, thereby allowing sister chromatid separation and the fulfillment of anaphase. The phosphorylation of Pds1 by Chk1 in budding yeast prevents its ubiquitylation and destruction, blocking kleisin cleavage and preventing sister chromatid separation (**Figure 16.10**).

Even a single HO-induced DSB is sufficient to trigger extended cell cycle arrest in haploid budding yeast that lack donor sequences for homologous recombinational repair. Cells that normally double every 2 hr arrest for 12 hr or more, before they "adapt" and the checkpoint is turned off, even though the broken DNA ends are still present and still undergoing 5′ to 3′ resection. These events can be seen by monitoring the Mec1 (ATR)-dependent phosphorylation of Rad53 (Chk2) (**Figure 16.11**).

How the DNA damage checkpoint is turned off remains an active area of investigation. Overexpression of budding yeast's Polo-like kinase, Cdc5, suppresses the checkpoint response but its precise targets are not known. Recovery from DSB-induced cell cycle arrest depends on PP2C phosphatases that dephosphorylate Rad53 and presumably other phosphorylated checkpoint proteins, but a number of other mutations also impair resumption of cell cycle progression. One example—deletion of

(A)

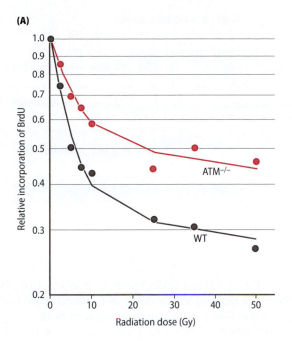

(B)

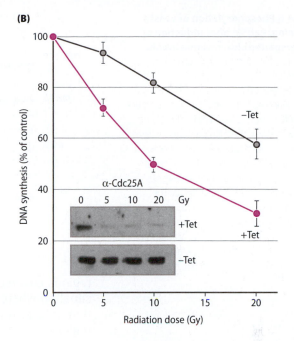

Figure 16.9 Radiation-resistant DNA synthesis in checkpoint-deficient cells. (A) Radio-resistant DNA synthesis, measured by BrdU incorporation, in wild-type cells (*gray circles*) and cells carrying an ATM mutation (*red circles*). (B) Radio-resistant DNA synthesis in human U20S cells carrying a conditionally expressed tetracycline-repressible Cdc25A gene. The inset shows that, at normal levels of expression, when only the normal Cdc25A gene is transcribed (+Tet), the abundance of Cdc25A was dramatically reduced after γ-irradiation. When Cdc25A is overexpressed (-Tet), Cdc25A persists and promotes radio-resistant DNA synthesis. (A, adapted from Gilad S, Chessa L, Khosravi R et al. [1998] *Am J Hum Genet* 62:551–561. B, adapted from Falck J, Mailand N, Syljuasen RG et al. [2001] *Nature* 410:842–847. With permission from Macmillan Publishers, Ltd.)

the resection-associated protein Sae2—is shown in Figure 16.11. Without Sac2, Rad53 phosphorylation persists indefinitely and securin remains phosphorylated and undegraded.

In fact the signals for these different checkpoints overlap. For example, the extent of the arrest of yeast cells in response to inducing a single irreparable DSB depends not only on the DNA damage response (DDR) but also on the SAC. In different ways, both the DDR and the SAC act to prevent the destruction of securin, which in turn prevents separase from cleaving the cohesin bands that must be removed before cells can separate their sister chromatids. Consequently cells arrest as large dumbbell-shapes with a single nucleus.

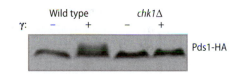

Figure 16.10 Checkpoint-mediated phosphorylation of Pds1 (securin) in budding yeast. A western blot using an anti-HA antibody that reacts with the tag on Pds1 shows the Chk1-dependent phosphorylation of Pds1 after γ-irradiation. (From Wang H, Liu D, Wang Y et al. [2001] *Genes Dev* 15:1361–1372. With permission from Cold Spring Harbor Laboratory Press.)

16.7 γ-H2AX IS IMPORTANT FOR SISTER CHROMATID REPAIR IN MAMMALS

In mammals the H2AX variant of histone H2A comprises about 10% of histone H2A. The γ-H2AX modification spreads over more than 1 Mb of sequences around a DSB. In budding and fission yeasts there is no special H2AX gene; the two copies of histone H2A encode an isoform that contains a terminal SQ site (H2A-S129 in budding yeast), so that nearly all the H2A in chromatin can be phosphorylated by ATM and ATR homologs. An important minor isoform, H2A.Z, is not modified. In budding yeast, γ-H2AX spreads strongly over 50 kb on either side of the DSB, but an increase in modification can be seen over several hundred kilobases.

The role of widespread modification of histones around a DSB is not well understood. There has been a suggestion that, in mammalian cells,

Figure 16.11 Phosphorylation of yeast's Chk2 homolog, Rad53, after induction of a single unrepaired chromosome break. An unrepaired DSB is induced by expression of the HO endonuclease at 0 hr. Rad53 hyperphosphorylation is evident at 1 hr but crescendos over the next several hours to produce a series of differently migrating phosphorylated species. As cells adapt the DNA damage checkpoint is turned off and Rad53 returns to being mostly dephosphorylated. This activation depends entirely on Mec1 and the course of modification is nearly wild type in *tel1Δ*. The absence of Sae2 results in cells that fail to adapt and thus Rad53 hyper phosphorylation persists for more than 24 hr. (From Clerici M, Mantiero D, Lucchini G & Longhese MP [2006] *EMBO Rep* 7:212–218. With permission from Macmillan Publishers, Ltd.)

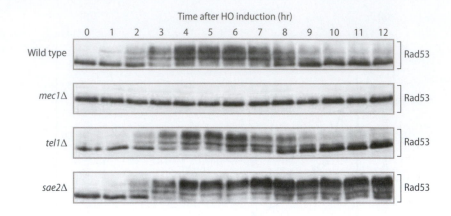

regions covered by γ-H2AX exhibit a decrease in transcription, but in a situation where there are defined endonuclease-induced DSBs transcription in fact prevents extensive γ-H2AX modification over the transcribed region. A similar conclusion is reached from studies of HO cleavage in yeast.

γ-H2AX might serve as a better "landing pad" for various DNA repair proteins, though how modifications hundreds of kilobases from the DSB would be advantageous in repair is not evident. One possibility is that it recruits proteins that are themselves ATM/ATR targets, thus accelerating the checkpoint response by increasing the local concentration of downstream targets. Cells lacking H2AX or carrying the H2AX-S139A mutation are sensitive to several DNA damaging agents, but there is not a dramatic effect in most homologous recombination assays. However, in one context γ-H2AX does play a key role in homologous recombination. Sister chromatid repair in mammalian cells is strongly inhibited in cells lacking H2AX or with the mutation of H2AX-S139A that cannot be phosphorylated.

γ-H2AX does play other roles in repair. One important process involves the 5'-to-3' resection of DSB ends that, in budding yeast, involves Fun30-dependent chromatin remodeling of nucleosomes. In yeast the absence of γ-H2AX results in faster 5'-to-3' resection, which can be explained by *in vitro* assays showing that there is a stronger affinity of Fun30 for reconstituted nucleosomes carrying H2A-S129 or H2A-S129A compared with the phosphomimetic form H2A-S129E. In mammals, the ability of the CtIP endonuclease to cleave RAG-mediated hairpin ends during immunoglobulin gene rearrangements in the G1 phase of the cell cycle is strongly inhibited by γ-H2AX, allowing the Artemis nuclease to create ends that are amenable to NHEJ.

16.8 THE DNA DAMAGE CHECKPOINT MODULATES DSB REPAIR IN MANY WAYS

In spite of the fact that many proteins involved in homologous recombination are phosphorylated by ATM and ATR, very few of these modifications have been shown to be critical for the function of the target protein. In fact in budding yeast, a *mec1Δ tel1Δ* strain (kept alive by overexpressing RNR genes) is surprisingly recombination competent, although repair (HO-mediated recombination) is about 40% as efficient as in the wild type. Some of the 60% lethality comes from the fact that cells do not arrest prior to mitosis, so that when mitosis does occur, one of the daughter cells may end up with the centromere-containing end of the DSB but fail to inherit the distal segment, making subsequent gene conversion or SSA impossible. If cells are arrested by nocodazole treatment to prevent mitosis these *mec1Δ tel1Δ* cells are even more proficient in DSB repair.

Nevertheless, some DDR-mediated phosphorylations are important. In budding yeast, during SSA, 3′-ended nonhomologous tails need to be clipped off and this requires the phosphorylation of the Slx4 protein by either Mec1 or Tel1. A phosphomimetic mutation Slx4-T113E is able to bypass this defect. The effects of other damage-induced phosphorylations are less profound in yeast. For example, the large- and middle-sized subunits of RPA are phosphorylated by Mec1; however, a mutant changing the principal site of Mec1-dependent modification of the large subunit, Rfa1-S178, to alanine had very minor phenotypes. Similarly, Rad55 is also phosphorylated but this change has no profound consequence on HO-mediated recombination. However, Rad55 lacking its phosphorylation sites (one of which is SQ that should be modified by Mec1 or Tel1) displays a very slow traversal of S phase under DNA-damaging conditions. This delay could reflect slower repair of replication-associated DNA damage, but more work will be required to sort this out.

16.9 FAILURES OF THE DNA DAMAGE RESPONSE CONTRIBUTE TO GENOME INSTABILITY

We began this book with a consideration of the amazing fact that as our cells have divided trillions of times, they have accurately retained their original karyotype, although they have undoubtedly acquired some small number of point mutations and possibly other rearrangements. In contrast are metastatic cancer cells that usually, but not always, exhibit astonishing genome instability (**Figure 16.12**). The cytological observation that nearly all cancer cells exhibit frequent changes in chromosome number and arrangement has been impressively extended by the deep sequencing of the genomes of many cancer cells. Genome sequencing has demonstrated in much greater detail that cancer cells have experienced hundreds, if not thousands, of simple mutations as well as numerous changes in gene copy number and many translocations, as illustrated in the example shown in **Figure 16.13**. Many of these changes are "passenger mutations" that do not contribute to cancer pathology but accumulate while a few "driver" mutations enable these cells to proliferate as they progress to full-blown cancers. As noted before, one apparent source of these mutations is a dramatic increase in the incidence of replication fork stalling and consequent breakage in these cells. This instability is caused by the expression of oncogenes and the loss of tumor suppressors, but we don't actually know why replication is so

16.12 Karyotypes of normal and cancer cells. (A) A normal diploid karyotype of a male, as revealed by SKY karyotyping, in which different fluorescently-tagged chromosome-specific DNA probes reveal each chromosome in a different color. (B) A multiply rearranged karyotype including several changes in chromosome number from a female with colorectal adenocarcinoma. (C) Karyotype from a woman with breast cancer. (Courtesy of Hesed Padilla-Nash, National Institutes of Health.)

(A)   (B)   (C)

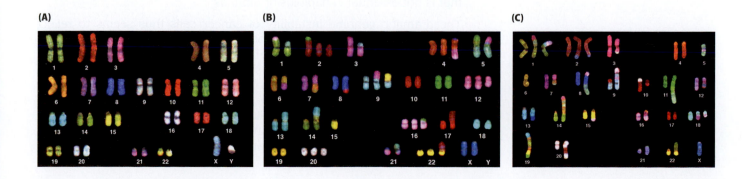

16.13 Structural sequence alterations identified by genomewide sequencing of a breast cancer cell. The outer circle shows the cytologically identified staining of bands for each chromosome, followed by a plot indicating the copy number alterations. The *green* bars in the center refer to smaller intrachromosomal changes such as duplications and inversions, whereas *pink* lines indicate interchromosomal translocations. In this sample, 13 chromosome arms had some complex chromosome aberrations; six of these had many rearrangements (these are in bold and marked with an asterisk). The two regions with most rearrangements were chromosome arm 7p and chromosome 15. (From Russnes HG, Vollan HK, Lingjaerde OC et al. [2010] *Sci Transl Med* 2:38ra47. With permission from the American Association for the Advancement of Science.)

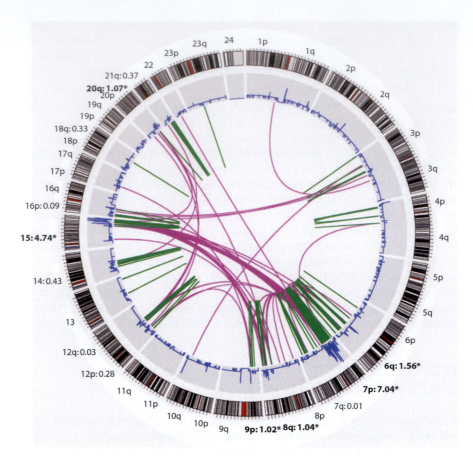

severely affected. Is there a dis-coordination between the initiation of replication and the synthesis of dNTPs or of histones or other chromatin proteins? Are helicases and other factors that help to replicate through hard-to-copy regions not transcribed or not post-translationally modified sufficiently to keep up with replication? These are among the many questions that need to be answered.

One other important feature of cancer cells is that they are almost always aneuploid. Very recent studies have shown that aneuploidy is, in nearly every case, detrimental to cells; but some aneuploidy combinations must enable cancer cells to evade elimination and to proliferate. Imbalances in gene copy numbers on different chromosomes place cells under many forms of stress, making them sensitive to many DNA damaging agents. But in what way these imbalanced cells are sensitive to DNA damage is still unexplored. Curiously, a fully euploid but tetraploid yeast cell, which should easily be able to tolerate chromosome losses because of only a small change in gene dosages, is unable to grow without the Rad52 gene that is not essential in haploids or diploids.

The dramatic genome instability of cancer cells thus seems to reflect two major failures: the absence of appropriate DNA damage response checkpoints and the failure to repair DSBs by the most accurate means—homologous recombination that should restore the karyotype even if sometimes associated with loss of heterozygosity. Cells lacking these most accurate ways of DSB repair are still able—through NHEJ or micro-homology-mediated end joining—to repair breaks, but at the huge cost of generating chromosome rearrangements. Indeed, the proteins of the DNA damage response can be regarded as important tumor suppressors.

16.10 CANCER CHEMOTHERAPIES EXPLOIT TARGETS IN MULTIPLE DNA REPAIR PATHWAYS

Beyond surgery, the main strategy for the control and elimination of cancer has been to inflict DNA damage, either in the form of ionizing radiation or by chemotherapy, on the premise that dividing cells will be selectively killed over nondividing cells. There is no strong evidence that cancer cells divide more rapidly than other cells that are cycling. Consequently, DNA damage impairs the growth of hair cells and weakens the immune system while it cripples cancer cells.

Ideally such treatment should make cancer cells more sensitive than other dividing cells because cancer cells usually have defects in one or more DNA repair pathway or in the DNA damage checkpoint. The cancer chemotherapeutic agents that are used cause many forms of DNA damage including double-strand breaks (bleomycin), inter- and intra-strand DNA cross-links (cisplatin or mitomycin C), and alkylated or oxidized bases such as 8-oxoguanine (nitrogen mustard or methotrexate). These lesions invoke the participation of different repair pathways, such as intra-strand cross-link repair or base-excision repair, leading to nicks, gaps, and breaks in the DNA. Thus cancer cells that have defects in a particular pathway should be susceptible to enhanced killing if they cannot repair lesions produced by therapeutic treatment.

In practice we know very little about how most chemotherapeutic agents kill. First of all, cells are resourceful and when one repair pathway is disabled, cells often repair damage in alternative ways. Indeed, the large numbers of nonreciprocal translocations seen in most cancer cells are the result of promiscuous nonhomologous end-joinings when normal, more accurate homologous recombinational repair processes are lacking. Thus, it seems that to selectively kill cancer cells, it will be necessary to block more than one repair process. This strategy is a form of *synthetic lethality* in which cells are only fatally compromised when two parallel repair pathways are disabled. This idea is manifest in the recent demonstrations that breast cancers in which either BRCA1 or BRCA2 are mutated can be selectively killed in the presence of a DNA damaging agent when an inhibitor of poly-ADP ribose polymerase (PARP) is concomitantly administered. PARP is required for base excision repair (BER) and its absence should cause an increase in single-stranded nicks that can be converted by replication into DSBs that in turn require the BRCA proteins for their repair. But it is worth noting that PARP also is needed for Ku- and Xrcc4-independent "alternative" end-joining (A-EJ). So whether it is PARP's role in BER or in A-EJ that is most important is not yet known.

Similarly, for several decades it was assumed that methotrexate killed dividing cancer cells by blocking the formation of folate, which in turn blocked the formation of deoxythymidine (dT); the reduction in dT would in turn prevent normal DNA replication. But recent work has shown that methotrexate also causes an increase in the oxidation of bases such as 8-oxoguanine, which should be removed primarily by BER. But oxidative DNA damage cannot only be repaired by BER; NER, mismatch repair, homologous recombination, and the translesion DNA polymerases all play a role. Consequently cells lacking two or more of these pathways show increased sensitivity. Thus, even though cells lacking the *MSH2* mismatch repair gene are sensitive to methotrexate, we really have not precisely defined what defects in DNA repair are most important in the action of a chemotherapeutic drug against a particular cancer.

In fact, in nearly all cases we do not know exactly why effective cancer chemotherapeutic agents actually overwhelm cancer cells more than their normal relatives. Both testicular cancer and one form of lung cancer are effectively cured by treatment with cisplatin, a drug that has little effect on many other forms of cancer. Cisplatin provokes intra- and inter-strand cross-links. But we do not know why these two types of cancer are so susceptible to this treatment. Nor do we know which repair pathways are active and which are inactivated.

Conversely, why are cancers in women with defects in BRCA1 or BRCA2 largely confined to breast and ovarian tissues (but not all tissue types)? Is it that in these estrogen-responsive tissues some other pathway is not working in breast tissue? Or is it that in most other tissues a homozygous defect in these two genes results in such poor growth that no cancers arise? Similarly, why should loss of Rb result in retinoblastoma and not many other forms of cancer? A detailed understanding of the arsenal of DNA repair and recombination pathways needs to be undertaken to give us a clearer insight into why certain chemotherapies work in specific cell types.

What I hope is clear is that the remarkable progress in understanding DSB repair processes from genetic, molecular biological, biochemical, and structural vantage points has made it possible to think much more clearly about each of the many DSB repair processes; and to see ways in which these pathways can be protected to prevent genome instability from arising; or to see ways that cells with defects in these pathways can be treated to eliminate them before they develop into full-blown cancers.

16.11 HOMOLOGOUS RECOMBINATION TURNS UP IN STEM CELL REPROGRAMMING

One of the biggest revelations of the twenty-first century was the discovery of relatively simple ways to reprogram differentiated cells into induced pluripotent stem cells (iPSCs). The addition of four transcription factors can drive cells back to a pluripotent state. Recently it was shown that this reprogramming is accompanied by the appearance of γ-H2AX foci that indicate some sort of DNA damage; possibly, but not certainly, DSBs. However, reprogramming is impaired in HR-defective cells lacking BRCA1 or BRCA2 function, or when Rad51 was knocked down by siRNA, or when Fanconi Anemia proteins were absent. It is possible that overexpression of Myc—one of the four transforming factors to drive cells into pluripotency—may drive initiation of S phase too strongly, creating replication stress and a need for homologous recombination. So another vista has opened up. Now we need to understand what roles homologous recombination plays during reprogramming.

SUMMARY

The DNA damage checkpoints play a fundamental role in preventing cells with broken chromosomes from undergoing mitosis to produce daughter cells with broken chromosomes. The DNA damage response also ensures that G1 cells that are damaged, or which inherit damage from the previous cell cycle, are blocked from proliferating and are often eliminated by apoptosis. The kinds of chromosome rearrangements one sees in cancer cells arise predominantly in cells that have lost both checkpoint control and the capacity to repair DSBs accurately by homologous recombination, leaving NHEJ mechanisms to produce cells with many chromosome rearrangements.

SUGGESTED READING

Bartek J, Bartkova J & Lukas J (2007) DNA damage signalling guards against activated oncogenes and tumour progression. *Oncogene* 26:7773–7779.

Ciccia A & Elledge SJ (2010) The DNA damage response: making it safe to play with knives. *Mol Cell* 40:179–204.

Davoli T & de Lange T (2011) The causes and consequences of polyploidy in normal development and cancer. *Annu Rev Cell Dev Biol* 27:585–610.

Dimitrova N, Chen YC, Spector DL & de Lange T (2008) 53BP1 promotes non-homologous end joining of telomeres by increasing chromatin mobility. *Nature* 456:524–528.

Goodarzi AA & Jeggo PA (2013) The repair and signaling responses to DNA double-strand breaks. *Adv Genet* 82:1–45.

Gobbini E, Cesena D, Galbiati A, et al. (2013) Interplays between ATM/Tel1 and ATR/Mec1 in sensing and signaling DNA double-strand breaks. *DNA Repair* (Amst). 12:791–799.

Halazonetis TD, Gorgoulis VG & Bartek J (2008) An oncogene-induced DNA damage model for cancer development. *Science* 319:1352–1355.

Harper JW & Elledge SJ (2007) The DNA damage response: ten years after. *Mol Cell* 28:739–745.

Harrison JC & Haber JE (2006) Surviving the breakup: the DNA damage checkpoint. *Annu Rev Genet* 40:209–235.

Lukas J, Lukas C, & Bartek J (2011) More than just a focus: The chromatin response to DNA damage and its role in genome integrity maintenance. *Nat Cell Biol* 13:1161–1169.

Matsuoka S, Ballif BA, Smogorzewska A et al. (2007) ATM and ATR substrate analysis reveals extensive protein networks responsive to DNA damage. *Science* 316:1160–1166.

Pfau SJ & Amon A (2012) Chromosomal instability and aneuploidy in cancer: from yeast to man. *EMBO Rep* 13:515–527.

INDEX

Note: abbreviations following page numbers are: B, box; F, figure; and T, table.